Antonio B. Nassar · Salvador Miret-Artés

Bohmian Mechanics, Open Quantum Systems and Continuous Measurements

Springer

Antonio B. Nassar
Department of Sciences
Harvard-Westlake School
and UCLA Extension
Los Angeles, CA
USA

Salvador Miret-Artés
Superior Council for Scientific Research
Institute of Fundamental Physics
Madrid
Spain

ISBN 978-3-319-53651-4 ISBN 978-3-319-53653-8 (eBook)
DOI 10.1007/978-3-319-53653-8

Library of Congress Control Number: 2017932636

Printed on acid-free paper

This Springer imprint is published by Springer Nature
The registered company is Springer International Publishing AG
The registered company address is: Gewerbestrasse 11, 6330 Cham, Switzerland

To Isabela, Clarissa and Priscilla

Antonio B. Nassar

and

To Virginia, Jessica, David, my parents and brother†

Salvador Miret-Artés

Preface

In standard quantum mechanics, the collapse of the wave function resulting of the interaction of a given system with a measuring apparatus is a clear unsatisfactory assumption. As is well known, this assumption, due to von Neumann, was established around 1932 by means of the so-called reduction (or collapse or projection) postulate. A unitary Schrödinger evolution can never lead to such a collapse and, therefore, at some point, the time evolution of a quantum system undergoing a measurement has to be suspended and replaced by a discontinuous, noncausal, nonlineal, nonunitary (but norm-preserving) and stochastic one. The collapse is postulated to happen in accordance with the well-known Born probability rule, with no dynamical theory behind such a mechanism. It is also very well known that the absence of macroscopic superpositions is at the heart of the measurement problem in quantum mechanics. Another aspect worth mentioning is the emergence of the classical features from the quantum world; in other words, the conversion of a coherent superposition into a mixture. The resulting decoherence process leads to the selection of one of the components of this mixture. This selection of one out of many alternatives still remains obscure. Thus, we still do not understand why the measurement process destroys the linear superposition of the initial states and why macroscopic objects are not found in superposed states. Even more, as asked by Bell, where is the quantum-classical dividing line? Mesoscopic systems are also ideal to address this fundamental question. In brief, too many open questions are even today waiting for some sort of satisfactory response.

Many physicists believe that there is a *measurement problem* in quantum mechanics. Even more, measurement definitely and unavoidably affects the dynamics of the measured system. In the Copenhagen interpretation of quantum theory, the world is divided into *system* and *apparatus*. In the simplest case, the system, which is quantum mechanical, is characterized by a wave function. This interpretation implies that the apparatus that performs and records measurements on the system must be classical. In practice, experimenters are always on the classical side of the line. As pointed out by Bell, the lack of clarity in regard to where the transition between the classical and quantum worlds is located is one aspect of the measurement problem. This problem represents one of the most important conceptual difficulties in quantum mechanics. The presence of a classical apparatus considerably affects the behavior of the observed quantum system and measurements typically fail to have outcomes of the sort the theory was created to explain. Another conceptual difficulty is that in a system under observation there are many degrees of freedom such that information can be lost in the couplings which may account for dissipation and decoherence. It appears that the measurement problem before alluded is becoming slowly a critical or singular quantum process (converted to a dynamical problem) rather than a conceptual issue. In other words, it is no longer considered as a conceptual or philosophical problem but a typical many-body problem (with all the computational issues involved in its resolution).

A measurement can be seen as a physical process by which the state of a quantum system influences somehow a classical variable of a macroscopic apparatus. This influence is usually described by an interaction between the system of interest (S) and the apparatus (A). Then, the combined quantum system SA has properties that S and A alone do not have and it is said that S and A are entangled by means of quantum correlations due to the quantum nonlocality. It is widely accepted that A can also be considered as an environment. In 1970, Zeh proposed to analyze the measuring process within the context of the theory of open quantum systems and see it as a typical many-body interaction problem. In 1984, Gisin carried out a theoretical connection between them. Furthermore, almost any interaction with an environment produces localization (or reduction) in position space. It is very often alluded that the system can not have a state on its own when interacting with its environment and, therefore, the only description in the standard quantum formalism is by means of the density matrix. The measurement process then implies the conversion of a coherent superposition into a mixture together with the selection of one of the components (or alternatives) of the mixture leading to a definite measurement readout. Thus, this decoherence process leads to the emergence of the classical world (classicality). Decoherence, which is ubiquitous in nature, is extremely effective and virtually impossible to escape (only irreversible when the environment is macroscopic). It is a pure quantum process and has no classical counterpart. Furthermore, entanglement is the key process underlying decoherence. This fundamental process has been and is still being studied from many, many points of view emphasizing its very different aspects and subtleties. The number of articles and books devoted to such an important and fundamental issue is extremely huge; to cite only some of them, we have the Ghirardi–Rimini–Weber model of continuous spontaneous collapse, the entropic-dynamic approach to quantum evolution and the Fisher information based derivation of the fundamental Lagrangians leading to relativistic wave equations. Several stochastic approaches are being used nowadays within the quantum theory which could be classified according to the following scheme: group or semigroup algebra, the influence functional, the restricted path integral, consistent histories, effective wave equations and quantum state diffusion equations. They are mathematically connected. Continuous quantum measurement theory is playing a fundamental role in this context since it provides new aspects and ingredients to the global quantum theory. For example, quantum transitions can be monitored as well as perturbed, altering the dynamics of the measured system. However, very few studies within the Bohmian framework can be found in literature. This approach should also be considered in the measuring process since the outcome of an experiment is very often the position of a particle. In Bohmian mechanics, particles are assumed to have well-defined positions in configuration space. Moreover, the description provided by this mechanics to the measurement process is entirely compatible with Born's rule in standard quantum mechanics.

Bohmian mechanics, which is also called the de Broglie–Bohm theory, the pilot-wave model, and the causal interpretation of quantum mechanics, was originated in the 1920s by Louis de Broglie, re-discovered and developed in 1952 by David Bohm. Its roots are also in the pioneering works of Madelung's hydrodynamic formulation of quantum mechanics. No one more eloquently than John Bell has championed this theory in recent decades. It should be remembered that Bell's inequalities were inspired by his reaction to the work by Bohm. Bell and Bernstein also argued convincingly that the de Broglie–Bohm interpretation of quantum mechanics should be part of any college curriculum on the subject. Tegmark and Wheeler also considered this theory as one of the most significant elements in the history of quantum mechanics. Comprehensive discussions of Bohmian mechanics can be found everywhere in the literature. This mechanics mainly arose as a result of the unsatisfactory interpretation of standard quantum mechanics, which claimed that the wave function provides the most general and complete physical information about a quantum system. This led to a very exciting, never-ending debate focussed on the completeness of the wave function and the quantum theory of measurement. Within Bohmian mechanics, a quantum system has a well-defined (in space and time) trajectory, namely a quantum or Bohmian trajectory; the evolution of this trajectory is determined by the wave function associated with the system. Quite recently, a revival of the debate about the role played by this mechanics in quantum physics can be found in the specialized literature. There are several groups for whom this theory constitutes the natural framework of quantum mechanics, whereas other groups consider it as an alternative and exact formulation able to characterize, interpret and predict quantum processes, standing on equal footing with the standard theory. In fact, it can also be considered as a picture of quantum mechanics at the same level as those due to Schrödinger, Heisenberg, Dirac (Interaction) and Feynman. As Aharanov says, this theory provides the same experimental predictions than quantum mechanics and it contains additional hidden variables which are nonlocal in time that account for the results of every measurement. In this sense, there is no way to decide from experiment which picture is prevalent. This formulation has also received an important impulse over the last 20 years from different communities, which translates into an impressive and fruitful theoretical development. Furthermore, from our own longstanding experience in the field, Bohmian mechanics can tackle any quantum problem as standard quantum mechanics can do it. Obviously, the amount of time and effort invested by many researchers in standard quantum mechanics completely outweigh that invested in Bohmian mechanics. However, we think that this situation will be corrected in the near future owing to the fact that this theory appears in more and more modern quantum mechanics books at the introductory level. We are going to deal with quantum or causal trajectories but they must not be confused with the same concept introduced by several authors in optics and quantum state diffusion which can be seen as the realizations of the wave function or state vector considered as an underlying stochastic process in Hilbert space. Unfortunately, the concept of quantum trajectory (stochastic or deterministic) is not unique and there is an important proliferation of quantum stochastic trajectories in different fields of research within the theory of open quantum systems as well as the Bohmian mechanics.

The scope of the book is to tackle such conceptual difficulties mentioned above. In particular, when certain conditions are met, an answer to the alluded challenge posed by Bell on the dividing line between the quantum and classical regimes in a measurement problem is also discussed and proposed. In Bell's opinion, the most important element in quantum theory was that the subjectivity of the orthodox version, the necessary reference to the observer, could be eliminated. The theoretical procedure followed here it is inspired from Mensky's theory about restricted path integrals, combined with the so-called Langevin–Schrödinger equation or Kostin equation (which can be seen as an effective wave equation) within the Bohmian framework. This new approach presents several advantages. First, it provides a causal aspect to the dynamics of the measuring process. Second, the proposed stochastic nonlinear approach is an alternative and complementary view of the continuous quantum measurement. Furthermore, in this context, it is put on equal footing the stochastic character coming from the presence of the environment and the phenomenological contribution coming from the measuring process. Third, the information due to the readout of the measurement is governing the equations of motion of Bohm through the quantum potential which is the nonlocal agent of this theory. Fourth, the correspondence principle, established in terms of the smallness of the quantum potential (that is, when it is approaching a zero value), provides a natural way to the emergence of the classical world. Fifth, the gradual decoherence process due to the interaction with the measuring device is introduced and described in terms of quantum stochastic trajectories; that is, in a continuos way with no collapse of the wave function. And sixth, following a dressing scheme, the quantum (stochastic) trajectory is always split into a classical trajectory plus a quantum or nonlocal term.

With this goal, and in order to be as self-contained as possible, although a minimal background on stochastic processes is assumed, this monograph is divided into four chapters. In Chap. 1, a brief historical description and short and updated account of Bohmian mechanics is provided for closed and open quantum systems; in particular, how open quantum systems and the measurement problem are typically considered in this formalism. This introductory chapter aims at providing the essential tools and background to better understand the rest of the monograph. It is clear that this mechanics is not the only one describing quantum mechanics in terms of trajectories but maybe it is becoming more and more known and widely used.

In Chap. 2, our purpose is to simply show that Bohmian mechanics is a powerful route to bring about new solutions to problems discussed by conventional quantum mechanical approaches, apart from allowing some striking correspondence between both frameworks. This is not carried out by means of a systematic exposition of conservative quantum processes but following some illustrations of solutions to some key quantum mechanical problems such as, for example, the so-called Ermakov–Bohm invariants, boundary conditions and uncertainty principle in tunneling, the quantum traversal time, Airy wave packets and Airy slits, the detection of inertial and gravitational masses with Airy wave packets, the geometric phase analyzing the Aharonov–Bohm effect and quantum vortices, the reformulation of the Gross–Pitaevskii equation within the hydrodynamical framework and ending with the well-known Caldirola–Kanai Hamiltonian for dissipative processes. In this dissipative scenario, the motion of a free particle, the quantum interference of two wave packets and the dynamics in a linear potential as well as the corresponding of a damped harmonic oscillator (within the underdamped, critically damped and overdamped regimes) are finally analyzed for ulterior references.

In the last decades, many nonlinear extensions of the Schrödinger equation have been proposed in literature either to explore the fundamental aspects of quantum mechanics, with the usual linear theory representing only a limiting case, or to describe open quantum systems. In Chap. 3, one of those nonlinear quantum equations is introduced and widely developed. For the description of nonconservative quantum systems, Kostin formulated in an heuristic way the so-called Schrödinger–Langevin (SL) equation or Kostin equation for the Brownian motion. This equation has been subsequently rederived, improved and extended for its use in numerous applications, mainly without including the noise term. Numerous features of this SL equation can be better revealed within the framework of de Broglie–Bohm (quantum hydrodynamical trajectory formulation) of quantum mechanics resulting the Schrödinger–Langevin–Bohm (SLB) equation. Within this formalism, several dissipative problems are presented and discussed: the Ramsauer–Townsend effect, the tunneling dynamics through a barrier, the plasma fluid formulation and the Lorentz–Abraham (extended electron) equation for a point-charge electron. These two last examples are also discussed in order to see the correspondence between classical and quantum dynamics. Very few applications of this SL equation are devoted to stochastic problems in the literature where the noise term needs to be included. The so-called Bohmian–Brownian motion is introduced in the context of surface diffusion with single adsorbates. An extension to interacting adsorbates is discussed within a simple, phenomenological model. Interestingly enough, this study leads us to quantum anomalous diffusion. Finally, a generalization of the SL equation is proposed for nonlinear dissipation.

Chapter 4 is devoted to the study of continuous or repeated (prolonged in time) measurements on a quantum system. This is a very active field since it is directly connected to the foundations of quantum mechanics, the theory of quantum measurements being one of its main topics. Following a phenomenological approach, a general theory for decoherence in the framework of restricted path integrals (RPI) has been proposed by Mensky. The corresponding propagator is modified according to the information provided by the measurement through the so-called quantum corridors, which correspond to different readouts of the measurement. The action uncertainty principle is then ruling the corresponding measurement process. The measured system is also considered in this theory as an open system since the back reaction of the environment, considered as a measuring apparatus, is taken into account implicitly. In this Chapter, the RPI formulation is briefly presented and discussed since our approach is inspired from Mensky's procedure. The linear time-dependent Schrödinger equation derived from this formulation is reached for a non-Hermitian (complex) effective Hamiltonian which takes into account the measurement readout. The RPI formulation is equivalent to a master equation, a special case of the Linblad equation, and to a stochastic Schrödinger equation. A nonlinear equation is proposed and discussed, within the Kostin framework, by extending Mensky's approach to be analyzed in terms of Bohmian trajectories, the so-called SLB equation for continuous measurement. When a time-dependent Gaussian shape is assumed for the probability density, soliton-like solutions are first analyzed in order to establish the dividing line between the quantum and classical trajectories, leading to the concept of Bohmian time. Afterwards, the solutions of the SLB equation for continuous measurement are analyzed in terms of mathematical stability for three simple cases, that is, the free particle and the linear and harmonic potentials. In particular, the continuous position measurement is seen that it is no longer governed by a standard continuity equation. The sign of the extra source/sink term appearing in this new equation is critical for the stability of the solutions. Two types of general solutions are found: (i) the stationary wave packet considered becomes unstable (that is, this solution is not a physically acceptable solution for that process since, as time evolves, the width of the corresponding wave packet becomes more delocalized) and (ii) the probability density becomes more and more localized at asymptotic times (the stationary wave packet is then an attractor of the corresponding width dynamics); in terms of trajectories, this analysis indicates that the Bohmian trajectories are not approaching the corresponding classical ones. These conclusions are also corroborated by carrying out an analysis of the entropy. Finally, it is also briefly discussed the question of coupling classical variables (apparatus) to quantum ones (system) in terms of Lyapunov exponents.

In the epilogue of this monograph some concluding remarks and future perspectives towards the establishment of a complete theory of continuous quantum measurement, within the Bohmian mechanics, are briefly presented and discussed.

This monograph is the result of more than 30 years working on trajectory-based formalisms; in particular, on Bohmian mechanics. Concerning citations, we have tried to furnish a historical development of the different topics presented here. However, to provide a selection of the very last references in a very active field is really difficult. We apologize to those who think they should be cited and are not. During this long but exciting time, we have benefitted from discussions with many colleagues from all over the world. In particular, we would like to acknowledge fruitful discussions and collaborations with P.T.S. Alencar, P. Bargueño, J.M.F. Bassalo, I. Besieris, M.S.D. Cattani, H. Peñate-Rodríguez, J. Margalef-Roig, R. Martínez-Casado, G. Rojas-Lorenzo, A.S. Sanz, E. Outerelo, E. Pollak and O. Roncero. ABN would like to thank all members of the Harvard-Westlake Science Department, the UCLA Extension Science Department and the College of the Canyons for the constant support and encouragement throughout the process of writing this monograph. Analogously, SMA would also like to thank all members (past and present) of the *Departamento de Física Atómica, Molecular y de Agregados* of the *Instituto de Física Fundamental* belonging to the *Centro de Física Miguel A. Catalán* (Consejo Superior de Investigaciones Científicas, CSIC) in Madrid. This work has been partially supported by the project FIS2014-52172-C2-1-P from the *Ministerio de Economia y Competitividad* (Spain).

Finally, we thank our Editor A. Lahee for her interest when we proposed this monograph as well as her patience when extending several times the deadline for finishing this project.

Los Angeles, USA
Madrid, Spain
December 2016

Antonio B. Nassar
Salvador Miret-Artés

Contents

Chapter 1
Historical and Introductory Account of Bohmian Mechanics

Abstract Bohmian mechanics, which is also called the de Broglie-Bohm theory, the pilot-wave model, and the causal interpretation of quantum mechanics, was originated in the 1920s by Louis de Broglie, re-discovered and developed in 1952 by David Bohm. No one more eloquently than John Bell has championed this theory in recent decades. Bell and Bernstein also argue convincingly that the de Broglie-Bohm interpretation of quantum mechanics should be part of any college curriculum on the subject. Comprehensive discussions of Bohmian mechanics can be found everywhere in the literature. In this introductory chapter, a brief historical description and short and updated account of this mechanics is provided for closed and open quantum systems; in particular, how the measurement problem is typically considered in this formalism. This chapter aims at providing the essential tools and background to better understand the rest of the monograph. It is clear that this mechanics is not the only one describing quantum mechanics in terms of trajectories but maybe it is becoming more and more known.

1.1 Introduction

John Bell was a theoretical physicist famous for his results in the field of foundations of quantum mechanics [1]. The famous *Bell inequalities* and similar results provide strict constraints on any future theory of quantum phenomena. His work is as monumental as the work of Kurt Gödel [2] on the *Incompleteness Theorems* in mathematics. Bell was a highly cultured and thoughtful thinker. His writings are still sending shock waves in the field of quantum mechanics. Bell presented his theory with extraordinary lucidity coupled with a vividness of expression. Tegmark and Wheeler have written an excellent and basic review about the 100 years of quantum mechanics [3].

Bell championed the theory put forth by de Broglie in 1927 and advanced by David Bohm in 1952 [4, 5], the so-called de Broglie-Bohm quantum theory, the pilot-wave model, the causal interpretation of quantum mechanics and Bohmian mechanics [6–14]. In fact, due to initial criticism of his theory, de Broglie abandoned his *guiding waves* project and so it lay abandoned until Bohm revived it. The particle motion

A.B. Nassar and S. Miret-Artés, *Bohmian Mechanics, Open Quantum Systems and Continuous Measurements*, DOI 10.1007/978-3-319-53653-8_1

dependens thus on the wave evolution with time, giving rise to quantum or Bohmian trajectories. The uncertainty principle in Bohmian mechanics is then understood from the initial conditions determining the trajectory, which are subject to the usual quantum mechanical uncertainties. The motions of the particles on their trajectories are deterministic, but which trajectory the particle is on is subject to probability. Bell was convinced that the de Broglie-Bohm interpretation of quantum mechanics should be part of any college curriculum on the subject. Bohmian mechanics stands out by being both historically influential and highly developed. Its position is further strengthened not only by Bell's theorem, which showed that all such theories have to be nonlocal as well as the experimental confirmation in 1981 due to Aspect et al. [15–17] and by recent experiments [18–20].

In the simplest pilot-wave theory, Bohmian mechanics, paradoxes of quantum mechanics can be resolved in a simple way – while the particles have positions, they move according to Schrödinger's equation. For example, one of the most famous paradoxes (in the standard interpretation) at the heart of quantum mechanics is the double-slit experiment. By using the concept of weak-value measurements introduced originally by Aharanov et al. [21],[1] recent experiments with single photons in a double-slit interferometer have revealed that single-particle trajectories can be processed in a indirect way and plotted. Weak measurements have then shown that it is possible to operationally define a set of trajectories for an ensemble of quantum particles. In this fashion, these results have reproduced those *predicted* by Bohmian mechanics (non-crossing trajectories) [19, 20, 22] in spite of the fact that this mechanics is developed for particles with mass. However, what qualifies some physical systems to play the role of a "meter"? After all, there is no reference to a measuring process in Schrödinger's equation. Besides, quantum mechanics should explain some aspects of macroscopic measurements. Bohm's ideas were applied to different prototypical models of quantum mechanics [7] during the late 1970s and particularly the 1980s and early 1990s. However, in the last 20 years or so, Bohmian mechanics has passed from being a mere way to formulate a quantum mechanics "without observers" [8, 9] to become a well-known (and increasingly accepted) theoretical framework used as a source for new quantum computational methods as well as quantum interpretations [12, 23, 24]. These two aspects of Bohmian mechanics are what Wyatt [12] has termed the *synthetic* and *analytic* approaches of this theory. The first approach essentially starts with the former numerical schemes developed to obtain quantum information without solving Schrödinger's equation, but its equivalent Bohmian counterparts [12, 25–27]. Since then, this computational branch of Bohmian mechanics has diversified into a myriad of numerical approaches, which can be summarized by the type of answer they try to give. For example, for wave packet propagation, different Lagrangian, Eulerian and combined Eulerian-Lagrangian algorithms have been developed [12]. Semiclassical initial value representation schemes based on Bohmian mechanics have been implemented [28–32].

[1]Following these authors, in principle it would be possible to measure transition probabilities in such a way that they would allow us to determine "any physical variable to a certain (even forbidden) value", e.g., canonically conjugate variables.

The purpose of avoiding and, therefore, solving the so-called *nodal problem* has led to schemes such as, for example, the bipolar ansatz [33–36]. The starting point for the second approach, on the other hand, based on obtaining the quantum trajectories from the wave function as a tool to interpret and simulate realistic experiments, can be established in the former studies of rare gas atom diffraction by metal surfaces [13, 14].

Bohmian mechanics allows us to understand and explain quantum systems in terms of the motion displayed (in configuration space) by a swarm of quantum trajectories. Each one of these trajectories represents the evolution in time of a particular initial state specified by a point on the configuration space associated with the system. Thus, unlike standard quantum mechanics, where the wave function determines the state of the system on the *whole* available configuration space, in Bohmian mechanics it is possible to follow one particular point of such a space. The time-evolution of this point is given according to some prescribed quantum laws of motion. The evolution of the trajectory ensemble is equivalent to the evolution of a quantum flow —this is precisely the viewpoint of quantum hydrodynamics. The Bohmian view does not invalidate at all other ways of understanding quantum systems; it only allows us to think of them on similar grounds as classical ones, i.e., using a similar intuitive scheme, which differs from a purely classical one precisely in the types of motion one can observe. However, unlike any classical approach to quantum mechanics, Bohmian mechanics is not an approximated theory, but an exact one. Furthermore, by looking at one of those diffraction experiments carried out with electrons [37] or atoms [38], one notices that a single measurement or detection is meaningless, many of them being necessary in order to visualize the diffraction pattern and then obtain information either about the diffracted particle or the diffracting object. In other words, individual particles behave like individual point-like particles, though their distribution displays a wave-like behavior, in accordance with Schrödinger's equation. It is therefore clear that ensemble properties need of an ensemble description, i.e., a density distribution function, whose role is played in quantum mechanics by the probability density or, at a more elementary level, the wave function. This is in agreement with Born's statistical interpretation of quantum mechanics. However, if (individual) particles are regarded as moving along single trajectories, are these trajectories the ones obtained from the guiding condition? Bohmian trajectories reproduce all the features of quantum mechanics and, therefore, one would be tempted to think that this is so. However, Bohmian equations are regarded as hydrodynamic equations, the corresponding trajectories obtained from the guiding condition should not be regarded as the trajectories pursued by real electrons, but rather as streamlines associated with the corresponding quantum fluid or paths along which quantum probability flows. Very interesting work is being addressed to several problems in quantum optics and atom interferometry by Sanz and coworkers [39–41] under this approach. Electrons may move or not like that, basically depending on the laminar or turbulent regimen displayed by the fluid, but surely not exactly as Bohmian trajectories. This description allows us to infer dynamical properties of the quantum fluid, which are usually hidden when studied by means of the wave function formalism. Therefore, Bohmian particles can be seen as the quantum equivalent of classical tracer particles in a fluid, so that they can help us to visualize its flow dynamics by moving along *streamlines*, i.e., lines along which the fluid current goes or energy is transported. For example, if the fluid is gaseous, one can use smoke; if it is a liquid, one can make use of tinny floating particles, e.g., pollen or charcoal dust, or another liquid, e.g., ink. In the case of the hydrodynamical approaches utilized to the universe dynamics in cosmology, the tracer particles can be stars, galaxies or clusters [13, 14].

The problem of many-body interactions —or, equivalently, many degrees of freedom— can be tackled from different points of view, since it appears in many different physical, chemical and biological contexts. For example, quantum chemistry constitutes nowadays one major field in importance and interest. In order to carry out the large-scale calculations and simulations, an increasing computational capacity of modern computers is critical. The corresponding theoretical analysis relies on considering the so-called *Born-Oppenheimer approximation*, which allows us to simplify the study of any molecular system (simple molecules, solid surfaces, polymeric chains, clusters, crystalline structures, etc.) by splitting it into its electronic and nuclear parts. By solving the Schrödinger equation associated with the electronic Hamiltonian, which is the ultimate goal of electronic structure methods, one determines electronic configuration of a (multi-electron) system, i.e., the way how electrons distribute throughout such a system. This knowledge will essentially determine its chemical properties, e.g., chemical bonding, intermolecular interactions, electronic charge distributions, dipole and multipole moments, or vibrational/rotational frequencies. Non–adiabatic problems have also been tackled within the Bohmian complex formalism [42, 43].

When the dimensionality of the problem considered increases, as happens with many systems of quantum chemistry and solid state physics, computational problems increase and start becoming unaffordable. To overcome this drawback, a series of approaches based on the density matrix theory [44] were also developed; for example, the independent electron approximation or Thomas-Fermi model [45, 46]. These models can be considered direct predecessors of the so-called *density functional theory* (DFT) [47], based on the principles established by Hohenberg, Kohn and Sham [48, 49]. Here we go from the Thomas-Fermi model [45, 46], where the electron kinetic energy of an atom is expressed as a functional of the atom electronic distribution, to the modern DFT [47], where the fundamental physical information about the molecular system is obtained from a single-particle density in a three-dimensional space, derived variationally either within a time-independent framework [47] (ground state) or a time-dependent one [50, 51] (excited states). The density functional theory is a very popular calculation method nowadays, not only in quantum chemistry, but also in solid state physics or condensed matter physics; in particular, due to its computational simplicity combined with its (numerical) accuracy. Accordingly, the relevant physical information about the ground state of a many-body system is obtained from single-particle densities in a three-dimensional space, which are determined variationally within a time-independent framework. Nonetheless, all practical applications of DFT rely on essentially uncontrolled approximations [52] and therefore the validity of this approach is conditioned to its ability to provide results sufficiently close to the experimental data.

Standard DFT is commonly applied to determine ground states in time-independent problems. However, its time-dependent generalization, the so-called time-dependent DFT (TD-DFT), are needed when processes and phenomena are intrinsically time-dependent [51]. This is the case, for example, of reactive and non-reactive scattering processes, or of atoms and molecules in laser fields, where the calculation of excited states is important. This approach was formerly started by Bartolotti [53–55] and Deb and Ghosh [56, 57], although the proofs of the fundamental theorems involved were later on provided by Runge and Gross [58]. One of these theorems corresponds to a version of the Hohenberg–Kohn theorem for time-dependent Schrödinger equation. Thus, as happens with standard DFT, TD-DFT can be started directly from the many-body time-dependent Schrödinger equation, the density being then determined from solving a set of time-dependent Schrödinger equations for single, non-interacting particles. A slightly extended version is based on Madelung's quantum hydrodynamic picture of wave mechanics [59] which has been widely exploited in the chemical physics community, giving rise to the so-called quantum hydrodynamic or quantum fluid DFT. According to Hirschfelder [60–62], quantum hydrodynamics facilitates the study of problems involving external electric and magnetic fields in molecular systems, since it contains in a natural way the concept of *equation of change* for any arbitrary quantum properties in configuration space. For instance, these equations can be used to study the energy flow from one part of a molecule to another, the nature of molecular collisions, or the magnetic properties of molecules. We would also like to notice an interesting relationship between this formulation and the weak measurement or weak value [21], proposed about twelve years after Hirschfelder's work. As mentioned above, experiments carried out by Kocsis et al. [19, 20] are not only in accordance to Bohmian mechanics but also to Bayesianism proposed by Wiseman [63]. As pointed out by Hiley [64], these weak measurements are not other thing that standard transition probabilities within the Bohmian scenario, i.e., just a particular type of equation of change, according to Hirschfelder.

On the other hand, strictly speaking, real physical systems do not exist in complete isolation in nature. All physical systems are open systems in the sense that the interaction with their environments can never be totally neglected. From its inception, the motion of particles in quantum mechanics was thought to happen in the presence of an environment and could be understood in terms of a stochastic process. In fact, a formal analogy between the Brownian motion and the Schrödinger equation was noticed by Fürth [65] in 1933. Afterwards, Fényes [66] and Weizel [67–69] developed this approach with more mathematical detail. The search for a stochastic support for quantum mechanics has taken place since the early 1950s. In particular, the first attempt in the Bohmian mechanics due to Bohm and Vigier [70] was reported in 1954. These authors assumed that the electron is a particle suspended in a Madelung fluid [59] (the hydrodynamical model of quantum mechanics was lately developed by Takabayasi [71, 72]) whose general motion is determined by the resolution of the Schrödinger equation, providing the probability density from the amplitude of the wave function and the deterministic contribution to the local velocity from its phase. The stochastic contribution to the local velocity is due to the random fluctuations arising from the presence of the fluid. In general, the aim was to show how close it is to classical theory of Brownian motion and Newtonan mechanics, and how the Schrödinger equation might have been discovered from this point of view. The theory was neatly established by Nelson [73] starting from a different point of view of those proposed by Fényes, Weizel, Kershaw [74], Comisar [75] and de la Peña [76]. After Nelson, every particle of a given mass, m, is subject to a Brownian motion with diffusion coefficient $\hbar/2m$ and no friction in order to preserve Galilean covariance. He explicitly wrote that his theory is not a causal theory and the physical interpretation is entirely classical. Olavo [77] also followed previous works but starting from a more axiomatic formulation. The main advantage of implementing this stochastic mechanics was to exploit the well developed mathematical methods of probability theory and stochastic processes. In a review written by Guerra [78] in 1981, one still can read "there is no stochastic interpretation of quantum mechanics different to the standard one... stochastic mechanics can be seen simply as a reformulation of it in terms of the language of stochastic processes". Grabert, Hänggi and Talkner analyzed the connection between both mechanics and concluded that quantum mechanics is not equivalent to a Markovian diffusion process [79]. Most of the work in this domain ignored Bohm's view of standard and stochastic quantum trajectories. In this respect, we can claim that the depart of Bohm and collaborators is again a breakthrough. Bohm and Hiley [10, 80] make clear that there is no way to avoid non-locality in the stochastic interpretation of quantum mechanics. After Vink [81], the causal and stochastic interpretations are two known examples of what is denominated beable interpretations as developed by Bell [1]. Lorenzen et al. [82] have recently extended Bell's beables to encompass dissipation, decoherence and the quantum-to-classical transition through quantum trajectories.

Concurrently, Nelson and de la Peña opened up an alternative route for deriving generalized nonlinear Schrödinger equations. Within the quantum hydrodynamical framework, Nassar [83, 84] proposed a generalized nonlinear equation containing some of the most well-known equations due to Kostin [85], Süssmann and Hasse [86], Bialynicki-Birula–Mycielski [87], Stocker-Albrecht [88] and Schuch–Chung–Hartmann [89, 90]. More recently, Olavo et al. [91] have provided an alternative way of combining Bohmian equations and the Langeving and Fokker–Planck equations and the Feynman propagator was analyzed for eight nonlinear Schrödinger equations [92] within this formalism. Garashchuk et al. [93] and Chou [94] applied the so-called Kostin equation to the symmetric and asymmetric double well potential for ground states. In the same framework, Nassar [95] also studied the linear Schrödinger equation within the Caldirola-Kanai model [96, 97]. A detailed analysis of quantum trajectories in the Caldirola-Kanai model has also been carried out by Sanz et al. [98].

The theory of open quantum systems has also been developed by different routes from the very beginning leading to different stochastic derivations of the Schrödinger equation. Traditionally, the so-called system-plus-reservoir methods are being largely used in this context, working either in the Schrödinger picture (density matrix formalism) or the Heisenberg picture of quantum mechanics (quantum Langevin equation). The reader interested in these different alternatives can look at, for example, Refs. [13, 14, 99–105]. In this context, the terminology of quantum trajectory is also used but it should not be confused with the corresponding Bohmian trajectory. When dealing with open quantum systems within the Bohmian formalism in the following chapters, these trajectories will be handled with care.

Dealing with measurements, the physical system to be measured together with the apparatus can also be considered as an open system [106]. In classical physics, measurements on a physical system are carried out without appreciably disturbing it. Positions and momenta can be empirically determined with arbitrary precision at the same time. In classical ontology terms, one assumes the existence of particles and fields to be essentially independent of the human observer. Moreover, following Holland [7], two successive observations of an apple, in a tree and then on the ground are connected to one another. In fact, they are causally connected. Non-disturbing measurements which simply reveal a preexisting reality cannot be sustained when we approach the quantum realm [10]. After Bohr, one cannot separate the behavior of the quantum objects from their interaction with the instruments. However, the key concept in this mechanics is the concept of a quantum state along the lines initiated by Dirac and von Neumann. After Dirac, each physical quantity is represented by an Hermitian operator which is called an observable. When this property is measured by an apparatus, the system is left with a wave function corresponding to an eigenfunction of this observable. Afterwards, we can measure the same observable again and again with the same result that the wave function does not change significantly (except for a phase factor). After von Neumann, the wave function provides the most complete possible description of what he called quantum reality (no further concepts, e.g., involving hidden variable could be introduced that would make possible a more detailed description of the state of the system than is afforded by the wave function). In 1955, von Neumann proposed to divide the world into two parts: an observed system and a system that may "observe", say a meter or a pointer. In other words, he was led to make a distinction between the quantum and classical levels. Between them, he said there was a cut or *a dividing line*. The location of the cut is to a large extent arbitrary. The apparatus must combine quantum and classical properties. The Schrödinger equation can not map a pure state into a mixture, or better said, a proper mixture killing the interference terms. Von Neumann also proposed a different evolution, discontinuous, noncausal and nonunitary, the so-called *projection postulate* or the hypothesis of *wave function collapse or reduction* which is an irreversible process and takes place instantaneously. Thus, the many-body Schrödinger theory is no longer applicable in the measuring process and this implies a massive incompleteness in the quantum mechanical treatment. No one knows when the collapse is supposed to occur, how long it takes or what brings it about. This process of reduction or collapse of the wave function is the simplest example of decoherence or conversion of a superposition state into one of the eigenstates of the observable measured with probability given by the squared modulus of the corresponding coefficient. The emergence of the classical world implies the phenomenon of decoherence, that is, the conversion of a coherent superposition of states into a mixture since superposition states are not observable. One of the components of the mixture has to be somehow selected to lead to a definite measurement readout. In other words, the measuring process is the selection of one out of many alternatives. Irreversibility is then settled.

Recently, Nassar [107] has proposed a different starting point within the Bohmian formalism when considering the measuring process as an open quantum dynamics. Following the phenomenological approach to continuous measurement by Mensky [108] which uses restricted path integrals, a new logarithmic nonlinear Schrödinger equation has been derived and analyzed in terms of quantum trajectories. A close connection between stochastic mechanics and the path integral formulation of quantum mechanics due to Feymann is noticeable. Furthermore, as an extension, the *stochastic weak measurement* is also introduced in this context following Hiley arguments [64]. As a further extension of this work, Nassar and Miret-Artés [109] and Bargueño and Miret-Artés [110] have derived new generalized Schrödinger-Langevin equations by considering continuous measurement for a dissipative and stochastic dynamics, respectively. Furthermore, Zander, Plastino and Díaz-Alonso [111] have investigated time-dependent solutions for the nonlinear equation proposed by Nassar and Miret-Artés.

This chapter is organized as follows. In Sect. 1.2, the main concepts and developments derived for closed or isolated quantum systems are briefly reviewed. In Sect. 1.3, the same is done but now for open quantum systems. Finally, in Sect. 1.4 some dynamical aspects of the measuring process are presented. This introductory chapter aims at providing the essential tools and background to better understand the rest of the monograph.

1.2 Closed Quantum Systems. Quantum Trajectories

1.2.1 Quantum Hydrodynamics Framework

In 1926, Madelung [59] provided an alternative interpretation to quantum mechanics which is known nowadays as *quantum hydrodynamics*. This interpretation is, for example, directly connected to some relevant phenomena in quantum mechanics, such as superconductivity [112] or Bose–Einstein condensation [113]. Furthermore, in chemical physics, it provides an ideal framework to understand and interpret quantum processes, going from chemical reactivity in collinear reactions [62, 114] to the understanding of molecular magnetic properties within a framework encompassing both electronic structure and topology [115–117].

In this framework, one starts by considering the wave function of a particle in the configuration space and in polar form such as

$$\Psi(\mathbf{r}, t) = \phi(\mathbf{r}, t)e^{iS(\mathbf{r},t)/\hbar} \tag{1.1a}$$

$$\rho(\mathbf{r}, t) = \phi(\mathbf{r}, t)^2 = \Psi^*(\mathbf{r}, t)\Psi(\mathbf{r}, t), \tag{1.1b}$$

$$\mathbf{J}(\mathbf{r}, t) = \rho(\mathbf{r}, t)\mathbf{v}(\mathbf{r}, t) = \phi^2(\mathbf{r}, t)\,\frac{\nabla_{\mathbf{r}} S(\mathbf{r}, t)}{m}, \tag{1.1c}$$

where $\rho(\mathbf{r}, t)$ is the probability density, $\mathbf{J}(\mathbf{r}, t)$ is the quantum probability density current and $\mathbf{v}(\mathbf{r}, t)$ is the velocity field, which describes the flow of the latter. In order to simplify the notation of equations, the arguments of the fields will be very often omitted and $\nabla_{\mathbf{r}}$ will be replaced simply by ∇ (for one or N particles). By substituting Eq. (1.1a) in the time dependent Schrödinger equation

$$i\hbar\frac{\partial\Psi(\mathbf{r}, t)}{\partial t} = H\Psi(\mathbf{r}, t) \tag{1.2}$$

where H is the Hamiltonian operator of a particle expressed as

$$H = K + V = -\frac{\hbar^2}{2m}\nabla^2 + V(\mathbf{r}) \tag{1.3}$$

K being the kinetic energy operator and V the interaction potential, the following two coupled equations are easily obtained

$$\frac{\partial\rho}{\partial t} + \nabla\cdot\mathbf{J} = 0, \tag{1.4a}$$

$$\frac{d\mathbf{v}}{dt} = \frac{\partial\mathbf{v}}{\partial t} + (\mathbf{v}\cdot\nabla)\mathbf{v} = -\frac{1}{m}\,\nabla(V + Q). \tag{1.4b}$$

The first equation is the so-called *continuity equation* and the second one is a Newtonian like equation in the quantum domain, or quantum Newton equation (or quantum Euler equation). In this last equation, the effective interaction potential consists of V and the so-called quantum potential Q expressed in terms of ρ as

$$Q = -\frac{\hbar^2}{2m}\frac{\nabla^2\rho^{1/2}}{\rho^{1/2}}. \tag{1.5}$$

Equations (1.4a) and (1.4b) constitute the formal basis of quantum hydrodynamics and have a direct correspondence with those of classical fluid mechanics, when m is identified with the mass of a piece of fluid separated from the rest by a closed surface, $m\rho$ is the fluid density and $\mathbf{v}$ is the velocity field of the flow (ideal classical fluid, that is, incomprenssible and nonviscous and thermal effects are not taken into account) [118]. The streamlines of the flow are lines along which the fluid current goes or energy is transported. Typically, some tracer particles are usually deposited on the fluid in order to visualize such streamlines. The tangents to the streamlines give the directions of the velocities of fluid particles at various points in space at a given instant and are perpendicular to the constant phase (or $S(\mathbf{r}, t) = constant$) lines. These streamlines are defined by the solution of the coupled differential equations

$$\frac{dx}{v_x(\mathbf{r}, t)} = \frac{dy}{v_y(\mathbf{r}, t)} = \frac{dz}{v_z(\mathbf{r}, t)}. \tag{1.6}$$

However, unlike classical fluids, quantum fluids correspond to probability flows with no material structure [119]. That is, they only characterize statistical events at each point in space and time, in spite of the fact that the time evolution of these events can be better understood when compared with the motion of ordinary fluids. Moreover, whereas the classical concept of fluid can be applied to describe the statistical behavior of a macroscopic ensemble of particles, in quantum mechanics it is applied to single particles.

When this formal similarity with the evolution of classical hydrodynamical flows is established, one is faced to two types of theoretical schemes: the *Eulerian* and *Lagrangian* schemes [120]. In the first scheme, the importance relies on the full quantum fluid rather than in the particular trajectory dynamics. Equations (1.4a) and (1.4b) are then integrated directly with the aid of the velocity field, $\nabla S/m$, but with no need to obtain any particular trajectory. Alternatively, one can also work following a *Lagrangian scheme*, i.e., considering a framework co-moving with the quantum fluid along trajectories. This is the typical scheme considered in order to benefit from the computational advantages of quantum trajectories. Thus, the first step is to pass from the Eulerian scheme to the Lagrangian one by means of the well-known time derivative or Lagrangian operator,

$$\frac{d}{dt} = \frac{\partial}{\partial t} + \mathbf{v} \cdot \nabla. \tag{1.7}$$

On the other hand, the classical Euler equation for the component v_i of $\mathbf{v}$ [121]

$$\rho\left[\frac{\partial v_i}{\partial t} + (\mathbf{v} \cdot \nabla)v_i\right] = \rho f_i + \frac{\partial}{\partial x_j}(-p\delta_{ij}), \tag{1.8}$$

where f_i is the external force acting on the fluid along the i–direction and p the fluid pressure, shows how the flow dynamics is determined by the influence of both, an external force and other internal one, given by $\rho^{-1}\partial(-p\delta_{ij})/\partial x_j$ and that depends on the fluid properties. Equation (1.4b) can then be rewritten in the form of Eq. (1.8) as

$$\rho\left[\frac{\partial v_i}{\partial t} + (\mathbf{v} \cdot \nabla)v_i\right] = \rho f_i + \frac{\partial T_{ij}}{\partial x_j}, \tag{1.9}$$

by defining the *quantum stress tensor* as

$$T_{ij} = \frac{\hbar^2}{4m^2}\,\rho\,\frac{\partial \ln \rho}{\partial x_{ij}}, \tag{1.10}$$

which is the quantum counterpart of the classical stress tensor $-p\delta_{ij}$, and whose explicit dependence on ρ can be easily obtained by expressing Q as

$$Q = -\frac{\hbar^2}{2m}\left[\frac{1}{2}\left(\nabla \ln \rho\right)^2 + \nabla^2 \ln \rho\right]. \tag{1.11}$$

In classical fluid dynamics, the Navier–Stokes equation [118] expresses the rate of change in the momentum density, which is defined as the linear momentum times the fluid density. In the quantum version, apart from the classical contributions, quantum contributions are clearly identified and are related to quantum stress and pressure.

As also happens with classical fluids, in quantum hydrodynamics one can also observe the presence of vortices and the corresponding associated (vortical) dynamics. The first theory on quantum vortices was first formulated by Dirac [122] in connection with the existence of magnetic monopoles. This theory has been described in detail in the literature [61, 123], finding some interesting applications [124]. More recent and explicit developments of the quantum theory of magnetic monopoles had led to a generalization of the concept of quantum vortex [125], the well-known Aharonov–Bohm effect [126] (see Chap. 2) being related to this generalization.

The conditions leading to the formation of quantum vortices can be obtained from the fact that the complex character of the wave function implies the multivaluedness of its phase

$$S'(\mathbf{r}, t) = S(\mathbf{r}, t) + 2\pi n\hbar, \qquad n = 0, \pm 1, \pm 2, \ldots \tag{1.12}$$

This multivaluedness can only take place at those points where $\rho = 0$ (nodal points or nodes, $\Psi = 0$), where the smoothness of the wave function disappears and the value of S may undergo discrete jumps. According to Eq. (1.12), under these conditions, $\mathbf{J}$ vanishes but not the velocity field $\mathbf{v}$. By inspecting the circulation of $\mathbf{v}$ along a closed path, $\mathcal{C}$, one finds that this magnitude is quantized,

$$\oint_{\mathcal{C}} d\mathbf{l}\cdot\mathbf{v} = \oint_{\mathcal{C}} d\mathbf{l}\cdot\frac{\nabla S}{m} = \frac{1}{m}\oint_{\mathcal{C}} dS = \frac{2\pi n\hbar}{m}, \tag{1.13}$$

n being an integer number. Applying Stoke's theorem to this result, it can be alternatively expressed as

$$\int_{\Sigma} d\mathbf{r}\cdot(\nabla \times \mathbf{v}) = \frac{2\pi n\hbar}{m}, \tag{1.14}$$

where Σ is the region enclosed by $\mathcal{C}$. This result indicates the appearance of vortices when $n \neq 0$, which happens only at those points where the wave function presents nodes. At these points, the streamlines will be closed paths around the nodes, which is consistent with the fact that the quantum density current vanishes at those points and the impossibility of passing through the regions where $\rho = 0$. Conversely, the velocity field $\mathbf{v}$ will be irrotational in those regions free of quantum vortices.

1.2.2 The Equations of Motion for One Particle

In 1952, Bohm [4, 5] proposed a different interpretation of quantum mechanics in terms of "hidden" (trajectories) variables and based on the quantum hydrodynamics of Madelung and de Broglie's ideas of a pilot wave. In configuration space, the fundamental Bohmian equations of motion for a particle are usually derived from the standard version of quantum mechanics through the transformation $(\Psi, \Psi^*) \rightarrow (\rho, S)$, where Ψ and Ψ^* are generally complex-valued functions or fields of the position ($\mathbf{r}$) and time (t), and ρ and S are also real-valued functions or fields of the same variables. More explicitly, in analogy to previous section, the same transformation relation between both types of fields for a particle of mass m is given by Eqs. (1.1a) and (1.1b) After introducing Eq. (1.1a) into the time-dependent Schrödinger equation (1.2), two real, coupled partial differential equations are obtained,

$$\frac{\partial \rho}{\partial t} + \frac{1}{m} \nabla(\rho \nabla S) = 0, \tag{1.15a}$$

$$\frac{\partial S}{\partial t} + \frac{(\nabla S)^2}{2m} + V_{\text{eff}} = 0, \tag{1.15b}$$

which come from the imaginary and real parts, respectively, of the resulting equation. The former is again the *continuity equation*, which accounts for the probability conservation, while the latter is the so-called *quantum Hamilton–Jacobi equation*, with

$$V_{\text{eff}}(\mathbf{r}, t) = V(\mathbf{r}) + Q(\mathbf{r}, t) \tag{1.16}$$

being an *effective* potential (notice the different way of expressing the same equations previously, Eqs. (1.4a) and (1.4b)). The last term in the right–hand side of Eq. (1.16) is again the *quantum potential* explicitly written now as

$$Q = \frac{\hbar^2}{4m} \left[\frac{1}{2} \left(\frac{\nabla \rho}{\rho} \right)^2 - \frac{\nabla^2 \rho}{\rho} \right], \tag{1.17}$$

which, as well as ρ, depends on both $\mathbf{r}$ and t. This term is regarded as a potential because, like V, also rules the quantum particle dynamics. However, unlike V, it is not a preassigned function. Moreover, its nature is fully quantum-mechanical due to its dependence on the quantum state via the evolution of ρ (state dependence). The quantum potential is not changed when the wave function is multiplied by an arbitrary constant, only its form (related to the curvature of the wave function) is important in the dynamics. Even more, again unlike V, this interaction does not necessarily fall off with the distance. Due to the fact that ρ describes statistically the evolution of a swarm of identical, non interacting particles, the dependence of Q on ρ means that the dynamics of a single particle from the swarm is going to be influenced by the behavior of the others. In other words, the quantum particle dynamics is *nonlocal*. Rather than being a particular feature of Bohmian mechanics, this property is inherent to quantum mechanics in general, which manifests through the kinetic energy operator, $K = -(\hbar^2/2m)\nabla^2$. Note that Q also arises from the action of this operator on Ψ after considering Eq. (1.1a)

$$K\Psi = -\frac{\hbar^2}{2m}\nabla^2\Psi = \frac{(\nabla S)^2}{2m} - \frac{\hbar^2}{2m}\frac{\nabla^2\rho^{1/2}}{\rho^{1/2}}, \tag{1.18}$$

and, therefore, it could also be associated with a sort of nonlocal kinetic energy [7].

Under the presence of an external potential, the de Broglie wavelength of the particle has to be modified due to the presence of the quantum potential. Thus, this variable is also state dependent,

$$\lambda_Q(\mathbf{r}, t) = \frac{h}{|\nabla S(\mathbf{r}, t)|} = \frac{h}{[2m(-\frac{\partial S(\mathbf{r},t)}{\partial t} - Q - V)]^{1/2}}. \tag{1.19}$$

In particular, when one is interested in calculating the turning points of a given problem, the quantum potential plays a very important role. In a certain sense, we have to differentiate the "classical" turning points (which are obtained from V) from the "quantum" turning points [127] (obtained from $V + Q$).

Paths along which quantum particles travel may be defined according to the so-called *guidance condition*

$$\mathbf{v} = \frac{\nabla S}{m} = \frac{\mathbf{J}}{\rho} = \frac{\hbar}{2im}\left[\frac{\nabla\Psi}{\Psi} - \frac{\nabla\Psi^*}{\Psi^*}\right], \tag{1.20}$$

with $\mathbf{v}(\mathbf{r}, t) = \dot{\mathbf{r}}$ being the local velocity field. The initial velocity of the particle is then determined by the initial field $\Psi(\mathbf{r}, 0)$. However, the resulting trajectories are of quantum nature and termed *quantum or Bohmian trajectories* This expression reminds us the so-called method of characteristics in the partial differential equations [128]. The integration of the guiding condition generates the corresponding characteristics perpendicular to the S-surface of constant phase (streamlines). Note that in standard quantum mechanics the system dynamics is only described by Ψ. However, in Bohmian mechanics, one focusses on the particular evolution in time of a given initial system configuration (initial condition), which gives rise to the corresponding Bohmian trajectories. This evolution is strongly determined by the wave function which acts like a field, apart from any other external field described by the potential V. The particle is then always accompanied by its quantum field or wave function, this combined system being causally determined. However, the converse is not always true. Imagine the wave is split up into a set of parts with no appreciable spatial overlap (nodal regions in-between). The particle can only follow one of them, the remaining waves being called *empty waves* [7].

After the quantum Newton equation given by Eq. (1.4b), one can find asymptotic regions where the external potential V is zero or negligible but the quantum potential is still active. In other words, the quantum force or the acceleration of the particle can still be important in a classical free particle regime. Furthermore, we can also imagine a situation where the classical force cancels out the quantum force. This only occurs in a stationary state (the wave function can be taken to be real) and, therefore, the particle is at rest (zero momentum) in a fixed position. Similarly, the tunneling effect can also be easily explained in terms of the quantum potential since it lowers the barrier height given by V, thus permitting particles to enter it.

Due to the linearity of the Schrödinger equation, the superposition principle holds. Thus, if ψ_1 and ψ_2 are solutions of that equation, its superposition

$$\Psi(\mathbf{r}, t) = c_1\psi_1(\mathbf{r}, t) + c_2\psi_2(\mathbf{r}, t) = c_1\left[\psi_1(\mathbf{r}, t) + \sqrt{\alpha}\psi_2(\mathbf{r}, t)\right] \tag{1.21}$$

is also a solution, with $\alpha = (c_2/c_1)^2$. There should also be an additional phase factor $e^{i\delta}$ multiplying ψ_2, but we will consider $\delta = 0$ here, for simplicity; the consideration of this phase is, on the contrary, important for some problems; in particular, as will be discussed in the next chapter, for the so-called Aharonov-Bohm effect. The particular interest in solutions like (1.21) relies on the fact that they express rather well the character of quantum interference [129], more apparent through the associated probability density,

$$\rho = |\Psi|^2 = c_1^2\left[\rho_1 + \alpha\rho_2 + 2\sqrt{\alpha}\sqrt{\rho_1\rho_2}\cos\varphi\right], \tag{1.22}$$

where both ψ_1 and ψ_2 are expressed in polar form ($\psi_j = \rho_j^{1/2}e^{iS_j/\hbar}$, $j = 1, 2$) and $\varphi = (S_2 - S_1)/\hbar$ with $\Psi = \rho^{1/2}e^{iS/\hbar}$. The presence of the oscillatory term in (1.22) constitutes an observable evidence of interference, since ψ_1 and ψ_2 form a *coherent superposition*. As can be noticed, ρ does not satisfy the superposition principle, as also happens with the associated probability density current,

$$\mathbf{J} = \frac{1}{m}\mathrm{Re}\left[\Psi^*\mathbf{p}\Psi\right] = -\frac{i\hbar}{2m}\left[\Psi^*\nabla\Psi - \Psi\nabla\Psi^*\right], \tag{1.23}$$

where $\mathbf{p} = -i\hbar\nabla$ is the momentum (vector) operator. Substituting Eq. (1.21) into (1.23), with again ψ_1 and ψ_2 expressed in polar form, yields

$$\mathbf{J} = \frac{c_1^2}{m}\Big[\rho_1 \nabla S_1 + \alpha\rho_2 \nabla S_2 + \sqrt{\alpha}\sqrt{\rho_1\rho_2}\nabla(S_1 + S_2)\cos\varphi + \hbar\sqrt{\alpha}(\rho_1^{1/2}\nabla\rho_2^{1/2} - \rho_2^{1/2}\nabla\rho_1^{1/2})\sin\varphi\Big]. \tag{1.24}$$

One can further proceed and determine probability streamlines (lines along which the probability flows), as in classical hydrodynamics. Now, from the guiding condition (1.20), well-defined (in space and time) quantum trajectories are obtained by integrating the equation of motion

$$\begin{aligned}\dot{\mathbf{r}} = \frac{\mathbf{J}}{\rho} = \frac{\nabla S}{m} &= -\frac{i\hbar}{2m}\frac{\Psi^*\nabla\Psi - \Psi\nabla\Psi^*}{\Psi^*\Psi} \\ &= \frac{1}{m}\frac{\rho_1\nabla S_1 + \alpha\rho_2\nabla S_2 + \sqrt{\alpha}\sqrt{\rho_1\rho_2}\nabla(S_1+S_2)\cos\varphi}{\rho_1 + \alpha\rho_2 + 2\sqrt{\alpha}\sqrt{\rho_1\rho_2}\cos\varphi} \\ &\quad + \sqrt{\alpha}\frac{\hbar}{m}\frac{(\rho_1^{1/2}\nabla\rho_2^{1/2} - \rho_2^{1/2}\nabla\rho_1^{1/2})\sin\varphi}{\rho_1 + \alpha\rho_2 + 2\sqrt{\alpha}\sqrt{\rho_1\rho_2}\cos\varphi},\end{aligned} \tag{1.25}$$

and where the initial momentum of particles is predetermined by the choice of the initial wave function and, therefore, by $\rho_0(\mathbf{r})$.

Gaussian wave packets are very helpful and easy to manipulate in theoretical treatments. Let us consider a free one-dimensional, spinless quantum system of mass m with its initial state being described (in configuration space) by

$$\Psi_0(x) = \left(\frac{1}{2\pi\sigma_0^2}\right)^{1/4} e^{-(x-x_0)^2/4\sigma_0^2 + ip_0(x-x_0)/\hbar}, \tag{1.26}$$

where x_0 and p_0 are the position and (translational or propagation) momentum of its center (in brief, the centroidal position and momentum), respectively, and σ_0 its initial spatial spreading. The exact time-dependent form of this wave function in free space $[V(x) = 0]$ can be readily obtained

$$\Psi_t(x) = \left(\frac{1}{2\pi\tilde{\sigma}_t^2}\right)^{1/4} e^{-(x-x_t)^2/4\sigma_0\tilde{\sigma}_t + i\,p_0(x-x_t)/\hbar + iE_0t/\hbar}, \tag{1.27}$$

where the centroidal position and momentum are given by $x_t = x_0 + (p_0/m)t$ and $p_t = p_0$, respectively, and correspond to a classical trajectory describing a rectilinear uniform motion. This means that the wave packet travels with a speed or *group velocity* $v_0 = p_0/m$ and has a total mechanical energy $E_0 = p_0^2/2m$. On the other hand, taking into account that

$$\tilde{\sigma}_t = \sigma_0 \left(1 + \frac{i\hbar t}{2m\sigma_0^2} \right), \tag{1.28}$$

the spreading of the wave packet at a time t is given by

$$\sigma_t = |\tilde{\sigma}_t| = \sigma_0 \sqrt{1 + \left(\frac{\hbar t}{2m\sigma_0^2} \right)^2}. \tag{1.29}$$

Therefore, the evolution of a wave packet is characterized by two types of motion: translational or extrinsic, and spreading or intrinsic [14].

By considering the wave packet as a whole rather than a superposition of plane waves, one can associate its spreading along time with a type of internal or intrinsic kinetic energy [129]. This can be seen when computing the expectation or average value of the total energy,

$$\bar{E} = \langle \hat{H} \rangle = \frac{p_0^2}{2m} + \frac{p_s^2}{2m}. \tag{1.30}$$

Here we observe a second contribution to the energy, which comes from averaging the quantum potential [129] and is directly connected with the spreading of the wave packet. Hence, we can we can define an effective *spreading momentum*,

$$p_s = \frac{\hbar}{2\sigma_0}. \tag{1.31}$$

It is interesting that this momentum also appears when one computes the wave-packet energy dispersion (variance),

$$\Delta E \equiv \sqrt{\langle \hat{H}^2 \rangle - \langle \hat{H} \rangle^2} = \sqrt{\frac{2p_s^2}{m}} \sqrt{\frac{p_0^2}{2m} + \frac{p_s^2}{4m}}. \tag{1.32}$$

The relationship between p_s and the wave-packet spreading becomes more apparent when (1.31) is analyzed in the light of Heisenberg's uncertainty principle: a spreading of the size of σ_0 gives rise to a spreading in momenta of the order of p_s. Actually, one could be tempted to think that classical-like regimes correspond to conditions where spreading momenta vanish.

Given these two types of motions governing the time-evolution of a wave packet, it is clear that the ratio between the corresponding velocities, v_t $(= p_t/m)$ and v_s $(= \hbar/2m\sigma_0)$, will play an important role in many physical processes such as, for example, interference and tunneling. From Eq. (1.29), one can always define the following time scale

$$\tau = \frac{2m\sigma_0^2}{\hbar} = \frac{\sigma_0}{v_s}, \tag{1.33}$$

which is associated with the relative spreading of the wave packet. Depending on the relationship between t and τ, we can identify three dynamical regimes [130]:

1. The very-short-time or *Ehrenfest-Huygens regime*, $t \lll \tau$, where the wave packet remains almost spreadless: $\sigma_t \approx \sigma_0$.
2. The short-time or *Fresnel regime*, $t \ll \tau$, where the spreading increases nearly quadratically with time: $\sigma_t \approx \sigma_0 + (\hbar^2/8m^2\sigma_0^3)t^2$.
3. The long-time or *Fraunhofer regime*, $t \gg \tau$, where the Gaussian wave packet spreads linearly with time: $\sigma_t \approx (\hbar/2m\sigma_0)t$.

This scheme results very useful in order to determine a way to elucidate which process, translational motion or spreading, is going to dominate the future evolution of the wave packet. Let us recast (1.29) in terms of v_s and consider that t is the time required for the centroidal position to cover a distance $d = v_0 t \approx \sigma_0$. Substituting the latter expression into (1.29), we have that

$$\sigma_t = \sigma_0 \sqrt{1 + \left(\frac{v_s}{v_0}\right)^2}. \tag{1.34}$$

Accordingly, using only information about the initial preparation of the wave packet, some information can be extracted about its subsequent dynamical behavior. More specifically, if $v_s \ll v_0$, the wave-packet spreading will be relatively slow, this being equivalent to the condition in time $t \ll \tau$. Furthermore, if $v_s \gg v_0$, the wave packet will spread very rapidly compared to its propagation along x, which corresponds to the time condition $t \gg \tau$.

From this brief wave packet analysis, and within the Bohmian formalism, the superposition principle in one dimension can be easily analyzed. Let us consider the interference between two identical Gaussian wave packets propagating in opposite directions at the same speed (by assuming $\alpha = 1$) [129]. In Fig. 1.1, the corresponding dynamics is illustrated showing quantum or Bohmian trajectories ((a) panel) for the coherent superposition of the two counter-propagating Gaussian wave packets with an initial velocity $v_0 = 10$ (in arbitrary units) and spreading motion with velocity $v_s = 1$ (also in arbitrary units). The (b)–(f) panels and transversal dashed lines correspond to different times in this coherent evolution. For simplicity in the discussion, we will refer to the two non-overlapping regions where ψ_1 and ψ_2 are initially placed as I (red plot) and II (black plot), respectively. If $t_{\max}^{\rm int}$ denotes the time at which interference is maximum (see Fig. 1.1d), then for $t \gg t_{\max}^{\rm int}$ ψ_1 will continue its evolution in II. Now, the wave packets (or, to be more precise, their probability densities) can be interpreted statistically as describing a swarm of identical, noninteracting quantum particles distributed accordingly. As also seen in Fig. 1.1a, Bohmian trajectories avoid crossing the line $x = 0$ at any time, this leading us to interpret the superposition principle in a different way. Quantum statistics retains the essence of (quantum) coherence and transmits it to the corresponding dynamics, leading to a zero velocity field at $x = 0$. Thus, a net flux of probability (or particles) crossing between regions I and II is not possible. In other words, trajectories starting in

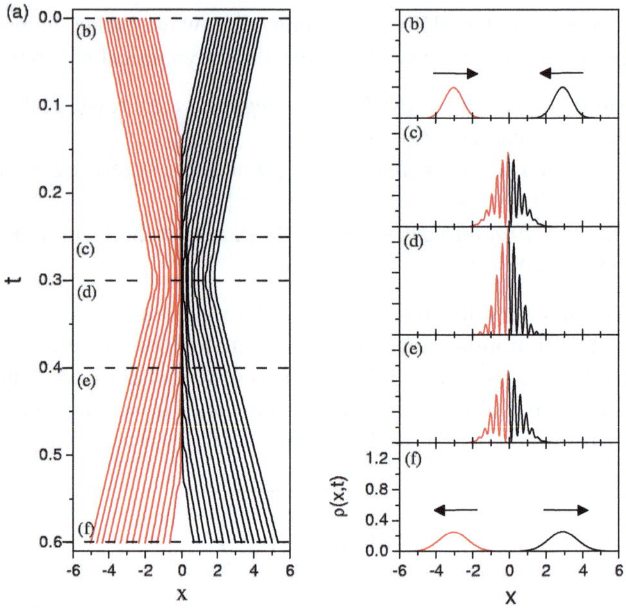

Fig. 1.1 **a** Bohmian trajectories associated with the coherent superposition of two counter-propagating Gaussian wave packets with $v_0 = 10$ and $v_s = 1$. Panels **b**–**f** represent snapshots illustrating the time-evolution of the superposition at the times indicated by the transversal dashed lines in panel (**a**)

region I, for example, will always keep moving inside this region. Correspondingly, the outgoing wave packet in region I (see Fig. 1.1f) accounts for the same swarm of trajectories associated with ψ_1 in Fig. 1.1b. This whole process can then be understood as a sort of bouncing motion of the wave packets once they have reached the intermediate position between them.

In spite of the fact that the two swarms of trajectories do not cross each other, the swarm associated initially with one of the wave packets behaves asymptotically as associated with the other one. Thus, from the standard picture provided by quantum mechanics for the superposition principle, the wave packets cross each other. However, from the Bohmian picture, the process should be understood as a transfer or interchange of probabilities from region I to region II, and vice versa, with the velocity field also changing its sign. Similarly to a particle-particle elastic scattering process, where particles exchange their momenta, here the swarms of particles exchange their probability distributions elastically.

On the other hand, nonlocality only disappears when $Q \equiv 0$, that is, when the particle dynamics becomes fully classical. In this case, the classical Hamilton–Jacobi equation is recovered and one has then to replace S by S_c. According to the *correspondence principle* stated by Bohr, quantum systems may behave classical-like under certain conditions, thus being describable by means of the classical mechanical laws. This is usually regarded as the *classical limit* of quantum mechanics. Usually, this limit relies on assuming that the value of a certain magnitude of interest becomes meaningless (e.g., $\hbar$) or, on the contrary, very large (e.g., the principal quantum number) —though not always different classicality criteria lead to the same limit [131, 132]. This limit can also be satisfied in Bohmian mechanics, for example, by increasing the mass m of the particles under consideration. However, though this may seem the correct way to operate, what one really observes is that this condition does not ensure the appearance of classical trajectories, as reported in atom–surface scattering [133]. Similarly, the quantum trajectories do not behave at all as their classical counterparts, for they still contain some nonlocal information (i.e., *coherence*) through ρ. Remember that regarding dynamical effects the "shape" of ρ is more relevant than its intensity. Thus, very small variations and tinny values of ρ may lead to very dramatic dynamical effects. This is in sharp contrast with other limits in physics, like the passage from relativistic to Newtonian mechanics, where a gradual, smooth transition is observed as particle velocities become much smaller than the speed of light. Ehrenfest's theorem helps us to establish certain criteria of classicality, e.g., obtaining the conditions for the center of a wave packet to move like a classical particle [134]. Such a wave packet can be interpreted as a swarm of non-interacting particles moving according to the motion given by

$$\frac{d\langle \mathbf{r} \rangle}{dt} = \frac{\langle \mathbf{p} \rangle}{m}, \tag{1.35a}$$

$$\frac{d\langle \mathbf{p} \rangle}{dt} = -\langle \nabla V_{\text{eff}} \rangle = -\langle \nabla V \rangle. \tag{1.35b}$$

where $\langle Q \rangle = 0$ at any time. This does not mean necessarily that Bohmian particles move like classical ones, but only on average (i.e., their distribution). A good example illustrating this fact is the one mentioned above on atom–surface scattering, where the mass of the incident particles is gradually increased [133]: the average distributions reproduce classical-like results, but quantum trajectories behave very differently with respect to their classical counterparts.

It is interesting to show very briefly that quantum trajectories can be interpreted as classical trajectories but "dressed" with a series of terms coming from quantum interference [135]. Ehrenfest's theorem may constitute a first step when trying to render some light on the transition to the classical limit. However, the WKB approximation results more insightful when dealing with Bohmian mechanics. Due to the expansion in series in terms of $\hbar$—though not always this is a good criterion— of the wave function, it yields a more direct correspondence between Bohmian and classical mechanics, i.e., to establish a closer connection between two trajectory-based formulations. Thus, as in optics [136], here one also proceeds with the ansatz

$$\Psi(\mathbf{r}, t) = e^{i\bar{S}(\mathbf{r},t)/\hbar}, \tag{1.36}$$

with $\bar{S}$ being a complex function that varies slowly in space. When (1.36) is substituted into the time-dependent Schrödinger equation, one obtains

$$\frac{\partial \bar{S}}{\partial t} + \frac{(\nabla \bar{S})^2}{2m} + V + \frac{\hbar}{2mi} \nabla^2 \bar{S} = 0. \tag{1.37}$$

In the classical limit $\hbar \to 0$, it can be assumed that $\bar{S}$ can be expanded as a series of $\hbar/i$,

$$\bar{S} = \sum_{n=0}^{\infty} \left(\frac{\hbar}{\mathrm{i}}\right)^n \bar{S}^{(n)}, \tag{1.38}$$

where the functions $\bar{S}^{(n)}$ are real. Inserting this series into (1.37), the following equation is reached

$$\sum_{n=0}^{\infty} \left(\frac{\hbar}{\mathrm{i}}\right)^n \frac{\partial \bar{S}^{(n)}}{\partial t} + \frac{1}{2m} \sum_{n=0}^{\infty} \left(\frac{\hbar}{i}\right)^n \sum_{k=0}^{n} \nabla \bar{S}^{(k)} \cdot \nabla \bar{S}^{(n-k)}$$
$$+ V + \frac{1}{2m} \sum_{n=0}^{\infty} \left(\frac{\hbar}{i}\right)^{n+1} \nabla^2 \bar{S}^{(n)} = 0. \tag{1.39}$$

The WKB approximation consists of solving order by order, in powers of $\hbar/i$, the coupled equations involved in (1.39). Thus, at zero order,

$$\frac{\partial \bar{S}^{(0)}}{\partial t} + \frac{(\nabla \bar{S}^{(0)})^2}{2m} + V = 0, \tag{1.40}$$

which is the classical Hamilton–Jacobi equation, with $\bar{S}^{(0)}(\mathbf{r}, t)$ being the classical action, S_{cl}. Note that although $\bar{S}$ is not a real function in general, $\bar{S}^{(0)}$ is real because (1.40) is real.

Regarding the remaining terms of Eq. (1.39), they lead us to a hierarchy of equations which couple the different higher orders of $\bar{S}$. These equations can be expressed in a general form as

$$\frac{\partial \bar{S}^{(n)}}{\partial t} + \frac{1}{2m} \sum_{k=0}^{n} \nabla \bar{S}^{(k)} \cdot \nabla \bar{S}^{(n-k)} + \frac{1}{2m} \nabla^2 \bar{S}^{(n-1)} = 0, \tag{1.41}$$

which couple the nth order with the remaining lower ones. Here, it is interesting to note that, since (1.41) is real as well as $\bar{S}^{(0)}$, all the remaining n orders will also be real.

If the wave function (1.36) is expressed in polar form, taking (1.38) into account,

$$\phi = \exp\left[\sum_{n=0}^{\infty}(-1)^n \hbar^{2n}\bar{S}^{(2n+1)}\right], \tag{1.42a}$$

$$S = \sum_{n=0}^{\infty}(-1)^n \hbar^{2n}\bar{S}^{(2n)}. \tag{1.42b}$$

Substituting Eq. (1.42b) into the guiding condition (1.20), one obtains the expression of the quantum trajectories within the WKB approach, i.e., in terms of $\hbar$,

$$\dot{\mathbf{r}} = \frac{1}{m}\sum_{n=0}^{\infty}(-1)^n \hbar^{2n}\nabla\bar{S}^{(2n)} = \dot{\mathbf{r}}_{\text{cl}} + \frac{1}{m}\sum_{n=1}^{\infty}(-1)^n \hbar^{2n}\nabla\bar{S}^{(2n)}, \tag{1.43}$$

where $\dot{\mathbf{r}}_{\text{cl}} = \nabla\bar{S}_t^{(0)}/m$ is the classical law of motion. Therefore, from (1.43) one can interpret quantum trajectories as classical trajectories "dressed" with a series of terms coming from quantum interference, showing the capital difference between both types of trajectories. This type of dressing scheme is going to be recurrent along this monograph. Moreover, also from (1.43) it is very apparent how classical mechanics underlies quantum mechanics and, therefore, how quantum phenomena will keep a reminiscence of a classical-like feature —which will be stronger as the classical limit is approached.

As we have previously mentioned, the phase function or field $S(\mathbf{r}, t)$ is multivalued. However, its gradient is a single-valued function of position; in other words, at each point in configuration space and at each instant of time, there is a unique tangent vector associated with the corresponding gradient. Thus, quantum trajectories can not cross each other or even touch (*noncrossing property*). A given space point may have more than one trajectory pass through it at different times. A single-valued trajectory field is called a *congruence* [7]. In classical mechanics, something similar can be found when dealing with the phase space although the evolution of the system is only determined by external fields or preassigned functions.

On the other hand, particles can not pass at all through nodal regions (nodes or points where $\Psi = 0$). This fact is closely related to the emergence of vortices or vortical dynamics. For example, within the field of surface physics, the presence of this vortical motion has been theoretically observed in the atom–surface scattering process with presence of impurities, where atoms may undergo a series of loops before they scape from the surface [137, 138].

Equations (1.15) form a closed set of coupled partial differential equations. Different numerical techniques have been suggested to solve this set of equations [12], which is analogous to those describing the evolution of classical hydrodynamical flows (see Sect. 1.2.1). Within the Lagrangian scheme, the trajectories are calculated one by one, obtaining ρ and S and, therefore, Ψ can be synthesized. Hence, from a computational point of view, this can be regarded as a "local" calculation. However, it is important to stress that this locality has nothing to do with the nonlocal dynamical behavior mentioned above. Note that the information related to the whole ensemble (though evaluated along one particular path) appears in a very precise and unambiguous manner in V_{eff}, as seen in Eq. (1.16), and influences the field S. Thus, it is always very important to distinguish between the locality of the calculations and the nonlocality inherent to the dynamical behavior of quantum particles [135].

By using the Lagrangian operator, Eq. (1.15b) becomes

$$\frac{dS}{dt} = \frac{1}{2} m\mathbf{v}^2 - V_{\text{eff}}, \tag{1.44}$$

where, as in classical mechanics, the time–derivative of the quantum action S is equal to a generalized quantum Lagrangian,

$$S[\mathbf{r}(t)] - S[\mathbf{r}(0)] = \int_0^t \mathcal{L}_Q[\mathbf{r}(t')]dt', \tag{1.45}$$

with $\mathcal{L}_Q \equiv m\mathbf{v}^2/2 - V_{\text{eff}}$. As is known, the wave formulation of quantum mechanics can also be derived from the (quantum) Lagrangian density [7, 139],

$$\mathcal{L}_q = \frac{i\hbar}{2}\left(\Psi^* \frac{\partial \Psi}{\partial t} - \frac{\partial \Psi^*}{\partial t}\Psi\right) - \frac{\hbar^2}{2}\nabla\Psi \cdot \nabla\Psi^* - V|\Psi|^2, \tag{1.46}$$

when the corresponding integral is required to be stationary with respect to variations in the complex-valued field variables Ψ and Ψ^*. Then, when variations are taken with respect to Ψ^*, the Euler–Lagrange equations yield the time-dependent Schrödinger equation (as well as its complex conjugate when variations are considered with respect to Ψ). Similarly, one can also proceed taking into account the polar form, Eq. (1.1a), which gives rise to the Lagrangian density [71, 72]

$$\begin{aligned}\mathcal{L}_q &= -\left[\frac{\partial S}{\partial t} + \frac{1}{2}(\nabla S)^2 + V\right]\rho - \frac{\hbar^2}{8}\left(\frac{\nabla\rho}{\rho}\right)^2\rho \\ &= -\left[\frac{\partial S}{\partial t} + \frac{1}{2}(\nabla S)^2 + \frac{1}{2}(\nabla\mathcal{K})^2 + V\right]\rho,\end{aligned} \tag{1.47}$$

where

$$\mathcal{K} \equiv \frac{\hbar}{2}\ln\rho \tag{1.48}$$

is a term, as we know, from which the quantum potential emerges. In this regard, note that it would be more appropriate to associate this term (and therefore the quantum potential) with a sort of inner kinetic energy, since Q does not appear explicitly in the Lagrangian density as the external potential V does. This is in correspondence with the previous analysis on the evolution of a wave packet which is ruled by two types of motions, one associated with its translation (and, therefore, with ∇S) and another one with its spreading (i.e., with $\nabla\mathcal{K}$). Furthermore, it is the presence of this term what makes quantum motion so different from the classical one, as can be readily seen when Eq. (1.47) is compared with its classical counterpart.

Finally, it is worth mentioning the so-called the wave equation of classical mechanics [140–144]. From the Lagrangian given by Eq. (1.46) and varying with respect to the complex fields Ψ^* and Ψ, the following nonlinear, time-dependent wave equation is reached

$$i\hbar\frac{\partial\Psi}{\partial t} = -\frac{\hbar^2}{2m}\nabla^2\Psi + V\Psi + \frac{\hbar^2}{2m}\frac{\nabla^2|\Psi|}{|\Psi|}\Psi. \tag{1.49}$$

The term which is responsible for the nonlinearity is proportional to the quantum potential (the last term on the right hand side of Eq. (1.49)); as mentioned by Holland [7], a nonlinear differential equation may be a limiting case of a linear equation, the time-dependent Schrödinger equation). Furthermore, a nonlinear quantum-classical transition equation has been proposed by introducing a degree of quantumness by Richardson et al. [145] which is governed by a simple parameter ϵ with $\epsilon \in [0, 1]$. The corresponding transition equation can then be written as

$$i\hbar\frac{\partial\Psi_\epsilon}{\partial t} = -\frac{\hbar^2}{2m}\nabla^2\Psi_\epsilon + V\Psi_\epsilon + (1-\epsilon)\frac{\hbar^2}{2m}\frac{\nabla^2|\Psi_\epsilon|}{|\Psi_\epsilon|}\Psi_\epsilon. \tag{1.50}$$

This transition is thus seen as a gradual process starting from $\epsilon = 0$ (classical regime) and going up to $\epsilon = 1$ (quantum regime). Several interesting applications with Gaussian wave packets dealing with this transition equation can be found in the literature [146–149]. However, no decoherence dynamics is established since we are dealing with closed quantum systems. The decoherence process emerges when interaction with an environment or an apparatus (open quantum systems) is present. This is the subject of the last two chapters of this monograph.

1.2.3 *Restricted Probability and Tubes*

As it can be inferred from Eq. (1.44), in the Lagrangian scheme it is also necessary the evaluation of the quantum potential along the trajectory. However, instead of integrating the continuity equation to obtain $\rho(t)$, one can alternatively proceed as in semiclassical mechanics [150] assuming that the solution of Eq. (1.15a) along the Bohmian trajectory $\mathbf{r}(t)$ is given by

$$\rho[\mathbf{r}(t)] = \left|\frac{\partial\mathbf{r}(0)}{\partial\mathbf{r}(t)}\right|\rho[\mathbf{r}(0)]. \tag{1.51}$$

This procedure has been used with practical computational purposes in the literature (see, for example, Refs. [28–31]), with the idea of applying Bohmian mechanics as a sort of quantum initial value representation numerical approach. Substituting Eq. (1.51) into (1.17), one finds

$$Q[\mathbf{r}(t)] = \frac{\hbar^2}{4m}\left[\frac{\partial \mathbf{r}(0)}{\partial \mathbf{r}(t)}\right]^2 \left\{ \frac{1}{2}\left[\frac{\nabla_{\mathbf{r}(0)}\rho[\mathbf{r}(0)]}{\rho[\mathbf{r}(0)]}\right]^2 - \frac{\nabla^2_{\mathbf{r}(0)}\rho[\mathbf{r}(0)]}{\rho[\mathbf{r}(0)]} \right\}, \tag{1.52}$$

where $\nabla_{\mathbf{r}(0)} \equiv \partial/\partial\mathbf{r}(0)$ is the action of the ∇–operator evaluated at $\mathbf{r}(0)$. In this way, $\rho[\mathbf{r}(t)]$ and $Q[\mathbf{r}(0)]$ can be both determined from $\rho[\mathbf{r}(0)]$. On the other hand, $\rho(\mathbf{r}, t)$ can be obtained by sampling the initial probability density with a sufficient number of initial conditions (trajectories). Of course, the numerical evaluation of the derivatives of ρ within this Lagrangian framework can lead to instabilities, as also happens in classical hydrodynamics, due to the approaching of the particle to nodal regions. Some alternative methods have been proposed in the literature to solve these drawbacks [151, 152].

When dealing with trajectories, a causal connection between the initial and the final states is readily established. Making use of the noncrossing property in configuration space, a kind of *probability tubes* can be defined [153]. The interest of these particular tubes arises from the fact that Bohmian mechanics is the only formulation of quantum mechanics that allows to specify such a connection in a unique and unambiguous way. Once the initial region Ω_0 is defined, one has also a clear and unambiguous prescription to follow its causal evolution throughout configuration space and a mapping $\Omega_t = \Phi_Q(\Omega_0)$ ('Q' here for 'quantum map') can also be established. Thus, trajectories distributed along the boundary $\partial\Sigma_0$ of Ω_0 will subsequently form the boundary $\partial\Sigma_t$ of Ω_t. These are called *separatrix trajectories* [10]. Furthermore, the deformation of the tubes with time is not completely arbitrary since the noncrossing rule of trajectories has to be always preserved.

A direct consequence of dealing with quantum probability tubes is the following: if we define a *partial* or *restricted probability* [154–156] as the fraction of the total probability that has ended up inside a region or domain $\mathcal{D}$ of the system configuration space,

$$\mathcal{P}_{\mathcal{D}}(\infty) \equiv \lim_{t\to\infty} \mathcal{P}_{\mathcal{D}}(t) = \lim_{t\to\infty} \int_{\mathcal{D}} \rho(\mathbf{r}, t) d\mathbf{r}, \tag{1.53}$$

it will remain constant in time whenever $\mathcal{D}$ corresponds to a probability tube. This means that, in principle, asymptotic probabilities like (1.53) can be specified from the initial state without any further calculation if we know: (i) the analytical form for the separatrix defining the initial boundary $\partial\Sigma_0$, and (ii) any *bifurcation* or *branching process* undergone by the probability tubes between t_0 and $t \to \infty$. The presence of branchings is actually a very important issue: any region Ω_t (including Ω_0) may consist of more than one separate subregions, which emerge or disappear along time.

Let $\mathcal{P}_{\Omega_t}(t)$ be a quantity describing the time-evolution of a certain probability of interest (e.g., a reaction probability, a transmittance, a cross-section, etc.) inside a region Ω_t of the corresponding configuration space. This quantity can be expressed in terms of the restricted probability as

$$\mathcal{P}_{\Omega_t}(t) \equiv \int_{\Omega_t} \rho(\mathbf{r}, t) d\mathbf{r}. \tag{1.54}$$

The variation of $\mathcal{P}_{\Omega_t}(t)$ with time inside Ω_t is

$$\frac{d\mathcal{P}_{\Omega_t}}{dt} = \int_{\Omega_t} \frac{\partial \rho}{\partial t} d\mathbf{r}, \tag{1.55}$$

although it can also be written as

$$\frac{d\mathcal{P}_{\Omega_t}}{dt} = -\int_{\Omega_t} (\nabla \cdot \mathbf{J}) \, d\mathbf{r} = -\int_{\partial \Sigma_t} \mathbf{J} \cdot d\mathbf{S}. \tag{1.56}$$

In the second equality of Eq. (1.56), which is a straightforward application of the divergence or Gauss-Ostrogradsky theorem, $d\mathbf{S}$ denotes a vector normal to a surface element $d(\partial \Sigma_t)$ of Ω_t and pointing outwards. By combining Eqs. (1.55) and (1.56), we find that the losses or gains of $\mathcal{P}_{\Omega_t}(t)$ inside Ω_t are described, respectively, by the outgoing or ingoing probability flux $\mathbf{J}$ through $\partial \Sigma_t$. This is a well-known result, which translates into the continuity equation when generalizing to the whole configuration space and making the flux to be independent of Ω_t. However, an additional interesting result which is worth mentioning is when the time-evolution of this region follows some particular rule, then one could keep the value of $\mathcal{P}_{\Omega_t}(t)$ constant all the way through.

From the above statements, an interesting consequence is easily deduced: any restricted probability can be determined directly from the initial state if the end points of the associated separatrix trajectories as well as any intermediate branching process are known. That is, in principle, one could determine or estimate final probabilities without carrying out the full calculation, only from the particular region covered by the initial wave function causally connected with the feature of interest [157]. Let us consider that in the restricted probability (1.53), the domain $\mathcal{D}$ corresponds to the region Ω_∞ which is the asymptotic extreme of a probability tube starting in Ω_0 at $t = 0$. By integrating back in time (i.e., considering the inverse mapping transformation $\Omega_0 = \Phi_Q^{-1}(\Omega_\infty)$), we find that

$$\mathcal{P}_{\Omega_\infty} = \int_{\Omega_\infty} \rho(\mathbf{r}, \infty) d\mathbf{r} = \lim_{t \to \infty} \int_{\Omega_t} \rho(\mathbf{r}, t) d\mathbf{r} = \lim_{t \to \infty} \mathcal{P}_{\Omega_t}(t). \tag{1.57}$$

The difference of this expression with respect to the guiding condition is that now we can keep track of the amount of probability going into $\mathcal{D} = \Omega_\infty$ by means of an unambiguous time-dependent relationship. But, since the probability inside the corresponding tube remains constant, we can just write down (1.57) as

$$\mathcal{P}_{\Omega_\infty} = \int_{\Omega_0} \rho(\mathbf{r}, 0) d\mathbf{r} = \mathcal{P}_{\Omega_0}(0). \tag{1.58}$$

The initial restricted probability $\mathcal{P}_{\Omega_0}$ can be computed from an appropriate sampling of (Bohmian) initial conditions (according to $\rho(\mathbf{r}, 0)$ inside Ω_0), as in classical mechanics. By proceeding in this way, the physical meaning of Eq. (1.58) becomes more apparent (and almost trivial): given a certain set of initial conditions enclosed in some region of the configuration space, their total number is conserved regardless of how the ensemble evolves. This is a result of *general* validity, which goes again beyond standard quantum mechanics, for it states that the probability within a certain region of the configuration space can be conveyed to another one causally connected, i.e., in an unambiguous way when following probability tubes.

The previous results lead us straightforwardly to establish a connection with the so-called Born rule [158–163]. This is readily seen as follows. If two arbitrary times, t and t' (we will assume $t' > t$), are considered, it follows from Eq. (1.58) that

$$\int_{\mathbf{r}(t')} \rho[\mathbf{r}(t')] d\mathbf{r}(t') = \int_{\mathbf{r}(t)} \rho[\mathbf{r}(t)] d\mathbf{r}(t). \tag{1.59}$$

On the other hand, because of the causal connection or mapping between $\mathbf{r}(t)$ and $\mathbf{r}(t')$ in Bohmian mechanics, one can also define a Jacobian

$$\mathcal{J}[\mathbf{r}(t)] = \frac{\partial \mathbf{r}(t')}{\partial \mathbf{r}(t)}. \tag{1.60}$$

This relation is equivalent to the one found in classical mechanics when solving the (classical) continuity equation [150, 164, 165], although in that case it includes the corresponding momenta, since it is defined in phase space. Thus, taking into account Eq. (1.59) and the connection between the layers defined by $d\mathbf{r}(t)$ and $d\mathbf{r}(t')$ enabled by the Jacobian

$$d\mathbf{r}(t') = |\mathcal{J}[\mathbf{r}(t)]| d\mathbf{r}(t), \tag{1.61}$$

the probability density evaluated along a quantum trajectory at a time t' is related through the inverse Jacobian transformation with its value at an earlier time t, as

$$\rho[\mathbf{r}(t')] = |\mathcal{J}[\mathbf{r}(t')]| \rho[\mathbf{r}(t)], \tag{1.62}$$

with $|\mathcal{J}[\mathbf{r}(t')]| = |\mathcal{J}[\mathbf{r}(t)]|^{-1}$. That is, Born's rule is preserved along time whenever the evolution of the probability $\rho[\mathbf{r}(t)]$ is monitored within the probability tube defined by the swarms of quantum trajectories $\mathbf{r}(t)$ and $\mathbf{r}'(t) = [\mathbf{r} + d\mathbf{r}](t)$,

$$\rho[\mathbf{r}(t')] d\mathbf{r}(t') = \rho[\mathbf{r}(t)] d\mathbf{r}(t). \tag{1.63}$$

In order to illustrate the notion of tubes, the scattering of a wave function off a rectangular barrier is a convenient example [153]. Let the initial wave function be an arbitrary coherent superposition of three Gaussian wave packets,

$$\Psi_0(x) = A_0 \sum_{i=1}^{3} c_i \psi_i(x), \tag{1.64}$$

where each ψ_i is described by

$$\psi_0(x) = \left(\frac{1}{2\pi\sigma_0^2}\right)^{-1/4} e^{-(x-x_0)^2/4\sigma_0^2 + ip_0(x-x_0)/\hbar}, \tag{1.65}$$

with x_0 and p_0 being respectively the (initial) position and momentum of the wave-packet centroid (i.e., $\langle\hat{x}\rangle = x_0$ and $\langle\hat{p}\rangle = p_0$), and σ_0 its initial spreading. Without loss of generality, the parameters chosen are: $(c_1, c_2, c_3) = (1.0, 0.75, 0.5)$, $(x_{0,1}, x_{0,2}, x_{0,3}) = (-10, -12, -9)$, $(p_{0,1}, p_{0,2}, p_{0,3}) = (10, 20, 15)$, $(\sigma_{0,1}, \sigma_{0,2}, \sigma_{0,3}) = (0.2, 1.6, 0.8)$, and $m = \hbar = 1$. The initial wave function Ψ_0 is hence normalized, this being denoted by the constant prefactor A_0 in Eq. (1.64). The square barrier is assumed to consist of a sum of two hyperbolic tangents,

$$V(x) = \frac{V_0}{2}\left\{\tanh\left[\alpha(x - x_-)\right] - \tanh\left[\alpha(x - x_+)\right]\right\}, \tag{1.66}$$

with $V_0 = 150$, $\alpha = 10$, and $x_\pm = \pm 2$. The quantum dynamics is carried out by means of a standard wave-packet propagation method, while the associated Bohmian trajectories are computed "on-the-fly", substituting the wave function resulting at each iteration into the guiding condition (1.20) and then integrating this equation of motion.

The initial and final probability densities are displayed in Fig. 1.2a (black and red solid lines, respectively); the barrier being plotted in the blue shadowed region. The system wave function is evolved until the probability within the intra-barrier region is negligible (our asymptotic time). In panel (b), the transmission (green dashed line), reflection (blue dash-dotted line) and intra-barrier (red dotted line) probabilities are displayed along time. In the calculation of these restricted probabilities, it is assumed that: $\mathcal{D}_\mathrm{T}$ is the region beyond the right-most barrier edge, $\mathcal{D}_\mathrm{I}$ is the region confined between the two barrier edges, and $\mathcal{D}_\mathrm{R}$ is the region to the left-most edge. As seen in the figure, after $t \approx 1.15$, we find that $\mathcal{P}_\mathrm{R} \approx 0$ and $\mathcal{P}_\mathrm{T}$ reaches its maximum (asymptotic) value, which already remains constant with time.

The initial wave function is then splitted up into reflected and transmitted, with Ω_0 encompassing the portion associated with the latter. The upper bound for this region can be the initial position, $x(0)$, of any trajectory on the right-most border of the initial probability density, for which $\rho(x(0)) \approx 0$. For the lower boundary, a search has to be done [157], so that it is ensured that the chosen trajectory is the last (or nearly the last) one in crossing the right-most barrier edge and not exhibiting a backwards motion. Determining this trajectory constitutes a major drawback, since Bohmian trajectories are not analytical in general, neither there is a simple, general

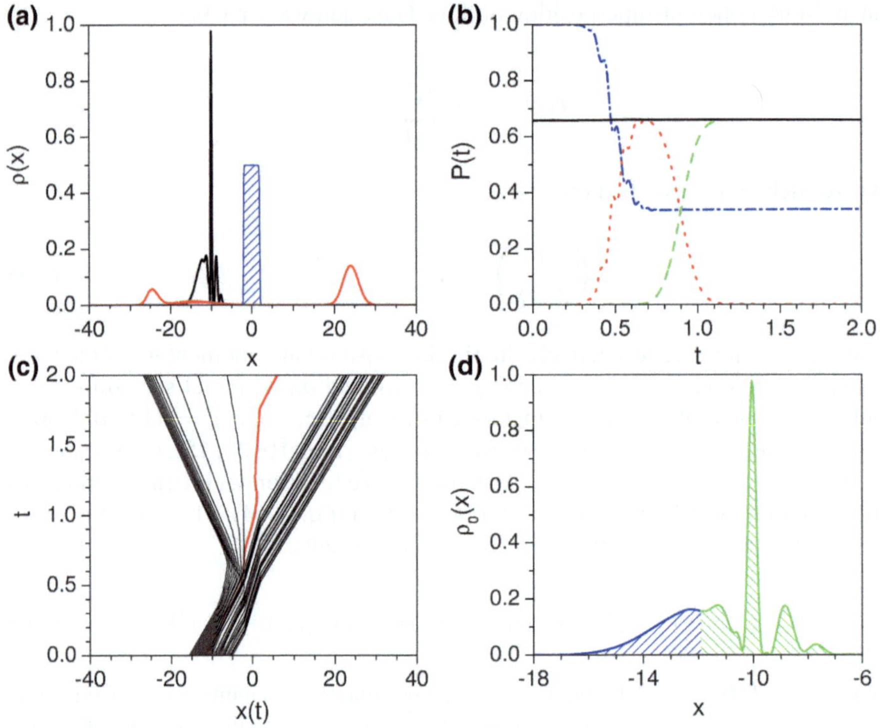

Fig. 1.2 **a** Initial (*black line*) and final (*red line*) probability densities in the problem of scattering off a nearly square barrier (*blue* shadowed region). *b* Time-dependence of the transmission (*green dashed line*), reflection (*blue dash-dotted line*) and intra-barrier (*red dotted line*) probabilities. The probability enclosed in Ω_t and obtained with the aid of the Bohmian calculation is displayed with *black solid line*. **c** Bohmian trajectories illustrating the process dynamics; the separatrix is denoted with *red thicker line*. **d** Splitting of the initial probability density according to the separatrix initial position. Only the *green* shadowed region (Ω_0) contributes to transmission

way to make an estimate [157]. This initial condition has to be then determined either from a series of sampling runs or just fixing the asymptotic value of the trajectory and running backwards in time the dynamics until $t = 0$.

A sampling set of Bohmian trajectories is shown in Fig. 1.2c, with their initial positions evenly distributed along the extension covered by the initial probability density. The red thicker line denotes the separatrix splitting the initial swarm into two groups of trajectories: those that will surmount the barrier (transmitted) and those that will bounce backwards (reflected). Accordingly, at any time t, the region Ω_t (i.e., the time-evolved of Ω_0) always confines trajectories that eventually become transmitted; $\partial\Sigma_t$ is determined by the positions at t of two trajectories, namely the separatrix and the rightmost one considered. The evolution of these two trajectories defines the corresponding transmission probability tube, along which all the transmitted probability density flows. Bearing this scheme in mind, it is now rather simple and straightforward to determine which part of the initial probability density contributes to tunneling transmission, denoted by the green shadowed area in panel (d). The integral over this area readily provides the value otherwise found from the asymptotic $\mathcal{P}_{\mathrm{T}}$ (see panel (b)). Actually, the evaluation of $\mathcal{P}_{\Omega_t}(t)$ at each time renders a constant value (see black solid line in panel (b)), thus proving the conservation of the probability inside Ω_t. Furthermore, this also proves our previous assertion that final probabilities can be, in principle, directly and unambiguously obtained from the initial state by means of Bohmian trajectories.

1.2.4 Local and Expectation Values

Though Bohmian mechanics allows us to describe the evolution of quantum processes and phenomena in terms of individual trajectories, it is clear that any observable will be extracted from a statistical treatment of the corresponding quantum trajectories. In other words, any observable will arise as a consequence of the counting of trajectory arrivals at a certain region, which is much connected with the way how experiments occur. For example, in a typical diffraction experiment, the diffraction pattern arises by counting (detecting) individual arrivals [37, 38, 166–168]. Similarly, when using Bohmian mechanics one can reproduce the experiment by counting the arrivals of quantum trajectories [121, 127]. Thus, taking this into account, the statistical nature of quantum mechanics arises in a very natural way, where expectation values are directly associated with average (ensemble) values.

In order to understand the relationship between the expectation value of a quantum operator and the statistical Bohmian description, suppose A is a Hermitian operator, which can be a function of the position and momentum operators, $\mathbf{r}$ and $\mathbf{p} = -i\hbar\nabla$, i.e., $A = A(\mathbf{r}, -i\hbar\nabla)$. The expectation value of this operator is defined as

$$\langle A \rangle = \langle \Psi | A | \Psi \rangle = \frac{\int \Psi^*(A\Psi) d\mathbf{r}}{\int \Psi^*\Psi d\mathbf{r}}, \tag{1.67}$$

where $\Psi(\mathbf{r}, t) = \langle \mathbf{r} | \Psi(t) \rangle$ is the wave function in the system configuration space and

$$A\Psi(\mathbf{r}, t) \equiv \langle \mathbf{r} | A \left(\int |\mathbf{r}'\rangle\langle\mathbf{r}'| d\mathbf{r}' \right) |\Psi(t)\rangle = \int A(\mathbf{r}, \mathbf{r}')\Psi(\mathbf{r}', t) d\mathbf{r}', \tag{1.68}$$

with $A(\mathbf{r}, \mathbf{r}') \equiv \langle \mathbf{r} | A | \mathbf{r}' \rangle$. Moreover, one can also consider the quantity

$$\mathcal{A} \equiv \frac{\mathrm{Re}\,\{\Psi^* A \Psi\}}{\Psi^*\Psi} \tag{1.69}$$

to represent the *local* value of the operator A, given in terms of the associated field function $A(\mathbf{r}, t)$ [7]. For example, the local field functions

$$\mathbf{r}(\mathbf{r},t) = \frac{\mathrm{Re}\,\{\Psi^*\mathbf{r}\Psi\}}{\Psi^*\Psi} = \mathbf{r}(t), \tag{1.70a}$$

$$\mathbf{p}(\mathbf{r},t) = \frac{\mathrm{Re}\,\{\Psi^*(-i\hbar\nabla)\Psi\}}{\Psi^*\Psi} = \nabla S \tag{1.70b}$$

$$E(\mathbf{r},t) = \frac{\mathrm{Re}\left\{\Psi^*\left(-\frac{\hbar^2}{2m}\nabla^2 + V\right)\Psi\right\}}{\Psi^*\Psi} = \frac{(\nabla S)^2}{2m} + V_{\mathrm{eff}}, \tag{1.70c}$$

provide us the position, momentum and energy of a Bohmian particle when they are evaluated along its trajectory.

If instead of a particle there is a statistical ensemble of them (or, equivalently, some set of initial conditions has to be sampled) distributed according to $\rho(\mathbf{r},t)$, the average value of A can be computed as

$$\bar{A}(t) = \int \rho(\mathbf{r},t)\, A(\mathbf{r},t)\, d\mathbf{r}. \tag{1.71}$$

In particular, we have

$$\bar{\mathbf{r}} = \int \rho\mathbf{r}d\mathbf{r} = \int \Psi^*\mathbf{r}\Psi d\mathbf{r} = \langle\mathbf{r}\rangle, \tag{1.72a}$$

$$\bar{\mathbf{p}} = \int \rho\nabla S d\mathbf{r} = \int \Psi^*(-i\hbar\nabla)\Psi d\mathbf{r} = \langle\mathbf{p}\rangle, \tag{1.72b}$$

$$\bar{E} = \int \rho\left[\frac{(\nabla S)^2}{2m} + V_{\mathrm{eff}}\right] d\mathbf{r} = \int \Psi^*\left[-\frac{\hbar^2}{2m}\nabla^2 + V\right]\Psi d\mathbf{r} = \langle H\rangle, \tag{1.72c}$$

which coincide with the corresponding expectation values obtained from standard quantum mechanics, this showing the equivalence at a predictive level of both approaches. Obviously, from a trajectory viewpoint, i.e., when the associated local field functions are evaluated along trajectories, Eqs. (1.72) read as

$$\bar{\mathbf{r}}_B = \frac{1}{N}\sum_{i=1}^{N} w_i\mathbf{r}_i(t), \tag{1.73a}$$

$$\bar{\mathbf{p}}_B = \frac{1}{N}\sum_{i=1}^{N} w_i\nabla S(\mathbf{r}_i(t)), \tag{1.73b}$$

$$\bar{E}_B = \frac{1}{N}\sum_{i=1}^{N} w_i\left\{\frac{[\nabla S(\mathbf{r}_i(t))]^2}{2m} + V_{\mathrm{eff}}(\mathbf{r}_i(t))\right\}, \tag{1.73c}$$

where N is the total number of trajectories considered, w_i is the associated weight —if the trajectories are initially sampled according to ρ_0, then $w_i = 1$ for all trajectories, otherwise $w_i \approx \rho(\mathbf{r}(t_0))$— and the subscript B means that these average values are computed from a sampling of Bohmian trajectories. As in classical statistical treatments, provided the sampling of initial conditions is properly carried out according to some initial distribution function (this role is played here by $\rho(0)$), in the limit $N \to \infty$, the quantities issued from Eqs. (1.73) will correspond with their quantum homologous equations (1.72), respectively. One readily notes that the uncertainty principle can be directly related to a statistical result instead of to an inherent impossibility to measure positions or momenta —the source for this impossibility would be rather associated with the way how things happen (interact) at quantum scales. In this sense, the inequality

$$\Delta r_i \Delta p_i \geq \frac{\hbar}{2} \tag{1.74}$$

expresses the relationship between two statistical quantities (in this case, position and momentum) in quantum mechanics, where

$$(\Delta r_i)^2 = \overline{r_i^2} - \overline{r_i}^2 \approx \overline{r_{B,i}^2} - \overline{r_{B,i}}^2, \tag{1.75a}$$

$$(\Delta p_i)^2 = \overline{p_i^2} - \overline{p_i}^2 \approx \overline{p_{B,i}^2} - \overline{p_{B,i}}^2, \tag{1.75b}$$

and $i = 1, 2, 3$.

It is clear from the above discussion that the role played by time in Bohmian mechanics and in quantum mechanics differs. Time in the quantum theory is not an observable, but a parameter, i.e., there is no time operator such that its eigenvalues provide us with some information (for example, about the time a particle needs to cross a barrier by tunnel effect, the time of a scattering process or the lifetime of a resonance phenomenon). Actually, when the time calculated through a given expression is compared to some experimental data value, things become more troublesome. This important issue has been considered in length in the literature [169–171], where several definitions of time can be found, such as dwell time, tunneling time, interaction time, arrival time, etc. This situation changes within the context of Bohmian mechanics, where the concept of arrival time is unambiguously defined because it is based on the concept of well-defined trajectories. Indeed, from the guidance condition or the trajectory, just by integration or by inspecting a graph [7], information about arrival times can be readily obtained with no need for a time operator.

1.2.5 *Complex Quantum Trajectories*

Bohmian mechanics arises when a transformation from the complex fields (Ψ, Ψ^*) to the real fields (ρ, S) is carried out. Alternatively, the wave function can be expressed in terms of a complex phase as in Eq. (1.36) and therefore there will be a one–to–one correspondence between Ψ and $\bar{S}$ —except for a constant $2\pi n\hbar$, as in Eq. (1.12). Of course, in the particular case

$$\bar{S}(\mathbf{r}, t) = S(\mathbf{r}, t) - i\hbar \ln \phi(\mathbf{r}, t) = S(\mathbf{r}, t) - \frac{i\hbar}{2} \ln \rho(\mathbf{r}, t), \tag{1.76}$$

standard Bohmian mechanics is recovered. After substitution of Eq. (1.36) into the time-dependent Schrödinger equation, one obtains

$$\frac{\partial \bar{S}}{\partial t} + \frac{(\nabla \bar{S})^2}{2m} + \bar{V}_{\text{eff}} = 0, \tag{1.77}$$

which is a complex quantum Hamilton–Jacobi equation that can be regarded as the time-dependent Schrödinger equation associated with a logarithmic wave function. In this equation,

$$\bar{V}_{\text{eff}} \equiv V - \frac{i\hbar}{2m} \nabla^2 \bar{S} \tag{1.78}$$

is now an effective complex potential, whose second component is a *complex quantum potential*,

$$\bar{Q} = -\frac{i\hbar}{2m} \nabla^2 \bar{S} = \frac{i\hbar}{2} \left[\left(\frac{\nabla \Psi}{\Psi} \right)^2 - \frac{\nabla^2 \Psi}{\Psi} \right]. \tag{1.79}$$

This (complex) quantum potential can also be referred to a sort of (complex) quantum kinetic energy, since

$$K\Psi = \frac{(\nabla \bar{S})^2}{2m} - \frac{i\hbar}{2m} \nabla^2 \bar{S}. \tag{1.80}$$

Actually, if Eq. (1.76) is substituted into the right–hand side of (1.80),

$$\begin{aligned} K\Psi &= \frac{1}{2m} \left\{ \left[(\nabla S)^2 - \frac{\hbar^2}{4} \left(\frac{\nabla \rho}{\rho} \right)^2 - \frac{i\hbar}{\rho} \nabla \rho \nabla S \right] \right. \\ &\quad \left. - i\hbar \left[\nabla^2 S - \frac{i\hbar}{2} \frac{\nabla^2 \rho}{\rho} + \frac{i\hbar}{2} \left(\frac{\nabla \rho}{\rho} \right)^2 \right] \right\} \\ &= \frac{(\nabla S)^2}{2m} + Q - \frac{i\hbar}{2m\rho} \nabla (\rho \nabla S). \end{aligned} \tag{1.81}$$

From the second equality it is very apparent that the real part of the complex kinetic energy contains the real quantum potential. Furthermore, its imaginary part provides us with information about the quantum flux conservation (in terms of the rate of change $\partial\rho/\partial t$).

In analogy to standard Bohmian mechanics, complex quantum trajectories can also be defined by analytic continuation of Eq. (1.77) to the complex plane, which is necessary for the corresponding equation of motion

$$\bar{\mathbf{v}} = \frac{\bar{\nabla}\bar{S}}{m} = \frac{\hbar}{im}\frac{\bar{\nabla}\bar{\Psi}}{\bar{\Psi}}, \tag{1.82}$$

to be self-consistent. If $\bar{S}$ is complex, $\bar{\mathbf{v}}$ has also to be complex-valued and, therefore, the integrated trajectory. This also implies that $\bar{S}$ (and eventually Ψ) will be evaluated along a complex trajectory $\mathbf{z}(t)$ rather than along a real coordinate $\mathbf{r}$ (this is the reason why $\bar{\Psi}$ and $\bar{\nabla}$ are used in Eq. (1.82), instead of Ψ and ∇), though its value will be (physically) meaningful only along the real axis. The relationship between the real Bohmian velocity and its complex counterpart,

$$\bar{\mathbf{v}} = \mathbf{v} - \frac{i\hbar}{2m}\frac{\nabla\rho}{\rho}, \tag{1.83}$$

can be followed from Eq. (1.76). Formerly, Rosen [143, 144, 172] considered this expression a sort of total quantum mechanical momentum (or velocity) field which explains why it is possible to observe non vanishing momenta in cases where the momentum ∇S vanishes. This effect would arise from the second term on the right–hand side of Eq. (1.83), which is assumed to be a "local" momentum assigned to the quantum mechanical field with which the particle interacts. In this way, Eq. (1.83) becomes the momentum that matches the momentum distributions provided by standard quantum mechanics, rather than the momentum ∇S from Bohmian mechanics, which is quite different [173–175]. This is consistent with the fact that Bohmian trajectories only carry information about the dynamics of the quantum flow, while complex quantum trajectories will also include information about the probability (as inferred from the analytic continuation of $\bar{S}$ from Eq. (1.76)). The dynamics in the complex configuration space thus explains in a natural way how to obtain the correct momentum distribution. This also explains why algorithms based on complex trajectories are quite stable and accurate.

In one dimension, it is clear that if $\bar{\mathbf{v}}$ is assumed to be complex, depending on the z–variable, while $\mathbf{v}$ and ρ depend on the real variable x, then (1.83) becomes an inequality. In the literature [176–179], (1.83) has been considered as an identity, particularly within the so-called *stochastic Bohmian mechanics* (see Sect. 1.3.1),[2] where the second term on the right–hand side is interpreted as a stochastic diffusive term. However, not only this is not mathematical consistent, but the inequality forbids to associate the real part of the complex quantum trajectories with the (real, standard) Bohmian trajectories, as assumed in the literature [176–181]. That is, assuming

[2]This terminology is again somehow confusing when alluding to open quantum systems within the Bohmian framework in the rest of this monograph.

$z(t) = z_r(t) + iz_i(t)$ describes the complex trajectory, with z_r and z_i being its real and imaginary parts, respectively, both being real functions, the equality $z_r(t) = x(t)$, where $x(t)$ would represent the standard Bohmian trajectory, does not hold as t goes on even if $z_0 = x_0$ at $t = 0$. Note that, from a strict Bohmian viewpoint, the equality $z_r(t) = x(t)$ means that the same (projected) z_r may present different (quantum) velocities, something contrary to what happens in standard Bohmian mechanics, where only one velocity can be associated with a space coordinate. Of course, only when one moves to the complex plane realizes that, effectively, that univaluedness still continues, except in those cases where a node of the wave function appears.

Once it is assumed that both terms in Eq. (1.83) depend on the complex $\mathbf{z}$–variable, and then $\mathbf{v}$ and ρ are some generalized complex Bohmian functions ($\tilde{\mathbf{v}}$ and $\tilde{\rho}$, respectively) depending on this variable instead of $\mathbf{r}$, it is instructive to apply $\bar{\nabla}$ on both sides, which yields

$$\bar{\nabla}\bar{\mathbf{v}} = \bar{\nabla}\tilde{\mathbf{v}} + \frac{i\hbar}{2m}\left[\left(\frac{\bar{\nabla}\tilde{\rho}}{\tilde{\rho}}\right)^2 - \frac{\bar{\nabla}^2\tilde{\rho}}{\tilde{\rho}}\right]. \tag{1.84}$$

As seen, the second term in the right–hand side reminds the functional dependence of Q on ρ in Eq. (1.17), except for a 1/2 factor inside the square bracket. In this sense, the effects of the (real) quantum potential (i.e., the nonlocality) are still present in the complex dynamics although the corresponding complex quantum potential, which is proportional to $\bar{\nabla}\bar{\mathbf{v}}$, might be relatively small (or even zero).

Equation (1.77) was formerly derived by Pauli during his studies on the quantum WKB approximation [182, 183]. However, the formalism based on the complex version of Bohmian mechanics is relatively recent, receiving much attention in these last years. The work developed here can also be placed within the framework of the analytic versus synthetic approaches previously mentioned. From the analytic viewpoint, with interpretative purposes, one of the trends followed is the one aimed at studying stationary states. As can be easily shown, the velocity field $\mathbf{v}$ vanishes when the wave function is described by energy eigenfunctions associated with zero angular momentum states (i.e., s–waves) and, therefore, the corresponding Bohmian particles will remain standing at their initial positions at any time. In order to overcome this problem, different time-independent quantum Hamilton–Jacobi formulations have been formulated in the literature. For example, Floyd [184–187] and Faraggi and Matone [188, 189] developed time-independent quantum Hamilton–Jacobi-like formulations starting from real bipolar ansatz, though not fully equivalence to standard quantum mechanics at a predictive level. However, later on, John [180, 181] proposed a time-dependent complex quantum trajectory formalism, namely the "modified de Broglie–Bohm approach to quantum mechanics", which has also been used and further developed by other authors [176–179, 190] to understand the problem of Bohmian stationarity in other kind of problems. On the other hand, also with interpretative purposes, Eq. (1.77) was found [121, 135, 150] when trying to discriminate the amount of "quantumness" implicit in Bohmian trajectories (i.e., how different they are with respect to their classical counterparts) within the framework of the semiclassical WKB approximation. Finally, a series of fundamental works can also be found in the literature dealing with the dynamics of problems in the continuum, such as interference [191–193], entanglement [177] or stochastic complex Bohmian mechanics [176, 179].

On the other hand, from the analytic viewpoint, Leacock and Padgett used [194, 195] the connection formula (1.36)—the same considered later by John— as an alternative way to tackle the problem of the calculation of stationary or bound states. This is formally equivalent to the more recent "Bohmian mechanics with complex action" developed by Tannor and coworkers [196–201] and the methodology developed by Wyatt and coworkers [202–205] for computational purposes without the need to solve the time-dependent Schrödinger equation. In principle, using analytical continuation it is possible to implement numerical codes that benefit from working in a complex configuration space—these advantages are similar to those that in electromagnetism, for example, lead to consider complex fields instead of real ones. More specifically, the main idea behind the development of computational tools based on the complex trajectory methodology is that the wave function on the whole real axis can be synthesized using the information transported by those particles crossing the real axis simultaneously. This allows us to define a curve, namely the *isochrone* [198, 205], which joins the specific initial positions of trajectories such that their crossing with the real axis occurs at the same time. Computationally, the main problem in dealing with the complex dynamics is locating isochrones, which is a problem similar to the root search problem in semiclassical mechanics [165]. In this regard, methods introduced in the latter may prove useful for the isochrone problem [205]. Actually, within this picture, three points are worth stressing. First, all the complex trajectories associated with an isochrone will reach the real axis simultaneously [204]. Second, the uniqueness in complex Bohmian mechanics arises from the bonds established by the initial real wave function, which is in the end the observable magnitude (through ρ). Therefore, though there might be many initial conditions leading to the same point on the real axis, only the isochrones connect the complex problem with the real one, thus establishing the same uniqueness observed in standard Bohmian mechanics. Third, a Bohmian trajectory does not correspond to a given complex trajectory or to the family of complex trajectories associated with a given isochrone, but to a family of complex trajectories such that their crossings with the real axes take place consecutively, one after the other. In other words, a Bohmian trajectory itself defines a family or set of complex quantum trajectories [191].

1.2.6 Many Particle Systems

Let us consider now two isolated quantum systems regardless of their size or properties where the total wave function describing them is a product state (factorizable) of the wave functions associated with each system. As already mentioned by Schrödinger [206, 207], as soon as these objects interact, the total wave function becomes non longer factorizable and they cannot be described as independent entities. The new quantum state becomes *entangled*, with the property that any quantum operation performed on one of the objects will have important implications on the other one independently of how far apart they are [15–17, 208]. This property has led to the well-known quantum information theory [209] as well as mechanisms, such as decoherence, which are used to explain the emergence of the classical world from quantum mechanics [104, 210, 211].

Before going to the implications of entanglement, we focus on a set of N independent particles of mass m, each one represented at a given time by a solution of the corresponding separated Schrödinger equation. The total N–body wave function in the configuration space can be expressed as a product of N single wave functions,

$$\begin{aligned}\Psi(\mathbf{r}_1, \mathbf{r}_2, \ldots, \mathbf{r}_N, t) &= \psi_1(\mathbf{r}_1, t)\psi_2(\mathbf{r}_2, t)\cdots\psi_N(\mathbf{r}_N, t)\\ &= \Pi_{i=1}^{N}\rho_i^{1/2}(\mathbf{r}_i, t)\, e^{iS_i(\mathbf{r}_i,t)/\hbar}.\end{aligned} \tag{1.85}$$

In this case, it is easy to show that from this total wave function, individual solutions of the Bohmian equations of motion are obtained according to

$$\mathbf{v}_i = \dot{\mathbf{r}}_i = \frac{\nabla_{\mathbf{r}_i} S_i(\mathbf{r}_i, t)}{m}, \tag{1.86}$$

since the quantum potential is also separable as a sum of partial quantum potentials,

$$Q = \sum_{i=1}^{N} Q_i = -\frac{\hbar^2}{2m}\sum_{i=1}^{N}\frac{\nabla^2\rho_i^{1/2}}{\rho_i^{1/2}}. \tag{1.87}$$

This implies that the Bohmian trajectories described by the different particles are independent and no information is transmitted among them. Of course, this also happens if instead of N different particles, N different degrees of freedom are considered. Typically, factorizability is closely connected to *distinguishability*. For example, this is the case of particles describable by means of a Maxwell–Boltzmann statistics. For two particles, their total wave function can be expressed as

$$\Psi(\mathbf{r}_1, \mathbf{r}_2, t) = \Phi_A(\mathbf{r}_1, t)\Phi_B(\mathbf{r}_2, t). \tag{1.88}$$

Usually, the particles constituting this kind of ensembles are non-interacting and they can be accounted for by only taking into account how they distribute. When the particles forming an ensemble do interact among themselves and become indistinguishable (i.e., truly quantum-mechanical), apart from their distribution it is also very important to know how they interact. In this case, even if they are very far apart, their quantum state cannot be described in a simple manner since it is an entangled state. This is the case of the Fermi–Dirac or Bose–Einstein statistics, which describe fermions or bosons, respectively. That is, if two of these particles are considered, their total wave function will be

$$\Psi(\mathbf{r}_1, \mathbf{r}_2, t) = N_\pm \left[\Phi_A(\mathbf{r}_1, t)\Phi_B(\mathbf{r}_2, t) \pm \Phi_A(\mathbf{r}_2, t)\Phi_B(\mathbf{r}_1, t)\right], \tag{1.89}$$

where $N_\pm$ is the normalization factor and the minus sign stands for fermions (wave function with odd parity under exchange of the positions of the particles) and the plus sign for bosons (even parity).

As is known in quantum statistical physics, spatial correlation exist even if the particles are supposedly noninteracting. This is due to the symmetry properties of the corresponding wave function. This correlation is significant if the interparticle distances in the whole system are comparable with the so-called mean thermal wavelength of the particles [212]

$$\lambda_T = \frac{h}{(2\pi m k_B T)^{1/2}}, \tag{1.90}$$

where k_B and T are the Boltzmann constant and the temperature, respectively. A positive and negative spatial correlations are displayed by bosons and fermions, respectively. Many times, an alternative way to express such correlations is by means of the so-called statistical potential which is defined as

$$V_s(r) = -k_B T \ln[1 \pm exp(-2\pi r^2/\lambda_T^2] \tag{1.91}$$

where r is the interparticle distance. For bosons (plus sign), the statistical potential is attractive whereas for fermions (minus sign), the corresponding potential is repulsive; decaying in both cases rapidly to zero when r becomes larger than λ_T.

In general, given an entangled state described by the global wave function

$$\Psi(\mathbf{r}_1, \mathbf{r}_2, \ldots, \mathbf{r}_N, t) = \phi(\mathbf{r}_1, \mathbf{r}_2, \ldots, \mathbf{r}_N, t)\, e^{iS(\mathbf{r}_1, \mathbf{r}_2, \ldots, \mathbf{r}_N, t)/\hbar}, \tag{1.92}$$

the corresponding quantum trajectories will be obtained by integrating the equations of motion

$$\mathbf{v}_i = \frac{\nabla_i S}{m}. \tag{1.93}$$

This scheme gives rise to a set of N equations coupled through the total phase S, where the evolution of a particle will be strongly nonlocally influenced by the other (apart from other classical like interactions through V). This entanglement [206, 207] becomes more apparent through the quantum potential,

$$Q = -\frac{\hbar^2}{2m}\sum_{i=1}^{N}\frac{\nabla_i^2\phi}{\phi}, \tag{1.94}$$

where $Q = Q(\mathbf{r}_1, \mathbf{r}_2, \ldots, \mathbf{r}_N, t)$, which is nonseparable and, therefore, strongly nonlocal. In other words, the usual hydrodynamical magnitudes described above become now many-body functions. Thus, for example, in order to calculate any trajectory of a given particle, the initial positions of all particles have to be specified. Instantaneous motion of any particle depends on the coordinates of all other particles at the same time. Disturbing a part of the system in a localized region, the configuration space of the whole system will respond instantaneously. The quantum force defined by the gradient of the quantum potential does not, in general, fall off with distance and interaction among particles can be rather strong even though the wave function amplitude is small. Different works in the literature have analyzed the trajectory correlation among entangled particles [213–217]. As Holland [7] mentioned, in this mechanics, the whole system is prior to its constituent particles displaying a radical, different behavior to the classical dynamics where the whole is the sum of the parts and their interacions. Furthermore, the symmetry properties of the wave function associated with a set of particles (bosons or fermions) implies clearly the introduction of forces between the particles making up the system. The correlations in their motions lead to quantum potentials distinct for each set of particles. Although identical particles are not distinguishable, they are distinguished by the continuity of their trajectories. In this sense, the separation or distance between two particles following two different trajectories has a precise meaning, at any instant of time. Furthermore, depending if we are dealing with a set of fermions or bosons the corresponding distance behaves with time differently.

Recently, the Bohmian formalism has been used to derive the ro-vibrational spectrum of diatomic molecules [218]. The same formalism was also used to devise some hybrid methods when systems of many degrees of freedom are considered [219, 220].

1.3 Open Quantum Systems. Quantum Stochastic Trajectories

1.3.1 *The Causal Stochastic Approach*

As has been already mentioned in the first section, Bohm and Vigier [10, 70] assumed that a particle could be thought to be suspended in a Madelung fluid. Their basic assumption was to express the velocity of a given particle in configuration space as

$$d\mathbf{v} = -\nabla(V + Q)dt + d\mathbf{W}(t) \tag{1.95}$$

where $d\mathbf{W}(t)$ is the increment vector (along the three directions) of the so-called Wiener stochastic process at time t which gives the random fluctuations (noise) of the velocity due to the presence of the environment.

This equation can provide us some sort of *quantum (Bohmian) stochastic trajectories*. In this approach, we are dealing with a statistical ensemble of particles with a probability distribution given by $P(\mathbf{r}, t)$. It is usually assumed that this probability distribution has nothing to do with the wave function. Only after the evolution in a long interval of time of the corresponding dynamics, $P(\mathbf{r}, t) \rightarrow \rho(\mathbf{r}, t) = |\Psi(\mathbf{r}, t)|^2$. One could suppose that the stochastic process could be come from some sort of vacuum fluctuations represented by a simple diffusion process [80]. The analysis of a Brownian particle in a gravitational field is simple and illustrative of some new concepts which should be introduced. If P is the probability density of particles, the probability current or flux and the conservation equation are given by

$$\begin{aligned} \mathbf{J} &= -D\nabla P \\ \frac{\partial P}{\partial t} &= -D\nabla\left(\frac{mg}{k_B T} zP + \nabla P\right), \end{aligned} \tag{1.96}$$

respectively, where D is the diffusion coefficient and k_B is the Boltzmann constant. The so-called osmotic velocity which gives us the drift component (due to the gravitational field) is written as

$$u = D\frac{mg}{k_B T} \tag{1.97}$$

and was introduced by Einstein. At equilibrium, $\partial P/\partial t = 0$, the well-known Boltzman factor is obtained,

$$P = Ae^{-mgz/k_B T} \tag{1.98}$$

where A is a constant. In other words, the particle is drifting downward in the gravitational field and the net upward diffusive movement is balanced to produce equilibrium.

In the quantum domain, we would like to have that the equilibrium state of the diffusion process corresponds to a probability density $P = |\psi|^2 = \rho$ and to a mean current $j = \rho\bar{v} = \rho\nabla S/m$. Such a state is a consistent possibility if the wave function of the particle, written in polar form as $\psi = \sqrt{\rho}\exp(iS/\hbar)$, satisfies the Schrödinger equation because this implies the continuity equation

$$\frac{\partial \rho}{\partial t} + \nabla.\mathbf{j} = 0. \tag{1.99}$$

This could be accomplished with a suitable choice of the osmotic velocity. The origin of this osmotic velocity is not coming from a force field but from different causes. It could be postulated a field such as

$$\mathbf{u} = D\frac{\nabla\rho}{\rho}. \tag{1.100}$$

The total current is thus written as

$$\mathbf{j} = \frac{P}{m}\nabla S + DP\frac{\nabla\rho}{\rho} - D\nabla P \tag{1.101}$$

and the continuity equation as

$$\frac{\partial P}{\partial t} + \nabla . \left(P \frac{\nabla S}{m} + DP \frac{\nabla \rho}{\rho} - D \nabla P \right) = 0. \tag{1.102}$$

Then, the systematic velocity is given by

$$\mathbf{v} = \frac{\nabla S}{m} + D \frac{\nabla \rho}{\rho} \tag{1.103}$$

where the mean velocity arises from the guidance condition and the osmotic velocity is derivable from the potential

$$V_o = D \, ln \rho \tag{1.104}$$

ρ being a solution of

$$\frac{\partial \rho}{\partial t} + \nabla . \left(\rho \frac{\nabla S}{m} \right) = 0. \tag{1.105}$$

The equilibrium position implies that $P = \rho$, that is, it means that the osmotic velocity is balanced by the diffusion current so that the mean velocity is $\nabla S/m$. The next important question is, any arbitrary probability distribution P will always approach ρ? Bohm and Vigier [70] have shown that for the general case P approaches ρ. The osmotic velocity is then constantly pushing the particle to the regions of highest $|\psi|^2$ and this explains why most particles are found near the maxima of the wave function. Moreover, as previously stated, for stationary states, $S = const$ and therefore $\mathbf{p} = \nabla S = 0$. When considering the stochastic contribution, the sistematic velocity is no longer zero.

As a last remark, notice that Eq. (1.83) issued from the complex Bohmian mechanics is reminiscent of Eq. (1.103) when speaking about stochastic Bohmian mechanics with an imaginary diffusion coefficient D.

1.3.2 Diffusion. Quantum Brownian Motion

One of the simplest integrable dissipative quantum systems that we may devise is the diffusion of a wave packet on a flat surface. This is also a very instructive problem regarding the effects of stochastic dynamics on wave packets as well as an alternative way to also introduce quantum stochastic trajectories in this framework [13, 14]. Obviously, some of the results derived in this Section go beyond this particular context. In particular, the decoherence process which will be widely analyzed in Chap. 4.

The quantum Langevin equations for the Heisenberg position operators (for convenience in the notation, the operators will not be denoted with the usual 'hat') read for Ohmic friction, γ, or linear spectral density as [13, 14, 100, 221, 222]

$$\begin{aligned}\ddot{x}(t) &= -\gamma\dot{x}(t) + \delta F_x(t),\\ \ddot{y}(t) &= -\gamma\dot{y}(t) + \delta F_y(t).\end{aligned} \tag{1.106}$$

Here, x and y denote the position operators accounting for the motion of a single adsorbate across a flat surface, $V(x, y) = 0$. Lattice vibrational effects due to the surface temperature are assumed to be well described by a Gaussian white noise (per mass unit), which interacts and affects importantly the wave packet dynamics representing the adsorbate. Accordingly, the noise fluctuations acting on each degree of freedom are given by $\delta F_i(t) = F_i(t) - \langle F_i \rangle$, where $i = x, y$. At high (surface) temperatures, we have $\beta^{-1} \gg \hbar\gamma$ or, equivalently, $\gamma^{-1} \gg t_c \sim \hbar\beta$, where $\beta = (k_B T)^{-1}$ and the coherence time, t_c, is commonly known as *thermal time* and gives us an idea of the time scales at which thermal effects start playing a role. Under these conditions, the noise can be assumed to be classical and, hence, its autocorrelation function at two different times is well described by a Dirac δ-function.

Consider an adparticle of mass m initially placed at a given position on the flat surface. This particle is represented initially by a Gaussian wave packet,

$$\psi(x, y, 0) = \frac{1}{\sqrt{2\pi\sigma_0^2}} e^{-x^2/4\sigma_0^2 - y^2/4\sigma_0^2}, \tag{1.107}$$

with initial width, σ_0, along both directions. The adparticle is assumed to be initially in equilibrium with the reservoir or heat bath (surface) at a temperature T, but weakly coupled to the environment, so that dissipation can be neglected. The role of the initial conditions has been very often discussed in the literature (see, for example, Refs. [223, 224]). After a time t, the probability distribution to find the particle at a given position (x, y) is given by

$$P(x, y, t) = |\langle\psi(x, y, t)\rangle|^2 = \frac{1}{2\pi w_x(t) w_y(t)} e^{-x^2/2w_x^2(t) - y^2/2w_y^2(t)}, \tag{1.108}$$

which arises after averaging the survival probability over a thermal (Maxwell-Boltzmann) distribution [225] and ψ is solution of the free particle time dependent Schrödinger equation (see previous Section). Note here that the interaction with the environment makes the quantum state describing the system to pass from pure to a statistical mixture. According to Ford et al. [226–228], this normal distribution is also associated with two measurements at two different times. For each degree of freedom, the overall time-dependent spreading of the distribution (1.108) can be recast as

$$w_i^2(t) = \sigma_0^2 + \sigma_i^2(t) + s_i(t), \tag{1.109}$$

with $i = x, y$. The quantum contribution to this spreading is given by

$$
\begin{aligned}
\sigma_x^2(t) &= -\frac{[x(0), x(t)]^2}{4\sigma_0^2}, \\
\sigma_y^2(t) &= -\frac{[y(0), y(t)]^2}{4\sigma_0^2},
\end{aligned}
\tag{1.110}
$$

while $s_i(t)$ is the mean-squared displacement (MSD) along the i-th direction, i.e.,

$$
\begin{aligned}
s_x(t) &= \langle\{x(t) - x(0)\}^2\rangle, \\
s_y(t) &= \langle\{y(t) - y(0)\}^2\rangle.
\end{aligned}
\tag{1.111}
$$

The quantum spreading depends on the commutator of the position operators at two different times, from which

$$
\sigma_i^2(t) = \frac{\hbar^2}{4m^2\sigma_0^2\gamma^2}\,\Phi^2(\gamma t), \tag{1.112}
$$

with $\Phi(\gamma t) = 1 - \exp(-\gamma t)$. For each degree of freedom, one obtains the same spreading because the initial width and the noise fluctuations are assumed to be the same in both directions (isotropic surface). The same happens with the MSD along each direction, which acquires the form

$$
s_i(t) = \frac{2\hbar\gamma}{\pi m}\, t^2 G(\gamma t; T), \tag{1.113}
$$

with

$$
G(\gamma t; T) = \int_0^\infty \frac{1 - \cos z}{z(z^2 + \gamma^2 t^2)} \coth\left(\frac{\hbar z}{2tk_B T}\right) dz. \tag{1.114}
$$

In our particular case, this G-function has an analytical solution, allowing us to express (1.113) as

$$
s_i(t) = \frac{2k_B T}{m\gamma}\left[t - \frac{1}{\gamma}\Phi(\gamma t)\right]. \tag{1.115}
$$

In the long-time or diffusion regime, $\gamma t \gg 1$, the quantum spreading becomes time-independent,

$$
\sigma^2(t \to \infty) \approx \frac{\hbar^2}{4m^2\sigma_0^2\gamma^2}, \tag{1.116}
$$

while the MSD is linear with time,

$$
s(t) \approx \frac{2k_B T}{m\gamma}\, t. \tag{1.117}
$$

Thus, Einstein's diffusion law is satisfied, with the diffusion constant given by

$$D = \frac{k_B T}{m\gamma} \tag{1.118}$$

in both directions. Accordingly, the overall time-dependent spreading can be expressed as

$$w^2(t) \approx \sigma_0^2 + \frac{\hbar^2}{4m^2\sigma_0^2\gamma^2} + \frac{2k_B T}{m\gamma}\, t. \tag{1.119}$$

In the short-time regime, $\gamma t \ll 1$, the quantum spreading (1.112) goes like t^2, according to

$$\sigma^2(t) \approx \frac{\hbar^2}{4m^2\sigma_0^2}\, t^2, \tag{1.120}$$

which corresponds to the wave packet spreading in the absence of dissipation. Analogously, the MSD also goes like t^2 according to

$$s(t) \approx \frac{k_B T}{m}\, t^2, \tag{1.121}$$

where the prefactor is the thermal velocity in two dimensions. Thus, the overall time-dependent spreading can be expressed in a more compact form as

$$w^2(t) \approx \sigma_0^2 + \langle v^2 \rangle t^2, \tag{1.122}$$

with

$$\langle v^2 \rangle = \frac{k_B T}{m} + \frac{\hbar^2}{4m^2\sigma_0^2}. \tag{1.123}$$

In this short-time regime (friction-free motion or ballistic regime), we have usually that $\langle v^2 \rangle t^2 \ll \sigma_0^2$, that is, the wave packet has not spread too much compared to its initial spreading which can be assumed to be arbitrary large.

At low surface temperatures, we have $\gamma^{-1} \ll t_c \sim \hbar\beta$. In this case, the noise autocorrelation function is complex [229] and depends on the ratio between the interval of the two times and t_c (colored noise). In general, this noise function will act like a driving force and the surface dynamics is better described within the generalized Langevin framework as

$$\begin{aligned} \ddot{x}(t) + \int_{-\infty}^{t} dt' \gamma_x(t - t')\dot{x}(t') &= \delta F_x(t), \\ \ddot{y}(t) + \int_{-\infty}^{t} dt' \gamma_y(t - t')\dot{y}(t) &= \delta F_y(t), \end{aligned} \tag{1.124}$$

where $\gamma_x(t)$ and $\gamma_y(t)$ now represent the time-dependent frictions or memory functions along each direction —a one-dimensional expression for the noise function can be found in the literature [222, 230]. If we assume again an isotropic surface and Ohmic friction, Eq. (1.124) reduce to (1.106). Thus, the corresponding quantum mechanical process is not a Markovian process in the customary sense of the term [222, 230]. The quantum spreading is then the same as before, i.e., given by (1.112), since it is independent of the surface temperature. However, the MSD along each direction, given by (1.113), now reads as

$$s_i(t) = \frac{2k_BT}{m\gamma}\left[t - \frac{1}{\gamma}\Phi(\gamma t)\right] + \frac{4}{\beta m}\sum_{n=1}^{\infty}\frac{\gamma - \nu_n - \gamma e^{-\nu_n t} + \nu_n e^{-\gamma t}}{\nu_n(\gamma^2 - \nu_n^2)}, \qquad (1.125)$$

with

$$\nu_n = \frac{2\pi n}{\hbar\beta} \qquad (1.126)$$

being the so-called *Matsubara frequencies*, which come from the Taylor series expansion of the coth-function in the integrand of (1.114). At high surface temperatures, Eq. (1.125) reduces to (1.115), while the sum over n plays an important role only at very low, but non-vanishing surface temperatures. Notice that in the limit of zero surface temperature, the sum vanishes —in this limit, the coth-function in (1.114) becomes unity and $G(\gamma t; 0)$ displays a different time-dependence. In this case ($T = 0$), the corresponding MSD expression reads as

$$s_i(t) = \frac{2\hbar}{\pi m\gamma}\left\{\gamma_E + \ln\gamma t - \frac{1}{2}\left[e^{\gamma t}\bar{E}i(-\gamma t) + e^{-\gamma t}Ei(\gamma t)\right]\right\}, \qquad (1.127)$$

where $\gamma_E = 0.577$ is Euler's constant, and $\bar{E}i(-\gamma t)$ and $Ei(\gamma t)$ are the exponential integrals [231]. The environment no longer transfers energy to the adparticle due to the zero-point motion of the surface oscillators.

A similar dynamical analysis can also be carried out in terms of the short- and long-time regimes. The quantum spreading in both dimensions remains the same and is given by (1.112). However, the MSD is different due to its temperature dependence. Thus, in the short-time regime, $\gamma t \ll 1$, we find

$$s(t) \approx \frac{k_BT}{m}t^2 + \frac{\hbar\gamma}{\pi m}t^2, \qquad (1.128)$$

which depends linearly on γ. Analogously, at zero surface temperature, we have

$$s(t) \approx \frac{\hbar\gamma}{\pi m}t^2\left(\frac{3}{2} - \gamma_E - \ln\gamma t\right), \qquad (1.129)$$

where we observe essentially the same friction and time dependence, since the presence of the friction in the log-function is much weaker. On the other hand, in the long-time regime, it can be shown that

$$s(t) \approx \frac{2k_B T}{m\gamma} t + \frac{2\hbar}{\pi m} \frac{1}{\gamma + \nu_1}, \tag{1.130}$$

which becomes

$$s(t) \approx \frac{2\hbar}{\pi m \gamma} (\gamma_E + \ln \gamma t) \tag{1.131}$$

at zero surface temperature.

From this behavior, at equilibrium or diffusive regime [80], it is possible to determine the corresponding quantum potential and then to derive the corresponding quantum stochastic trajectories issued from the quantum Newton-like equation of motion

$$m\mathbf{a} = -\nabla(V + Q), \tag{1.132}$$

where $\mathbf{a}$ is the acceleration, V is the interaction potential between the adsorbate and the surface (in the free wave-packet problem, $V = 0$), and the quantum potential is given by Eq. (1.17). According to Eq. (1.108), $P(x, y, t)$ is identified with the probability density $\rho(x, y, t)$ displaying the total spreading of the wave packet with time. Only the spreading due to MSD depends explicitly on the surface temperature. However, the friction coefficient describing the interaction with the environment or bath is present in both the quantum and MSD spreadings. At very short times, in the ballistic regime, the stochastic dynamics is frictionless, but temperature dependent, as previously shown. On the contrary, in the diffusion regime, the corresponding dynamics depends on both, friction and temperature. Direct time integration of Newton's equation of motion gives the corresponding quantum stochastic trajectories. Very important implications such as the decoherence process and the violation of the

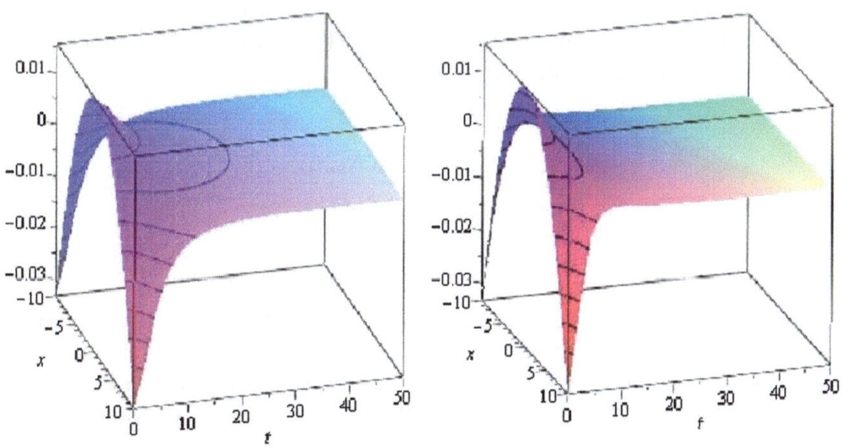

Fig. 1.3 Quantum potential for a friction $\gamma = 0.2$ and two surface temperatures: $T = 50$ K (*left*) and $T = 300$ K (*right*). The associated Gaussian wave packet describes a particle characterized by $m = 1$ and $\sigma_0 = 2$ ($\hbar = 1$)

noncrossing rule are worth mentioning. These points will be analyzed and discussed under a different perspective in the following chapters.

As an illustration, in Fig. 1.3 the quantum potential associated with a Gaussian wave packet with $m = 1$ and $\sigma_0 = 2$ ($\hbar = 1$) is displayed for a friction $\gamma = 0.2$ and two surface temperatures: $T = 50$ K (left) and $T = 300$ K (right). For simplicity, only the x-direction is plotted. The stronger variations of the quantum potential takes place at short times and large distances. The MSD is linear with the temperature as well as the total spreading of the Gaussian distribution.

This simple analysis can be extended to a many particle system. The non-local effects are now not only due to the quantum potential but also due to the osmotic velocity. This point is crucial to the stochastic methods derived from Nelson's approach.

It should be taken into account that when speaking about quantum stochastic trajectories in the Bohmian framework, there are several routes as the one followed in this Section. After Bohm and Vigier, the more formal approach is when one is starting with a Langevin-like equation such as Eq. (1.95). In any case, the osmotic and systematic velocities, given by Eqs. (1.100) and (1.103) respectively, can straightforwardly be obtained for the Brownian motion in the diffusive regime. In Chap. 3, a different approach is proposed and developed.

1.3.3 Stochastic Weak Values

As pointed out by Aharonov [232] the initial state of a closed system does not determine its final state, complete initial and final boundary conditions must be imposed. Thus, for an initial boundary condition (before the measurement) we are pre-selecting an incident wave function while for a final boundary condition we are post-selecting a final wave function. Both boundary conditions define what is called a pre- and post-selected ensemble, both wave functions being non-orthogonal. In this sense, the weak value of an operator A, is defined as

$$\langle A \rangle_w = \frac{\langle \psi_{final} | A | \psi_{initial} \rangle}{\langle \psi_{final} | \psi_{initial} \rangle}. \tag{1.133}$$

Thus, weak values reduce to expectation values when both wave functions coincide. A measurement of the observable A yielding $\langle A \rangle_w$ is a *weak* measurement.

Weak measurements are playing a very important role in measuring very tiny, but important effects [19, 233]. Most people think that this type of measurements will be important to shed some light to fundamental problems in quantum mechanics. As shown by Hiley [64], weak values which can be seen as complex, transition probability amplitudes are related to Bohm momentum, the Bohm energy, and the quantum potential. In one dimension, for example, the expectation value of the momentum operator is given by

$$\langle\psi(t)|p|\psi(t)\rangle = \int \rho(x,t)\frac{\langle x|p|\psi(t)\rangle}{\langle x|\psi(t)\rangle}dx \tag{1.134}$$

where in atomic units

$$\langle x|p|\psi(t)\rangle = \int \langle x|p|x'\rangle\langle x'|\psi(t)\rangle dx' = -i\nabla_x\psi(x,t). \tag{1.135}$$

By writting as usual the wave function in polar form, the weak value obtained from measuring the position (post-selection) after the momentum is pre-selected leads to

$$\frac{< x|p|\psi(t) >}{< x|\psi(t) >} = \nabla_x S(x,t) - i\frac{\nabla_x\rho(x,t)}{2\rho(x,t)} \tag{1.136}$$

where it is a complex quantity where the real part is the Bohm momentum and the imaginary part is identical to the so-called osmotic velocity.

As we have also seen in open quantum systems, the osmotic velocity comes directly from an underlying diffusion motion (as introduced by Nelson in his stochastic theory [73]). This osmotic velocity is related to the probability density. Thus, when we are dealing with quantum stochastic trajectories, the weak values are also depending on the friction coefficient and the surface temperature. Further investigation along these lines are crucial to better understand future experiments.

1.3.4 *The Density Matrix*

When one has an ensemble of particles at different positions but the same state or wave function, we talk about a pure state, or better said a pure ensemble. On the contrary, if the states or wave function are different, a mixed state or ensemble is often used.

Concerning statistical properties of the quantum systems, two types of averaging is usually considered [10]. The first one is over the particle positions and the second one, over the distribution of wave functions. If the wave function is represented in terms of an orthonormal basis set, for example, eigenfunctions of the energy, we can always write

$$\Psi(\mathbf{r}) = \sum_{E,j} c_{Ej}\psi_{E,j}(\mathbf{r}) \tag{1.137}$$

where E runs over the energy eigenvalues and the index j over all the remaining quantum numbers. The complex coefficients c can also be expressed in polar form, $c_{Ej} = R_{Ej}exp[i\phi_{Ej}]$. Considering an ensemble of wave functions also means to have a probability distribution of amplitudes and phases. Thus, a given point in the space of wave functions is given by all c_{Ej}'s or all R_{Ej}s and ϕ_{Ej}'s. The element of volume of this space can then be written as

$$d\Omega = \prod_{E,j} dp_{Ej} d\phi_{Ej} \tag{1.138}$$

where $p_{Ej} = R_{Ej}^2$. The mean number, dZ, of systems in this element of volume is given by

$$dZ = P(...R_{Ej}...; ...\phi_{Ej}...)d\Omega \tag{1.139}$$

P being the probability density of systems in this wave function space. The first averaging is then, for a given wave function ψ and physical operator O,

$$\langle O \rangle = \int \sum_{E,j} \sum_{E',j'} c_{Ej}^* c_{E'j'} \psi_{E,j}^*(\mathbf{r}) O \psi_{E'j'}(\mathbf{r}) d\mathbf{r} \tag{1.140}$$

and the second averaging can be expressed as

$$\begin{aligned} \langle\langle O \rangle\rangle = & \int \int \sum_{E,j} \sum_{E',j'} P(...R_{Ej}...R_{E'j'}...; ...\phi_{Ej}...\phi_{E'j'}...) \\ & . \ R_{Ej} R_{E'j'} e^{-i\phi_{Ej}} e^{i\phi_{E'j'}} \psi_{E,j}^*(\mathbf{r}) O \psi_{E'j'}(\mathbf{r}) d\mathbf{r} d\Omega \\ = & \ Tr(\rho O) \end{aligned} \tag{1.141}$$

where the usual density matrix is defined as

$$\begin{aligned} \rho_{EjE'j'} = & \int P(...R_{Ej}...R_{E'j'}...; ...\phi_{Ej}...\phi_{E'j'}...) \\ & . \ R_{Ej} R_{E'j'} e^{-i\phi_{Ej}} e^{i\phi_{E'j'}} d\Omega. \end{aligned} \tag{1.142}$$

It is clear that the density matrix determines the statistical results of all measurements in standard quantum mechanics [104, 105].

The density matrix does not determine the probability distribution P unambiguously. In other words, physical averages are extremely insensitive to the detail of the P-distribution. In fact, for a given density matrix, a very wide range of P-distributions is possible. A good illustration of this fact is the following. Let us consider a general wave function expressed as

$$\Psi(\mathbf{r}) = \sum_j R_j e^{i\phi_j} \psi_j(\mathbf{r}) \tag{1.143}$$

then the density matrix in the configuration space is given by

$$\rho(\mathbf{r},\mathbf{r}') = \int \sum_{j,j'} P(...R_j...R_{j'}...;...\phi_j...\phi_{j'}...)$$
$$\cdot\ R_j R_{j'} e^{i(\phi_j-\phi_{j'})}\psi_j^*(\mathbf{r})\psi_{j'}(\mathbf{r}')d\Omega. \tag{1.144}$$

Now, if we assume that the P-distribution is independent on the phases and these are random (random phase approximation),

$$\rho(\mathbf{r},\mathbf{r}') = \int \sum_{j} P(...R_j...R_{j'}...)R_j^2\psi_j^*(\mathbf{r})\psi_j(\mathbf{r}')d\Omega$$
$$= \sum_j \langle R_j^2\rangle \psi_j^*(\mathbf{r})\psi_j(\mathbf{r}'). \tag{1.145}$$

If we further assume that the P-distribution is a symmetric function of the R_j's, $\langle R_j^2\rangle = \langle R^2\rangle$ and then

$$\rho(\mathbf{r},\mathbf{r}') = \sum_j \langle R^2\rangle \psi_j^*(\mathbf{r})\psi_j(\mathbf{r}') \tag{1.146}$$

where a very wide range of P-functions can lead to the above density matrix.

Finally, in order to also extract some useful information about the system of interest, one usually computes its associated reduced density matrix by tracing out over the total density matrix, $\hat{\rho}_t$, over the environment degrees of freedom. In the configuration representation and for an environment constituted by N particles, the system reduced density matrix is obtained after integrating $\hat{\rho}_t \equiv |\Psi\rangle_{t\,t}\langle\Psi|$ over the $3N$ environment degrees of freedom, $\{\mathbf{r}_i\}_{i=1}^N$,

$$\tilde{\rho}(\mathbf{r},\mathbf{r}',t) = \int \langle \mathbf{r},\mathbf{r}_1,\mathbf{r}_2,\ldots\mathbf{r}_n|\Psi(t)\rangle\langle\Psi(t)|\mathbf{r}',\mathbf{r}_1,\mathbf{r}_2,\ldots\mathbf{r}_n\rangle\ d\mathbf{r}_1 d\mathbf{r}_2\cdots d\mathbf{r}_n. \tag{1.147}$$

The system (reduced) quantum density current can be derived from the following expression

$$\tilde{\mathbf{J}}(\mathbf{r},t) \equiv \frac{\hbar}{m}\ \mathrm{Im}[\nabla_{\mathbf{r}}\tilde{\rho}(\mathbf{r},\mathbf{r}',t)]\Big|_{\mathbf{r}'=\mathbf{r}}, \tag{1.148}$$

which satisfies the continuity equation

$$\dot{\tilde{\rho}} + \nabla\tilde{\mathbf{J}} = 0. \tag{1.149}$$

In (1.149), $\tilde{\rho}$ is the diagonal element (i.e., $\tilde{\rho} \equiv \tilde{\rho}(\mathbf{r},\mathbf{r},t)$) of the reduced density matrix and gives the measured intensity [234].

Taking into account (1.148) and (1.149), it is possible to define the velocity field, $\dot{\mathbf{r}}$, associated with the (reduced) system dynamics as

$$\tilde{\mathbf{J}} = \tilde{\rho}\dot{\mathbf{r}}, \tag{1.150}$$

which is analogous to the Bohmian velocity field. Now, from (1.150), a new class of quantum trajectories is define, which are the solutions to the equation of motion

$$\dot{\mathbf{r}} \equiv \frac{\hbar}{m} \frac{\mathrm{Im}[\nabla_{\mathbf{r}}\tilde{\rho}(\mathbf{r}, \mathbf{r}', t)]}{\mathrm{Re}[\tilde{\rho}(\mathbf{r}, \mathbf{r}', t)]}\bigg|_{\mathbf{r}'=\mathbf{r}}. \tag{1.151}$$

These new trajectories are the so-called *reduced quantum trajectories* [235, 236], which are only related to the system reduced density matrix. As shown in [235], the dynamics described by (1.151) leads to the correct intensity (whose time-evolution is described by (1.149)) when the statistics of a large number of particles is considered. Moreover, it is also straightforward to show that (1.151) reduces to the well-known expression for the velocity field in Bohmian mechanics when there is no interaction with the environment.

The formalism of the density matrix (or reduced density matrix) is widely used in open quantum systems dynamics through master equations or, in general, when the decoherence process is also taken into account due to the presence of an apparatus [104, 105].

1.4 Dynamical Aspects of the Measuring Process

The idea behind the measuring process is to deal with devices to be considered as quantum systems. This view entails an arbitrary division of the world into classical and quantum parts. As mentioned at the beginning of this chapter, the issue of a dividing line is critical in this context. The equation of motion of Heisenberg is very advantageous because both the classical and quantum equations of motion for a given observable coincide, leading to a better understanding due to the analogy with the classical evolution in time [232]. Here, on the contrary, we are going to briefly discuss the measuring process from a dynamical perspective in the Schrödinger picture of quantum mechanics, more adequate for the Bohmian formalism. In any case, quantum measurements preserve the Heisenberg uncertainty relations (see Chap. 4).

Following von Neumann, let A be the Hermitian operator that is going to be measured and A_n and $\psi_n(\mathbf{x})$ its eigenvalues and eigenfunctions in the configuration space, respectively (for simplicity, a discrete and non-degenerate spectrum is assumed). The wave function of the pointer or apparatus with a large number of coordinates, represented globally by $\mathbf{y}$, is assumed to be described by a wave packet $\chi(\mathbf{y})$. The initial wave function of the composite system is then written as (the state preparation) [7, 10]

$$\Psi_i(\mathbf{x}, \mathbf{y}) = \chi(\mathbf{y}) \sum_n c_n \psi_n(\mathbf{x}). \tag{1.152}$$

If the interaction between the observed system and the pointer is very short, say T (impulsive measurement), and so strong that throughout that time the total Hamiltonian is only governed by the interaction term, the time-dependent Schrödinger equation can be written as .

$$i\hbar\frac{\partial\Psi(\mathbf{x},\mathbf{y},t)}{\partial t} = H_I\Psi(\mathbf{x},\mathbf{y},t) = i\hbar\kappa A\frac{\partial\Psi(\mathbf{x},\mathbf{y},t)}{\partial\mathbf{y}} \tag{1.153}$$

which can be easily solved for the period of the interaction to give

$$\Psi(\mathbf{x},\mathbf{y},t) = \sum_n c_n\psi_n(\mathbf{x})\chi(\mathbf{y}-\kappa t A_n) \tag{1.154}$$

where κ is the strength of the coupling. During this period of time, the various components of the total wave function will overlap and interfere being non-factorizable. The total wave function at the end of the impulse is written as

$$\Psi_f(\mathbf{x},\mathbf{y},T) = \sum_n c_n\psi_n(\mathbf{x})\chi(\mathbf{y}-\kappa T A_n) \tag{1.155}$$

consisting of wave packets with centers given by $\kappa T A_n$. Under the condition

$$\kappa T \Delta A_n \gg 1 \tag{1.156}$$

where ΔA_n is the change of A_n for the different values of n, which means that the interaction is strong enough, the wave packets multiplying the different eigenstates of A will be distinct and non-overlapping (in order to prevent any influence on the observed system and extract unambiguous information) and evolve freely lately. In other words, to each wave packet of the pointer there will correspond an eigenfunction and eigenvalue of the system since both are correlated; the corresponding widths being much less than the separation between eigenvalues (complete measurement). Thus, when the wave packets cease to overlap, the interference terms are negligible and

$$|\Psi_f(\mathbf{x},\mathbf{y},T)|^2 \approx \sum_n |c_n|^2|\psi_n(\mathbf{x})|^2|\chi(\mathbf{y}-\kappa T A_n)|^2 \tag{1.157}$$

If the apparatus registers the value $\mathbf{y}$ (the registration step), according to Born's postulate, the total wave function will collapse to one of the many components or products $\psi_n(\mathbf{x})\chi(\mathbf{y}-\kappa T A_n)$, $|c_n|^2$ being the probability of the measurement A_n. A typical example of this type of measurements is given by the well-known Stern–Gerlach experiment where hydrogen atoms pass through an inhomogeneous magnetic field in the z-direction. The interaction Hamiltonian is then given by the gradient of the magnetic field times the z-component of the angular momentum operator. After the interaction, the three wave packets will eventually separate allowing their detection in a screen.

From a Bohmian perspective [7, 10], during the interaction time, the different wave packets strongly interfere and the quantum potential develops a quite complicated and rapidly varying space-time structure. When the interaction is over, the different wave packets or channels start to be well defined and the system will enter one of them, remaining from then on, since the probability of being between wave packets is zero. The rest of the wave packets will not contribute to the quantum potential (or to the guiding condition) acting on the the system and can therefore be ignored. They can be regarded as constituting inactive or physically ineffective information (the so-called empty waves). This fact is behind the possibility of a set of clearly distinct results of a quantum measurement and no mention to the collapse of the wave function is needed. The measuring process can be also considered as a quantum transition, from Ψ_i to one of the final states (or channels) explicitly written in Ψ_f with probability $|c_n|^2$. Therefore, in principle, there is no need to place any arbitrary cut between the quantum and classical worlds. In Chap. 4, this dividing line in the Bohmian formalism will be widely developed.

In Bohmian mechanics, the position of the system plays a key role since it is an intrinsic property. It is intrinsic in a double sense. First, it is defined in a way that is conceptually independent on the wave function or, in other words, it is not inherently dependent on the overall context, unlike the actual momentum of the sytem which depends on the total wave function (in particular, on the gradient of the wave function phase). And second, the position can be measured without being altered. Let us analyze the particular case of $A \equiv x$ when the interaction Hamiltonian is given by

$$H_I = \kappa x p_y. \tag{1.158}$$

Due to the fact that x is a continuous variable, the initial wave function can be written as

$$\Psi_i(x, y) = \int dx' c(x')\delta(x - x')\chi_0(y) \tag{1.159}$$

and the final wave function as

$$\Psi_f(x, y, T) = \int dx' c(x')\delta(x - x')\chi_0(y - \kappa x' T). \tag{1.160}$$

If δy is the width of the wave packet χ_0, the corresponding impulsive measurement will make possible an observable distinction between values of x' differing by $\delta x = \delta y / \kappa T$, the only change being the shifting of the argument of χ_0 (the x position is not altered and the corresponding momentum is also unchanged if χ_0 is assumed to be real). However, other properties are changed due to the fact that the total wave function has been modified.

On the other hand, for measuring properties other than position, the situation is quite different. Let us assume that we want to measure the momentum, that is, $A \equiv p_x$. Following the same procedure as before, the interaction Hamiltonian is written as

$$H_I = \kappa p_x p_y, \tag{1.161}$$

the initial wave function can be written as

$$\Psi_i(x, y) = \int dp_x c(p_x) e^{ip_x x} \chi_0(y), \tag{1.162}$$

and the final wave function as

$$\Psi_f(x, y, T) = \int dp_x c(p_x) e^{ip_x x} \chi_0(y - \kappa p T). \tag{1.163}$$

Again, the corresponding measurement will make possible an observable distinction between values of p_x differing by $\delta p_x = \delta y / \kappa T$ in momentum space and, in position space, $\delta x = 1/\delta p_x$. The wave packets do not interfere because they correspond to non-overlapping states of the apparatus variable. The initial and final true momenta may have different values. The final value depends not only of the total wave function but also on the details of the initial conditions of both the particle and the apparatus. After the standard interpretation of quantum mechanics, this fact can be considered contradictory because $[H_I, p_x] = 0$, so that one would have expected it to be a constant of motion. In Bohmin mechanics, the momentum operator is related to the particle momentum only when the wave function is an eigenfunction (a plane wave) of this operator. As mentioned before, this also means that the momentum is not an intrinsic property unlike the position. It is inherently context dependent so that it involves the participation of the measuring apparatus and the quantum system as a whole. This fact implies that the mean value of p^2, for example, has an extra term with respect to the standard result of quantum mechanics [7, 10]. The same is true for higher moments of the momentum.

The next straightforward question is what happens when trying to measure position and momentum together. The interaction Hamiltonian is now written as

$$H_I = \kappa(x p_y + p_x p_z), \tag{1.164}$$

where y and z are the coordinates of two independent meters corresponding to commuting operators. After the impulsive interaction, the three interacting systems become evidently correlated. As may be expected, the outcome of the joint measurement does not allow one to infer values for the particle position or momentum [7].

This theoretical treatment should be changed when noncanonical physical observables are considered such as velocity and other time derivatives of canonical observables [232].

In Chap. 4, the measurement problem is studied under a different and new point of view. The so-called continuous quantum measurement on a given system surrounded by an environment is tackled from a nonlinear wave equation.

References

1. Bell, J.S.: Speakable and Unspeakable in Quantum Mechanics. Cambridge University Press, Cambridge (1987)
2. Dawson Jr., J.W.: Logical dilemmas: The life and work of Kurt Gödel. Wellesley, Massachusetts (1997)
3. Tegmark, M., Wheeler, J.A.: 100 years of quantum misteries, pp. 72-79. Scientific American (2001)
4. Bohm, D.: A suggested interpretation of the quantum theory in terms of "hidden" variables. I. Phys. Rev. **85**, 166–179 (1952)
5. Bohm, D.: A suggested interpretation of the quantum theory in terms of "hidden" variables. II. Phys. Rev. **85**, 180–193 (1952)
6. Bernstein, J.: More about Bohm's Quantum. Am. J. Phys. **79**, 601–606 (2011)
7. Holland, P.R.: The Quantum Theory of Motion. Cambridge University Press, Cambridge (1993)
8. Goldstein, S.: Quantum theory without observers - Part I. Phys. Today **51**(3), 42–46 (1998)
9. Goldstein, S.: Quantum theory without observers - Part II. Phys. Today **51**(4), 38–42 (1998)
10. Bohm, D., Hiley, B.J.: The Undivided Universe. Routledge, London (1993)
11. Dürr, D., Teufel, S.: Bohmian Mechanics. Springer, Heidelberg (2009)
12. Wyatt, R.E.: Quantum Dynamics with Trajectories: Introduction to Quantum Hydrodynamics. Springer, New York (2005)
13. Sanz, A.S., Miret-Artés, S.: A trajectory description of quantum processes. I. Fundamentals. Lect. Notes Phys. **850**, 1–299 (2012)
14. Sanz, A.S., Miret-Artés, S.: A trajectory description of quantum processes. II. Applications. Lect. Notes Phys. **831**, 1–333 (2014)
15. Aspect, A., Grangier, P., Roger, G.: Experimental tests of realistic local theories via Bell's theorem. Phys. Rev. Lett. **47**, 460–463 (1981)
16. Aspect, A., Grangier, P., Roger, G.: Experimental realization of Einstein-Podolsky-Rosen-Bohm gedankenexperiment: a new violation of Bell's inequalities. Phys. Rev. Lett. **49**, 91–94 (1982)
17. Aspect, A., Dalibard, J., Roger, G.: Experimental test of Bell's inequalities using time-varying analyzers. Phys. Rev. Lett. **49**, 1804–1807 (1982)
18. Braverman, B., Simon, C.: Proposal to observe the nonlocality of Bohmian trajectories with entangled photons. Phys. Rev. Lett. **110**(1–5), 060406 (2013)
19. Kocsis, S., Braverman, B., Ravets, S., Stevens, M.J., Morin, R.P., Shalm, L.K., Steinberg, A.M.: Observing the average trajectories of single photons in a two-slit interferometer. Science **332**, 1170–1174 (2011)
20. Mahler, D.H., Rozema, L., Fisher, K., Vermeyden, L., Resch, K.J., Wiseman, H.M., Steinberg, A.: Experimental nonlocal and surreal Bohmian trajectories. Sci. Adv. **2**, 1–7 (2016)
21. Aharonov, J., Albert, D.Z., Vaidman, L.: How the result of a measurement of a component of the spin of a 1/2-spin particle can turn out to be 100. Phys. Rev. Lett. **60**, 1351–1354 (1988)
22. Sanz, A.S., Davidovic, M., Bozic, M., Miret-Artés, S.: Understanding interference experiments with polarized light through photon trajectories. Ann. Phys. (N.Y.) **325**, 763–784 (2010)
23. Chattaraj, P.K. (ed.): Quantum Trajectories. CRC Press, Taylor and Francis, NY (2010)
24. Oriols, X., Mompart, J. (eds.): Applied Bohmian Mechanics: From Nanoscale Systems to Cosmology. Pan Standford Publishing, Singapore (2011)
25. Weiner, J.H., Askar, A.: Particle method for the numerical solution of the time-dependent Schrödinger equation. J. Chem. Phys. **54**, 3534–3541 (1971)
26. Dey, B.K., Askar, A., Rabitz, H.A.: Multidimensional wave packet dynamics with the fluid dynamical formulation of the Schrödinger equation. J. Chem. Phys. **109**, 8770–8782 (1998)
27. Sales Mayor, F., Askar, A., Rabitz, H.A.: Quantum fluid dynamics in the Lagrangian representation and applications to photodissociation problems. J. Chem. Phys. **111**, 2423–2435 (1999)

28. Zhao, Y., Makri, N.: Bohmian versus semiclassical description of interference phenomena. J. Chem. Phys. **119**, 60–67 (2003)
29. Makri, M.: Forward-backward quantum dynamics for time correlation functions. J. Phys. Chem. A **108**, 806–812 (2004)
30. Liu, J., Makri, N.: Monte Carlo Bohmian dynamics from trajectory stability properties. J. Phys. Chem. A **108**, 5408–5416 (2004)
31. Bittner, E.R.: Quantum initial value representations using approximate Bohmian trajectories. J. Chem. Phys. **119**, 1358–1364 (2003)
32. Garashchuk, S., Rassolov, V.A.: Semiclassical dynamics with quantum trajectories: formulation and comparison with the semiclassical initial value representation propagator. J. Chem. Phys. **118**, 2482–2490 (2003)
33. Poirier, B.: Reconciling semiclassical and Bohmian mechanics. I. Stationary States. J. Chem. Phys. **121**, 4501–4515 (2004)
34. Trahan, C., Poirier, B.: Reconciling semiclassical and Bohmian mechanics. II. Scattering states for discontinuous potentials. J. Chem. Phys. **124**, 034115(1–18) (2006)
35. Trahan, C., Poirier, B.: Reconciling semiclassical and Bohmian mechanics. III. Scattering states for continuous potentials. J. Chem. Phys. **124**, 034116(1–14) (2006)
36. Poirier, B.: Reconciling semiclassical and Bohmian mechanics. V. Wavepacket dynamics. J. Chem. Phys. **128**, 164115(1–15) (2008)
37. Tonomura, A., Endo, J., Matsuda, T., Kawasaki, T., Ezawa, H.: Demonstration of single-electron buildup of an interference pattern. Am. J. Phys. **57**, 117–120 (1989)
38. Shimizu, F., Shimizu, K., Takuma, H.: Double-slit interference with ultracold metastable neon atoms. Phys. Rev. A **46**, R17–R20 (1992)
39. Sanz, A.S.: Particles, waves and trajectories: 2010 years after Young's experiment. J. Phys. Conf. Ser. **504**(1–14), 012028 (2014)
40. Sanz, A.S.: Investigating puzzling aspects of the quantum theory by means of its hydrodynamic representation. Found. Phys. **45**, 1153–1165 (2015)
41. Luis, A., Sanz, A.S.: Nonclassical polarization dynamics in classical-like states. Phys. Rev. A **92**, 023832(1–7) (2015)
42. Zamstein, N., Tannor, D.J.: Non-adiabatic molecular dynamics with complex quantum trajectories.I. The diabatic representation. J. Chem. Phys. **137**, 22517 (1–6) (2012)
43. Zamstein, N., Tannor, D.J.: Non-adiabatic molecular dynamics with complex quantum trajectories. II. The adiabatic representation. J. Chem.Phys. **137**, 22A518 (1–6) (2012)
44. Blum, K.: Density Matrix Theory and Applications. Plenum Press, New York (1981)
45. Thomas, L.H.: The calculation of atomic fields. Proc. Cambridge Phil. Soc. **23**, 542–548 (1927)
46. Fermi, E.: Un metodo statistico per la determinazione di alcune priopriettà dell'atomo. Rend. Accad. Naz. Lincei **6**, 602–607 (1927)
47. Koch, W., Holthausen, M.C.: A Chemist's Guide to Density Functional Theory, 2nd edn. Wiley-VCH, Weinheim (2001)
48. Hohenberg, P., Kohn, W.: Inhomogeneous electron gas. Phys. Rev. B **136**, 864–871 (1964)
49. Kohn, W., Sham, L.J.: Self-consistent equations including exchange and correlation effects. Phys. Rev. A **140**, 1133–1138 (1965)
50. Marques, M.A.L., Gross, E.K.U.: Time-dependent density functional theory. Ann. Rev. Phys. Chem. **55**, 427–455 (2004)
51. Botti, S., Schindlmayr, A., Del Sole, R., Reining, L.: Time-dependent density-functional theory for extended systems. Rep. Prog. Phys. **70**, 357–407 (2007)
52. See, for example: Proceedings of the VIth International Conference on the Applications of Density Functional Theory. Paris, France, 29 Aug.-1 Sept., 1995. Int. J. Quantum Chem. **61**, 181–196 (1997)
53. Bartolotti, L.J., Mollmann, J.C.: 4th Order time-dependent variation perturbation-theory based on the hydrodynamic analogy. Mol. Phys. **38**, 1359–1365 (1979)
54. Bartolotti, L.J.: Time-dependent extension of the Hohenberg-Kohn-Levy energy-density functional. Phys. Rev. A **24**, 1661–1667 (1981)

55. Bartolotti, L.J.: Time-dependent Kohn-Sham density-functional theory. Phys. Rev. A **26**, 2243–2244 (1982)
56. Ghosh, S.K., Deb, B.M.: Quantum fluid dynamics of many-electron systems in three-dimensional space. Int. J. Quantum Chem. **22**, 871–888 (1982)
57. Deb, B.M., Ghosh, S.K.: Schrödinger fluid dynamics of many-electron systems in a time-dependent density-functional framework. J. Chem. Phys. **77**, 342–348 (1982)
58. Runge, E., Gross, E.K.U.: Density-functional theory for time-dependent systems. Phys. Rev. Lett. **52**, 997–1000 (1984)
59. Madelung, E.: Quantentheorie in hydrodynamischer form. Z. Phys. **40**, 322–326 (1926)
60. Hirschfelder, J.O., Christoph, A.C., Palke, W.E.: Quantum mechanical streamlines. I. Square potential barrier. J. Chem. Phys. **61**, 5435–5456 (1974)
61. Hirschfelder, J.O., Goebel, C.J., Bruch, L.W.: Quantized vortices around wavefunction nodes. II. J. Chem. Phys. **61**, 5456–5459 (1974)
62. Hirschfelder, J.O., Tang, K.T.: Quantum mechanical streamlines. III. Idealized reactive atom-diatomic molecule collision. J. Chem. Phys. **64**, 760–786 (1976)
63. Wiseman, H.M.: Grounding Bohmian mechanics in weak values and Bayesianism. New. J. Phys. **9**, 165(1–12) (2007)
64. Hiley, B.J.: Weak values:Approach through the Cliffordand Moyal algebras. J. Phys. Conf. Ser. **361**, 012014(1-11) (2012)
65. Fürth, R.: Über einige Beziehungen zwischen klassischer Statistik und Quantenmechanik. Z. Phys. **81**, 143–162 (1933)
66. Fényes, I.: Eine wahrscheinlichkeitstheoretische Begründung und Interpretation Der Quantenmechanik. Z. Phys. **132**, 81–106 (1952)
67. Weizel, W.: Ableitung der Quantentheorie aus einem klassischen, kausal determinierten Modell. Z. Physik **134**, 264–285 (1953)
68. Weizel, W.: Ableitung der Quantentheorie aus einem klassischen Modell. II. Z. Phys. **135**, 270–273 (1953)
69. Weizel, W.: Ableitung der Quantenmechanischen Wellengleichung des Mehrteilchensystems aus einem klassischen Modell. Z. Phys. **136**, 582–604 (1953)
70. Bohm, D., Vigier, J.P.: Model of the causal interpretation of quantum theory in terms of a fluid with irregular fluctuations. Phys. Rev. **96**, 208–216 (1954)
71. Takabayasi, T.: On the formulation of quantum mechanics associated with classical pictures. Prog. Theor. Phys. **8**, 143–182 (1952)
72. Takabayasi, T.: Remarks on the formulation of quantum mechanics with classical pictures and on relations between linear scalar fields and hydrodynamical fluids. Prog. Theor. Phys. **9**, 187–222 (1953)
73. Nelson, E.: Derivation of the Schrödinger equation from Newtonian mechanics. Phys. Rev. **150**, 1079–1085 (1966)
74. Kershaw, D.: Theory of hidden variables. Phys. Rev. **136**, B1850–B1856 (1964)
75. Comisar, G.G.: Brownian motion of nonrelativistic quantum mechanics. Phys. Rev. **138**, B1332–B1337 (1965)
76. de la Peña-Auberbach, L.: New formulation of stochastic theory and quantum mechanics. J. Math. Phys. **10**, 1620–1630 (1969)
77. Olavo, L.S.F.: Foundations of quantum mechanics: connection with stochastic processes. Phys. Rev. A **61**, 052109(1–14) (2000)
78. Guerra, F.: Structural aspects of stochastic mechanics and stochastic field theory, Phys. Rep. **77**, 263-312 (1981). The same volume **77** is entirely devoted to *New Stochastic Methods in Physics*, edited by C. DeWitt-Morette and K. D. Elworthy
79. Grabert, H., Hänggi, P., Talkner, P.: Is quantum mechanics equivalent to a classical stochastic process? Phys. Rev. A **19**, 2440–2445 (1979)
80. Bohm, D., Hiley, B.J.: Non-locality and locality in the stochastic interpretation of quantum mechanics. Phys. Rep. **172**, 93–122 (1989)
81. Vink, J.C.: Quantum mechanics in terms of discrete beables. Phys. Rev. A **48**, 1808–1818 (1993)

82. Lorenzen, F., de Ponte, M.A., Moussa, M.H.Y.: Extending Bell's beables to encompass dissipation, decoherence and the quantum-to-classical transition through quantum trajectories. Phys. Rev. A **80**, 032101(1–8) (2009)
83. Nassar, A.B.: Derivation of a generalized nonlinear Schrödinger-Langevin equation. Phys. Lett. A **109**, 1–3 (1985)
84. Nassar, A.B.: Time-dependent invariant associated to nonlinear Schrödinger-Langevin equation. J. Math. Phys. **27**, 2949–2952 (1986)
85. Kostin, M.D.: On the Schrödinger-Langevin equation. J. Chem. Phys. **57**, 3589–3591 (1972)
86. Hasse, R.W.: On the quantum mechanical treatment of dissipative systems. J. Math. Phys. **16**, 2005–2011 (1975)
87. Bialynicki-Birula, I., Mycielski, J.: Nonlinear wave mechanics. Ann. Phys. **100**, 62–93 (1976)
88. Stocker, W., Albrecht, K.: A formalism for the construction of quantum friction equations. Ann. Phys. **117**, 436–446 (1979)
89. Schuch, D., Chung, K.M., Hartmann, H.: Nonlinear Schrödinger-type field equation for the description of dissipative systems. I. Derivation of the nonlinear field equation and one-dimensional example. J. Math. Phys. **24**, 1652–1660 (1984)
90. Schuch, D., Chung, K.M., Hartmann, H.: Nonlinear Schrödinger-type field equation for the description of dissipative systems. III. Frictionally damped free motion as an example for an aperiodic motion. J. Math. Phys. **25**, 3086–3092 (1984)
91. Olavo, L.S.F., Laps, L.C., Figueiredo, A.: Foundations of quantum mechanics: the Langevin equation for QM. Ann. Phys. **327**, 1391–1407 (2012)
92. Bassalo, J.M.F., da Silva, D.G., Nassar, A.B., Cattani, M.S.D.: The Feynman's propagators for non-linear Schrödinger equations. J. Adv. Math. Appl. **1**, 1–30 (2012)
93. Garashchuk, S., Dixit, V., Gu, B., Mazzuca, J.: The Schrödinger equation with friction from the quantum trajectory perspective. J. Chem. Phys. **138**, 054107(1–7) (2013)
94. Chou, C.-C.: Trajectory approach to the Schrödinger-Langevin equation with linear dissipation for ground states. Ann. Phys. **362**, 57–73 (2015)
95. Nassar, A.B.: Ermakov and non-Ermakov systems in quantum dissipative models. J. Math. Phys. **27**, 755–758 (1986)
96. Caldirola, P.: Forze non conservative nella meccanica quantistica. Nuovo Cimento **18**, 393–400 (1941)
97. Kanai, E.: On the quantization of the dissipative systems. Prog. Theor. Phys. **3**, 440–442 (1948)
98. Sanz, A.S., Martinez-Casado, R., Peñate-Rodriguez, H.C., Rojas-Lorenzo, G., Miret-Artés, S.: Dissipative Bohmian mechanics within the Caldirola-Kanai framework: a trajectory analysis of wave-packet dynamics in viscid media. Ann. Phys. **347**, 1–20 (2014)
99. Percival, I.C.: Quantum State Diffusion. Cambridge University Press, Cambridge (1998)
100. Weiss, U.: Quantum Dissipative Systems. World Scientific, Singapore (1999)
101. Breuer, H.P., Petruccione, F.: The Theory of Open Quantum Systems. Oxford University Press, Oxford (2002)
102. Gardiner, C.W., Zoller, P.: Quantum Noise. Springer Complexity, Berlin (2004)
103. Razavy, M.: Classical and Quantum Dissipative Systems. Imperial College Press, London (2005)
104. Joos, E., Zeh, H.D., Kiefer, C., Giulini, D., Kupsch, J., Stamatescu, I.-O.: Decoherence and the Appearance of a Classical World in Quantum Theory, 2nd edn. Springer, Heidelberg (2003)
105. Schlosshauer, M.: Decoherence and the Quantum-to-Classical Transition. The Frontier Collection. Springer, Heidelberg (2007)
106. Gisin, N.: Quantum measurement and stochastic processes. Phys. Rev. Lett. **52**, 1657–1660 (1984)
107. Nassar, A.B.: Quantum trajectories and the Bohm time constant. Ann. Phys. **331**, 317–322 (2013)
108. Mensky, M.B.: Quantum Measurements and decoherence. Springer-Science, Heidelberg (2000)

109. Nassar, A.B., Miret-Artés, S.: Dividing line between quantum and classical trajectories in a measurement problem: Bohmian time constant. Phys. Rev. Lett. **111**, 150401(1–4) (2013)
110. Bargueño, P., Miret-Artés, S.: The generalized Schrödinger-Langevin equation. Ann. Phys. **346**, 59–65 (2014)
111. Zander, C., Plastino, A.R., Díaz-Alonso, J.: Wave dynamics for a non-linear Schrödinger equation describing continuous position measurements. Ann. Phys. **362**, 36–56 (2015)
112. London, F.: Planck's constant and low temperature transfer. Rev. Mod. Phys. **17**, 310–320 (1945)
113. Fetter, A.L., Svidzinsky, A.A.: Vortices in a trapped dilute Bose-Einstein condensate. J. Phys. Condens. Matter **13**, R135–R194 (2001)
114. McCullough, E.A., Wyatt, R.E.: Dynamics of the collinear $H + H_2$ reaction. II. Energy analysis. J. Chem. Phys. **54**, 3592–3600 (1971)
115. Gomes, J.A.N.F.: Topological elements of the magnetically induced orbital current densities. J. Chem. Phys. **78**, 4585–4591 (1983)
116. Lazzeretti, P.: Ring currents. Prog. Nuc. Mag. Res. Spect. **36**, 1–88 (2000)
117. Pelloni, S., Lazzeretti, P., Zanasi, R.: Topological models of magnetic field induced current density field in small molecules. Theor. Chem. Acc. **123**, 353–364 (2009)
118. Landau, L.D., Lifschitz, E.M.: Fluid Mechanics. Pergamon Press, Oxford (1959)
119. Bialynicki-Birula, I., Cieplak, M., Kaminski, J.: Theory of Quanta. Oxford University Press, Oxford (1992)
120. Spurk, J.H.: Fluid Dynamics. Springer, Berlin (1997)
121. Sanz, A.S., Borondo, F., Miret-Artés, S.: Particle diffraction studied using quantum trajectories. J. Phys. Condens. Matter **14**, 6109–6145 (2002)
122. Dirac, P.A.M.: Quantised singularities in the electromagnetic field. Proc. Roy. Soc. Lond. A **133**, 60–72 (1931)
123. Riess, J.: Nodal structure, nodal flux fields, and flux quantization in stationary quantum states. Phys. Rev. D **2**, 647–653 (1970)
124. Wu, H., Sprung, D.W.L.: Inverse-square potential and the quantum vortex. Phys. Rev. A **49**, 4305–4311 (1994)
125. Bialynicki-Birula, I., Bialynicki-Birula, Z.: Magnetic monopoles in the hydrodynamic formulation of quantum mechanics. Phys. Rev. D **3**, 2410–2412 (1971)
126. Aharonov, Y., Bohm, D.: Significance of electromagnetic potentials in the quantum theory. Phys. Rev. **115**, 485–491 (1959)
127. Sanz, A.S., Borondo, F., Miret-Artés, S.: Causal trajectories description of atom diffraction by surfaces. Phys. Rev. B **61**, 7743–7751 (2000)
128. Courant, R., Hilbert, D.: Methods of Mathematical Physics, vol. 2. Wiley, New York (1966)
129. Sanz, A.S., Miret–Artés, S.: A trajectory-based understanding of quantum interference. J. Phys. A **41**, 435303(1–23) (2008)
130. Sanz, A.S., Miret-Artés, S.: Quantum phase analysis with quantum trajectories: a step towards the creation of a Bohmian thinking. Am. J. Phys. **80**, 525–533 (2011)
131. Liboff, R.L.: The correspondence principle revisited. Phys. Today **37**, 50–55 (1984)
132. Berry, M.V.: Quantum chaology, not quantum chaos. Phys. Scr. **40**, 335–336 (1989)
133. Sanz, A.S., Borondo, F., Miret-Artés, S.: On the classical limit in atom-surface diffraction. Europhys. Lett. **55**, 303–309 (2001)
134. Ehrenfest, P.: Bemerkung über die angenäherte Gültigkeit der klassischen Mechanik innerhalb der Quantenmechanik. Z. Phys. **45**, 455–457 (1927)
135. Sanz, A.S., Miret-Artés, S.: Aspects of nonlocality from a quantum trajectory perspective: a WKB approach to Bohmian mechanics. Chem. Phys. Lett. **445**, 350–354 (2007)
136. Elmore, W.C., Heald, M.A.: Physics of Waves. Dover Publications, New York (1985)
137. Sanz, A.S., Borondo, F., Miret-Artés, S.: Quantum trajectories in atom-surface scattering with single adsorbates: the role of quantum vortices. J. Chem. Phys. **120**, 8794–8806 (2004)
138. Sanz, A.S., Borondo, F., Miret–Artés, S.: Role of quantum vortices in atomic scattering from single adsorbates. Phys. Rev. B **69**, 115413(1–5) (2004)
139. Goldstein, H.: Classical Mechanics. Addison-Wesley, Reading, MA (1980)

140. Schiller, R.: Quasi-classical theory of the nonspinning electron. Phys. Rev. **125**, 1100–1108 (1962)
141. Schiller, R.: Quasi-classical transformation theory. Phys. Rev. **125**, 1109–1115 (1962)
142. Schiller, R.: Quasi-classical theory of the spinning electron. Phys. Rev. **125**, 1116–1123 (1962)
143. Rosen, N.: The relation between classical and quantum mechanics. Am. J. Phys. **32**, 597–600 (1964)
144. Rosen, N.: Quantum particles and classical particles. Found. Phys. **16**, 687–700 (1986)
145. Richardson, C.D., Schlagheck, P., Martin, J., Vandewalle, Bastin, T.: Nonlinear Schrödinger wave equation with linear quantum behavior. Phys. Rev. A **89**, 032118(1–4) (2014)
146. Haug, K., Metiu, H.: A test of the possibility of calculating absorption spectra by mixed quantum-classical methods. J. Chem. Phys **97**, 4781–4791 (1992)
147. Hagelstein, P.L., Senturia, S.D., Orlando, T.P.: Introductory Applied Quantum and Statistical Mechanics. Wiley, Hoboken (2004)
148. Chou, C.-C.: Trajectory description of the quantum-classical transition for wave packet interference. Ann. Phys. **371**, 437–459 (2016)
149. Chou, C.-C.: Quantum-classical transition equation with complex trajectories. Int. J. Quantum Chem. **116**, 1–11 (2016)
150. Sanz, A.S., Miret-Artés, S.: Selective adsorption resonances: Quantum and stochastic approaches. Phys. Rep. **451**, 37–154 (2007)
151. Babyuk, D., Wyatt, R.E.: Coping with the node problem in quantum hydrodynamics: the covering function method. J. Chem. Phys. **121**, 9230–9238 (2004)
152. Garashchuk, S., Rassolov, V.A.: Modified quantum trajectory dynamics using a mixed wave function representation. J. Chem. Phys. **120**, 8711–8715 (2004)
153. Sanz, A.S., Miret-Artés, S.: On the unique mapping relationship between initial and final quantum states. Ann. Phys. **339**, 11–21 (2013)
154. Sanz, A.S., Miret-Artés, S.: Quantum trajectories in elastic atom-surface scattering: threshold and selective adsorption resonances. J. Chem. Phys. **122** 014702(1–12) (2005)
155. Sanz, A.S., Giménez, X., Bofill, J.M., Miret-Artés, S.: Understanding chemical reactions within a generalized Hamilton-Jacobi framework. Chem. Phys. Lett. **478**, 89–96 (2009); (Erratum, Chem. Phys. Lett. **488**, 235–236 (2010))
156. Sanz, A.S., López-Durán, D., González-Lezana, T.: Investigating transition state resonances in the time domain by means of Bohmian mechanics: the F+HD reaction. Chem. Phys. **399**, 151–161 (2012)
157. Sanz, A.S., Miret-Artés, S.: Setting-up tunneling conditions by means of Bohmian mechanics. J. Phys. A: Math. Theor. **44**, 485301(1–17) (2011)
158. Born, M.: Zur Quantenmechanik der Stoßvorgänge. Z. Phys. **37**, 863–867 (1926)
159. Born, M.: Quantenmechanik der Stoßvorgänge. Z. Phys. **38**, 803–827 (1926)
160. Born, M.: Das Adiabatenprinzip in der Quantenmechanik. Z. Phys. **40**, 167–192 (1926)
161. Zurek, W.H., Wheeler, J.A.: Quantum Theory of Measurement. Princeton University Press, Princeton, NJ (1983)
162. Landsman, N.P.: Born rule and its interpretation. In: Greenberger, D., Hentschel, K., Weinert, F. (eds.) Compendium of Quantum Physics: Concepts, Experiments. History and Philosophy., pp. 64–70. Springer, Berlin (2009)
163. Brumer, P., Gong, J.: Born rule in quantum and classical mechanics. Phys. Rev. A **73**, 052109(1–4) (2006)
164. Gutzwiller, M.C.: Chaos in Classical and Quantum Mechanics. Sprigner, New York (1990)
165. Guantes, R., Sanz, A.S., Margalef-Roig, J., Miret-Artés, S.: Atom-surface diffraction: a trajectory description. Surf. Sci. Rep. **53**, 199–330 (2004)
166. Dimitrova, T.L., Weis, A.: The wave-particle duality of light: a demonstration experiment. Am. J. Phys. **76**, 137–142 (2008)
167. Dimitrova, T.L., Weis, A.: Lecture demonstrations of interference and quantum erasing with single photons. Phys. Scr. **T135**, 014003(1–4) (2009)
168. Dimitrova, T.L., Weis, A.: Single photon quantum erasing: a demonstration experiment. Eur. J. Phys. **31**, 625–637 (2010)

169. Landauer, R., Martin, Th: Barrier interaction time in tunneling. Rev. Mod. Phys. **66**, 217–228 (1994)
170. Muga, J.G., Sala Mayato, R., Egusquiza, I. (eds.): Time in Quantum Mechanics. Lecture Notes in Physics, vol. 734, 2nd edn. Springer, Berlin (2008)
171. Muga, J.G., Ruschhaupt, A., Campo, A. (Eds.): Time in Quantum Mechanics, Lecture Notes in Physics **789**, Springer, Berlin (2009). Vol. 2
172. Ghosh, S.K., Deb, B.M.: Densities, density-functionals and electron fluids. Phys. Rep. **92**, 1–44 (1982)
173. Belinfante, F.J.: A Survey of Hidden-Variables Theories. Pergamon, New York (1973)
174. Sanz, A.S., Miret–Artés, S.: A causal look into the quantum Talbot effect. J. Chem. Phys. **126**, 234106(1–11) (2007)
175. Davidović, M., Arsenović, D., Bozić, M., Sanz, A.S., Miret–Artés, S.: Should particle trajectories comply with the transverse momentum distribution? Eur. Phys. J.–Special Topics **160**, 95–104 (2008)
176. Yang, C.-D.: Quantum dynamics of hidrogen atom in complex space. Ann. Phys. **319**, 399–443 (2005)
177. Yang, C.-D.: Wave-particle duality in complex space. Ann. Phys. **319**, 444–470 (2005)
178. Yang, C.-D.: Modeling quantum harmonic oscillator in complex domain. Chaos, Solitons Fractals **30**, 342–362 (2006)
179. Yang, C.-D.: Quantum motion in complex space. Chaos, Solitons Fractals **33**, 1073–1092 (2007)
180. John, M.V.: Modified de Broglie-Bohm approach to quantum mechanics. Found. Phys. Lett. **15**, 329–343 (2002)
181. John, M.V.: Probability and complex quantum trajectories. Ann. Phys. **324**, 220–231 (2010)
182. Pauli, W.: Die allgemeine Prinzipien der Wellenmechanick, in Handbuch der Physik, Geiger, H., Scheel, K., (eds.), vol. 24, part 1, 2nd Ed., Springer, Berlin (1933)
183. Gottfried, K.: Quantum Mechanics, vol. 1. W.A. Benjamin, New York (1966)
184. Floyd, E.R.: Modified potential and Bohm's quantum-mechanical potential. Phys. Rev. D **26**, 1339–1347 (1982)
185. Floyd, E.R.: Arbitrary initial conditions of nonlocal hidden-variables. Phys. Rev. D **29**, 1842–1844 (1984)
186. Floyd, E.R.: Interference, reduced action, and trajectories. Found. Phys. **37**, 1386–1402 (2007)
187. Floyd, E.R.: Welcher Weg? A trajectory representation of a quantum Young's diffraction experiment. Found. Phys. **37**, 1403–1420 (2007)
188. Faraggi, A.E., Matone, M.: The equivalence principle of quantum mechanics: uniqueness theorem. Phys. Lett. B **437**, 369–380 (1998)
189. Faraggi, A.E., Matone, M.: The equivalence postulate of quantum mechanics. Int. J. Mod. Phys. A **15**, 1869–2017 (2000)
190. Barker–Jarvis, J., Kabos, P.: Modified de Broglie approach applied to the Schrödinger and Klein–Gordon equation. Phys. Rev. A **68**, 042110(1–8) (2003)
191. Sanz, A.S., Miret-Artés, S.: Interplay of causticity and vorticality within the complex quantum Hamilton-Jacobi formalism. Chem. Phys. Lett. **458**, 239–243 (2008)
192. Chou, C.-C., Sanz, A.S., Miret–Artés, S., Wyatt, R.E.: Hydrodynamic view of wave–packet interference: quantum caves. Phys. Rev. Lett. **102**, 250401(1–4) (2009)
193. Chou, C.-C., Sanz, A.S., Miret-Artés, S., Wyatt, R.E.: Quantum interference within the complex quantum Hamilton-Jacobi formalism. Ann. Phys. **325**, 2193–2211 (2010)
194. Leacock, R.A., Padgett, M.J.: Hamilton-Jacobi theory and the quantum action variable. Phys. Rev. Lett. **50**, 3–6 (1983)
195. Leacock, R.A., Padgett, M.J.: Hamilton-Jacobi/action-angle quantum mechanics. Phys. Rev. D **28**, 2491–2502 (1983)
196. Tannor, D.J.: Introduction to Quantum Mechanics: A Time Dependent Perspective. University Science Press, Sausalito (2006)
197. Boiron, M., Lombardi, M.: Complex trajectory method in semiclassical of wave packets. J. Chem. Phys. **108**, 3431–3444 (1998)

198. Goldfarb, Y., Degani, I., Tannor, D.J.: Bohmian mechanics with complex action: A new trajectory formulation of quantum mechanics. J. Chem. Phys. **125**, 231103(1–4) (2006)
199. Sanz, A.S., Miret–Artés, S.: Comment on Bohmian mechanics with complex action: A new trajectory-based formulation of quantum mechanics [J. Chem. Phys. **125**, 231103 (1–4) (2006)]. J. Chem. Phys. **127**, 197101(1–3) (2007)
200. Goldfarb, Y., Degani, I., Tannor, D.J.: Reply to "Comment on "Bohmian mechanics with complex action: a new trajectory-based formulation of quantum mechanics" [J. Chem. Phys. **125**, 231103(1–4) (2006)]". J. Chem. Phys. **127**, 197101(1–3) (2007)
201. Goldfarb, Y., Tannor, D.J.: Interference in Bohmian mechanics with complex action. J. Chem. Phys. **127**, 161101(1–4) (2007)
202. Chou, C.-C., Wyatt, R.E.: Computational method for the quantum Hamilton–Jacobi equation: bound states in one dimension. J. Chem. Phys. **125**, 174103(1–10) (2006)
203. Chou, C.-C., Wyatt, R.E.: Computational method for the quantum Hamilton–Jacobi equation: one–dimensional scattering problems. Phys. Rev. E **74**, 066702(1–9) (2006)
204. Chou, C.-C., Wyatt, R.E.: Quantum trajectories in complex space. Phys. Rev. A **76**, 012115 (1–14) (2007)
205. David, J.K., Wyatt, R.E.: Barrier scattering with complex-valued quantum trajectories: taxonomy and analysis of isochrones. J. Chem. Phys. **128**, 094102(1–9) (2008)
206. Schrödinger, E.: Discussion of probability relation between separated systems. Proc. Camb. Phil. Soc. **31**, 555–563 (1935)
207. Schrödinger, E.: Probability relations between separated systems. Proc. Camb. Phil. Soc. **32**, 446–452 (1936)
208. Kwiat, P.G., Mattle, K., Weinfurter, H., Zeilinger, A., Sergiemko, V., Shih, Y.: New high-intensity source of polarization-entangled photon pairs. Phys. Rev. Lett. **75**, 4337–4341 (1995)
209. Nielsen, M., Chuang, I.: Quantum Computation and Quantum Information. Cambridge University Press, Cambridge (2000)
210. Zurek, W.H.: Decoherence and the transition from quantum to classical. Phys. Today **44**, 36–44 (1991)
211. Joos, E., Zeh, H.D.: The emergence of classical properties through interaction with the environment. Z. Phys. B **59**, 223–243 (1985)
212. Pathria, K.E.: Statistical Mechanics. Pergamon Press, New York (1972)
213. Dewdney, C.: Nonlocally corelated trajectories in 2-particle quantum mechanics. Found. Phys. **18**, 867–886 (1988)
214. Lam, M.M., Dewdney, C.: Locality and nonlocality in correlated two-particle interferometry. Phys. Lett. A **150**, 127–135 (1990)
215. Guay, E., Marchildon, L.: Two-particle interference in standard and Bohmian quantum mechanics. J. Phys. A **36**, 5617–5624 (2003)
216. Na, K., Wyatt, R.E.: Quantum hydrodynamic analysis of decoherence: quantum trajectories and stress tensor. Phys. Lett. A **306**, 97–103 (2002)
217. Na, K., Wyatt, R.E.: Quantum hydrodynamic analysis of decoherence. Phys. Scr. **67**, 169–180 (2003)
218. Yang, C.-D., Weng, H.-J.: State-dependent molecular dynamics. Molecules **19**, 16122–16145 (2014)
219. Gindensperger, E., Meier, C., Beswick, J.A.: Mixing quantum and classical dynamics using Bohmian trajectories. J. Chem. Phys. **113**, 9369–9372 (2000)
220. Gindensperger, E., Meier, C., Beswick, J.A.: Quantum-classical including continuum states using quantum trajectories. J. Chem. Phys. **116**, 8–13 (2002)
221. Ford, G.W., Kac, M.: On the quantum Langevin equation. J. Sta. Phys. **46**, 803–810 (1987)
222. Ford, G.W., Lewis, J.T., O'Connell, R.F.: Quantum Langevin equation. Phys. Rev. A **37**, 4419–4428 (1988)
223. Hakim, V., Ambegaokar, V.: Quantum theory of a free particle interacting with a linearly dissipative environment. Phys. Rev. A **32**, 423–434 (1985)
224. Ford, G.W., O'Connell, R.F.: Exact solution of the Hu-Paz-Zhang master equation. Phys. Rev. D **64**, 105020(1–13) (2001)

225. Ford, G.W., O'Connell, R.F.: Wave packet spreading: temperature and squeezing effects with applications to quantum measurement and decoherence. Am. J. Phys. **70**, 319–324 (2002)
226. Ford, G.W., Lewis, J.T., O'Connell, R.F.: Quantum measurement and decoherence. Phys. Rev. A **64**, 032101(1–4) (2001)
227. Ford, G.W., O'Connell, R.F.: Decoherence without dissipation. Phys. Lett. A **286**, 87–90 (2001)
228. Ford, G.W., O'Connell, R.F.: Decoherence at zero temperature. J. Opt. B: Quantum Semiclass. Opt. **5**, S609–S612 (2003)
229. Ingold, G.-L.: Path integrals and their application to dissipative quantum systems. Lect. Notes Phys. **611**, 1–53 (2002)
230. Ford, G.W., O'Connell, R.F.: Canonical commutator and mass renormalization. J. Stat. Phys. **57**, 803–810 (1989)
231. Grandshteyn, I.S., Ryzhik, I.M.: Table of Integrals Series and Products, vol. 7. American Press, New York (2007)
232. Aharonov, Y., Rohrlich, D.: Quantum Paradoxes. Wiley-VCH, Weinheim (2005)
233. Lundeen, J.S., Sutherland, B., Patel, A., Stewart, C., Bamber, C.: Direct measurement of the quantum wavefunction. Nature **474**, 188–191 (2011)
234. Sanz, A.S., Borondo, F., Bastians, M.: Loss of coherence in double–slit diffraction experiments. Phys. Rev. A. **71**, 042103(1–7) (2005)
235. Sanz, A.S., Borondo, F.: A quantum trajectory description of decoherence. Eur. Phys. J. D **44**, 319–326 (2007)
236. Sanz, A.S., Borondo, F.: Contextuality, decoherence and quantum trajectories. Chem. Phys. Lett. **478**, 301–306 (2009)

Chapter 2
Some Selected Applications of Bohmian Mechanics

Abstract In this chapter, our purpose is to simply show that Bohmian mechanics is a powerful route to bring about new solutions to problems discussed by conventional quantum mechanical approaches, apart from allowing some striking correspondence between both frameworks. This goal is carried out by choosing some key quantum mechanical problems in the framework of Bohmian mechanics such as, for example, the so-called Ermakov–Bohm invariants, boundary conditions and uncertainty principle in tunneling, the quantum traversal time, Airy wave packets and Airy slits, the detection of inertial and gravitational masses with Airy wave packets, the geometric phase analyzing the Aharonov–Bohm effect and quantum vortices, the reformulation of the Gross–Pitaevskii equation within the hydrodynamical framework and, finally, the study of simple dissipative dynamics by using the well-known Caldirola-Kanai Hamiltonian. In this dissipative scenario, the motion of a free particle, the quantum interference of two wave packets and the dynamics in a linear potential as well as the corresponding of a damped harmonic oscillator (within the underdamped, critically damped and overdamped regimes) are finally analyzed for ulterior references.

2.1 Introduction

Recently, Bernstein has presented a pedagogically clear and historically brilliant review of the Bohmian theory [1] showing that it is sharp where the usual one is fuzzy and general where the usual one is special. Bell and Bernstein argued convincingly that the de Broglie-Bohm interpretation of quantum mechanics should be known from the very beginning and part of any college curriculum on the subject.

In this theory, the wave function provides only a partial description of the system. The description is completed by the specification of the actual positions of the particles which evolve according to the so-called guidance condition or guiding equation, Eq. (1.20). This equation provides the velocity of the particles in terms of the wave function. However, the trajectory of a particle is not at all classical; it is instead determined by the structure of the associated quantum wave which guides or pilots the particle along its path.

A.B. Nassar and S. Miret-Artés, *Bohmian Mechanics, Open Quantum Systems and Continuous Measurements*, DOI 10.1007/978-3-319-53653-8_2

Comprehensive discussions of Bohmian mechanics can be found everywhere in the literature [2–6]. In this Chapter, we want to show that Bohmian mechanics is a powerful route to bring about new solutions to problems discussed by conventional quantum mechanical approaches, apart from allowing some striking correspondence between both frameworks. This goal is carried out by choosing some key quantum mechanical problems in the framework of Bohmian mechanics such as, for example, the so-called Ermakov–Bohm invariants, Airy wave packets and Airy slits, the geometric phase analyzing the Aharonov–Bohm effect and quantum vortices, reformulating the Gross–Pitaevskii equation within the hydrodynamical framework and, finally, the study of simple dissipative dynamics by using the well-known Caldirola-Kanai Hamiltonian. In this dissipative scenario, the motion of a free particle, the quantum interference of two wave packets and the dynamics in a linear potential as well as the corresponding of a damped harmonic oscillator (within the underdamped, critically damped and overdamped regimes) are finally analyzed for ulterior references.

2.2 Ermakov–Bohm Invariants for the Time-Dependent Harmonic Oscillator

A paradigmatic problem in classical and quantum mechanics is the one-dimensional harmonic oscillator. After Schrödinger developed wave mechanics, he tried very hard to make it into a classical theory. The wave packets describing the simple harmonic oscillator with constant frequency seemed like a good candidate. It is well known that the ground state of the oscillator has the minimum possible uncertainty, which suggests that wave packets associated with this state might show some classical behavior. A quite interesting extension of this problem is within the theory of invariants (constant of motion) for the time-dependent harmonic oscillator (TDHO) discovered by Ermakov [7–9]: it consists of demonstrating that for the corresponding equation of motion

$$\ddot{q}(t) + \omega^2(t) q(t) = 0 \tag{2.1}$$

there exists a time-dependent invariant

$$I = \frac{1}{2} \left[\left(\dot{q}(t)\delta(t) - \dot{\delta}(t) q(t) \right)^2 + \left(\frac{q(t)}{\delta(t)} \right)^2 \right] \tag{2.2}$$

provided that

$$\ddot{\delta}(t) + \omega^2(t)\delta(t) = \frac{\hbar^2}{4m^2\delta^3(t)}, \tag{2.3}$$

where $q(t)$, $\delta(t)$ and $\omega(t)$ represent the wave packet center of mass, width and frequency of the TDHO, respectively. Interestingly enough, this time-dependent invariant emerges naturally within Bohmian mechanics. It constitutes a natural connection between the classical to the quantum regime.

To demonstrate such an invariant, the associated Schrödinger equation can be written as

$$i\hbar\frac{\partial\psi(x,t)}{\partial t} = -\frac{\hbar^2}{2m}\frac{\partial^2\psi(x,t)}{\partial x^2} + \frac{1}{2}m\omega^2(t)x^2\psi(x,t), \tag{2.4}$$

where m is the mass of the TDHO. Equation (2.4) can now be solved via the Bohmian formalism. To this end, the wave function is again expressed in polar form as

$$\psi(x,t) = \phi(x,t)\exp(iS(x,t)/\hbar). \tag{2.5}$$

Now, after substitution of Eq. (2.5) into (2.4), we obtain the following differential equation in partial derivatives

$$i\hbar\left[\frac{\partial\phi}{\partial t} + \frac{i}{\hbar}\frac{\partial S}{\partial t}\phi\right] =$$

$$= -\frac{\hbar^2}{2m}\left\{\left[\frac{\partial^2\phi}{\partial x^2} - \frac{\phi}{\hbar^2}\left(\frac{\partial S}{\partial x}\right)^2\right] + \frac{i}{\hbar}\left[2\frac{\partial S}{\partial x}\frac{\partial\phi}{\partial x} + \frac{\partial^2 S}{\partial x^2}\right]\right\} + \frac{1}{2}m\omega^2x^2\phi.$$

The equation above can be as usual separated into real and imaginary parts to give

$$\frac{\partial v}{\partial t} + v\frac{\partial v}{\partial x} = -\frac{1}{m}\frac{\partial}{\partial x}(Q+V), \tag{2.6}$$

and

$$\frac{\partial\rho}{\partial t} + \frac{\partial}{\partial x}(\rho v) = 0, \tag{2.7}$$

with the following definitions for the quantum hydrodynamical density

$$\rho(x,t) = \phi^2(x,t), \tag{2.8}$$

the velocity

$$v = \frac{1}{m}\frac{\partial S}{\partial x}, \tag{2.9}$$

the quantum potential

$$Q = -\frac{\hbar^2}{2m\phi}\frac{\partial^2 \phi}{\partial x^2}, \tag{2.10}$$

and the interaction potential

$$V = \frac{1}{2}m\omega^2(t)x^2. \tag{2.11}$$

Equation (2.6) is an Euler-type equation describing trajectories of a fluid particle, with momentum $p = mv$, whereas Eq. (2.7) describes the evolution of the quantum fluid density ρ. This density is interpreted as the probability density of a particle being actually present within a specific region. Such a particle follows a definite space-time trajectory that is determined by its wave function through an equation of motion in accordance with the initial position, formulated in a way that is consistent with the Schrödinger time evolution. An essential and unique feature of the quantum potential is that the force arising from it is unlike a mechanical force of a wave pushing on a particle with a pressure proportional to the wave intensity.

By assuming now that the wave packet is initially centered at $x = 0$ and given by a Gaussian function, $\rho(x,0) = \left[2\pi\delta^2(0)\right]^{-1/2}\exp\left[-x^2/2\delta^2(0)\right]$ then ρ vanishes for $|x| \to \infty$ at any time and we may rewrite

$$\rho(x,t) = |\psi(x,t)|^2 = \left[2\pi\delta^2(t)\right]^{-1/2}\exp\left(-\frac{[x-q(t)]^2}{2\delta^2(t)}\right), \tag{2.12}$$

$\delta(t)$ being the total width of the Gaussian wave packet and $q(t)$ a classical trajectory. Equation (2.12) can be readily used to demonstrate that

$$\int_{-\infty}^{+\infty} \left([x-q(t)]^2\right)\rho(x,t)dx = \delta^2(t). \tag{2.13}$$

Substitution of Eq. (2.12) into (2.7) yields

$$\frac{\partial \rho}{\partial t} = \left[-\frac{\dot{\delta}}{\delta} + \frac{(x-q)}{\delta^2}\dot{q} + \frac{1}{\delta^3}(x-q)^2\dot{\delta}\right]\rho, \tag{2.14}$$

and

$$\frac{\partial(\rho v)}{\partial x} = \left(\frac{\dot{\delta}}{\delta}\right)\rho + \left[\left(\frac{\dot{\delta}}{\delta}\right)(x-q) + \dot{q}\right]\left(-\frac{(x-q)}{\delta^2}\right)\rho, \tag{2.15}$$

which implies that

$$v(x,t) = \left(\frac{\dot{\delta}}{\delta}\right)(x-q) + \dot{q}\,. \tag{2.16}$$

Analogously, substitution of Eq. (2.16) into (2.6) yields

$$\left(\ddot{\delta}(t) + \omega^2(t)\delta - \frac{\hbar^2}{4m^2\delta^3(t)}\right)(x-q)^1 + (\ddot{q} + \omega^2(t)q)(x-q)^0 = 0, \tag{2.17}$$

leading to Eqs. (2.1) and (2.3).

In 1880, Ermakov [10] was the first to describe a relationship between nonlinear differential equations of second order such as Eq. (2.3) and a particular type of the linear equation (2.1). At the beginning of the thirties, Milne [11] developed a method quite similar to the WKB technique where the same nonlinear equation found by Ermakov occurred, and applied it successfully to several model problems in quantum mechanics. Further, in 1950, the solution to this nonlinear differential equation was given by Pinney [12]. It is worth pointing out that the term on the right side of Eq. (2.3) depends on $\hbar$ in contrast to a similar nonlinear equation studied by Ermakov, Milne and Pinney which contains an arbitrary constant. This $\hbar$-term emerges naturally from the presence of the quantum potential Q given by Eq. (2.10).

By eliminating $\omega^2(t)$ between the two equations above, one obtains the invariant in Eq. (2.2). Inasmuch as this invariant connects the classical and quantum regimes, we call it an *Ermakov–Bohm invariant*. Further, from Eqs. (2.6) and (2.7), we can construct the relations for the energy density U, energy density flux $\mathcal{Q}$, momentum J and momentum flux P, respectively

$$\frac{\partial U}{\partial t} + \frac{\partial \mathcal{Q}}{\partial x} = 0, \tag{2.18}$$

$$\frac{\partial J}{\partial t} + \frac{\partial P}{\partial x} + \frac{\rho}{m}\frac{\partial V}{\partial x} = 0, \tag{2.19}$$

where

$$U = \rho\left(\frac{1}{2}mv^2 + V + Q\right), \tag{2.20}$$

$$\mathcal{Q} = vU + \frac{\hbar^2}{2m^2}\left[\sqrt{\rho}\frac{\partial^2\sqrt{\rho}}{\partial x\partial t} - \frac{\partial\sqrt{\rho}}{\partial t}\frac{\partial\sqrt{\rho}}{\partial x}\right], \tag{2.21}$$

$$J = \rho v \tag{2.22}$$

and

$$P = \rho v^2 - \frac{\hbar^2}{4m^2}\left[\frac{\partial^2\rho}{\partial x^2} - \frac{1}{\rho}\left(\frac{\partial\rho}{\partial x}\right)^2\right]. \tag{2.23}$$

Next, the wave packet dynamics will be completely determined by the solutions to Eqs. (2.3) and (2.1), which describe the time evolution of the width and center of the packet.

Integration of Eqs. (2.3) and (2.1) are subject to the general initial conditions [13]

$$\delta(0) = \delta_0\ ,\ \dot{\delta}(0) = \mathrm{d}_0, \tag{2.24}$$

and

$$q(0) = q_0\ ,\ \dot{q}(0) = V_0. \tag{2.25}$$

In the following, we show that the general solution to the quantum Eq. (2.3) can be obtained with the help of just one particular solution to the classical Eq. (2.1). To this end, let us now substitute

$$\delta(t) = r(\theta)\ \alpha(t) \tag{2.26}$$

where we define [13]

$$d\theta = dt/\alpha^2(t) \tag{2.27}$$

and substitute into Eq. (2.26) to obtain the first Ermakov–Bohm invariant of motion

$$I_1 = [r\ (\theta)]^2 + \frac{\hbar^2}{4m^2}\left(\frac{1}{r(\theta)}\right)^2, \tag{2.28}$$

provided that

$$\ddot{\alpha} + \omega^2(t)\alpha = 0\ . \tag{2.29}$$

Upon a second integration Eq. (2.28) can be recast with the help of Eq. (2.26) as

$$\delta^2(t) = \left(\frac{\hbar^2}{4m^2 I_1} + I_1 I_2^2\right)\ \alpha_1^2(t) + I_1\alpha_2^2(t) + 2I_1I_2\alpha_1(t)\alpha_2(t) \tag{2.30}$$

where α_1 and α_2 are two independent solutions to the classical Eq. (2.29). The two constants I_1 and I_2 represent two Ermakov–Bohm invariants of motion of the problem. In addition, the two independent solutions to Eq. (2.29) can be obtained in general from just one particular solution to the same equation, namely,

$$\alpha_1(t) \equiv \alpha(t) \tag{2.31}$$

and

$$\alpha_2(t) \equiv \alpha(t) \int^t \frac{dt}{\alpha^2(t)}. \tag{2.32}$$

If initial conditions are imposed as follows

$$\alpha(0) = 1 \; and \; \dot{\alpha}(0) = 0 \tag{2.33}$$

we are led to

$$\alpha_1(0) = 1, \; \alpha_2(0) = 0, \; \dot{\alpha}_1(0) = 0, \; and \; \dot{\alpha}_2(0) = 1. \tag{2.34}$$

These initial conditions allow us to find that the two *Ermakov–Bohm* invariants of motion are given by:

$$I_1 = d_0^2 + (\delta_0^2/\tau^2), \tag{2.35}$$

and

$$I_2 = \delta_0 d_0/[d_0^2 + (\delta_0^2/\tau^2)]. \tag{2.36}$$

Thus, the complete dynamics of the TDHO can be found with the help of Eqs. (2.31), (2.32), (2.35) and (2.36): the generalized squeezed states for the TDHO wave packet is finally written as

$$\delta(t) = \delta_0 \, \alpha(t) \left\{ 1 + \left(\frac{2\varsigma}{\tau}\right) \left[\int^t \frac{dt}{\alpha^2(t)} \right] + \left(\frac{(1+\varsigma^2)}{\tau^2}\right) \left[\int^t \frac{dt}{\alpha^2(t)} \right]^2 \right\}^{1/2} \tag{2.37}$$

where $\tau = 2m \; \delta_0^2/\hbar$, and $\varsigma = 2m \; \delta_0 d_0/\hbar$. In turn, the packet center of gravity evolves according to

$$q(t) = \alpha(t) \left\{ q_0 + V_0 \int^t \frac{dt}{\alpha^2(t)} \right\}. \tag{2.38}$$

Now, the full wave packet can be written in its final, general form as

$$\psi(x,t) = \left(2\pi\delta^2(t)\right)^{-1/4} \exp\left[-\frac{[x-q(t)]^2}{4\delta^2(t)} + \frac{i}{\hbar}\left[S_1(x,t) + S_0(t)\right]\right] \quad (2.39)$$

where

$$S_1(x,t) = \frac{m\dot{\delta}(t)}{2\delta(t)}[x-q(t)]^2 + m\dot{X}(t)[x-q(t)] \quad (2.40)$$

and

$$S_0(t) = \int_0^t dt' \left(\frac{1}{2}m\dot{q}^2(t') - \frac{1}{2}m\Omega^2 q^2(t') - \frac{\hbar^2}{4m\delta^2(t')}\right). \quad (2.41)$$

The wave packet surrounds the position of the classical particle and the center of gravity of the packet follows the classical trajectory: its time evolution is completely determined by the quantum and classical solutions (2.37) and (2.38), respectively.

At the moment of observation, the packet is moving with an initial velocity V_0 and spreading with an initial rate d_0. The associated variance of x [squared uncertainty] can be written as $[\Delta x\,(t)]^2 = \delta^2\,(t)$, which exhibits generalized squeezed states for the TDHO. Above all, our method gives a general solution to the quantum TDHO from just one particular solution $\alpha(t)$ to the classical time-dependent oscillator (2.29) in terms of two Ermakov–Bohm invariants of motion [13].

Finally, the associated Bohmian trajectories of an evolving *ith* particle of the ensemble with an initial position x_{oi} can be calculated via Eq. (2.16)

$$\dot{x}_i(t) = \dot{q}(t) + [x(t) - q(t)]\frac{\dot{\delta}(t)}{\delta(t)} \quad (2.42)$$

or

$$\int_0^t \frac{d}{dt'}\left(\ln[x_i(t') - q(t')]\right) dt' = \int_0^t \frac{d}{dt'}[\ln \delta(t')]dt' \quad (2.43)$$

leading to

$$x_i(t) = q(t) + (x_{0i} - q_0)\frac{\delta(t)}{\delta_0}, \quad (2.44)$$

where $\delta(t)$ and $q(t)$ are given by (2.37) and (2.38), respectively. Furthermore, Eqs. (2.16) and (2.44) also reproduce the dressing scheme of the quantum trajectory (velocity and position, respectively) mentioned in Chap. 1. In this scheme, the quantum or nonlocal contribution is governed by the time evolution of the Gaussian wave packet width. This pattern is going to be also reproduced along this chapter as well as in the remaining chapters dealing with open quantum systems.

2.3 Boundary Conditions in Tunneling

A commonly used assumption in quantum mechanics [14–17] is that the boundary conditions at a surface where the potential undergoes a finite jump reduce to the requirement that both the wave function ψ and its derivative $(\partial\psi/\partial x)$ be continuous at the surface. It is illustrative to show through Bohmian mechanics how more general boundary conditions follow from the continuity of mass, momentum and energy densities. With this new boundary conditions, a novel approach to tunneling through sharp-edged potential barriers is presented.

Let us consider the dynamics of a quantum particle described by the coupled hydrodynamical equations (2.6) and (2.7), as previously defined. The momentum density ρv appearing in the hydrodynamical equations can be the real part of a more general quantum mechanical local momentum field P defined from the momentum-density operator as

$$P = \frac{\hbar}{i}\psi^* \frac{\partial\psi}{\partial x} = m\rho\,(v + iu)\,, \tag{2.45}$$

where $u = -(\hbar/2m\rho)(\partial\rho/\partial x)$. So, it follows that the boundary conditions for the continuity of mass, momentum and energy are [18]

$$\rho,\, \rho v,\, \rho u,\, \rho\left(\frac{1}{2}mv^2 + V + Q\right). \tag{2.46}$$

In terms of the wave function, these boundary conditions become $(\psi^*\psi)$, $[\psi^*(\partial\psi/\partial x)]$ and $\partial S/\partial t = -(\frac{1}{2}mv^2 + V + Q)$.

Next, let us apply these boundaries conditions to the following quantum tunneling problem. Consider a particle with incident energy E striking a potential barrier of height V and width L: $V(x) = V$ for $0 < x < L$ and zero elsewhere. In what follows $a, b, c, d, \alpha, \beta$ and δ are constants to be determined and $k^2 = 2mE/\hbar^2$ and $\bar{q}^2 = 2m(V - E)/\hbar^2$. In a polar form, the wave function for $x < 0$ (incident region 1) is

$$\psi_1(x, t) = \sqrt{\rho_1}\exp(iS_1/\hbar), \tag{2.47}$$

where – detailed calculations can be found in Ref. [19] and Sect. 2.5,

$$\rho_1 = 1 + a^2 + 2a\cos(2kx - \alpha) \tag{2.48}$$

and

$$S_1/\hbar = -\omega t + \frac{\alpha}{2} + \tan^{-1}\left[\frac{1-a}{1+a}\tan\left(kx - \frac{\alpha}{2}\right)\right]. \tag{2.49}$$

For $0 < x < L$ (tunneling region 2), the wave function reads

$$\psi_2(x,t) = \sqrt{\rho_2}\exp(iS_2/\hbar) \tag{2.50}$$

where

$$\rho_2 = \frac{1}{\bar{q}}\left[c^2 e^{2\bar{q}x} + d^2 e^{-2\bar{q}x} + 2dc\cos(\beta - \delta)\right] \tag{2.51}$$

and

$$S_2/\hbar = -\omega t + \frac{\beta+\delta}{2} + \tan^{-1}\left[\frac{ce^{\bar{q}x} - de^{-\bar{q}x}}{ce^{\bar{q}x} + de^{-\bar{q}x}}\tan\left(\frac{\beta-\delta}{2}\right)\right]. \tag{2.52}$$

For $x > L$ (transmission region 3), the wave function is given by

$$\psi_3(x,t) = \sqrt{\rho_3}\exp(iS_3/\hbar), \tag{2.53}$$

where

$$\rho_3 = b \tag{2.54}$$

and

$$S_3/\hbar = -\omega t + kx + \beta, \tag{2.55}$$

The boundary conditions from Eq. (2.46) where the potential undergoes a finite jump read

$$\rho_1(0) = \rho_2(0), \tag{2.56}$$

$$\rho_2(L) = \rho_3(L), \tag{2.57}$$

$$\rho_1{}'(0) = \rho_2{}'(0), \tag{2.58}$$

$$\rho_2{}'(L) = \rho_3{}'(L), \tag{2.59}$$

$$\rho_1(0)v_1(0) = \rho_2(0)v_2(0), \tag{2.60}$$

$$\rho_2(L)v_2(L) = \rho_3(L)v_3(L), \tag{2.61}$$

$$\left(\frac{\partial S_1}{\partial t}\right)_0 = \left(\frac{\partial S_2}{\partial t}\right)_0, \tag{2.62}$$

$$\left(\frac{\partial S_2}{\partial t}\right)_L = \left(\frac{\partial S_3}{\partial t}\right)_L. \tag{2.63}$$

By applying the above boundary conditions on the wave functions (2.47), (2.50) and (2.53), we then obtain

$$1 + a^2 + 2a\cos\alpha = \frac{c^2 + d^2 + 2cd\cos(\beta - \delta)}{\bar{q}}, \tag{2.64}$$

$$\frac{c^2 e^{2\bar{q}L} + d^2 e^{-2\bar{q}L} + 2cd\cos(\beta - \delta)}{\bar{q}} = b^2, \tag{2.65}$$

$$2ak\sin\alpha = (c^2 - b^2), \tag{2.66}$$

$$c = de^{-2\bar{q}L}, \tag{2.67}$$

$$1 - a^2 = \frac{2d^2 e^{-2\bar{q}L}\sin(\beta - \alpha)}{k}, \tag{2.68}$$

$$\frac{2d^2 e^{-2\bar{q}L}\sin(\beta - \alpha)}{k} = b^2. \tag{2.69}$$

From Eqs. (2.65) and (2.67), we have

$$b^2 = \frac{2d^2 e^{-2\bar{q}L}[1 + \cos(\beta - \alpha)]}{\bar{q}}, \tag{2.70}$$

which combined to Eq. (2.69) yields

$$\tan\left(\frac{\beta - \delta}{2}\right) = \frac{k}{\bar{q}}, \tag{2.71}$$

$$\sin(\beta - \delta) = \frac{2k\bar{q}}{\bar{q}^2 + k^2}, \tag{2.72}$$

$$cos(\beta - \delta) = \frac{\bar{q}^2 - k^2}{\bar{q}^2 + k^2}. \tag{2.73}$$

Equations (2.69) and (2.73) allow us to write Eq. (2.70) as

$$b^2 = \left(\frac{4\bar{q}}{\bar{q}^2 + k^2}\right) d^2 e^{-2\bar{q}L}, \tag{2.74}$$

which, in turn, combined with Eqs. (2.67) and (2.73), reduces Eq. (2.64) to

$$1 + a^2 + 2a\cos\alpha = b^2 \left(\frac{\bar{q}^2 - k^2}{2\bar{q}^2}\right) \left(1 + \frac{\bar{q}^2 + k^2}{\bar{q}^2 - k^2} \cosh 2\bar{q}L\right). \tag{2.75}$$

Equations (2.68) and (2.69) imply that

$$a^2 = 1 - b^2 \tag{2.76}$$

which inserted into Eq. (2.75) gives

$$a\cos\alpha = b^2 \left(1 + \left[\frac{\bar{q}^2 + k^2}{2\bar{q}^2}\right] \sinh^2 \bar{q}L\right) - 1. \tag{2.77}$$

By the same procedure above, Eq. (2.66) can be rewritten as

$$a\sin\alpha = -\left[\frac{\bar{q}^2 + k^2}{4k\bar{q}}\right] b^2 \sinh 2\bar{q}L. \tag{2.78}$$

Combination of Eqs. (2.76), (2.77) and (2.78) leads to

$$b^{-2} = \frac{\left[1 + \left(\frac{\bar{q}^2+k^2}{2\bar{q}^2}\right) \sinh^2 \bar{q}L\right]^2 + \left(\frac{\bar{q}^2+k^2}{4k\bar{q}}\right) \sinh 2\bar{q}L}{\left[1 + \left(\frac{\bar{q}^2+k^2}{2\bar{q}^2}\right) \sinh^2 \bar{q}L\right]}. \tag{2.79}$$

Using the identity $\sinh^2 2\bar{q}^2 L = 4(\sinh^2 \bar{q}L + \sinh^4 \bar{q}L)$ and after dividing the numerator by the denominator of Eq. (2.79), we arrive at the known result

$$b^{-2} = 1 + \left(\frac{\bar{q}^2 + k^2}{2k\bar{q}}\right)^2 \sinh^2 \bar{q}L. \tag{2.80}$$

2.4 Quantum Potential and Uncertainty Principle in Tunneling

Quantum tunneling is an important effect in many physical phenomena, such as the rate of nuclear fusion, many chemical reactions, and a lot of technology, for example, scanning tunneling microscopy. The basic physics of fusion has been known for some time, and a key element of understanding it is quantum tunneling. Nuclei have a positive electric charge, and since like charges repel, there is an energy barrier to be overcome. Once the barrier is overcome, the strong nuclear force takes over. One way of overcoming the barrier is quantum tunnelling. There is even a possibility that the universe itself might at some point tunnel through to a lower energy state, depending on what the mass of the Higgs boson is exactly.

The role of the quantum potential on the tunneling of a Gaussian wave packet through a rectangular potential barrier is studied here by solving numerically the time-dependent Schrödinger equation via a computer simulation [20–23]. The algorithm used was originally developed by Goldberg et al. [21] and further developed by Koonin [20]. It is assumed that the barrier was confined in a finite region of a one-dimensional space (0, L). At x = 0 and L, the wave function is set equal to zero, corresponding to a rigid barrier. The initial wave packet is established in the region to the left of the barrier, just far enough so that initially it does not interact or overlap the barrier. The time-dependent Schrödinger equation is solved numerically to determine the wave function at each time step, with small time increments and parameters used are the same as in Koonin's work [20]. As the time evolves, the wave packet travels to the right, and interacts with the barrier; part of the packet is reflected and part tunnels through, and then emerges from the other side of the barrier.

The familiar picture of a Gaussian wave packet prior to tunneling through a rectangular potential barrier is depicted in Fig. 2.1. The factors δ_i, L, V_0. represent the initial width of the packet, the width of the barrier and the barrier potential energy, respectively.

By analyzing the role of the quantum potential, we demonstrate an unfamiliar aspect of the tunneling rate as the initial wave packet width increases. We begin by recalling that the total energy of the quantum particle is given by: $E_T = \frac{1}{2}mv^2 + V + Q$, where m, v, $V = V_0$ and Q are the mass, velocity, potential-barrier height and

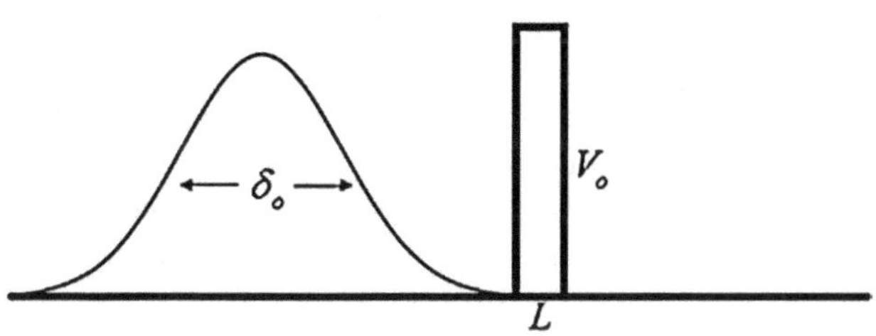

Fig. 2.1 Gaussian wave packet prior to the tunneling through a rectangular potential barrier

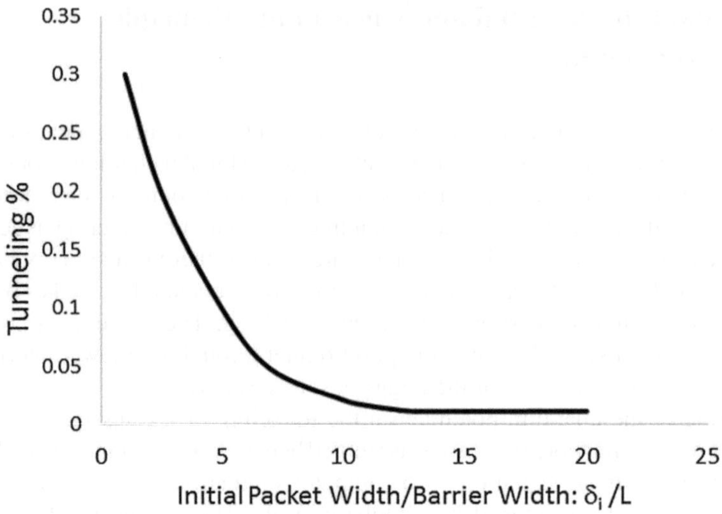

Fig. 2.2 Tunneling versus initial wave packet width

the quantum potential, respectively. In the tunneling case of an initially free-particle wave packet through a potential barrier, the wave packet width before the packet strikes the barrier must have a minimum initial value so that the energy uncertainty $\frac{1}{2}mv_i^2 + Q_i < V_0$ needs to be fulfilled in order to guarantee tunneling through and not hopping over the barrier. The quantum potential, for a free particle, decreases as δ_i increases. This fact can be verified for a free-particle where

$$Q_i = \frac{\hbar^2}{4m\delta_i^2[1+(t/\tau)^2]}\left[1 - \frac{[x - x_0 - v_i t]^2}{2\delta_i^2[1+(t/\tau)^2]}\right] \tag{2.81}$$

with $\tau \equiv 2m\delta_i^2/\hbar$. Choosing $\hbar = m = 1$, $E_i/V_0 = 0.8$, $(2mV_0/\hbar)^{1/2}L = 3.1$ and $\delta_i/L = 1$, Fig. 2.2 above indicates that initially the tunneling rate decreases as the incident wave packet width increases until it reaches a plateau (at $\delta_i/L \simeq 13$)– due to the uncertainty principle, for a minimum (or greater) value of the incident wave packet width the tunneling remains the same, i.e., the quantum potential is reduced to a minimum (since it depends inversely on the width of the packet). Consequently, above a critical initial wave packet width the tunneling rate is no longer dependent of the width of the wave packet. This point has been overlooked in the literature in light of another issue: the tunneling time. Many authors have advocated that the tunneling time through an opaque barrier is independent of the thickness of the barrier (the Hartman effect) [22, 24]. The quantum potential and the uncertainty principle then demand that the spatial extent of the packet be much greater than the barrier width. As a result, the duration of the tunneling event will be the temporal extent of the wave packet, assumed propagating at its initial velocity [25].

2.5 Scattering via Invariants of Bohmian Mechanics

We present here a new application of the theory of invariants to the problem of scattering within Bohmian mechanics. Within this formalism [26], we derive a formula for the transmission coefficient of a steady flux of fluid particles scattered by an arbitrary shaped potential well, and discuss the possibility of exact and approximate solutions for some particular cases.

The quantum potential is a responsible, crucial determining factor for the unbroken transition from microscopic to macroscopic levels. Quantum-wave phenomena, such as diffraction and interference, are precluded from a scattering process in the semiclassical regime [14] and therefore it becomes necessary to improve such approximations if the de Broglie wavelength is comparable with the potential dimensions. In the framework of Bohmian mechanics, this implies that the quantum potential can still be significant and its complete neglect is not desirable. So, let us consider again the dynamics of a quantum-fluid particle described previously by the pair of coupled hydrodynamical equations (2.6) and (2.7).

The stationary solution to the coupled Eqs. (2.6) and (2.7) ($\dot{\rho} = \dot{v} = 0$) can be found by expressing

$$\psi(x, t) = \phi(x) \exp\left(i \left[S(x) - Et\right]/\hbar\right) \tag{2.82}$$

which substituted into Eqs. (2.6) and (2.7) gives

$$\phi'(x)S'(x) + S''(x)\phi(x) = 0, \tag{2.83}$$

$$\phi''(x) + \lambda^2(x)\phi(x) = \left[\frac{S'(x)}{\hbar}\right]^2 \phi(x) \tag{2.84}$$

where for an arbitrarily shaped potential well $V(x)$, we have

$$\lambda^2(x) = 2m[E + V(x)]/\hbar^2. \tag{2.85}$$

Integration of Eq. (2.83) implies that

$$\rho(x)S'(x) = I_1, \tag{2.86}$$

which is the *first invariant*: the fluid-particle density current ($J(x) = \rho(x)v(x)$) is spatially conserved. Then, by inserting Eq. (2.86) into (2.84), and defining the dimensionless quantity

$$\tilde{\phi}(x) = \sqrt{\frac{k}{I_1}}\phi(x), \tag{2.87}$$

we have

$$\tilde{\phi}''(x) + \lambda^2(x)\tilde{\phi}(x) = \frac{\hbar^2 k^2}{\tilde{\phi}^3(x)}, \tag{2.88}$$

which is of the form of the so-called Ermakov-Pinney equation [10–12]. Noticeably, the term on the right side of Eq. (2.88) depends on $\hbar$ in contrast to the similar nonlinear equation studied by Ermakov, Milne and Pinney which contains an arbitrary constant and analyzed previously. This $\hbar$-term emerges naturally from the presence of the quantum potential. A complete solution to this equation can be found by successive transformations. By following Reid and Ray [27], we express $\tilde{\phi}(x) = r(\theta)\alpha(x)$ and $d\theta = dx/\alpha^2(x)$, yielding $\ddot{r}(\theta) = k^2\hbar^2/r^3(\theta)$ and the second invariant

$$I_2 = [\dot{r}(\theta)]^2 + \frac{k^2\hbar^2}{r^2} \tag{2.89}$$

or

$$I_2 = \left[\tilde{\phi}\alpha' - \tilde{\phi}'\alpha\right]^2 + \frac{k^2\hbar^2}{\left(\alpha/\tilde{\phi}\right)^2}, \tag{2.90}$$

provided that α obeys

$$\alpha''(x) + \lambda^2(x)\alpha(x) = 0. \tag{2.91}$$

The overdot denotes differentiation with respect to θ. Upon integration, Eq. (2.90) leads to

$$\tilde{\rho}(x) = \left(\frac{k^2\hbar^2}{I_2} + I_2 I_3^2\right)\alpha_1^2(x) + I_2\alpha_2^2(x) + 2I_2 I_3 \alpha_1(x)\alpha_2(x) \tag{2.92}$$

where

$$\alpha_1(x) \equiv \alpha(x) \tag{2.93}$$

and

$$\alpha_2(x) \equiv \alpha(x)\int^x \frac{dx}{\alpha^2(x)}. \tag{2.94}$$

Because α_1 and α_2 are linearly independent functions and, therefore, cannot be zero simultaneously, $\tilde{\rho}(x)$ can never be zero. As justified in Lutzky's work [28], Eqs. (2.93) and (2.94) can be expressed as

$$\alpha_1(x) = \tilde{\phi}_P(x) cos\beta(x) \tag{2.95}$$

and

$$\alpha_2(x) = [\tilde{\phi}_p(x)/k] sin\beta(x), \tag{2.96}$$

where $\tilde{\phi}_P(x)$ is *any* particular solution to (2.88) and

$$\beta(x) = k \int^x \frac{dx}{\tilde{\rho}_P(x)}, \tag{2.97}$$

such that we obtain after some manipulations

$$\tilde{\rho}(x) = \tilde{\rho}_P(x) \left\{ a + \sqrt{a^2 - 1} \cos[2(\beta(x) - b)] \right\}, \tag{2.98}$$

where a and b are two (redefined) constants, which will be determined below. An advantage of having α_1 (and α_2) expressed in terms of $\tilde{\rho}_P$ is that while α_1 may vary rapidly (through $\cos\beta$), $\tilde{\rho}_P$ can be a slowly varying function, and therefore its corresponding equation is more suitable for approximations: $\tilde{\rho}_P$ can also be seen as a modulation of the total amplitude $\tilde{\rho}$.

Moreover, it is worth noting that a complete solution to (2.88) can be obtained from any (just one) particular solution to another equation of the same form as that of (2.88). This is the central point towards an exact and/or approximate solution without having to neglect its quantum potential term.

For a bound-state energy case, after some manipulations, we find that condition

$$\oint mv(x,t)dx = 2\pi n\hbar \tag{2.99}$$

is sufficiently satisfied if $\Delta\beta = n\pi$,

Now, consider the scattering of a steady flux of particles with energy E ($k^2 = 2mE/\hbar^2$) by an arbitrarily shaped potential well $V(x) = V$ for $0 \leq x \leq L$ and 0 elsewhere. For an incident flux of particles from the left ($x < 0$), with wave function given by

$$\psi_L(x) = e^{ikx} + Ae^{-ikx}, \tag{2.100}$$

and a transmitted flux to the right ($x > L$), with wave function

$$\psi_R(x) = Be^{ikx}, \tag{2.101}$$

the boundary conditions are now matched as follows at $x = 0$

$$1 + A = \phi(0)e^{iS(0)/\hbar}, \tag{2.102}$$

$$ik(1 - A) = \left(\phi'(0) + i\frac{S'(0)\phi(0)}{\hbar}\right) e^{iS(0)/\hbar}; \tag{2.103}$$

at $x = L$

$$\phi(L)e^{iS(L)/\hbar} = Be^{ikL}, \tag{2.104}$$

$$\left(\phi'(L) + i\frac{S'(L)\phi(L)}{\hbar}\right) e^{iS(L)/\hbar} = ikBe^{ikL}. \tag{2.105}$$

Eliminating A between Eqs. (2.102) and (2.103), the real and imaginary parts of the resulting expression are

$$2k \sin\left(\frac{S(0)}{\hbar}\right) = \phi'(0), \tag{2.106}$$

$$2k \cos\left(\frac{S(0)}{\hbar}\right) = \phi(0)\left(k + \frac{S(0)}{\hbar}\right). \tag{2.107}$$

Likewise, elimination of B between Eqs. (2.104) and (2.105) leads to

$$\frac{S'(L)}{\hbar} = k \Rightarrow \tilde{\rho}(L) = 1, \tag{2.108}$$

$$\tilde{\rho}'(L) = 0. \tag{2.109}$$

Now, with the help of Eq. (2.108), the scattering transmission coefficient stems from Eq. (2.105)

$$T = |B|^2 = \rho(L) = I_1/k. \tag{2.110}$$

Squaring Eqs. (2.106) and (2.107) and adding, one finds

$$4k^2 = [\phi'(0)]^2 + \phi^2(0)\left(k + \frac{S'(0)}{\hbar}\right)^2. \tag{2.111}$$

With the help of Eqs. (2.86) and (2.87), we recast Eq. (2.111) as

$$T = 4k^2(C_1 + C_2 + C_3)^{-1} \tag{2.112}$$

where

$$C_1 = 2k^2 - k\sqrt{a^2-1}\frac{\tilde{\rho}_P'(0)}{\tilde{\rho}_P(0)}\sin(2[\beta(0)-b]), \tag{2.113}$$

$$C_2 = \frac{k^2}{\tilde{\rho}_P(0)}\frac{1+(a^2-1)\sin^2(2[\beta(0)-b])}{a+\sqrt{a^2-1}\cos(2[\beta(0)-b])}, \tag{2.114}$$

$$C_3 = \tilde{\rho}_P(0)\left(k^2 + \frac{[\tilde{\rho}_P'(0)]^2}{4\tilde{\rho}_P^2(0)}\right)\left(a+\sqrt{a^2-1}\cos(2[\beta(0)-b])\right), \tag{2.115}$$

where the constants a and b can be determined through the boundary conditions (2.108) and (2.109), namely,

$$a = \frac{\tilde{\rho}_P(L)}{2}\left(1 + \frac{1}{\tilde{\rho}_P^2(L)} + \frac{[\tilde{\rho}_P'(L)]^2}{4k^2\tilde{\rho}_P^2(L)}\right), \tag{2.116}$$

$$b = \beta(L) - \frac{1}{2}\sin^{-1}\left(\frac{\tilde{\rho}_P'(L)}{2k\sqrt{a^2-1}\tilde{\rho}_P(L)}\right), \tag{2.117}$$

and $\beta(x)$ is given by Eq. (2.97). All quantities above are given in terms of $\tilde{\rho}_P(x)$, and, therefore, any (just one!) particular solution for $\tilde{\rho}_P$ completes the problem. A large class of analytic, closed-form solutions for $\tilde{\rho}_P$, for different choices of $\lambda(x)$, can be found in the literature [26].

2.5.1 *Quantum Traversal Time*

The question of the duration of a tunneling process of a particle through a potential barrier has been debated considerably in the literature [19]. Let us consider the dynamics of a quantum particle described by the coupled hydrodynamical equations (2.6) and (2.7) recast in the form [19]

$$\frac{\partial u}{\partial t} = -\frac{\partial}{\partial x}(vu) - \frac{\hbar}{2m}\frac{\partial^2 v}{\partial x^2} \tag{2.118}$$

and

$$\frac{\partial v}{\partial t} = -\frac{1}{m}\frac{\partial V}{\partial x} - v\frac{\partial v}{\partial x} + u\frac{\partial u}{\partial x} + \frac{\hbar}{2m}\frac{\partial^2 u}{\partial x^2} \tag{2.119}$$

such that for each quantum state with the wave function given by Eq. (2.5) we associate the velocity (2.9) and

$$u = (\hbar/2m)(\partial \ln \rho/\partial x). \tag{2.120}$$

We consider a particle with energy E scattering (tunneling) by a static potential well of depth V_o (or barrier) for $0 \leq x \leq L$ and 0 elsewhere [19]. The particle's forward (backward) drift velocity $v_+(v_-)$ can be written as the sum of (difference) of the current velocity v and the osmotic velocity u: $v_\pm = v \pm u$. Both the forward and backward velocities of the particle contribute to the traversal kinetic energy

$$KE = (m/2)(v^2 + u^2) = (m/2)[(v_+^2 + v_-^2)/2] \tag{2.121}$$

where

$$v_\pm(x) = \frac{\hbar}{m} \frac{\left[2k(q/k)^2 \pm q[1-(q/k)^2]\sin 2q(x-L)\right]}{\left[[1+(q/k)^2] - [1-(q/k)^2]\cos 2q(x-L)\right]}. \tag{2.122}$$

For tunneling ($E < V_o$), we replace $q = \sqrt{2m(V_o + E)/\hbar}$ by $i\bar{q} = i\sqrt{2m(V_o - E)/\hbar}$ and $k = \sqrt{2mE/\hbar}$. We define for the stationary regime the quantum traversal time across a square barrier of width L as

$$\tau = \int_0^L \frac{dx}{\sqrt{(2/m)KE}}. \tag{2.123}$$

With the help of Sect. 2.5, we find that for the thin-barrier tunneling problem, the traversal time is given by $\tau = (m/\hbar k)L$, whereas for the opaque-barrier case, $\tau = (m/\hbar\bar{q})L$ [19].

In contrast, for an opaque rectangular barrier, the stationary-phase method [24] yields a traversal time that is independent of the barrier width. This result depends strongly on the form of the wave packet, and is sound only if the wave packet is characterized by a narrow momentum distribution. If a wave packet with a wide momentum distribution strikes a barrier, the transmitted wave packet will exhibit a distribution displaced to higher momenta since the high-energy components of the wave packet tunnel more easily. Thus, the transmitted wave packet moves faster and the reflected wave packet moves slower than the incident one. In contrast, our traversal time definition is that, while working in the configuration space, we avoid the restricted condition on the initial momentum distribution width. As remarked by Büttiker [29], the stationary-phase method does not distinguish between particle which at the end stay in the interaction region and are subsequently reflected and those that are transmitted. This yields the average dwell time of a particle in the barrier and not the traversal time, if most particles are reflected.

2.6 Wave Propagator for the Guiding Wave Function

Via the de Broglie-Bohm approach to quantum mechanics, we develop here a protocol to obtain a propagator for the guiding wave function as defined by the integral equation

$$\psi(x,t) = \int_{-\infty}^{+\infty} dx_0 K(x,x_0,t)\psi(x_0,0). \tag{2.124}$$

The primary concept is introduced that a particle has a definite path which is determined by a suitable equation of motion and that this path is fundamentally affected by a guiding wave function [30]. Accordingly, the connection between the particle and wave properties can be obtained by writing the wave function in the polar form as in Eq. (2.5), and from Eqs. (2.6) and (2.7), the paths of a particle with velocity is given by

$$\frac{dx}{dt} = v(x,t)|_{x=x(t)} = \left.\frac{1}{m}\frac{\partial S}{\partial x}\right|_{x=x(t)} \tag{2.125}$$

as in Eq. (2.9) and subject to an arbitrary external potential V and the quantum potential Q. With the help of Eq. (2.125), we can readily obtain

$$\frac{\partial S}{\partial t} + \frac{1}{2m}\left(\frac{\partial S}{\partial x}\right)^2 + V + Q = 0, \tag{2.126}$$

where

$$\frac{d}{dt} = \frac{\partial}{\partial t} + v\frac{\partial}{\partial x} \tag{2.127}$$

is the hydrodynamical derivative. Equation (2.126) has the form of Newton's second law, in which the particle is subject to a quantum potential Q in addition to the classical potential V. A set of paths is obtained by considering the case when the amplitude of the wave function is a slowly varying function of position. In what follows, we develop a protocol to obtain a propagator for the wave function by retaining explicitly some of the features of the quantum potential. This procedure attempts to generalize that developed by Feynman and Hibbs [31], since their procedure is viewed as a method for obtaining the wave function from the set of classical paths, for which $Q = 0$.

We investigate the quantum hydrodynamical evolution of the Gaussian guiding wave packet given by Eq. (2.12). We now expand $S(x,t)$, $V(x,t)$ and Q around $q(t)$ up to second order

$$S(x,t) = S[q(t)] + S'[q(t),t][x - q(t)] + \frac{S''[q(t)]}{2}[x - q(t)]^2, \tag{2.128}$$

$$V(x,t) = V[q(t)] + V'[q(t),t][x - q(t)] + \frac{V''[q(t)]}{2}[x - q(t)]^2, \tag{2.129}$$

$$Q(x,t) = Q[q(t)] + Q'[q(t),t][x - q(t)] + \frac{Q''[q(t)]}{2}[x - q(t)]^2. \tag{2.130}$$

Next, substituting Eq. (2.12) into (2.7), we find again Eq. (2.16). Now, a connection to Eqs. (2.16) and (2.125) can be established by collecting terms in $[x - q(t)]^0$ and $[x - q(t)]^1$

$$S'[x(t),t] = m\dot{q}(t), \tag{2.131}$$

$$S''[x(t),t] = m\frac{\dot{\delta}(t)}{\delta(t)}. \tag{2.132}$$

Then, by substituting Eq. (2.12) and the previous Taylor expansions into Eq. (2.126) and again collecting terms in $[x - q(t)]^0$ and $[x - q(t)]^1$, we have

$$\dot{S}_0 = \frac{1}{2}m\dot{q}^2(t) - V[q(t),t] - \frac{\hbar^2}{4m\delta^2(t)}, \tag{2.133}$$

$$\ddot{q}(t) = -\frac{1}{m}V'[q(t),t], \tag{2.134}$$

$$\ddot{\delta}(t) + \frac{1}{m}V''[q(t),t]\delta(t) = \frac{\hbar^2}{4m^2\delta^3(t)}, \tag{2.135}$$

where have denoted $\dot{S}_0 \equiv S[q(t),t]$. It is worth noticing the presence of the quantum potential in the last terms of Eqs. (2.133) and (2.135). These equations have the initial conditions

$$q(0) = x_0,\ \dot{q}(0) = v_0, \tag{2.136}$$

$$\delta(0) = \delta_0,\ \dot{\delta}(0) = 0, \tag{2.137}$$

and

$$S_0(0) = mv_0x_0. \tag{2.138}$$

Now, the wave packet can be fully written as

$$\psi(x,t) = [2\pi\delta^2(t)]^{-1/4} \exp\left[\left(\frac{im\dot{\delta}(t)}{2\hbar\delta(t)} - \frac{1}{4\delta^2(t)}\right)[x-q(t)]^2\right]$$
$$\times \exp\left[\frac{im\dot{q}(t)}{\hbar}[x-q(t)] + \frac{imv_0x_0}{\hbar}\right]$$
$$\times \exp\left[\frac{i}{\hbar}\int_0^t dt' \left(\frac{1}{2}m\dot{q}^2(t) - V[q(t)] - \frac{\hbar^2}{4m\delta^2(t)}\right)\right]. \quad (2.139)$$

Next, we turn to finding the propagator $K(x, x_0, t)$ as defined by the integral Eq. (2.124). Let us first define the normalized quantity

$$\Phi(v_0, x, t) = (2\pi\delta_0^2)^{1/4}\psi(v_0, x, t), \quad (2.140)$$

which satisfies the completeness relation

$$\int_{-\infty}^{+\infty} dv_0 \Phi^*(v_0, x, t)\Phi(v_0, x', t) = \frac{2\pi\hbar}{m}\delta(x-x') \quad (2.141)$$

where $\delta(x-x')$ denotes the well-known delta function. From Eq. (2.7), it follows that

$$\frac{\partial(\Phi^*\psi)}{\partial t} + \frac{\partial(\Phi^*\psi v)}{\partial x} = 0 \quad (2.142)$$

which after integration yields

$$\frac{\partial}{\partial t}\int_{-\infty}^{+\infty} dx \Phi^*\psi = 0, \quad (2.143)$$

whence

$$\int_{-\infty}^{+\infty} dx' \Phi^*(v_0, x', t)\psi(x', t) = \int_{-\infty}^{+\infty} dx_0 \Phi^*(v_0, x_0, t)\psi(x_0, t). \quad (2.144)$$

Multiplying Eq. (2.144) by $\Phi^*(v_0, x, t)$, integrating with respect to v_0, and using Eq. (2.141), we have

$$\psi(x,t) = (m/2\pi\hbar) \int_{-\infty}^{+\infty} dv_0 \Phi(v_0, x, t) \int_{-\infty}^{+\infty} dx_0 \Phi^*(v_0, x_0, 0)\psi(x_0, 0), \quad (2.145)$$

whence the propagator reads

$$K(x, x_0, t) = (m/2\pi\hbar) \int_{-\infty}^{+\infty} dv_0 \Phi(v_0, x, t)\Phi^*(v_0, x_0, 0). \quad (2.146)$$

With the help of Eqs. (2.136)–(2.140), we have explicitly

$$K(x, x_0, t) = \frac{m}{2\pi\hbar} \int_{-\infty}^{+\infty} dv_0 \left(\frac{\delta(t)}{\delta_0}\right)^{-1/2} \exp\left[\left(\frac{im\dot{\delta}(t)}{2\hbar\delta(t)} - \frac{1}{4\delta^2(t)}\right)[x - q(t)]^2\right]$$
$$\times \exp\frac{i}{\hbar}\left[m\dot{q}(t)[x - q(t)] + \int_0^t dt' \left(\frac{1}{2}m\dot{q}^2(t') - V[q(t'), t'] - \frac{\hbar^2}{4m\delta^2(t')}\right)\right]. \quad (2.147)$$

Equation (2.147) shows that the quantum-mechanical information given by the propagator can be found in the guiding wave function. Besides, the quantum propagator can be viewed as an expansion of the guiding wave function over the v_0-space (and not in just over the ordinary position space). It is a merit of the formalism of the de Broglie-Bohm quantum theory to bring this out so explicitly that it cannot be ignored.

2.7 Bohmian Trajectories of Airy Packets

The discovery of Berry and Balazs in 1979 [32] that the Schrödinger equation for a free particle of mass m allows a non-dispersive and accelerating Airy wave packet solution has taken the folklore of quantum mechanics by surprise. They have shown that for the Schrödinger equation

$$i\hbar\frac{\partial\psi(x,t)}{\partial t} = -\frac{\hbar^2}{2m}\frac{\partial^2\psi(x,t)}{\partial x^2}, \quad (2.148)$$

a unique solution is given by

$$\psi(x,t) = Ai^2\left[\frac{B}{\hbar^{2/3}}\left(x - \frac{B^3t^2}{4m^2}\right)\right]\exp\left[i\left(\frac{B^3t}{2m\hbar}\right)\left(x - \frac{B^3t^2}{6m^2}\right)\right]. \quad (2.149)$$

This is easily verified by direct substitution and use of the Airy function's differential equation [33]. Here Ai is the Airy function and B is an arbitrary constant. Airy packets continue to propagate without spreading even when a spatially uniform and time-varying force F(t) acts. Over the years, this intriguing class of wave packets has sparked a considerable resurgence on research on diffraction theory and experiments on a classical analogue in optics of nondiffracting beams [34–42]. Despite its numerous applications in electrodynamics, optical theory, solid state physics, radiative transfer, semiconductors in electric fields, Airy functions have found to be a distinctive solution in quantum mechanics but only for force-free and for linear potentials [42–44].

Within the Bohmian mechanics framework, new features of Airy wave packet solutions to Schrödinger equation with time-dependent quadratic potentials can be introduced. In particular, we provide some insights to the problem by calculating its Bohmian trajectories. It is shown that by using general space-time transformations, these trajectories can display a variety of cases depending upon the initial position of the individual particle in the Airy wave packet. These results suggest some mathematical similarities between Schřodinger equation and the paraxial equation of diffraction and pave the way toward the discovery of new experimental observations notwithstanding the fact that the paraxial equation has solutions in terms of Airy functions with complex arguments. Numerous experimental configurations of optics and atom physics have shown that the dynamics of Airy beams depends significantly on initial parameters and configurations of the experimental set-up.

The simplest problem is defined by two evolution equations: the Schrödinger equation with time-dependent quadratic and linear potentials for the wave function $\psi(x,t)$

$$i\hbar\frac{\partial\psi(x,t)}{\partial t} = -\frac{\hbar^2}{2m}\frac{\partial^2\psi(x,t)}{\partial x^2} + \left(\frac{1}{2}m\omega^2(t)x^2 - F(t)x\right)\psi(x,t) \quad (2.150)$$

and the first-order guiding equation for $x(t)$

$$\dot{x}_i(t) = \frac{\hbar}{m}Im\left(\frac{\partial}{\partial x}\log\psi(x,t)\right)\bigg|_{x=x_i(t)} \quad (2.151)$$

which constitutes the simplest first-order evolution equation for the position of the particle that is compatible with the Galilean and time-reversal covariance of the Schrödinger evolution. In Eq. (2.150), $\omega(t)$ and $F(t)$ represent the time-dependent harmonic-oscillator frequency and linear force, respectively. Thus, the main objective is to solve Eq. (2.151) for the Bohmian trajectories of an evolving *ith* particle of the Airy wave packet ensemble with an initial position x_{0i}.

A general solution to Eq. (2.150) can be found through a proper time rescaling of the space variables and introducing new times [45–47]. This new group of transformations reduce Eq. (2.150) to the case of the problem of a free-particle type motion. Although different techniques have been used, the scale and phase transformations presented here yield a simpler physical meaning to the mathematical protocol [48]. To this end, the extended space-time transformations is introduced as

$$\psi(x,t) = \frac{1}{\sqrt{\delta(t)}} \exp\left(\frac{i\phi_1(x,t)}{\hbar}\right) \psi_1(x,t), \tag{2.152}$$

$$x' = \frac{1}{\delta(t)} (x - q(t)), \tag{2.153}$$

and

$$t' = \int_0^t \frac{d\tau}{\delta^2(\tau)}. \tag{2.154}$$

These transformations represent a scale and phase transformation on the wave function and a scale transformation on space and time along with a space translation. In particular, Eq. (2.153) is a Galilean-type transformation. After lengthy but straightforward calculations is found that Eq. (2.150) reduces to

$$i\hbar \frac{\partial \psi_1(x',t')}{\partial t'} = -\frac{\hbar^2}{2m} \frac{\partial^2 \psi_1(x',t')}{\partial x'^2} + f_1(t')\psi_1(x',t') \tag{2.155}$$

where

$$f_1(t') = \frac{m}{2}[\dot{\delta}^2 q^2 + \delta^2 \dot{q}^2 - 2q\dot{q}\delta\dot{\delta}]. \tag{2.156}$$

By performing the phase change

$$\psi_1(x',t') = \psi_2(x',t') \exp\left(-\frac{i}{\hbar} \int_0^{t'} f_1(t'')dt''\right), \tag{2.157}$$

Equation (2.155) reduces further to

$$i\hbar \frac{\partial \psi_2(x',t')}{\partial t'} = -\frac{\hbar^2}{2m} \frac{\partial^2 \psi_2(x',t')}{\partial x'^2}. \tag{2.158}$$

Thus, the general solution to Eq. (2.150) then reads

$$\psi(x,t) = \frac{1}{\sqrt{\delta(t)}} Ai \left\{ \frac{B}{\hbar^{2/3}} \left[\frac{[x-q(t)]}{\delta(t)} - \frac{B^3}{4m^2} \left(\int_0^t \frac{d\tau}{\delta^2(\tau)} \right)^2 \right] \right\} \exp \frac{i}{\hbar} \phi_{total}(x,t), \tag{2.159}$$

where

$$\phi_{total} = \phi_1(x,t) + \phi_2(x,t) + \phi_3(x,t) \tag{2.160}$$

and

$$\phi_1 = -\int_0^{t'} f_1(\tau) d\tau, \tag{2.161}$$

$$\phi_2 = \frac{mx^2\dot{\delta}(t)}{2\delta} - \left[\frac{mq(t)\dot{\delta}(t)}{\delta(t)} - m\dot{q}(t) \right] x, \tag{2.162}$$

$$\phi_3 = \frac{B^3}{2m} \left(\int_0^t \frac{dt'}{\delta^2(t')} \right) \left[\frac{x-q(t)}{\delta(t)} - \frac{B^3}{6m} \left(\int_0^t \frac{dt'}{\delta^2(t')} \right)^2 \right], \tag{2.163}$$

$$f_1 = \frac{1}{2} m \left[\dot{\delta}^2 q^2 + \delta^2 \dot{q}^2 - 2q\dot{q}\delta\dot{\delta} \right]. \tag{2.164}$$

In turn, the auxiliary functions $q(t)$ and $\delta(t)$ obey

$$\ddot{q}(t) + \omega^2(t) q(t) = \frac{F(t)}{m} \tag{2.165}$$

and

$$\ddot{\delta}(t) + \omega^2(t)\delta(t) = 0. \tag{2.166}$$

Now, with the help of Eqs. (2.160), (2.151) yields

$$\begin{aligned} \dot{x}_i(t) &= \frac{1}{m} \left(\frac{\partial \phi_{total}(x,t)}{\partial x} \right) \Bigg|_{x=x_i(t)} \\ &= \dot{q}(t) + [x_i(t) - q(t)] \frac{\dot{\delta}(t)}{\delta(t)} + \frac{B^3}{2m^2\delta(t)} \int_0^t \frac{dt'}{\delta^2(t')}. \end{aligned} \tag{2.167}$$

Equation (2.167) can be integrated as follows. First, we recast it as

$$\frac{[\dot{x}_i(t) - \dot{q}(t)]\,\delta(t) - [x_i(t) - q(t)]\,\dot{\delta}(t)}{\delta^2(t)} = \frac{B^3}{2m^2\delta^2(t)} \int_0^t \frac{dt'}{\delta^2(t')} \tag{2.168}$$

or

$$\frac{d}{dt}\left[\frac{x_i(t) - q(t)}{\delta(t)}\right] = \frac{B^3}{2m^2\delta^2(t)} \int_0^t \frac{dt'}{\delta^2(t')} \tag{2.169}$$

which upon integration becomes

$$x_i(t) = q(t) + (x_{0i} - q_0)\frac{\delta(t)}{\delta_0} + \frac{B^3}{2m^2}\delta(t) \int_0^t \frac{dt'}{\delta^2(t')} \int_0^{t'} \frac{d\tau}{\delta^2(\tau)}, \tag{2.170}$$

where $q_0 = q(0)$ and $\delta_0 = \delta(0)$. Thus, Eq. (2.170) gives the associated Bohmian trajectories of the Airy wave packet that satisfies Eq. (2.150). In particular, if $\delta(t) = 1$, $\omega(t) = 0$, Eq. (2.159) yields the result found by Berry and Balazs in Eq. (2.148) [32]. Furthermore, Eq. (2.170) displays still a myriad of nontrivial Airy packet trajectories. For example, if x_{0i} is positive, then the particles distributed in the right half of the initial ensemble are accelerated whereas the particles distributed in the left half of the initial ensemble are decelerated. Besides, Eq. (2.170) implies that deviations from classical trajectories $\Delta x_i(t) = x_i(t) - X(t)$ are entirely dependent on the solution of Eq. (2.166) (which is a generalization of the Mathieu and Hill-Poschl-Teller equations) [13, 49, 50]. Two independent solutions to Eq. (2.166) can be obtained in general from just one particular solution to the same equation, namely,

$$\delta_1(t) \equiv \delta(t) \tag{2.171}$$

and

$$\delta_2(t) \equiv \delta(t) \int^t \frac{dt'}{\delta^2(t')}. \tag{2.172}$$

If initial conditions are imposed as follows: $\delta(0) = 1$ and $\dot{\delta}(0) = 0$ we are led to the orthogonal conditions $\delta_1(0) = 1$, $\delta_2(0) = 0$, $\dot{\delta}_1(0) = 0$ and $\dot{\delta}_2(0) = 1$. This constitutes a large set of general, nontrivial Bohmian trajectories associated to the Airy wave packet subject to time-dependent quadratic potentials.

These new features are worth introducing to the subject's theoretical folklore in light of the fact that the evolution of a quantum mechanical Airy wave packet, governed by the Schrödinger equation, is analogous to the propagation of a finite energy Airy beam satisfying the paraxial equation notwithstanding the fact that the paraxial equation has solutions in terms of Airy functions with complex arguments. As previously mentioned, numerous experimental configurations of optics and atom physics have shown that the dynamics of Airy beams depends significantly on initial parameters and configurations of the experimental set-up. The use of Airy beams for particle manipulation in nonlinear media remains a topic of intense theoretical and experimental research [33, 34]. The results discussed here could be extended in order to be applicable to physically realizable situations.

2.8 Airy and Gaussian Slits. Quantum Focal Point

It is well known that the Schrödinger equation describing a free particle can exhibit a remarkable, nonspreading wave packet solution in terms of an Airy function [32]. A remarkable feature of Airy packets is their ability to remain diffraction-free over long distances while they tend to freely accelerate during propagation in the absence of any external potential. They constitute nontrivial, unusual solutions to Schrödinger equation describing a free particle that remains invariant with time.

Another remarkable feature of Airy packets is worth discussing. We show that Young's two-slit quantum interference of two initially separated Airy wave packets (or *Airy slits*) yields a prominent focal point in the interference pattern. This point of maximum intensity resembles Fresnel's focal point in optics. We also show that Airy slits display a more robust interference pattern when compared to the case of two Gaussian slits. The results discussed here could be tested experimentally since Airy beams have been realized in both one- and two-dimensional configurations [34].

For comparison, let us recall that it is well-known that the solution for the free-particle Schrödinger equation [14–17]

$$i\hbar\frac{\partial\psi(x,t)}{\partial t} = -\frac{\hbar^2}{2m}\frac{\partial^2\psi(x,t)}{\partial x^2}, \tag{2.173}$$

can be described by a Gaussian wave packet

$$\psi_{Gauss}(x,t) = A(t)\exp\left[-\frac{[x-q(t)]^2}{4\tilde{\delta}(t)\delta(0)} + i\frac{mv[x-q(t)]}{\hbar} + i\frac{Et}{\hbar}\right] \tag{2.174}$$

where

$$A(t) = \left[2\pi\tilde{\delta}^2(t)\right]^{-1/4}, \tag{2.175}$$

the complex time dependent spreading is

$$\tilde{\delta}(t) = \delta(0)\left(1 + \frac{i\hbar t}{2m\delta^2(0)}\right), \tag{2.176}$$

where

$$\delta(t) = \delta(0)\sqrt{1 + \left(\frac{\hbar t}{2m\delta^2(0)}\right)^2}, \tag{2.177}$$

$$q(t) = q(0) + vt \tag{2.178}$$

and

$$E = \frac{1}{2}mv^2. \tag{2.179}$$

On the other hand, as seen in the previous Section, Berry and Balazs [32] showed that for the free-particle Schrödinger equation (2.173) a unique solution is given by Eq. (2.149). Now the initial total wave function can be written as

$$\psi(x,0) = N\left[\psi_1\left(x - \frac{d}{2}, 0\right) + \psi_2\left(x + \frac{d}{2}, 0\right)\right] \tag{2.180}$$

where N is a normalization constant. The functions $\psi_{1,2}$ correspond to two widely separated wave packets (Gaussian or Airy slits), each located $x = \pm d/2$ so that the wavepackets are separated by a distance d. The general solution to Eq. (2.173) is given by [16]

$$\psi(x,t) = \sqrt{\frac{m}{2\pi i\hbar t}}\int_{-\infty}^{\infty} dx' \exp\left\{-\frac{m(x-x')^2}{2i\hbar t}\right\}\psi(x',0). \tag{2.181}$$

For comparison, the absolute squared value of expression above is plotted in Figs. 2.3, 2.4, 2.5 and 2.6 for the Young's two-slit quantum interference of two initially separated wave packets (Airy and Gaussian wave packets, respectively). It is clear seen a focal point in the corresponding interference patterns. This point of maximum intensity is analogous to Fresnel focal point in optics.

It is clearly observed that Airy slits display a more robust interference pattern and exhibit a unique point of maximum intensity in the preceding part of the pattern. These results could be easily tested experimentally in light of the successful experimental realization of finite energy Airy beams (one and two-dimensional) obtained by Siviloglou et al. in 2007 [34]. These new features are worth introducing to the subject's theoretical folklore in light of the fact that the evolution of a quantum mechanical Airy wave packet governed by the Schrödinger equation is analogous

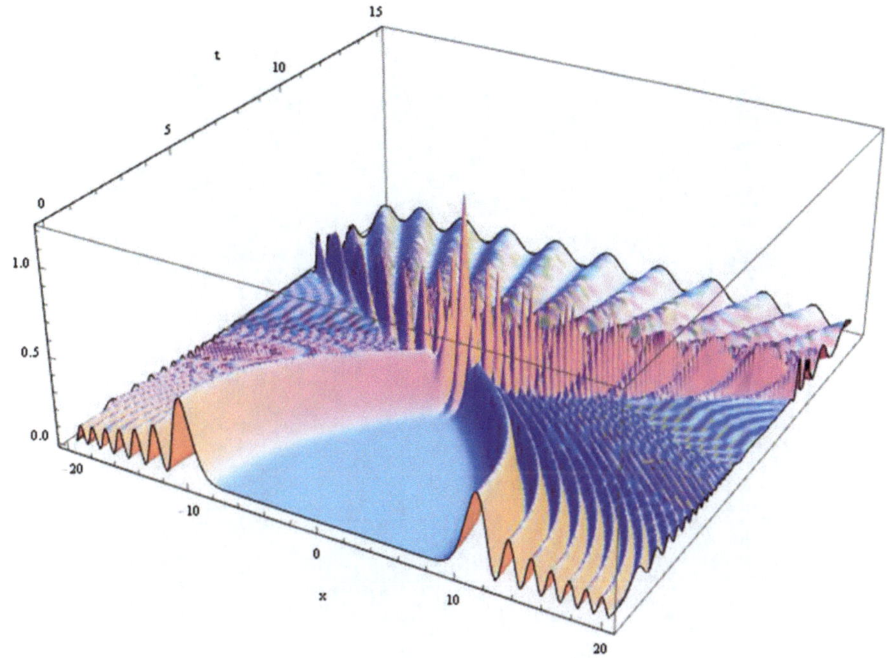

Fig. 2.3 Interference of two initially, widely separated Airy wave packets

to the propagation of a finite energy Airy beam satisfying the paraxial equation. Numerous experimental configurations of optics and atom physics have shown that the dynamics of Airy beams depends significantly on initial parameters and configurations of the experimental set-up.

2.9 Detection of Inertial and Gravitational Masses with Airy Wave Packets

The problem of how the inertial and gravitational mass enter in classical and quantum mechanics is of utmost interest. The first investigations on the equivalence of inertial and gravitational mass relied on pendulum experiments and can be traced back to Newton and Bessel [51]. Tests of higher accuracies were realized by the classical torsion balance experiments of Eötvös [52] and Roll et al. [53].

In quantum mechanics, tests of the universality of free fall have recently been carried out with atom interferometry in the context of the coherence of an atom laser or in connection with the so-called atom trampoline, or also known as the quantum bouncer. Matter wave interferometry with freely falling ^{85}Rb and ^{87}Rb isotopes has

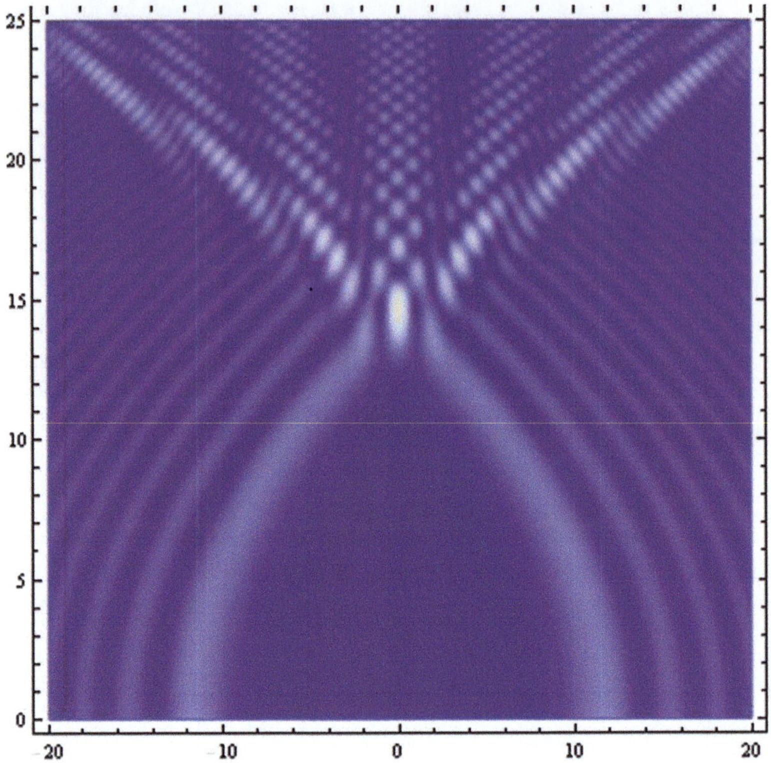

Fig. 2.4 Density plot of the interference of two initially, widely separated Airy wave packets: Quantum focal point

already been performed, and several other experiments worldwide using different species of atoms are in preparation [54–67].

In particular, the universality of the free fall problem has been reviewed and addressed for a quantum mechanical particle in a linear gravitational potential within the Wigner phase space formulation by Kajari et al. [54]. Their conclusion is three-fold: (i) The quantum dynamics reduces to classical dynamics and therefore can only involve the ratio of the inertial mass m_i and the (passive) gravitational mass m_g, (ii) the spatial modulation of the energy eigenfunctions depends on the third root of the product of the two masses, and (iii) the energy eigenvalues of the gravitational atom trampoline are proportional to $(m_g^2/m_i)^{1/3}$.

Here we show that the dynamics of a quantum particle in a linear gravitational potential can be readily described using Airy wave packets within the framework of Bohmian mechanics. It is clearly shown that the dynamics of a particle in a linear gravitational potential involves the gravitational mass (m_g) and the inertial mass (m_i) by the third root of the product of the two masses. This noteworthy feature may offer another avenue in quantum tests of the universality of free fall and experiments

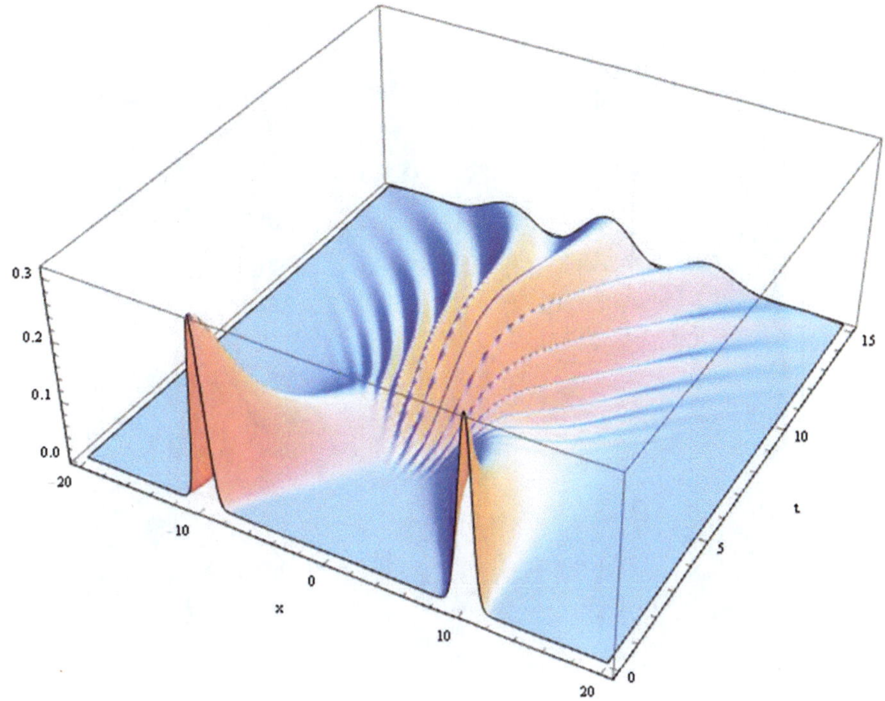

Fig. 2.5 Interference of two initially, widely separated Gaussian wave packets

capable of measuring the Airy-function shaped probability density would yield information about $\sqrt[3]{m_i m_g}$. Over the years, this intriguing class of Airy wave packets has sparked a considerable resurgence on research on diffraction theory and experiments on a classical analogue in optics of nondiffracting beams [33–38, 40]. Despite its numerous applications in electrodynamics, optical theory, solid state physics, radiative transfer, semiconductors in electric fields, Airy functions have found to be a distinctive solution in quantum mechanics not only for force-free and for linear potentials but for a large class of time-dependent quadratic potentials (see Sect. 2.7). The solution to the problem is sought in terms of Airy wave packets within the framework of Bohmian mechanics. Also, we show that the wave packet remains nonspreading and nonaccelerating if the well-known Berry-Balazs constant B is given specifically by $B = \left(2m_i m_g g\right)^{1/3}$. Therefore, B is not devoid of any physical meaning – a fact overlooked in the literature.

Our problem is defined by two evolution equations: the Schrödinger equation with a linear gravitational potential for the wave function $\psi(x,t)$

$$i\hbar\frac{\partial\psi(x,t)}{\partial t} = -\frac{\hbar^2}{2m_i}\frac{\partial^2\psi(x,t)}{\partial x^2} - (m_g g x)\psi(x,t) \tag{2.182}$$

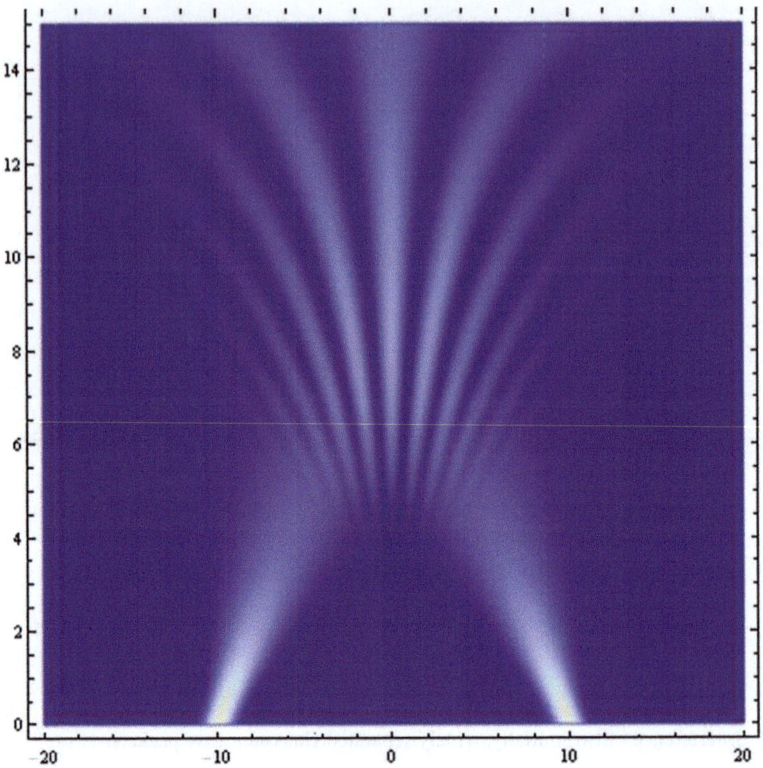

Fig. 2.6 Density plot of the interference of two initially, widely separated Gaussian wave packets: Quantum focal point

and the first-order de Broglie guiding equation for $x(t)$

$$\dot{x}_i(t) = \frac{\hbar}{m_i} Im \left(\frac{\partial}{\partial x} \log \psi(x,t) \right) \Bigg|_{x=x_i(t)} \tag{2.183}$$

which constitutes the simplest first-order evolution equation for the position of the particle that is compatible with the Galilean and time-reversal covariance of the Schrödinger evolution. We now solve Eqs. (2.182) and (2.183). To this end, a solution to Eq. (2.182) can be again found by introducing the space-time transformation

$$\psi(x,t) = \exp\left(\frac{i\phi_1(x',t)}{\hbar}\right)\psi_1(x',t), \tag{2.184}$$

$$x' = x - q(t), \tag{2.185}$$

and

$$t' = t. \tag{2.186}$$

These transformations represent a scale and phase transformation on the wave function and a scale transformation on space and time along with a space translation. Experiments have verified the validity of these transformation laws for noninertial frames also in the quantum limit. Equation (2.185) is a Galilean-type transformation which reduces Eq. (2.182) to

$$i\hbar\frac{\partial\psi_1(x',t')}{\partial t'} = -\frac{\hbar^2}{2m_i}\frac{\partial^2\psi_1(x',t')}{\partial x'^2} + f_1(t')\psi_1(x',t') \tag{2.187}$$

where

$$f_1(t') = \frac{1}{2}m_i\dot{q}^2. \tag{2.188}$$

By performing the phase change

$$\psi_1(x',t') = \psi_2(x',t')\exp\left(-\frac{i}{\hbar}\int_0^{t'} f_1(t'')dt''\right), \tag{2.189}$$

Equation (2.187) reduces further to

$$i\hbar\frac{\partial\psi_2(x',t')}{\partial t'} = -\frac{\hbar^2}{2m_i}\frac{\partial^2\psi_2(x',t')}{\partial x'^2}. \tag{2.190}$$

Thus, the wave packet solution to Eq. (2.182) then reads

$$|\psi_2(x,t)| = Ai\left\{\frac{B}{\hbar^{2/3}}\left[[x-q(t)] - \frac{B^3t^2}{4m_i{}^2}\right]\right\}, \tag{2.191}$$

with

$$f_1(t) = \frac{1}{2}\frac{m_g^2}{m_i}g^2t^2, \tag{2.192}$$

and

$$m_i\ddot{q}(t) = -m_g g. \tag{2.193}$$

Now, the Bohmian trajectories are found to be

$$x_i(t) = q(t) + (x_{oi} - q_o) + \frac{B^3t^2}{2m_i^2}. \tag{2.194}$$

Thus, we can summarize our main results as follows

$$|\psi(x,t)|^2 = Ai^2\left[\frac{B}{\hbar^{2/3}}\left(x - \frac{1}{2}\frac{m_g}{m_i}gt^2 + \frac{B^3t^2}{4m_i^2}\right)\right] \tag{2.195}$$

and

$$x_i(t) = -\frac{1}{2}\frac{m_g}{m_i}gt^2 + (x_{oi} - q_o) + \frac{B^3t^2}{4m_i{}^2}. \tag{2.196}$$

Further, we find that if B is given by the third root of the product of the inertial and gravitational mass, namely

$$B = \sqrt[3]{2m_i m_g g}, \tag{2.197}$$

the non-spreading Airy wave packet is also stationary $[x_i(t) = (x_{oi} - q_o)]$ in the uniform gravitational field.

The transformation to the free fall system changes the argument of the wave function by introducing a phase factor. This means that the laws of quantum physics are the same in a frame with gravitational potential $-m_g gx$ as in a corresponding frame lacking this potential but having acceleration $-g$ instead. Experiments have verified the validity of these classical transformation laws for noninertial frames also in the quantum limit [57, 63].

It has been assumed so far in the literature that B is a completely arbitrary constant devoid of any special physical meaning. Classically, when the inertial mass m_i and the gravitational mass m_g are equated, the mass drops out of Newton's equations of motion, implying that particles of different mass with the same initial conditions follow the same trajectories. However, in Schrödinger's equation (2.182) the masses do not cancel. Now, Eq. (2.183) defines the Bohmian trajectories of an evolving *ith* particle of the wave packet ensemble with an initial position x_{0i}. In general, the Bohmian trajectories imply mass-dependent differences in motion, except when the well-known Berry-Balazs constant B is given by $B = \left(2m_i m_g g\right)^{1/3}$. For atoms from a condensate ^{87}Rb [62, 63] this quantity can be related to an inverse scale related to the wave number $k \approx 3.3 \times 10^6\,\mathrm{m}^{-1}$ (k has the same physical units as the familiar wave vector of a plane wave) [54–61, 63–67].

$$B = h^{2/3}k \approx 7.34 \times 10^{-17}(\mathrm{mkg/s})^{1/3}. \tag{2.198}$$

This result might offer another avenue in quantum tests of the universality of free fall. The use of Airy beams for particle manipulation in nonlinear media remains a topic of intense theoretical and experimental research [33, 34]. Above all, experiments capable of measuring the Airy-function shaped probability density would yield information about $\sqrt[3]{m_i m_g}$.

2.10 The Geometric Phase

The time evolution of a quantum system is governed by differential equations, namely the Schrödinger equation, the Heisenberg or interaction picture equation of motion in the standard approach of quantum mechanics. The Hamiltonian operator plays an important role in this dynamics. Usually, this Hermitian linear operator is considered to be time-independent. However, there are many physical situations where the corresponding Hamiltonians are time-dependent due to the fact that some parameters are determined by external or environmental factors varying with time. The so-called geometric phase is a very important aspect of this type of dynamics [68–71]. The classical counterpart of the phase of the wave function is the phase of the quasi-periodic motion and is called Hannay's angle.

Around thirty years ago, Berry [72] noticed that the standard description of adiabatic processes in quantum mechanics missed something important. We should remember that the notion of adiabaticity lies on the border of statics and dynamics; in other words, dynamical effects are taken into account but in the limit of infinitely slow changes. If a set of parameters are varied adiabatically during the time evolution, a *cyclic variation* (or periodical evolution in time) of such parameters leads to a change of the wave function by an additional phase factor, being completely ignored more than half a century. This phase factor, apart from the usual dynamical phase, contains a purely geometric part independent on the duration of such an evolution. Berry's phase was derived in the adiabatic formalism and, therefore, this phase could only be an approximation of the true quantum phase. The latter was introduced for general unitary cyclic evolution by Aharonov and Anandan [73] and subsequently generalized to arbitrary, not necessarily unitary or cyclic, evolutions by Samuel and Bhandari [74]. In 1928, Fock [75] showed that such a phase could be redefined in non-cyclic evolutions and thought to be unimportant. Bortolotti [76] and Rytov [77] papers ignored by the optical community were pioneers in the rotation of the polarized vector of light travelling along the coiled ray. It turns out to be that the corresponding rotation can be simply interpreted in terms of the geometric phase. In 1956, Pancharatnam discovered an analog to the geometric phase when using polarized light [78] by defining the relative phase between two light beams in different polarization states. The same phase was reported in molecular physics when dealing with the Jahn-Teller effect by Longuet-Higgins et al. [79]. Aharonov and Bohm showed the importance of the electromagnetic vector potential in the interference pattern of electrons and the shift observed with respect to a normal interference pattern was coined the Aharonov–Bohm effect [80]. The first derivation of a geometric phase and the corresponding gauge potential is due to Mead and Truhlar when studying the well-known chemical reaction $H + H_2 \rightarrow H_2 + H$ [81]. Nowadays, geometric phases are been observed in many different fields going from magnetic resonance to condensed matter systems such as, for example, the anomalous Hall effect. The dynamics of quantized vortices is also a field where such a phase is relevant.

Soon after Berry reported his adiabatic geometric phase, Simon [82] realized that it could be interpreted as the holonomy of a fiber bundle and the corresponding gauge potential played the role of a connection in this fiber bundle. It seems like this mathematical theory was waiting for nice and important application in quantum phase factors susceptible to be measured in interference experiments. Mukunda and Simon [83] proposed a new approach consisting of deriving the geometric phase from the so-called quantum kinematic approach instead of the Schrödinger equation.

The geometric phase has also been considered within the Bohmian formalism [84–86] and quantum vortices can be regarded as a manifestation of it.

2.10.1 The Aharonov–Bohm Effect

As is well known [2, 87], the Schrödinger equation is form invariant with respect to a gauge transformation, where the wave function undergoes a suitable unitary transformation. Unlike classical physics, the electromagnetic potentials (scalar, A_0, and vector, $\mathbf{A}$, potentials) play a fundamental role in quantum processes involving charge interactions. In fact, both potentials enter in that equation instead of the electric and magnetic field strengths themselves.

The wave function of a particle with mass m and charge e in the presence of an electromagnetic field is governed by the Schrödinger equation

$$i\hbar\frac{\partial\Psi}{\partial t} = \left\{-\frac{\hbar^2}{2m}\left[\nabla - \frac{ie}{\hbar c}\mathbf{A}\right]^2 + eA_0 + V\right\}\Psi \tag{2.199}$$

where c is the speed of light. Under the gauge transformation in the configuration space

$$\Psi'(\mathbf{r}, t) = e^{ie\varphi(\mathbf{r},t)/\hbar}\Psi(\mathbf{r}, t) \tag{2.200}$$

φ being an arbitrary single-valued function of $\mathbf{x}$ and t, the particle is governed by the same Schrödinger equation by replacing $\Psi \to \Psi'$, $\mathbf{A} \to \mathbf{A}' = \mathbf{A} - \nabla\varphi$ and $A_0 \to A_0' = A_0 + (1/c)\partial\varphi/\partial t$.

Analogously, in the Bohmian formalism, it is a quite straightforward exercise to show that the probability density, the velocity field and the quantum potential are gauge invariant quantities (as well as the quantum trajectories) unlike the total energy [2]. The particle momentum field is given by

$$\mathbf{p}(\mathbf{r}, t) = \nabla S - \frac{e}{c}\mathbf{A} \tag{2.201}$$

and the quantum Newton equation (or quantum Lorentz equation) can be expressed as

$$m\frac{d\mathbf{v}}{dt} = \frac{e}{c}\mathbf{v} \times (\nabla \times \mathbf{A}) - \nabla Q \tag{2.202}$$

when no electric field is present. Again, there is a quantum force even when the magnetic field is zero. The paths are no longer orthogonal to the surface $S = constant$

$$\oint_C dS = \oint_C \mathbf{p}.d\mathbf{r} + \frac{e}{c}\oint_C \mathbf{A}.d\mathbf{r} = n\hbar \qquad (2.203)$$

where n is an integer number and the electromagnetic flux is expressed as

$$\Phi = \int_S \mathbf{B}.d\mathbf{S} = \oint_C \mathbf{A}.d\mathbf{r} \qquad (2.204)$$

along the circuit or closed path C encircling the surface **S**. The magnetic and electric fields are given by $\mathbf{B} = \nabla \times \mathbf{A}$ and $\mathbf{E} = -\nabla A_0 - (1/c)\partial\mathbf{A}/\partial t$, respectively.

The so-called Aharonov–Bohm (AB) effect consists of the influence of the vector potential on charged particles in interference patterns. In particular, consider a collimated electron beam from a source reaching a beam splitter. The two resulting beams go round a line of flux and are recombined again to meet in a screen. When the flux is zero, a certain interference pattern is observed. However, when the flux is not zero, the corresponding interference pattern is shifted with respect to the original one. This line of flux can be seen as an infinite cylindrical solenoid. A steady current in the solenoid generates a flux given by Eq. (2.204), the magnetic field vanishes outside but the vector potential is not zero. Following Holland [2], the total wave function when the field is switched on is given by

$$\Psi = N[\Psi_A + \Psi_B e^{ie\Phi/\hbar c}]e^{(ie/\hbar c)\int_A \mathbf{A}.d\mathbf{r}} \qquad (2.205)$$

since each wave is multiplied by a phase factor due to the open path going from the beam splitter to the screen. The closed path C is formed by the sum of path A and path B and the total flux, Φ, is expressed by Eq. (2.203). The normalization factor N is chosen depending on the type of slit assumed (Gaussian, etc.). The total intensity is then given by

$$|\Psi|^2 = N^2\{|\Psi_A|^2 + |\Psi_B|^2 + 2|\Psi_A||\Psi_B|cos[(S_A - S_B)/\hbar - (e\Phi/c\hbar)]\} \qquad (2.206)$$

where, as usual, the total wave function as well as the partial waves have been expressed in polar form. The observable AB shift is given by $e\Phi/c\hbar$, the total flux being a gauge invariant quantity. The AB phase is a particular example of Berry's geometric phase; namely, a topological phase. This phase is independent of the details of the dynamics and is even insensitive to the shape of the close path C. The latter argument is precisely the main feature distinguishing topological phases from the more general geometric phases.

The so-called bound-state Aharonov–Bohm effect [88] has also been studied in terms of Bohmian trajectories [86]. In this problem, a particle with a given mass is confined to a motion on a ring of a finite radius around a magnetic flux. This flux is concentrated in a line through the origin perpendicular to the plane of the ring.

Twenty five years after the prediction of the AB effect, Aharonov and Casher [89] predicted a *dual* effect, the so-called Aharonov-Casher effect, the particle is neutral but has a magnetic moment, and the tube contains a line of charge. Experiments in neutron [90], vortex [91], atom [92], and electron [93] interferometry bear out the prediction of Aharonov and Casher.

2.10.2 *Quantum Vortices*

Consider a wave function in the three-dimensional configuration space and an arbitrary open or closed path $\mathbf{r}(t)$ on this space. The variation of the wave function along this path as time evolves originates a one-dimensional curve labelled by t. The inner product of this funtion at two different times is a complex number and the total phase, measuring the relative phase of the final point with respect to the initial point of the curve, is given by its argument. This total phase, $\varphi_{tot} = arg[(\psi(t_1), \psi(t_2)]$, is defined modulo 2π and, after Aharonov and Anandan [73], is expressed as the sum of the so-called dynamical phase and the geometric phase. The dynamical phase measures the phase change locally accumulated along the path and is defined in Bohmian mechanics as [86]

$$\varphi_{dyn} = \frac{1}{\hbar} \int_{t_1}^{t_2} \frac{dS(\mathbf{r}(t), t)}{dt}. \tag{2.207}$$

Therefore, the geometric phase is then written as

$$\varphi_g = \frac{1}{\hbar}[S(\mathbf{r}(t_2), t_2) - S(\mathbf{r}(t_1), t_1)] - \varphi_{dyn} \tag{2.208}$$

where the total phase is now expressed in terms of the difference of the wave function phases. The single valuedness of the wave function implies that the phase difference at the final position of the path must be zero or $2\pi n$, n being an integer number. Thus, $\varphi_g = 2\pi n$ along an arbitrary path (closed or open). Even more, for stationary states, this phase along a closed path is proportional to the circulation integral

$$\varphi_g = \frac{1}{\hbar} \oint_C \mathbf{p}.d\mathbf{r}. \tag{2.209}$$

This integral is quantized to be an integer multiple of $\hbar$. Thus, quantum vortices can be regarded as a manifestation of the geometric phase [86].

2.11 Bohmian Formulation of the Gross–Pitaevskii Equation

The so-called Gross–Pitaevskii Equation has been successfully used to discuss the behavior of Bose-Einstein condensates up to first order at zero temperature [94–97]

$$i\,\hbar\,\frac{\partial\Psi(x,\,t)}{\partial t} \;=\; -\,\frac{\hbar^2}{2\,m}\,\frac{\partial^2\Psi(x,\,t)}{\partial x^2} \;+\; V\,\Psi(x,\,t) \;-\; G\mid\Psi(x,\,t)\mid^2\Psi(x,\,t). \tag{2.210}$$

This equation is effectively a mean-field approximation for the interparticle attractive interactions where m is an effective mass of the system, V is a constant potential and G a parameter that regulates the strength of the nonlinearity. It has a well-known solution in terms of a soliton

$$\Psi(x,\,t) \;=\; \left(\frac{k}{2}\right)^{1/2}\; sech\,[k\,(x\;-\;x_o\;-v\,t)]\;\times \tag{2.211}$$

$$\times\; exp\left[\frac{i\;m\;v}{\hbar}\,(x\;-\;x_o\;-\;v'\,t)\right]. \tag{2.212}$$

where

$$k \;=\; \frac{m\;G}{2\;\hbar^2} \tag{2.213}$$

and

$$v' \;=\; \frac{V_\epsilon\;+\;m\;v^2/2\;-\;\hbar^2\,k^2/2}{m\;v}. \tag{2.214}$$

An additional, attractive characteristic of this equation can be demonstrated within the framework of Bohmian mechanics. In particular, we find a most remarkable solution: existence of exact soliton-like solutions of Gaussian form (gausson). To this end, from the polar form of the wave function given by Eq. (2.5), we have that

$$\frac{\partial\rho}{\partial t}+\frac{\partial}{\partial x}\,(\rho v)=0, \tag{2.215}$$

and

$$\frac{\partial v}{\partial t}+v\frac{\partial v}{\partial x}=-\frac{1}{m}\frac{\partial}{\partial x}(Q+V_{gp}), \tag{2.216}$$

where ρ, v, Q are defined as usual and $V_{gp}=-G\phi^2(x,t)$ is the so-called Gross–Pitaevskii potential. With the help of Eq. (2.216), it is readily obtained that

$$\frac{\partial S}{\partial t} + \frac{1}{2m}\left(\frac{\partial S}{\partial x}\right)^2 + V + Q + V_{gp} = 0. \tag{2.217}$$

A semiclassical solution to the Gross–Pitaevskii equation in terms of a Gaussian wave packet is easily reached. Thus, if we start again from Eq. (2.12), $S(x, t)$, V_{gp} and Q can be again expanded around $q(t)$ up to second order according to Eqs. (2.128), (2.129) and (2.130).

Next, substituting Eq. (2.12) into (2.215), we again find Eq. (2.16) for the velocity. Now, as previously carried out in Sect. 2.6, by collecting terms in $[x - q(t)]^0$ and $[x - q(t)]^1$, we have

$$S'[x(t), t] = m\dot{q}(t), \tag{2.218}$$

$$S''[x(t), t] = m\frac{\dot{\delta}(t)}{\delta(t)}. \tag{2.219}$$

Then, by substituting Eq. (2.12) and the previous Taylor expansions (2.128), (2.129) and (2.130) into (2.217) and again collecting terms in $[x - q(t)]^0$ and $[x - q(t)]^1$, it is reached that

$$\dot{S}_0 = \frac{1}{2}m\dot{q}^2(t) - \frac{\hbar^2}{4m\delta^2(t)} + \frac{G}{\sqrt{2\pi}\delta(t)}, \tag{2.220}$$

$$\ddot{q}(t) = 0, \tag{2.221}$$

$$\ddot{\delta}(t) = \frac{\hbar^2}{4m^2\delta^3(t)} - \frac{G}{\sqrt{2\pi}m\delta^2(t)}, \tag{2.222}$$

where have denoted $\dot{S}_0 \equiv \dot{S}[q(t), t]$. It is worth noticing the presence of the quantum and Gross–Pitaevskii potentials in the last terms of Eqs. (2.220) and (2.222). For linear and quadratic potentials, we note a remarkable feature: the existence of exact soliton-like solutions of Gaussian form (gaussons). We find a non-spreading width if

$$\delta_0 = \frac{\sqrt{2\pi}\hbar^2}{4mG}. \tag{2.223}$$

2.12 Dissipation in the Caldirola-Kanai Model

An effective description of dissipation can be achieved from explicitly time-dependent Lagrangians and/or Hamiltonians, thus avoiding to deal with the environment degrees of freedom. This approach allows one to preserve the canonical formalism, which can be a good starting point to find out the quantum analogue of the corresponding dissipative dynamics [98, 99]. The so-called Caldirola-Kanai model [100, 101] can be considered a Hamiltonian formulation of the Langevin equation with zero fluctuations, that is, the Brownian-like thermal fluctuations due to the environment are neglected and the system undergoes a gradual decay until its total energy is completely and irreversibly lost by dissipation (leading to the stopping of the corresponding dynamics). This model can provide us a nice illustration of the dynamics of a particle on a quantum viscid media.

Consider first the classical equation of motion for a damped particle of mass m under the action of a one-dimensional external potential $V(x)$ and a mean friction γ. The corresponding Langevin equation can be written as

$$m\ddot{x} + m\gamma\dot{x} + V'(x) = 0, \tag{2.224}$$

where overdots on the position variable indicate the order of the total time derivatives, and primes for different orders of derivation of the external potential with respect to the position variable. Now, by multiplying this equation by $e^{\gamma t}$, we have

$$\frac{d}{dt}\left(me^{\gamma t}\dot{x}\right) + V'(x)e^{\gamma t} = 0. \tag{2.225}$$

If we consider the change of variable $X = x$ and

$$P \equiv me^{\gamma t}\dot{x} = pe^{\gamma t}, \tag{2.226}$$

where $p = m\dot{x}$ is the physical momentum, we readily notice that (2.225) is just the Lagrange equation satisfied by the time-dependent Lagrangian function

$$L = \frac{P^2}{2m}\, e^{-\gamma t} - V(X)e^{\gamma t}, \tag{2.227}$$

where P plays the role of a canonical momentum with $P = \partial L/\partial \dot{X}$. This equation allows us to obtain straightforwardly the corresponding Hamiltonian function

$$H = \dot{X}P - L = \frac{P^2}{2m}\, e^{-\gamma t} + V(X)e^{\gamma t}, \tag{2.228}$$

which is a function of the canonical variables X and P satisfying the Hamilton equations

$$\dot{X} = \frac{\partial H}{\partial P}, \qquad \dot{P} = -\frac{\partial H}{\partial X}. \tag{2.229}$$

Thus, the energy of the classical system can be expressed as

$$E = \frac{p^2}{2m} + V(x) = He^{-\gamma t}. \tag{2.230}$$

which exhibits the typical exponential decay due to the dissipative dynamics.

The quantum analog of (2.228) can be now obtained by considering the (canonical) momentum operator $P = -i\hbar\partial/\partial X$ $(= -i\hbar\partial/\partial x)$ leading to the standard commutation rule, $[X, P] = i\hbar$. Notice that the same commutation relation for the physical operators of position and momentum is written as $[x, p] = i\hbar\, e^{-\gamma t}$. The quantum Hamiltonian operator reads as

$$H = -\frac{\hbar^2}{2m}\, e^{-\gamma t}\, \nabla_X^2 + V(X)e^{\gamma t}, \tag{2.231}$$

leading to the generalized, dissipative Schrödinger equation

$$i\hbar\frac{\partial \Psi}{\partial t} = -\frac{\hbar^2}{2m}\, e^{-\gamma t}\nabla_x^2\Psi + V(x)e^{\gamma t}\Psi \tag{2.232}$$

in terms of the physical coordinate. Thus, as long as the momentum operator is not acting, the use of the wave equation in the physical coordinate is formally correct.

In the Bohmian formalism [102], the polar form of the wave function is again used, $\Psi(x,t) = \rho^{1/2}(x,t)e^{iS(x,t)/\hbar}$. When substituting into (2.232), the following coupled partial differential equations are reached

$$\frac{\partial \rho}{\partial t} + \nabla_x J = 0, \tag{2.233}$$

$$\frac{\partial S}{\partial t} + \frac{1}{2m}\nabla_x^2 S\, e^{-\gamma t} + V_{\text{eff}}(x,t) = 0, \tag{2.234}$$

where $J = (1/m)\rho\nabla_x S e^{-\gamma t}$ is the associated probability density current and $V_{\text{eff}}(x,t) = V(x)e^{\gamma t} + Q(x,t)$ is an effective potential, which includes the quantum potential

$$Q = -\frac{\hbar^2}{2m}\frac{\nabla_x^2\rho^{1/2}}{\rho^{1/2}}\, e^{-\gamma t} = -\frac{\hbar^2}{4m}\left[\frac{\nabla_x^2\rho}{\rho} - \frac{1}{2}\left(\frac{\nabla_x\rho}{\rho}\right)^2\right] e^{-\gamma t}. \tag{2.235}$$

The time-dependent decaying factor, $e^{-\gamma t}$, also manifests in the corresponding dissipative Bohmian trajectories, which are derived from the equation of motion

$$\dot{x} = \frac{J}{\rho} = \frac{\nabla_x S}{m}\, e^{-\gamma t} = \frac{\hbar}{2mi\rho}\left(\Psi^*\nabla_x\Psi - \Psi\nabla_x\Psi^*\right)e^{-\gamma t}. \tag{2.236}$$

In this expression, the Bohmian momentum also holds a similar relationship to (2.226), i.e., $P_B = \nabla_x S = me^{\gamma t}\dot{X}$, with $X(t) = x(t)$ being the corresponding Bohmian trajectory.

For simple problems such as, for example, free propagation, interference dynamics, linear potentials or the harmonic oscillator, analytical solutions are available. Usually, the initial wave function is described by a Gaussian wave packet written as [102]

$$\Psi(X,t) = e^{(i/\hbar)\left[\alpha_t(X-X_t)^2+P_t(X-X_t)+f_t\right]}, \tag{2.237}$$

where the subscript t is used to denote explicit time-dependence of the parameters characterizing the wave function (the subscript 0 is found when $t = 0$). If the wave function (2.237) is normalized, then

$$f_0 = \frac{i\hbar}{4}\ln\left[\frac{\pi\hbar}{2\mathrm{Im}\{\alpha_0\}}\right], \tag{2.238}$$

with $\mathrm{Im}\{\alpha_0\} \neq 0$. In such a case, the position and momentum expectation values (and therefore the wave-packet centroid) follow a classical trajectory, i.e., $\langle\hat{X}\rangle(t) = X_t$ and $\langle\hat{P}\rangle(t) = P_t$, with (X_t, P_t) being obtained by integrating the Hamilton equations of motion (2.229), i.e.,

$$\dot{X}_t = \frac{P_t}{m}\,e^{-\gamma t}, \tag{2.239}$$

$$\dot{P}_t = -\frac{\partial}{\partial X_t}V(X_t)e^{\gamma t}. \tag{2.240}$$

Moreover, the physical dispersion of the wave packet is found to be

$$\Delta x(t) = \Delta X(t) = \sqrt{\langle X^2\rangle(t) - [\langle X\rangle(t)]^2} = \sqrt{\frac{\hbar}{4\mathrm{Im}\{\alpha_t\}}} = \sigma_t, \tag{2.241}$$

where σ_t is the instantaneous wave-packet spreading. The expectation value of its (also physical) energy,

$$\bar{E} = \langle\hat{H}\rangle e^{-\gamma t} = \frac{P_t^2}{2m}\,e^{-2\gamma t} + V_t + \frac{\hbar}{2m}\frac{|\alpha_t|^2}{\mathrm{Im}\{\alpha_t\}}\,e^{-2\gamma t} + \frac{\hbar V_t''}{8\mathrm{Im}\{\alpha_t\}}, \tag{2.242}$$

where the primes mean derivation with respect to X. Clearly, two contributions are present: one coming from the translational motion along the classical centroidal trajectory (E_t) and another one related to the wave-packet spreading. This second contribution contains information about the spatial variations of both the wave packet and the potential.

Concerning the shape parameters α_t and f_t, their equations of motion can be readily obtained as follows. Let us recast $V(X)$ as a Taylor series expansion around the centroidal position X_t up to the second order, i.e.,

$$V(X) = V_t + V_t'(X - X_t) + \frac{1}{2}\,V_t''(X - X_t)^2. \tag{2.243}$$

By substituting Eq. (2.237) and the expansion (2.243) into the dissipative Schrödinger equation (2.232), and then collecting the coefficients associated with the same power of $(X - X_t)$, it is easily found that

$$\dot{\alpha}_t = -\frac{2\alpha_t^2}{m} e^{-\gamma t} - \frac{1}{2} V_t'' e^{\gamma t}, \tag{2.244}$$

$$\dot{f}_t = \frac{i\hbar\alpha_t}{m} e^{-\gamma t} + L_t, \tag{2.245}$$

with L_t given by (2.227) evaluated along the classical trajectory (X_t, P_t) —since this term is only time-dependent and does not have any space dependance, it will not influence the Bohmian dynamics and therefore we will not consider its explicit functional form. Thus, integrating the set of coupled ordinary differential equations (2.239), (2.240), (2.244) and (2.245), the wave function (2.237) is completely specified at any time. By substituting this wave function into the Bohmian equation of motion (2.236), the following expression for the velocity is reached

$$\dot{X} = \left[\frac{P_t}{m} + \frac{2\text{Re}\{\alpha_t\}}{m} (X - X_t)\right] e^{-\gamma t}, \tag{2.246}$$

which after integration renders the corresponding dissipative Bohmian trajectory $x(t) = X(t)$. From Eq. (2.246), it is clearly seen that if the second term vanishes, the Bohmian trajectory exactly coincides with the classical one given by (2.240). Accordingly, the condition to recover the classical equations does not require necessarily of the limiting procedure of $\hbar \to 0$, but that the wave packet remains relatively localized (i.e., $1/\text{Re}\{\alpha_t\} \to 0$), in agreement with the hypothesis of Ehrenfest's theorem. Furthermore, if $\text{Im}\{\alpha_t\} \to 0$ and the wave function becomes a pure phase factor (except for some time-dependent norm coming from f_t), quantum motion is still present through both P_t and $\text{Re}\{\alpha_t\}$.

2.12.1 Free Gaussian Wave Packet

If the initial wave function is considered to be a Gaussian wave packet [102]

$$\Psi(x, 0) = \left(\frac{1}{2\pi\sigma_0^2}\right)^{1/4} e^{-(x-x_0)^2/4\sigma_0^2 + ip_0(x-x_0)/\hbar}, \tag{2.247}$$

then the initial conditions for the above Gaussian parameters and variables are $X_0 = x_0$, $P_0 = p_0$, $\alpha_0 = i\hbar/4\sigma_0^2$, and $f_0 = (i\hbar/4)\ln(2\pi\sigma_0^2)$. The corresponding initial probability distribution is given by

$$\rho(x,0) = \frac{1}{\sqrt{2\pi\sigma_0^2}} e^{-(x-x_0)^2/2\sigma_0^2}, \tag{2.248}$$

which provides us the spatial region for the initial conditions of the Bohmian trajectories.

After integration of the aforementioned equations of motion, we have that

$$x_t = x_0 + \frac{p_0}{m\gamma}(1 - e^{-\gamma t}), \tag{2.249}$$

$$p_t = p_0 e^{-\gamma t}, \tag{2.250}$$

$$\alpha_t = \frac{\alpha_0}{1 + (2\alpha_0/m\gamma)(1 - e^{-\gamma t})} \tag{2.251}$$

$$f_t = \frac{i\hbar}{4} \ln\left[\frac{\pi\hbar}{2\mathrm{Im}\{\alpha_0\}}\right] + \frac{i\hbar}{2} \ln\left[1 + \frac{2\alpha_0}{m}\left(\frac{1 - e^{-\gamma t}}{\gamma}\right)\right] + S_{\mathrm{cl},t}, \tag{2.252}$$

where

$$S_{\mathrm{cl},t} = \int_0^t L_{t'} dt' \tag{2.253}$$

is the associated classical action. This term only adds a time-dependent phase factor, which does not play any role in the Bohmian dynamics (its gradient vanishes). Equation (2.251) can be alternatively expressed in terms of an effective time-dependent spreading

$$\tilde{\sigma}_t = \sigma_0 \left[1 + \frac{i\hbar}{2m\sigma_0^2}\left(\frac{1 - e^{-\gamma t}}{\gamma}\right)\right], \tag{2.254}$$

which is connected to α_t by the simple relationship $\alpha_t = i\hbar/4\sigma_0\tilde{\sigma}_t$. Accordingly, Eq. (2.252) can be recast in a simpler form,

$$f_t = \frac{i\hbar}{4} \ln(2\pi\tilde{\sigma}_t^2) + S_{\mathrm{cl},t}. \tag{2.255}$$

It is worth mentioning that, in terms of the canonical variables, the functional form of this wave packet is the same as that displayed by a standard free wave packet if t is replaced by

$$\tau = \frac{1 - e^{-\gamma t}}{\gamma} \tag{2.256}$$

in the former; in other words, a time contraction coming from the dissipative process and associated with an eventual freezing displayed by the wave packet.

In the free dispersion limit, it is straightforward to show that (2.237) approaches the well-known solution for the free Gaussian wave packet

$$\Psi(x,t) = \left[\frac{1}{2\pi(\tilde{\sigma}_t^0)^2}\right]^{1/4} e^{-(x-x_t)^2/4\sigma_0\tilde{\sigma}_t^0 + ip_0(x-x_t)/\hbar + iEt/\hbar}, \tag{2.257}$$

where $x_t = x_0 + (p_0/m)t$ provides the instantaneous centroidal position, $E = p_0^2/2m$ is its total mechanical energy, and

$$\sigma_t^0 = |\tilde{\sigma}_t^0| = \sigma_0 \sqrt{1 + \left(\frac{\hbar t}{2m\sigma_0^2}\right)^2} \tag{2.258}$$

is its spreading along time, with $\tilde{\sigma}_t^0 = \sigma_0 \left[1 + (i\hbar t/2m\sigma_0^2)\right]$. However, with friction, the wave packet undergoes asymptotically (i.e., for $t \to \infty$) a damping in its propagation (according to Eq. (2.249), stopping at the position $x_\infty = x_0 + (p_0/m\gamma)$) as well as in its spreading, described by

$$\sigma_t = |\tilde{\sigma}_t| = \sigma_0 \sqrt{1 + \left(\frac{\hbar}{2m\sigma_0^2}\right)^2 \left(\frac{1 - e^{-\gamma t}}{\gamma}\right)^2}. \tag{2.259}$$

Notice that asymptotically, $\sigma_\infty = \sigma_0 \sqrt{1 + (\hbar/2m\gamma\sigma_0^2)^2}$.

The corresponding Bohmian trajectories, which are obtained after integration of the equation of motion (2.246), are given by

$$x(t) = x_t + \frac{\sigma_t}{\sigma_0}[x(0) - x_0], \tag{2.260}$$

which is formally equivalent to the expression that one obtains for the free, frictionless case (see Chap. 1). In the case with friction, Eq. (2.260) reaches the asymptotic limit

$$x(t \to \infty) = x_0 + \frac{p_0}{m\gamma} + \sqrt{1 + \left(\frac{\hbar}{2m\gamma\sigma_0^2}\right)^2}\,[x(0) - x_0]. \tag{2.261}$$

Thus, the wave packet becomes localized: motionless and with the spreading being frozen. For strong friction, it becomes essentially parallel to the classical or centroidal trajectory, since the time-dependence vanishes very quickly and, therefore, σ_t becomes a constant value very rapidly. To some extent, this is a step towards the classicality of the quantum system without appealing to the more standard limiting procedure of $\hbar \to 0$.

In Fig. 2.7 a series of wave-packet properties and Bohmian trajectories are shown for different values of the friction coefficient: $\gamma = 0.025$ (black solid line), $\gamma = 0.1$ (blue dashed line), and $\gamma = 0.5$ (red dash-dotted line). The numerical simulations have been carried out with the following initial conditions: $\sigma_0 = 1$, $x_0 = 0$ and $p_0 = 2.5$ (atomic units). In the upper row, we show the average position (a), (spatial) dispersion (b), and energy (c) as a function of time. To compare with, we have also included the frictionless values, denoted with the gray dotted line. The energy

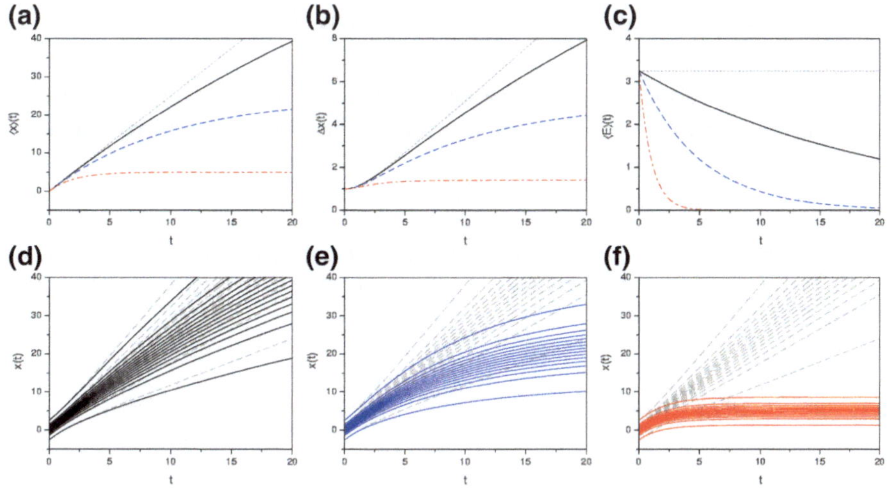

Fig. 2.7 *Top* Average position (**a**), dispersion (**b**), and energy (**c**) for a Gaussian wave packet in free space affected by friction. Different values for the friction coefficient have been considered: $\gamma = 0.025$ (*black solid line*), $\gamma = 0.1$ (*blue dashed line*), and $\gamma = 0.5$ (*red dash-dotted line*). To compare with, the frictionless case ($\gamma = 0$) has also been included and is denoted with the *gray dotted line*. The value of the other parameters considered in these simulations were: $x_0 = 0$, $p_0 = 2.5$ ($E_0 = 3.25$), $\sigma_0 = 1$, $m = 1$, and $\hbar = 1$. *Bottom* Bohmian trajectories associated with the three dissipative cases considered atop: **d** $\gamma = 0.025$, **e** $\gamma = 0.1$, and **f** $\gamma = 0.5$. Again, to compare with, the trajectories for the frictionless case (with the same initial conditions) have also been included in each panel (*gray dashed lines*). The initial positions have been distributed according to the initial Gaussian probability density

expectation value is given by

$$\bar{E} = \left(\frac{p_0^2}{2m} + \frac{\hbar^2}{8m\sigma_0^2} \right) e^{-2\gamma t}, \tag{2.262}$$

i.e., it is suppressed at twice the rate γ. Bohmian trajectories illustrating these dissipative cases are displayed in the three lower panels, from left to right: (d) $\gamma = 0.025$, (e) $\gamma = 0.1$, and (f) $\gamma = 0.5$. The trajectories for the frictionless case have also been included in each panel (gray dashed lines). To better appreciate the different effect of the wave-packet spreading and how it is suppressed as γ increases, 15 trajectories are chosen with initial positions distributed according the initial Gaussian probability density. Thus, for small frictions, it is observed that trajectories starting in the "wings" of the wave packet will increase faster their distance with respect to the centroid than those closer to the latter. Moreover, this distance will be faster for trajectories starting behind the centroid than in front of it due to the larger relative difference between their associated velocities. This effect is, however, damped as the friction coefficient increases, thus producing a smaller separation among trajectories and eventually a freezing of their position for times larger than γ^{-1}. These frozen positions are given by Eq. (2.261).

2.12.2 Interference of Two Gaussian Wave Packets

Let us consider now the interference problem of two free wave packets. As is known, a general solution to this problem can be expressed as a coherent superposition,

$$\Psi = \mathcal{N}(\Psi_1 + \Psi_2), \tag{2.263}$$

with $\mathcal{N}$ being the overall norm factor (it is assumed that each Gaussian wave packet, given by Eq. (2.257), is normalized). As it can be shown, the associated Bohmian equation of motion reads as [102]

$$\begin{aligned}
\dot{x} &= \frac{\rho_1}{\rho}\,\dot{x}_1 + \frac{\rho_2}{\rho}\,\dot{x}_2 + \frac{\hbar}{2mi\rho}\left(\Psi_1^*\partial_x\Psi_2 - \Psi_2\partial_x\Psi_1^*\right)e^{-\gamma t} \\
&+ \frac{\hbar e^{-\gamma t}}{2mi\rho}\left(\Psi_2^*\partial_x\Psi_1 - \Psi_1\partial_x\Psi_2^*\right)e^{-\gamma t} \\
&= \frac{\rho_1}{\rho}\left[\frac{p_{t,1}}{m} + \frac{2\mathrm{Re}\{\alpha_{t,1}\}}{m}\,(x - x_{t,1})\,e^{-\gamma t}\right] \\
&+ \frac{\rho_2}{\rho}\left[\frac{p_{t,2}}{m} + \frac{2\mathrm{Re}\{\alpha_{t,2}\}}{m}\,(x - x_{t,2})\,e^{-\gamma t}\right] \\
&+ 2\cos\xi_{12}\,\frac{\sqrt{\rho_1\rho_2}}{\rho}\left[\frac{p_{t,1} + p_{t,2}}{2m} + \left[\frac{\mathrm{Re}\{\alpha_{t,1}\}(x - x_{t,1}) + \mathrm{Re}\{\alpha_{t,2}\}(x - x_{t,2})}{m}\right]e^{-\gamma t}\right] \\
&- 2\sin\xi_{12}\,\frac{\sqrt{\rho_1\rho_2}}{\rho}\left[\frac{\mathrm{Im}\{\alpha_{t,1}\}(x - x_{t,1}) - \mathrm{Im}\{\alpha_{t,2}\}(x - x_{t,2})}{m}\right]e^{-\gamma t},
\end{aligned} \tag{2.264}$$

where $\dot{x}_i$ refers to the equation of motion for the ith wave packet and $\xi_{12} = (S_1 - S_2)/\hbar$, being ρ_i and S_i the probability density and real phase associated with the ith wave packet when it is expressed in polar form. As it can be seen, this expression simply reflects the sum of two separate fluxes, each one associated with one wave packet, plus another one coming from their interference.

Depending on the value given to the parameters of each wave packet, one can generate different dynamics. For example, if they have the same width ($\alpha_{t,1} = \alpha_{t,2} = \alpha_t$) and are located at symmetric positions with respect to $x = 0$ ($x_{t,1} = x_0 = -x_{t,2}$), then (2.264) simplifies to

$$\begin{aligned}
\dot{x} &= \frac{\rho_1}{\rho}\left[\frac{p_{t,1}}{m} + \frac{2\mathrm{Re}\{\alpha_t\}}{m}\,(x - x_0)e^{-\gamma t}\right] + \frac{\rho_2}{\rho}\left[\frac{p_{t,2}}{m} + \frac{2\mathrm{Re}\{\alpha_t\}}{m}\,(x + x_0)e^{-\gamma t}\right] \\
&+ 2\cos\xi_{12}\,\frac{\sqrt{\rho_1\rho_2}}{\rho}\left[\frac{p_{t,1} + p_{t,2}}{2m} + \frac{2\mathrm{Re}\{\alpha_t\}}{m}\,xe^{-\gamma t}\right] \\
&+ 2\sin\xi_{12}\,\frac{\sqrt{\rho_1\rho_2}}{\rho}\left[\frac{2\mathrm{Im}\{\alpha_t\}}{m}\,x_0e^{-\gamma t}\right].
\end{aligned} \tag{2.265}$$

This situation can simulate the two slit experiment when each wave packet is associated with one of the diffracted beams in a viscid medium. In such a case, if the longitudinal propagation (parallel to the diffractive screen) is slower than the perpendicular one (in the direction of the diffracted beam), both motions can be decoupled and treated as two independent one-dimensional propagations, one longitudinal and the other translational, the latter being well accounted for by a plane wave. Now, if one assumes that $p_{0,1} = p_{0,2} = 0$, i.e., there is no translational motion along the longitudinal direction, but only spreading of the two wave packets [103], the above expression (2.265) will read as

$$\begin{aligned}\dot{x} = & \frac{\rho_1}{\rho}\left[\frac{2\mathrm{Re}\{\alpha_t\}}{m}(x - x_0)\, e^{-\gamma t}\right] + \frac{\rho_2}{\rho}\left[\frac{2\mathrm{Re}\{\alpha_t\}}{m}(x + x_0)\, e^{-\gamma t}\right] \\ & + 2\cos\xi_{12}\frac{\sqrt{\rho_1\rho_2}}{\rho}\left[\frac{2\mathrm{Re}\{\alpha_t\}}{m} x e^{-\gamma t}\right] \\ & + 2\sin\xi_{12}\frac{\sqrt{\rho_1\rho_2}}{\rho}\left[\frac{2\mathrm{Im}\{\alpha_t\}}{m} x_0 e^{-\gamma t}\right]. \end{aligned} \tag{2.266}$$

Results after integration of this equation of motion for different values of γ (the same values as in the case of the free wave packet treated above) can be seen in Fig. 2.8. As it can be seen through the sets of trajectories selected, the friction of the medium leads to the localization of the two wave packets by gradually "freezing" them rather than by annihilating the coherence between them due to the interaction with an environment. In other words, the reason why interference is not observed in a quantum viscid medium is because the wave packets cannot be seen each other, rather than because the destruction (over time) of their mutual coherence.

Several types of interference dynamics can be devised by changing the initial values of the parameters involved in this superposition problem [102].

2.12.3 Linear Potential

For a linear interaction potential like a gravitational or an electric field, the classical solutions are also readily obtained [102, 104]. Thus, if $V(x) = -max$ (a being a positive parameter), we have that

$$x_t = x_0 + \frac{p_0}{m}\left(\frac{1 - e^{-\gamma t}}{\gamma}\right) + a\left(\frac{\gamma t - 1 + e^{-\gamma t}}{\gamma^2}\right), \tag{2.267}$$

$$p_t = p_0 e^{-\gamma t} + ma\left(\frac{1 - e^{-\gamma t}}{\gamma}\right), \tag{2.268}$$

and α_t and γ_t keep the same functional form as in the free damped case (although L_t varies for the latter due to the presence of a nonzero potential function), because only second-order derivatives influence the evolution of those parameters (through V_t'', as seen in Eq. (2.244)).

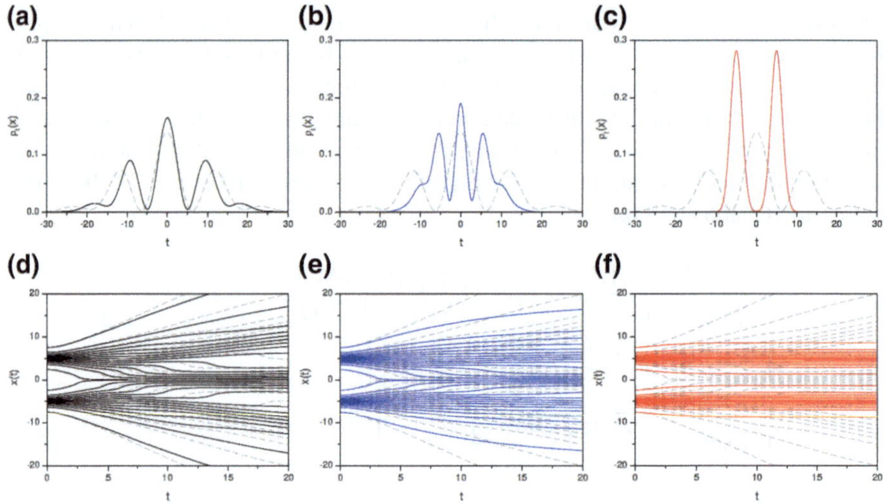

Fig. 2.8 *Top* Final probability density (at $t = 20$) for a coherent superposition of two Gaussian wave packets in free space affected by friction. The same friction values as in Fig. 2.7 have been considered: **a** $\gamma = 0.025$ (*black solid line*), **b** $\gamma = 0.1$ (*blue solid line*), and **c** $\gamma = 0.5$ (*red solid line*). To compare with, the frictionless case ($\gamma = 0$) has also been included in each panel (*gray dashed line*). The value of the other parameters considered in these simulations were: $x_{0,1} = 5 = -x_{0,2}$, $p_{0,1} = p_{0,2} = 0$ ($E_0 = 3.25$), $\sigma_{0,1} = \sigma_{0,2} = 1$, $m = 1$, and $\hbar = 1$. *Bottom* Bohmian trajectories associated with the three dissipative cases considered atop: **d** $\gamma = 0.025$, **e** $\gamma = 0.1$, and **f** $\gamma = 0.5$. Again, to compare with, the trajectories for the frictionless case (with the same initial conditions) have also been included in each panel (*gray dashed lines*). The initial positions have been distributed according to the initial Gaussian probability density

The frictionless limit $\gamma \to 0$ leads us to the well-known expressions for uniform accelerated motion, with $x_t = x_0 + (p_0/m)t + (a/2)t^2$ and $p_t = p_0 + mat$. The wave packet also approaches the expression corresponding to the solution to ramp-like potentials [104], equal to (2.257), except for the different functional form of E_t ($E_t = p_0^2/2m - max_0$) and an extra term coming from the classical action. Regarding the long-time limit for finite friction (i.e., for $t \gg \gamma^{-1}$), we find that while the wave packet freezes its spreading, as in the previous example, it still keeps moving due to the constant limit momentum, $p_\infty = ma/\gamma$. Accordingly, the centroid of the wave-packet displays a uniform motion described by

$$x_{t\to\infty} = x_0 + \frac{p_0}{m\gamma} - \frac{a}{\gamma^2} + \frac{a}{\gamma}\, t. \tag{2.269}$$

With γ, the transition from a uniformly accelerated motion, displayed by the wave-packet centroid, to a uniform rectilinear one, as a consequence of the damping, becomes more apparent. The expression for the average energy is given by

$$\bar{E} = \frac{p_0^2}{2m}\, e^{-2\gamma t} - max_0 - p_0 a\gamma \left(\frac{1-e^{-\gamma t}}{\gamma}\right)^2 + \frac{ma^2}{2}\left(\frac{3 - 2\gamma t - 4e^{-\gamma t} + e^{-2\gamma t}}{\gamma^2}\right) + \frac{\hbar^2}{8m\sigma_0^2}\, e^{-2\gamma t}. \tag{2.270}$$

In the limit $\gamma t \gg 1$, this expression approaches

$$\bar{E} = -max_0 - \frac{p_0 a}{\gamma} + \frac{ma^2}{2\gamma^2}\,(3 - 2\gamma t)\,. \tag{2.271}$$

Unlike the example of the free wave packet, here the energy does not approach zero asymptotically, but it continues decreasing below it as the wave packet slides downhill (unless some additional constraint is imposed in the dynamics to avoiding this situation). Nevertheless, although there is no suppression of the motion, spatial localization is still present.

On the other hand, due to the fact that α_t does not depend on the potential function, the functional form displayed by the Bohmian trajectories is exactly the same as in the free case, except for the fact that these trajectories contain information about the acceleration undergone by the centroid (this information is transmitted through x_t). Now, once the exponentials have vanished, all these trajectories evolve parallel one another, just like classical trajectories under similar circumstances.

2.12.4 Damped Harmonic Oscillator

In a frictionless situation, the wave functions for a harmonic potential given by $V(x) = m\omega_0^2 x^2/2$ are expressed as

$$\Phi_n(x,t) = N_n e^{-(m\omega_0/2\hbar)x^2 - i(n+1/2)\omega_0 t} H_n(\sqrt{m\omega_0/\hbar}x), \tag{2.272}$$

with energies $E_n = (n + 1/2)\hbar\omega_0$, and where H_n is the Hermite polynomial of degree n and $N_n = (1/\sqrt{2^n n!})(\pi\hbar/m\omega_0)^{-1/4}$ is the normalization constant. These energy values can be obtained by employing the same method used above to derive the analytical expression of the time-evolved wave packets [105]. Analogously, it can also be used to determine the dissipative counterpart of Eq. (2.272) [102],

$$\begin{aligned}\Phi_n(x,t) &= N_n e^{-(m\Omega/2\hbar)(1+i\gamma/2\Omega)e^{\gamma t}x^2 - i(n+1/2)\Omega t + \gamma t/4} H_n(\sqrt{m\Omega/\hbar e^{\gamma t/2}}x) \\ &= N_n e^{-(m\Omega/2\hbar)e^{\gamma t}x^2 - i(n+1/2)\Omega t - i(m\gamma/4\hbar)e^{\gamma t}x^2 + \gamma t/4} H_n(\sqrt{m\Omega/\hbar e^{\gamma t/2}}x).\end{aligned} \tag{2.273}$$

However, contrary to Eqs. (2.272), (2.273) describes a *quasi-stationary state*. As formerly shown by Vandyck [106], these states, at each time, are eigenstates of the dissipative Schrödinger equation and eventually collapse to zero. Of course, these states are only defined for $\omega_0 > \gamma/2$.

Concerning the associated Bohmian trajectories, from Eq. (2.273), we have that

$$\dot{x} = -\frac{\gamma}{2}\,x, \tag{2.274}$$

which after integration renders

$$x(t) = x(0)e^{-\gamma t/2}. \tag{2.275}$$

That is, regardless of the eigenstate considered, any trajectory falls down to the bottom of the potential at the same rate, $\gamma/2$, and therefore merging asymptotically at $x = 0$. The reason for such a behavior is that the model is fully dissipative and there is no possibility for a feedback with an environment, as happens when a Brownian-like motion is assumed. In this latter case, the stochastic fluctuations accounting for the feedback with a surrounding medium would be enough to sustain a dynamical regime (even if stationary) and avoid its full collapse.

We are going to consider here the case of a coherent (or minimum uncertainty) Gaussian wave packet initially centered around the turning point $x = x_0$ ($p_0 = 0$). At any subsequent time, this wave packet is described by

$$\Psi(x,t) = \left(\frac{1}{2\pi\sigma_0^2}\right)^{1/4} e^{-(x-x_t)^2/4\sigma_0^2 + ip_t(x-x_t)/\hbar - i\omega_0 t/2 + ip_t x_t/\hbar}, \tag{2.276}$$

with $x_t = x_0 \cos\omega_0 t$ and $p_t = -m\omega_0 x_0 \sin\omega_0 t$. The corresponding probability density reads as

$$|\Psi(x,t)| = \frac{1}{\sqrt{2\pi\sigma_0^2}}\, e^{-[x - x_0\cos(\omega_0 t)]^2/2\sigma_0^2}, \tag{2.277}$$

where $\sigma_0^2 = \hbar/(2m\omega_0)$ for the wave packet (2.276) to be coherent (otherwise, the wave packet will keep its Gaussian shape and will display an oscillating variation of its width or "breathing" as it moves back and forth between the two turning points). The corresponding quantum action is

$$S(x,t) = -\frac{1}{2}\,\hbar\omega_0 t - \frac{m\omega_0}{4}\left[4xx_0\sin(\omega_0 t) - x_0^2\sin(2\omega_0 t)\right], \tag{2.278}$$

which leads to Bohmian trajectories oscillating with the same frequency as x_t, but around their initial position

$$x(t) = [x(0) - x_0] + x_0\cos(\omega_0 t). \tag{2.279}$$

As also happens in classical mechanics, in order to proceed analytically in the dissipative case, it is important to distinguish three cases depending on whether ω_0 is larger than, equal to or smaller than $\gamma/2$, i.e., if we have underdamped oscillatory motion, critically damped motion, or overdamped motion, respectively. These situations lead to the following solutions for the centroidal trajectory:

$$\omega_0 > \gamma/2 \Longrightarrow \begin{cases} x_t = \left(\dfrac{\omega_0}{\Omega}\right) x_0 e^{-\gamma t/2} \cos(\Omega t - \varphi) \\ p_t = -m \left(\dfrac{\omega_0^2}{\Omega}\right) x_0 e^{-\gamma t/2} \sin \Omega t \end{cases}, \tag{2.280}$$

$$\omega_0 = \gamma/2 \Longrightarrow \begin{cases} x_t = x_0 \left(1 + \dfrac{\gamma}{2} t\right) e^{-\gamma t/2} \\ p_t = -m x_0 \left(\dfrac{\gamma}{2}\right)^2 e^{-\gamma t/2} \end{cases}, \tag{2.281}$$

$$\omega_0 < \gamma/2 \Longrightarrow \begin{cases} x_t = \left(\dfrac{\omega_0}{\Gamma}\right) x_0 e^{-\gamma t/2} \cosh(\Gamma t + \phi) \\ p_t = -m \left(\dfrac{\omega_0^2}{\Gamma}\right) x_0 e^{-\gamma t/2} \sinh \Gamma t \end{cases}, \tag{2.282}$$

where $\Omega = \sqrt{\omega_0^2 - (\gamma/2)^2}$, $\varphi = (\tan)^{-1}(\gamma/2\Omega)$, $\Gamma = i\Omega$, and $\phi = (\tanh)^{-1}(\gamma/2\Gamma)$. The relationship between ω_0 and γ also influences the calculation of α_t and f_t. In the case of α_t, the equation of motion to be solved is

$$\dot{\alpha}_t = -\frac{2\alpha_t^2}{m} e^{-\gamma t} - \frac{1}{2} m\omega_0^2 e^{\gamma t}, \tag{2.283}$$

which can be conveniently rearranged by introducing the change $\alpha_t = g_t e^{\gamma t}$. This leads to the equation of motion

$$\dot{g}_t = -\frac{2}{m} \left[g_t^2 + \frac{m\gamma}{2} g_t + \left(\frac{m\omega_0}{2}\right)^2 \right] = -\frac{2}{m} (g_t - g_+)(g_t - g_-), \tag{2.284}$$

which does not contain exponential terms, and where

$$g_\pm = \frac{m}{2} \left[-\frac{\gamma}{2} \pm \sqrt{\left(\frac{\gamma}{2}\right)^2 - \omega_0^2} \right]. \tag{2.285}$$

From the latter expression, it is now clear how the three cases of damped motion also rule the behavior of α_t, although there are some physical subtleties to take into account. If $\omega_0 \neq \gamma/2$, the general solution for g_t reads as

$$g_t = \frac{g_+(g_0 - g_-)e^{\beta t} - g_-(g_0 - g_+)e^{-\beta t}}{(g_0 - g_-)e^{\beta t} - (g_0 - g_+)e^{-\beta t}}, \tag{2.286}$$

where $\beta = \sqrt{(\gamma/2)^2 - \omega_0^2}$ and g_0 is the initial condition. Otherwise, if $\omega_0 = \gamma/2$ (critically damped motion), we have $g_+ = g_-$ and Eq. (2.284) becomes

$$\dot{g}_t = -\frac{2}{m}(g_t - g_s)^2, \tag{2.287}$$

with $g_s = -m\gamma/4$. The integration over time of this equation of motion yields

$$g_t = g_s + \frac{g_0 - g_s}{1 + (g_0 - g_s)(2t/m)}. \tag{2.288}$$

If we now assume that our initial wave packet is (2.276) and consider the initial condition $g_0 = \alpha_0 = \alpha_t = im\omega_0/2$, then g_t becomes an oscillatory function of time for $\omega_0 > \gamma/2$, since $\beta = i\Omega$. Otherwise, it becomes a monotonically decreasing function of time, with the asymptotic limits $g_\infty = g_s$ if $\omega_0 = \gamma/2$, or $g_\infty = g_+$ (with $\beta = \Gamma$) if $\omega_0 < \gamma/2$.

Notice in Eq. (2.286) that g_t remains constant with time if we assume that g_0 is equal to either g_+ or g_-, which can be inferred either from Eq. (2.286) or also directly from Eq. (2.284) by setting $\dot{g}_t = 0$. This means $\alpha_t = g_+ e^{\gamma t}$, if $g_0 = g_+$, or $\alpha_t = g_- e^{\gamma t}$, if $g_0 = g_-$. In order to choose the appropriate solution, we consider the fact that in the limit $\gamma \to 0$ the chosen solution has to approach the non-dissipative value, i.e., $\alpha_t \to im\omega_0/2$, which only happens if $g_0 = g_+$. Therefore, we have that

$$\alpha_t = \frac{im\Omega}{2}\left(1 + \frac{i\gamma}{2\Omega}\right)e^{\gamma t}. \tag{2.289}$$

It should also be stressed that in the limit of vanishing friction, this value approaches the frictionless one mentioned above. More importantly, α_t has a real and an imaginary part, and therefore the wave function will be properly normalized. This can be readily seen by integrating the equation of motion for f_t, which leads to

$$f_t = \frac{i\hbar}{4}\ln\left(\frac{\pi\hbar}{m\Omega}\right) - \frac{\hbar\Omega}{2}\left(1 + \frac{i\gamma}{2\Omega}\right)t + S_{\mathrm{cl},t}. \tag{2.290}$$

Substituting all these parameters in the expression for the wave function, it reads as

$$\Psi(x,t) = \left(\frac{1}{2\pi\sigma_t^2}\right)^{1/4} e^{-(x-x_t)^2/4\sigma_t^2 + ip_t(x-x_t)/\hbar - i\Omega t/2 - (im\gamma/4\hbar)e^{\gamma t}(x-x_t)^2 + iS_{\mathrm{cl},t}/\hbar}, \tag{2.291}$$

with the width being $\sigma_t = e^{-\gamma t/2}\sqrt{\hbar/2m\Omega}$. As can be noticed, the first three arguments in the exponential of Eq. (2.291) are identical to those in Eq. (2.276), but replacing ω_0 by Ω. The wave function is properly normalized, but its width decreases exponentially with time according to

$$\Delta x = \sigma_t e^{-2\gamma t}. \tag{2.292}$$

Thus, it eventually collapses over the center of the harmonic well as the whole wave packet follows the motion of a damped harmonic oscillator. Along this transit, the system energy is also exponentially lost according to

$$\bar{E} = \frac{1}{2} m\omega_0^2 x_0^2 \left(\frac{\omega_0}{\Omega}\right)^2 \left[1 - \frac{\gamma}{2\omega_0} \sin(2\Omega t - \varphi)\right] e^{-\gamma t} + \frac{1}{2} \omega_0 \hbar \left(\frac{\omega_0}{\Omega}\right) e^{-\gamma t}. \tag{2.293}$$

It is worth stressing that under underdamped conditions, α_t, given by Eq. (2.289), is a complex function, which makes the wave function (2.291) to display a vanishing Gaussian shape plus a phase factor as the system oscillates, as seen in Eq. (2.291). Critically damped and overdamped conditions imply that the motion amplitude of the system exhibits a monotonic decrease to zero, with no oscillations. In the present context, where we are seeking for solutions such that $\dot{g}_t = 0$, this now translates into a rather puzzling quantum behavior. In order to analyze it, let us start from the overdamped condition, $\omega_0 < \gamma/2$. Equation (2.286) is still valid, although we have $g_\pm = \pm(m\Gamma/2)(1 \mp \gamma/2\Gamma)$. Between these two solutions, we choose g_+, because it is consistent with Eq. (2.289) and also because it corresponds to the long-time limit of g_t in this case (i.e., for $\beta = \Gamma$). Accordingly, the spreading factor will read as

$$\alpha_t = \frac{m\Gamma}{2}\left(1 - \frac{\gamma}{2\Gamma}\right) e^{\gamma t}, \tag{2.294}$$

which is a pure real function. Thus,

$$f_t = f_0 + \frac{i\hbar\Gamma}{2}\left(1 - \frac{\gamma}{2\Gamma}\right) t + \mathcal{S}_{\mathrm{cl},t}, \tag{2.295}$$

where f_0 is now left as a free parameter, although its imaginary part has to be such that the initial plane wave is still normalized. Notice that in this case the corresponding wave function is written now as

$$\Psi(x,t) = e^{-(im\Gamma/2\hbar)(1+\gamma/2\Gamma)e^{\gamma t}(x-x_t)^2 + ip_t(x-x_t)/\hbar + \Gamma t(1+\gamma/2\Gamma)/2 + if_0 + i\mathcal{S}_{\mathrm{cl},t}/\hbar}, \tag{2.296}$$

which is a pure phase factor multiplied by a diverging time-dependent exponential. It is interesting to emphasize that, although the wave function itself diverges, the associated Bohmian trajectories are well-defined and approach asymptotically the classical overdamped centroid, x_t (see below).

Regarding the critically damped motion, if we consider here the initial value $g_0 = g_s$, the stationary solution is $g_t = g_s = -m\gamma/4$, and therefore

$$\alpha_t = -\frac{m\gamma}{4} e^{\gamma t}. \tag{2.297}$$

As it can be seen, this value of α_t allows us to connect smoothly those two previously obtained in underdamped and overdamped cases as the friction γ is gradually increased.

In spite of the type of motion displayed in each one of the regimes discussed above, the Bohmian equation of motion for the trajectories can be recast in the same form for all of them,

$$\frac{d(x-x_t)}{x-x_t} = -\frac{\bar{\gamma}}{2}\,dt, \tag{2.298}$$

where $\bar{\gamma} = \gamma$ for underdamped and critically damped motions, and $\bar{\gamma} = \gamma - 2\Gamma$ for overdamped motions. As it can be noticed, this equation looks exactly the same as Eq. (2.274) for the eigenstates, although replacing x by $x - x_t$. Thus, the solutions,

$$x(t) = x_t + [x(0) - x_0]e^{-\bar{\gamma}t/2}, \tag{2.299}$$

are also very similar. In the case of Eq. (2.275), since the associated wave function is a quasi-eigenstate, the trajectory approaches asymptotically the origin or, in other words, it falls down to the bottom of the well. In the case of Eq. (2.299), and differently to what we have seen before, any trajectory will eventually coalesce with the centroidal one. This is an interesting case where the Bohmian non-crossing rule seems to be violated. Actually, the evolution of the trajectories is in compliance with the shrinking undergone by the wave function in the damped oscillatory regime and its evanescent nature in the overdamped regime. These trajectories, described generically by Eq. (2.299), have the following analytical form

$$x(t) = \begin{cases} \left\{[x(0) - x_0] + x_0\left(\frac{\omega_0}{\Omega}\right)\cos(\Omega t - \varphi)\right\}e^{-\gamma t/2}, & \omega_0 > \gamma/2 \\ \left\{[x(0) - x_0] + x_0\left(1 + \frac{\gamma t}{2}\right)\right\}e^{-\gamma t/2}, & \omega_0 = \gamma/2 \\ \left\{[x(0) - x_0]e^{\Gamma t} + x_0\left(\frac{\omega_0}{\Gamma}\right)\cosh(\Gamma t + \phi)\right\}e^{-(\gamma-2\Gamma)t/2}, & \omega_0 < \gamma/2 \end{cases}, \tag{2.300}$$

where the first expression looks very similar to that for the unperturbed harmonic oscillator, with the exception of the overall exponentially decaying factor and the dephasing in the cosine function.

References

1. Bernstein, J.: More about Bohm's quantum. Am. J. Phys. **79**, 601–606 (2011)
2. Holland, P.R.: The Quantum Theory of Motion. Cambridge University Press, Cambridge (1993)
3. Goldstein, S.: Quantum theory without observers - Part I. Phys. Today, **51**(3), 42–46 (1998); Part II. Phys. Today **51**(4), 38–42 (1998)

4. Wyatt, R.E.: Quantum Dynamics with Trajectories: Introduction to Quantum Hydrodynamics. Springer, New York (2005)
5. Sanz, A.S., Miret-Artés, S.: A trajectory description of quantum processes. Part I. Fundamentals. Lect. Notes Phys. **850**, 1–299 (2012)
6. Sanz, A.S., Miret-Artés, S.: A trajectory description of quantum processes. Part II. Applications. Lect. Notes Phys. **831**, 1–333 (2014)
7. Nassar, A.B.: Time-dependent harmonic oscillator: An Ermakov-Nelson process. Phys. Rev. A **32**, 1862–1863 (1985) (See references there in)
8. Lewis, H.R.: Classical and quantum systems with time-dependent harmonic-oscillator-type Hamiltonians. Phys. Rev. Lett. **18**, 510–512 (1967)
9. Ray, J.R.: Quantum invariants. Phys. Rev. **28**, 2603–2605 (1983)
10. Ermakov, V.P.: Second-order differential equations. Conditions of complete integrability, Univ. Izv. Kiev Ser. III **9** 1–25 (1880)
11. Milne, W.E.: The numerical determination of characteristic numbers. Phys. Rev. **35**, 863–867 (1930)
12. Pinney, E.: The nonlinear differential equation $y'' + p(x)y + cy^{-3} = 0$. Proc. Am. Math. Soc. **1**, 681–681 (1950)
13. Nassar, A.B.: New quantum squeezed states for the time-dependent harmonic oscillator. J. Opt. B: Quantum Semiclassical Opt. **3**, S226–S228 (2002) (See references therein)
14. Davydov, A.S.: Quantum Mechanics, 2nd edn. Pergamon Press, Oxford (1965)
15. Merzbacher, E.: Quantum Mechanics, 2nd edn. Wiley, New York (1970)
16. Schiff, L.I.: Quantum Mechanics, 3rd edn. McGraw-Hill, New York (1970)
17. French, A.P., Taylor, E.F.: Quantum Mechanics. MIT Series. W. W. Norton, New York (1978)
18. Nassar, A.B.: Boundary conditions in tunneling via quantum hydrodynamics. NASA Conf. Publ. **3197**, 149–154 (1993)
19. Nassar, A.B.: Quantum traversal time. Phys. Rev. A **38**, 683–687 (1988)
20. Koonin, S.E.: Computational Physics. Benjamin/Cummings Publication, New York (1985) (See in part B, example 7)
21. Goldberg, A., Schey, H.M., Schwartz, J.L.: Computer-generated motion pictures of one-dimensional quantum-mechanical transmission and reflection phenomena. Am. J. Phys. **35**, 177–186 (1967)
22. Huang, Z.H., Cutler, P.H., Feuchtwang, T.E., Good Jr., R.H., Kazes, E., Nguyen, H.Q., Park, S.K.: Computer simulation of a wave packet tunneling through a square barrier. IEEE **36**, 2665–2670 (1989)
23. https://phet.colorado.edu/en/simulation/quantum-tunneling
24. Hartman, T.E.: Tunneling of a wave packet. J. Appl. Phys. **33**, 3427–3433 (1962)
25. Winful, H.G.: Delay time and the Hartman effect in quantum tunneling. Phys. Rev. Lett. **91**, 260401(1–4) (2003)
26. Nassar, A.B.: Scattering via invariants of quantum hydrodynamics. Phys. Lett. **146**, 89–92 (1990)
27. Reid, J.R., Ray, J.R.: Ermakov systems, nonlinear superposition, and solutions of nonlinear equations of motion. J. Math. Phys. **21**, 1583–1587 (1980)
28. Lutzky, M.: Noether's theorem and the time-dependent harmonic oscillator. Phys. Lett. A **68**, 3–4 (1978)
29. Büttiker, M.: Larmor precession and the traversal time for tunneling. Phys. Rev. **27**, 6178–6188 (1983)
30. Nassar, A.B., Bassalo, J.M.F., Alencar, P.T.S., Cancela, L.S.G.: Wave propagator via quantum fluid dynamics. Phys. Rev. **56**, 1230–1233 (1997)
31. Feynman, R.P., Hibbs, A.R.: Quantum Mechanics and Path Integrals. McGraw-Hill, New York (1965)
32. Berry, M.V., Balazs, N.L.: Nonspreading wave packets. Am. J. Phys. **47**, 264–267 (1979)
33. Vallée, O., Soares, M.: Airy Function and Applications to Physics. World Scientific, Hackensack (2004)

34. Siviloglou,G.A., Broky, J., Dogariu, A., Christodoulides, D.N.: Observation of accelerating Airy beams. Phys. Rev. Lett. **99**, 213901(1–4) (2007) (See references therein)
35. Stützle, R., Gø:bel, M.C., Hø:rner, Th., Kierig, E., Mourachko, I., Oberthaler, M.K., Fedorov, M.V., Yakovlev, V.P., van Leeuwen, K.A.H., Schleich, W.P.: Observation of nonspreading wave packets in an imaginary potential. Phys. Rev. Lett. **95**, 110405(1–4) (2005)
36. Gutiérrez-Vega, J.C., Iturbe-Castillo, M.D., Chávez-Cerda, S.: Alternative formulation for invariant optical fields: Mathieu beams. Opt. Lett. **25**, 1493–1495 (2000)
37. Bandres, M.A., Gutiérrez-Vega, J.C., Chavez-Cerda, S.: Parabolic nondiffracting optical wave fields. Opt. Lett. **29**, 44–46 (2004)
38. Durnin, J.: Exact solutions for nondiffracting beams. I. The scalar theory. J. Opt. Soc. Am. A **4**, 651–654 (1987)
39. Durnin, J., Miceli, J.J., Eberly, J.H.: Diffraction-free beams. Phys. Rev. Lett. **58**, 1499–1501 (1987)
40. Durnin, J., Miceli, J.J., Eberly, J.H.: Comment on Bessel and Gaussian beams. Phys. Rev. Lett. **66**, 838–838 (1991)
41. Besieris, I.M., Shaarawi, A.M.: Accelerating Airy wave packets in the presence of quadratic and cubic dispersion. Phys. Rev. E **78**, 046605(1–6) (2008)
42. Besieris, I.M., Shaarawi, A.M., Ziolkowski, R.W.: Nondispersive accelerating wave packets. Am. J. Phys. **62**, 519–521 (1994)
43. Unnikrishnan, K., Rau, A.R.P.: Uniqueness of the Airy packet in quantum mechanics. Am. J. Phys. **64**, 1034–1036 (1996)
44. Greenberger, D.M.: Nonspreading wave packets. Am. J. Phys. **48**, 256–256 (1980)
45. Nassar, A.B., Bassalo, J.F., Alencar, P.T.S.: Dispersive Airy packets. Am. J. Phys. **63**, 849–852 (1995)
46. Ray, J.R.: Exact solutions to the time-dependent Schrödinger equation. Phys. Rev. A **26**, 729–733 (1982)
47. Burgan, J.R., Feix, M.R., Fijalkow, E., Munier, A.: Solution of the multidimensional quantum harmonic oscillator with time-dependent frequencies through Fourier, Hermite and Wigner transforms. Phys. Lett. A **74**, 11–14 (1979)
48. These space-time transformations used here are a generalization of those found in references [45, 46]. However, Equations (2.166) and (2.167) have clear physical meaning; they are essential in determining the trajectories of the Airy packet
49. Ruby, L.: Applications of the Mathieu equation. Am. J. Phys. **64**, 39–44 (1996)
50. Nassar, A.B., Machado, F.L.A.: Solvable Hill and Pø:schl-Teller equations. Phys. Rev. A **35**, 3159–3160 (1987)
51. Bessel, F.W.: Versuche ber die Kraft, mit welcher die Erde Krper von verschiedener Beschaffenheit anzieht. Ann. Phys. **101**, 401–417 (1832)
52. Eötvös, R.V., Pekár, D., Fekete, E.: Beitrge zum Gesetze der Proportionalitt von Trgheit und Gravitt. Ann. Phys. **373**, 11–66 (1922)
53. Roll, P.G., Krotkov, R., Dicke, R.H.: The equivalence of inertial and passive gravitational mass. Ann. Phys. (N.Y.) **26**, 442–517 (1964)
54. Kajari, E., Harshman, N.L., Rasel, E.M., Stenholm, S., Süssmann, G., Schleich, W.P.: Inertial and gravitational mass in quantum mechanics. Appl. Phys. B **100**, 4360 (2010) (See references therein)
55. Colella, R., Overhauser, A.W., Werner, S.A.: Observation of gravitationally induced quantum interference. Phys. Rev. Lett. **34**, 1472–1474 (1975)
56. Greenberger, D.M., Overhauser, A.W.: Coherence effects in neutron diffraction and gravity experiments. Rev. Mod. Phys. **51**, 43–78 (1979)
57. Bonse, U., Wroblewski, T.: Measurement of neutron quantum interference in noninertial frames. Phys. Rev. Lett. **51**, 1401–1404 (1983)
58. Peters, A., Chung, K.Y., Chu, S.: Measurement of gravitational acceleration by dropping atoms. Nature **400**, 849–852 (1999)
59. Müller, H., Peters, A., Chu, S.: A precision measurement of the gravitational redshift by the interference of matter waves. Nature **463**, 926–930 (2010)

60. Lämmerzahl, C.: On the equivalence principle in quantum theory. Gen. Relativ. Gravit. **28**, 1043–1070 (1996)
61. Viola, L., Onofrio, R.: Testing the equivalence principle through freely falling quantum objects. Phys. Rev. D **55**, 455–462 (1997)
62. Hughes, K.J., Burke, J.H.T., Sackett, C.A.: Suspension of atoms using optical pulses, and application to gravimetry. Phys. Rev. Lett. **102**, 150403(1–4) (2009)
63. Fray, S., Alvarez Diez, C., Hüsch, T.W., Weitz, M.: Atomic interferometer with amplitude gratings of light and its applications to atom based tests of the equivalence principle. Phys. Rev. Lett. **93**, 240404(1–4) (2004)
64. Kasevich, M., Chu, S.: Atomic interferometry using stimulated Raman transitions. Phys. Rev. Lett. **67**, 181–184 (1991)
65. Aminoff, C.G., Steane, A.M., Bouyer, P., Desbiolles, P., Dalibard, J., Cohen-Tannoudji, C.: Cesium atoms bouncing in a stable gravitational cavity. Phys. Rev. Lett. **71**, 3083–3086 (1993)
66. Ovchinnikov, Yu.B., Manek, I., Grimm, R.: Surface trap for Cs atoms based on evanescent-wave cooling. Phys. Rev. Lett. **79**, 2225–2228 (1997)
67. Wallis, H., Dalibard, J., Cohen-Tannoudji, C.: Sisyphus cooling of a bound atom. Appl. Phys. B **54**, 407–419 (1992)
68. Shapere, A., Wilczek, F. (eds.): Geometric Phases in Physics. World Scientific, Singapore (1989)
69. Markowski, B., Vinitskii, S.I.: Topological Phases in Quantum Theory. World Scientific, Singapore (1989)
70. Bohm, A., Mostafazadeh, A., Koizumi, H., Niu, Q., Zwanziger, J.: The Geometric Phase in Quantum Systems. Springer, Heidelberg (2003)
71. Chruscinski, D., Jamiolkpwski, A.: Geometric Phases in Classical and Quantum Mechanics. Birkhäuser, Boston (2004)
72. Berry, M.V.: Quantal phase factors accompanying adiabatic changes. Proc. Roy. Soc. Lond. A **392**, 45–57 (1984)
73. Aharonov, Y., Anandan, A.: Phase change during a cyclic quantum evolution. Phys. Rev. Lett. **58**, 1593–1596 (1987)
74. Samuel, J., Bhandari, R.: General setting for Berry's phase. Phys. Rev. Lett. **60**, 2339–2342 (1988)
75. Fock, V.: ber die Beziehung zwischen den Integralen der quantenmechanischen Bewegungsgleichungen und der Schrdingerschen Wellengleichung. Z. Phys. **49**, 323–338 (1928)
76. Bortolotti, F.: Sulle rappresentazioni conformi, e su di una interpretazione fisica del parallelismo di Levi-Civita. Rend. R. Naz. Lincei **IV**, 552–556 (1926)
77. Rytov, S.M.: On the transition from wave to geometric optics. Dokl. Akad. Nauk. USSR **18**, 263–266 (1938)
78. Pancharatnam, S.: Generalized theory of interference and its applications. Proc. Ind. Acad. Sci. Ser. A **44**, 247–262 (1956)
79. Longuet-Higgins, H.C., Öpik, U., Pryce, M.H.L., Sack, R.A.: Studies of the Jahn-Teller effect. II. The dynamical problem. Proc. Roy. Soc. Lond. A **244**, 1–16 (1958)
80. Aharonov, Y., Bohm, D.: Significance of electromagnetic potentials in the quantum theory. Phys. Rev. **115**, 485–491 (1959)
81. Mead, C.A., Truhlar, D.G.: On the determination of Born-Oppenheimer nuclear motion wave functions including complications due to conical intersections and identical nuclei. J. Chem. Phys. **70**, 2284–2296 (1979)
82. Simon, B.: Holonomy, the quantum adiabatic theorem, and Berry's phase. Phys. Rev. Lett. **51**, 2167–2170 (1983)
83. Mukunda, N., Simon, R.: Quantum kinematic approach to the geometric phase I. General formalism. Ann. Phys. (NY) **228**, 205–268 (1993)
84. Sjöqvist, E., Carlsen, H.: Geometric phase, quantum measurements, and the de Broglie-Bohm model. Phys. Rev. A **56**, 1638–1641 (1997)
85. Parmenter, R.H., Valentine, R.W.: Properties of the geometric phase of a de Broglie-Bohm causal quantum mechanical trajectory. Phys. Lett. A **219**, 7–14 (1996)

86. Chou, C.-C., Wyatt, R.E.: Geometric phase in Bohmian mechanics. Ann. Phys. (NY) **325**, 2234–2250 (2010)
87. Bohm, D., Hiley, B.J.: The Undivided Universe. Routledge, London (1993)
88. Olariu, S., Popescu, I.I.: The quantum effects of electromagnetic fluxes. Rev. Mod. Phys. **57**, 339–436 (1985)
89. Aharonov, Y., Casher, A.: Topological quantum effects for neutral particles. Phys. Rev. Lett. **53**, 319–321 (1984)
90. Cimmino, A., Opat, G.I., Klein, A.G., Kaiser, H., Werner, S.A., Arif, M., Clothier, R.: Observation of the topological Aharonov-Casher phase shift by neutron interferometry. Phys. Rev. Lett. **63**, 380–383 (1989)
91. Elion, W.J., Wachters, J.J., Sohn, L.L., Mooij, J.D.: Observation of the Aharonov-Casher effect for vortices in Josephson-junction arrays. Phys. Rev. Lett. **71**, 2311–2314 (1993)
92. Sangster, K., Hinds, E.A., Barnett, S.M., Riis, E.: Measurement of the Aharonov-Casher phase in an atomic system. Phys. Rev. Lett. **71**, 3641–3644 (1993)
93. Konig, M., Tschetschetkin, A., Hankiewicz, E.M., Sinova, J., Hock, V., Daumer, V., Schäfer, M., Becker, C.R., Buhmann, H., Molenkamp, L.W.: Direct observation of the Aharonov-Casher phase. Phys. Rev. Lett. **96**, 076804(1–4) (2006)
94. Pitaesvskii, L.P.: Vortex lines in an imperfect bose gas. Soviet Phys. JETP **13**, 451–454 (1961)
95. Gross, E.P.: Structure of a quantized vortex in boson systems. Nuovo Cimento **20**, 454–457 (1961)
96. Landau, L.D., Lifshitz, E.M., Pitaevskii, L.P.: Statistical Physics: Theory of the Condensed State. Course of Theoretical Physics. Butterworth-Heinemann, Oxford (1980)
97. Nassar, A.B., Bassalo, J.M.F., Alencar, P.T.S., de Souza, J.F., de Oliveira, J.E., Cattani, M.: Gaussian solitons in nonlinear Schrödinger equation. Il Nuovo Cimento **117**, 941–946 (2002)
98. Goldstein, H.: Classical Mechanics. Addison-Wesley Publishing Company, Reading (1980)
99. Razavy, M.: Classical and Quantum Dissipative Systems. Imperial College Press, London (2005)
100. Caldirola, P.: Forze non conservative nella meccanica quantisitica. Nuovo Cimento **18**, 393–400 (1941)
101. Kanai, E.: On the quantization of the dissipative systems. Prog. Theor. Phys. **3**, 440–442 (1948)
102. Sanz, A.S., Martinez-Casado, R., Peñate-Rodriguez, H.C., Rojas-Lorenzo, G., Miret-Artés, S.: Dissipative Bohmian mechanics within the Caldirola-Kanai framework: a trajectory analysis of wave-packet dynamics in viscid media. Ann. Phys. **347**, 1–20 (2014)
103. Sanz, A.S., Miret–Artés, S.: A trajectory-based understanding of quantum interference. J. Phys. A **41**, 435303(1–23) (2008)
104. Sanz, A.S., Miret-Artés, S.: Setting-up tunneling conditions by means of Bohmian mechanics. J. Phys. A: Math. Theor. **44**, 485301(1–17) (2011)
105. Heller, E.J.: Time-dependent approach to semiclassical dynamics. J. Chem. Phys. **62**, 1544–1555 (1975)
106. Vandyck, M.A.: On the damped harmonic oscillator in the de Broglie-Bohm hidden-variable theory. J. Phys. A: Math. Gen. **27**, 1743–1750 (1994)

Chapter 3
Bohmian Stochastic Trajectories

Abstract In the last decades, many nonlinear extensions of the Schrödinger equation have been proposed in literature either to explore the fundamental aspects of quantum mechanics, with the usual linear theory representing only a limiting case, or to describe open quantum systems. For the description of nonconservative quantum systems, Kostin formulated in an heuristic way the so-called Schrödinger–Langevin (SL) equation or Kostin equation, for the Brownian motion. This equation has been subsequently rederived, improved and extended for its use in numerous applications, mainly without including the noise term. Numerous features of this SL equation can be better revealed within the framework of de Broglie–Bohm (quantum hydrodynamical trajectory formulation) of quantum mechanics. Within this formalism, several dissipative problems are presented and discussed: the Ramsauer–Townsend effect, the tunneling dynamics through a barrier, the plasma fluid formulation and the Lorentz–Abraham (extended electron) equation for a point-charge electron. These two last examples are also discussed in order to see the correspondence between classical and quantum dynamics. Very few applications of this SL equation are devoted to stochastic problems in the literature where the noise term needs to be included. The so-called Bohmian–Brownian motion is introduced in the context of surface diffusion with single adsorbates. An extension to interacting adsorbates is discussed within a simple, phenomenological model. Interestingly enough, this study leads us to quantum anomalous diffusion. The harmonic motion is also briefly considered in order to compare with the same open dynamics under the presence of a continuous quantum measurement (Chap. 4). Finally, a generalization of the SL equation is proposed for nonlinear dissipation.

3.1 Introduction

The study of open quantum systems constitutes a very broad and active field of research within quantum mechanics. As it is well known, strictly speaking, real physical systems do not exist in complete isolation in Nature. All physical systems are open systems since the interaction with their environment can never be totally neglected. The energy transfer from the system to the environment is termed quantum

A.B. Nassar and S. Miret-Artés, *Bohmian Mechanics, Open Quantum Systems and Continuous Measurements*, DOI 10.1007/978-3-319-53653-8_3

relaxation or *damping*. If there is no chance for the energy to move backwards into the system, the unidirectional energy flow into the reservoir is then called quantum *dissipation*. On short timescales, the distinction between quantum relaxation and dissipation is obviously unclear. Only when the environment has a small number of degrees of freedom, the energy moves backwards into the system; this phenomenon is called a *recurrence* This interaction also gives rise to a strong quantum correlation or *entanglement* between the system and environment which eventually leads the former to become a statistical mixture (decoherence). The so-called theory of open quantum systems [1–5] encompasses a series of formalisms and approaches developed to deal with the corresponding quantum dynamics. This line of research can be considered as an interdisciplinary field, where very broad branches of physics, chemistry and biology meet together to describe processes that are ubiquitous in nature [1–3, 6–8]. Nevertheless, finding good quantum analogues of classical dissipative systems constitutes a very difficult task and it still remains an important topic because of some problems such as, for example, commutation rules, time ordering and symmetrization can not be neglected in the quantum world.

Irreversible quantum processes take place when dealing with extended systems as environments or, more precisely, with systems displaying an infinite number of degrees of freedom (photon or phonon fields, for example). Quantum noise has its origin in our inability to specify each of all the infinite modes of a given field when considering an environment around a system; noise appears when the reaction of the field back on the system is not neglected. Under certain conditions the duration of the reservoir correlations is very short compared to the dynamical evolution of the system. This leads to a total memory loss of the bath dynamics that gives rise to a subsequent irreversible loss of coherence and energy (or population) relaxation in the system. This is called a *Markovian regime*. Within this regime, the time-evolution of the system does only depend on the present state of the system; this is called a *Markovian process*. As will be seen, when this happens, the system dynamics can be characterized by (relatively) simple Markovian master equations, where one does not need to take into account the reservoir dynamics and its effects on the system are described by means of certain operators. Quantum stochastic methods in the Markovian and non-Markovian regimes have been widely developed in quantum optics [7]. In particular, the non-Markovian dynamics of open quantum systems is being developed very fast in the last years. It has been finally established that there is no quantum Onsager regression theorem and the correct generalization of the Onsager hypothesis is the fluctuation-dissipation theorem [9].

In analogy to open classical systems, three main different approaches to deal with quantum dissipative/stochastic dynamics can also be considered here. First, effective

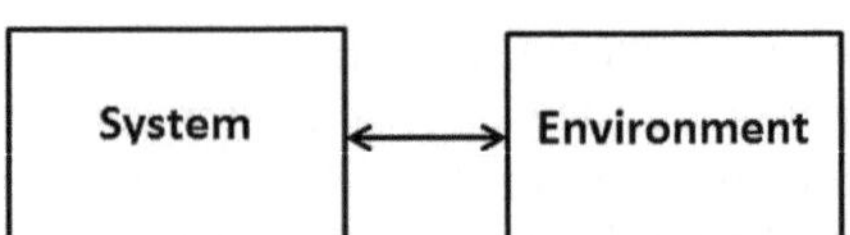

Fig. 3.1 Schematic view of the coupling of a system surrounded by its environment

time-dependent Hamiltonians. The Caldirola–Kanai model is a good example which has been developed shortly in Chap. 2. Second, the system–plus–environment model Hamiltonian where a schematic view is displayed in Fig. 3.1. The whole system (physical system and environment or thermal bath) is considered as an isolated system [2, 3]. Both system and reservoir are in continuous interaction and the effects —quantum coherence loss or decoherence, population transfer, and/or (system–environment) energy exchange— arising from that interaction will depend to a greater or a lesser extent on the coupling strength and its intrinsic nature. Under these conditions, one speaks about the environment induced decoherence. An interesting and very important process is vibrational dephasing of small molecules immersed in a rare gas in liquid phase [10–15]. Fluctuations occur because of random collision events. Due to its interaction with the environment, the system usually behaves quite different with respect to its behavior in isolation. Its time-evolution is not unitary and therefore cannot be described in terms of the Schrödinger equation. However, the powerful standard quantum-mechanical theoretical formalisms developed are at our complete disposal. Dissipation can be described following one of the three standard pictures of quantum mechanics: Schrödinger, Heisenberg, Dirac (interaction). In the first and third pictures, one always tries to find master equations that account for the time-evolution of the so-called reduced system, where the bath degrees of freedom have been traced out. As has been mentioned previously, among this type of equations of motion, the simplest class is that of Markovian character where one assumes that the bath has no memory and the time-evolution of the reduced density matrix depends only on its present time. The Lindblad equation plays a fundamental role in this context. On the contrary, non-Markovian approaches lead to a loss of analytical and conceptual simplicity as well as larger computational times. An alternative way to the density matrix formalism also comes from the path integral formulation of quantum mechanics (fourth picture of quantum mechanics). Many efforts are being addressed along this way with also great success. If the Heisenberg picture is followed, the quantum (generalized or standard) Langevin equation is reached (needless to say that all formulations are equivalent). Essentially, this procedure consists of replacing the reservoir by damping terms in the Heisenberg equations of motion of a conservative system and then adding random forces as driving terms that give rise to fluctuations over the system. These stochastic quantum formalisms are being addressed to more and more complex systems. Finally, if the wave function or vector state is considered as a stochastic process in Hilbert space, the corresponding formulation is given in terms of a time-dependent density functional, the dynamics being described no longer by a master equation but by the Itô stochastic differential equation. The realizations of the underlying stochastic processes are also called quantum trajectories, which have not to be confused with the same denomination used in Bohmian mechanics [4, 5]. And third, nonlinear Hamiltonians. Nonlinear wave mechanics has been sought since the inception of the original linear Schrödinger equation. It constitutes one among the most investigated topics in quantum mechanics. These generalizations emerge from the study of important phenomena occurring in condensed matter physics, for instance, in the propagation of light pulse in an optical fiber and propagation in a nonlinear birefringent medium. Among the many nonlinear variants of Schrödinger equations proposed in literature [16–25], some of them are applied in order to save the superposition principle of the linear quantum mechanics, or propose the introduction of homogeneous nonlinear terms in order to save, only partially, the superposition principle. The problem of experimental verification has also been studied [26, 27] – and the parameter that defines the coupling constant for nonlinear perturbations of the Schrödinger equation has been estimated. The main problem in applications of results obtained for evaluating the coupling constant is that research in the conventional nonlinear wave mechanics has been mainly concentrated on the equation with log-nonlinearity. As we show in what follows, the conceptual understanding of these equations can be better suited in the framework of Bohmian mechanics and some correspondence between classical and quantum quantities can be carried out [21–24]. The Feynman propagator for eight nonlinear Schrödinger equations within this context have recently been analyzed [28]. Many nonlinear equations have soliton solutions characterized by widths which do not spread in time; in particular, as has been mentioned in Chap. 2, when dealing with the Gross–Pitaevskii equation (see Sect. 2.11). Optical solitons are thought to be very important in optical transmission systems since the soliton pulse can maintain its wave form over long distances. Finally, we should also mention that when considering the measuring process (see Chap. 4), nonlinear equations also play an important role. In any case, without and with the presence of a measurement, the perspective of an open dynamics has to be kept in mind. However, the causal stochastic approach described in Chap. 1, Sect. 1.3.1, is not going to be followed from now on.

In this Chapter, dissipation and stochasticity are discussed within the fifth picture of quantum mechanics due to de Broglie and Bohm (as mentioned in the Preface). Section 3.2 is devoted to the introduction of the so-called nonlinear Schrödinger–Langevin (SL) or Kostin equation for a Brownian particle. This equation has been subsequently rederived, improved and extended for its use in numerous applications, mainly without including the noise term. In Sect. 3.3, this SL equation is derived within the Bohmian formalism, the SLB equation. In this hydrodynamical framework, a Gaussian probability density is considered for potentials up to second order in a Taylor expansion. In particular, for a harmonic potential, the so-called Pinney equation but with an additional friction term is finally obtained for its time spreading. In Sect. 3.4, the Ramsauer–Townsend effect in a dissipative medium is studied and analyzed in very detail by means of the SL and SLB equations. This effect tells us that under certain incident conditions for energy and a given width of the well depth, the reflection coefficient is totally suppressed. Section 3.5 presents the tunneling dynamics through a dissipative barrier. In the next two sections, Sects. 3.6 and 3.7, the plasma fluid formulation and the Lorentz–Abraham equation for a point-charge electron within Bohmian mechanics are also discussed in order see the correspondence between classical and quantum dynamics. Very few applications of this SL equation are devoted to stochastic problems in the literature where the noise term needs to be included. In Sect. 3.8, the so-called Bohmian–Brownian motion is introduced in the context of surface diffusion with single adsorbates. An extension to interacting adsorbates is discussed within a simple, phenomenological model. Interestingly enough, this study leads us to quantum anomalous diffusion. The harmonic motion is also considered in order to compare with the same open dynamics under the presence of a continuous quantum measurement (Chap. 4). Finally, Sect. 3.9 is devoted to present and discuss a generalization of the SL equation for nonlinear dissipation.

3.2 The Schrödinger–Langevin Equation

In the sixties, Senitzky [29] and Ford et al. [30] showed that a system of coupled harmonic oscillators could model a heat bath. The corresponding Brownian motion was then studied both classically and quantum-mechanically. The corresponding Langevin equation was obtained as a result of reducing the dimensionality of the full problem by tracing out over the bath variables. The heat bath was thus only described by two parameters: the friction coefficient and the temperature through a random force. This random force was shown to be a Gaussian stochastic process. The variables in the quantum Langevin equation were assumed to be operators with order preserved according to the Heisenberg picture. Different orderings are available, such as the symmetric rule, Weyl's rule, etc., in order to construct Hermitian operators. The normal product of the random force was shown to be a Gaussian process, but not Markovian. This point is critical when compared to the classical counterpart. For an Ohmic or constant friction coefficient, the classical Gaussian process is also Markovian. The denomination of Ohmic friction comes from the fact that for a constant field of force F, the average value of the momentum reads like Ohm's law from the Langevin equation, i.e., $\langle p\rangle = \lambda^{-1}F$, where the current described by $\langle p\rangle$ is proportional to the field applied, F, and proportionally inverse to the resistance or friction coefficient.

Kostin [16, 17] derived heuristically the so-called Schrödinger–Langevin (SL) equation for a Brownian particle of mass m subject to an external potential V and interacting with a thermal bath from the corresponding quantum Langevin equation written as

$$m\ddot{\mathbf{r}} = -\gamma\dot{\mathbf{r}} + \mathbf{F}_r(t) + \nabla V(\mathbf{r}) \tag{3.1}$$

where the time random force $\mathbf{F}_r$ is assumed to arise from a random potential linearly dependent on the particle position, V_r. In three dimensions, this equation reads as

$$\begin{aligned} i\hbar\,\frac{\partial\Psi(\mathbf{r},t)}{\partial t} &= -\frac{\hbar^2}{2m}\nabla^2\Psi(\mathbf{r},t) \\ &\quad + [V(\mathbf{r}) + V_r(\mathbf{r},t) + V_D(\mathbf{r},t) + G(t)]\,\Psi(\mathbf{r},t), \end{aligned} \tag{3.2}$$

where γ is the friction coefficient. The random potential is given by

$$V_r(\mathbf{r}, t) = -\mathbf{r}.\mathbf{F}_r(t), \tag{3.3}$$

the so-called damping potential by

$$V_D(\mathbf{r}, t) = -\frac{i\hbar\gamma}{2} \ln\left(\frac{\Psi(\mathbf{r}, t)}{\Psi^*(\mathbf{r}, t)}\right) \tag{3.4}$$

and

$$G(t) = \frac{i\hbar\gamma}{2} \int \Psi^*(\mathbf{r}, t) \ln\left(\frac{\Psi(\mathbf{r}, t)}{\Psi^*(\mathbf{r}, t)}\right) \Psi(\mathbf{r}, t) d\mathbf{r} \tag{3.5}$$

is a time dependent function resulting from the average value of V_D by integration with respect to the position variable. It is easy to show that the norm of the corresponding wave function is conserved and the expectation value of the corresponding nonlinear Hamiltonian is, as usual, the sum of the kinetic and potential energies at any time. The rate of change in energy along time is given by $-\gamma\langle p^2\rangle/m$. This $G(t)$ function can be removed from Eq. (3.2) by introducing the transformation

$$\Psi(\mathbf{r}, t) = e^{if(t)}\phi(\mathbf{r}, t) \tag{3.6}$$

leading to

$$\begin{aligned} i\hbar \frac{\partial\phi(\mathbf{r}, t)}{\partial t} &= -\frac{\hbar^2}{2m}\nabla^2\phi(\mathbf{r}, t) \\ &+ \left[V(\mathbf{r}) + V_r(\mathbf{r}, t) + \frac{i\hbar\gamma}{2} \ln\left(\frac{\phi(\mathbf{r}, t)}{\phi^*(\mathbf{r}, t)}\right)\right]\phi(\mathbf{r}, t), \end{aligned} \tag{3.7}$$

the $f(t)$ function being given by

$$f(t) = -\hbar^{-1}e^{-\gamma t} \int_0^t e^{\gamma s} G(s) ds. \tag{3.8}$$

The motion of a Brownian particle is strictly described by the nonlinear equation (3.7) when $V = 0$. It is easy to show that with the new transformation

$$\phi(\mathbf{r}, t) = e^{iS(\mathbf{r},t)/\hbar} \tag{3.9}$$

and looking for solutions of the type given by

$$S(\mathbf{r}, t) = \mathbf{r}.\mathbf{p}(t) + c(t) \tag{3.10}$$

the corresponding quantum Langevin equation is again obtained, the solution being

$$\mathbf{p}(t) = \mathbf{p}(0)e^{-\gamma t} + e^{-\gamma t} \int_0^t F_r(s)e^{\gamma s} ds. \tag{3.11}$$

where the time-dependent function $c(t)$ is given by

$$c(t) = c(0)e^{-\gamma t} - \frac{1}{2m} e^{-\gamma t} \int_0^t |\mathbf{p}(s)|^2 e^{\gamma s} ds. \tag{3.12}$$

We should emphasize that because of the nonlinearity of the SL equation, the superposition principle does not hold. Even more, the general solution of the corresponding equation can not be found and the uniqueness of the solution obtained can not be ascertained. The nonlinear SL wave equation has been received special attention and applied to several problems such as, for example, the damped harmonic oscillator and the motion of a charged particle in the presence of damping while it is moving in an external electromagnetic field [8]. It is also true that some drawbacks are also present when dealing with this equation. For example, the solutions for a damped harmonic oscillator contain the undamped frequency instead of the reduced or renormalized frequency (see below). The analysis of the solutions of the SL equation has been carried out usually without considering the stochastic term, analytically and numerically [8, 31]. For a free particle represented by a plane wave, the SL equation and the corresponding quantum Langevin equation lead to the same solution. However, Messer [32] has studied the evolution of a Gaussian wave function in the free and harmonic potentials and shown that both equations are not strictly equivalent. Very recently, the thermal relaxation given by the SL has been analyzed with white and colored noises (see also Ref. [33]) for an harmonic oscillator [31].

Another heuristic derivation of the equation governing the probability density evolution for the Brownian motion is due to Tsekov [34–36]. In terms of the probability density, the theory becomes nonlinear. Furthermore, the corresponding equations were completed by taking into account thermodynamical functions. A similar Kostin frictional term is found. This procedure is quite similar to that developed in the next section within the Bohmian formalism.

On the other hand, Hasse [37] showed that Kostin's Hamiltonian is a special case of a more general nonlinear Hamiltonian,

$$H = T + V + \gamma W, \tag{3.13}$$

where T and V are the usual kinetic and potential energy operators, and W satisfies the requirements: (i) $\langle W \rangle = 0$ and (ii) $\langle p \rangle = \partial W / \partial q$ according to Ehrenfest's theorem. The second requirement can also be fulfilled for a number of different W operators. For example, Süssmann's Hamiltonian [4] is a special case, originally found by this author in a completely empirical way when studying the force-free motion of wave packets traveling along classical damped paths. W can also be expressed as

$$W = \int_0^q \dot{q}^n dq, \tag{3.14}$$

which, with the usual quantization rule for the momentum, becomes a differential operator of order n. The corresponding Hamiltonian is, in general, complex and non-Hermitian. For linear damping ($n = 1$), the interaction potential becomes complex. This potential, called the *optical potential*, is in general nonlocal and has been widely used in atomic and nuclear physics (for example, in Feshbach's theory). Velocity dependent interactions have also been introduced in nuclear scattering and band theory of solids.

Several alternatives routes to derive the SL or Kostin equation can also be found in the literature (see, for example, Ref. [38]). For nonrandom potentials, the method of Pauli can be used to find the damping potential of the Schrödinger equation which makes the quantum analog of the corresponding classical equations. One can also follow the so-called Schrödinger's procedure to generate a nonlinear wave equation. Without loss of generality, for one dimension, the Hamilton–Jacobi equation for the action S and momentum $p = \partial S/\partial x$ is

$$\frac{\partial S}{\partial t} + H(x, \partial S/\partial x, t) = 0. \tag{3.15}$$

By introducing the wave function Ψ through the relation

$$S(x,t) = -i\hbar \ln \Psi(x,t), \tag{3.16}$$

the continuity equation for $\rho = \Psi^* \Psi$ reads as

$$\frac{\partial \rho}{\partial t} + \frac{1}{m} \nabla\left(\rho \, \mathrm{Re}\{p\}\right) = 0, \tag{3.17}$$

where Re$\{p\}$ stands for the real part of p, which is in general a complex quantity if Ψ is complex, such that

$$\begin{aligned} \mathrm{Re}\{p\} &= -\frac{i\hbar}{2m} \frac{\partial \ln\left(\Psi/\Psi^*\right)}{\partial x}, \\ \mathrm{Im}\{p\} &= -\frac{i\hbar}{2m} \frac{\partial \ln \rho}{\partial x}, \end{aligned} \tag{3.18a}$$

with the mean value of the imaginary part always vanishing. Thus, the expression for p does again fulfill Hasse's second requirement. The so-called Kostin *energy dissipation operator* can then be expressed as

$$W = -\frac{i\hbar\gamma}{2}\left[\ln\left(\frac{\Psi}{\Psi^*}\right) - \left\langle \ln\left(\frac{\Psi}{\Psi^*}\right)\right\rangle\right]. \tag{3.19}$$

The density ρ verifies the reversible continuity equation for a system displaying damping which follows an irreversible dynamics. This contradiction was avoided by Schuch et al. [39–41] by introducing a diffusion term in the continuity equation arriving at the Fokker–Planck equation,

$$\frac{\partial \rho}{\partial t} + \frac{1}{m}\,\nabla(\rho\ \mathrm{Re}\{p\}) - D\,\frac{\partial^2 \rho}{\partial x^2} = 0, \tag{3.20}$$

D being the diffusion coefficient and with the additional condition

$$-\frac{D}{\rho}\frac{\partial^2 \rho}{\partial x^2} = \gamma(\ln \rho - \langle \ln \rho \rangle) \tag{3.21}$$

to be satisfied in order to achieve separation of the two equations for the amplitudes Ψ and Ψ^*. The mean value on the right–hand side guarantees the normalization. Equations (3.20) and (3.21) can also be seen as a continuity equation with a source term. The so-called logarithmic nonlinear Schrödinger equation is now written as

$$i\hbar\,\frac{\partial \Psi}{\partial t} = [H - i\hbar\gamma(\ln \rho - \langle \ln \rho \rangle)]\,\Psi, \tag{3.22}$$

where Hasse's second requirement comes from considering the irreversible diffusion term. As is well known, the damping motion described by the Fokker–Planck equation can always be expressed in terms of its corresponding Langevin equation [42]. This equation will be again discussed in Chap. 4 but in a different context, the continuous quantum measurement.

Interestingly enough, for the dissipative time dependent harmonic oscillator, Schuch, Chung and Hartmann [40, 41] proposed the following nonlinear Schrödinger equation

$$i\hbar\,\frac{\partial \Psi}{\partial t} = \left[-\frac{\hbar^2}{2m}\frac{\partial^2}{\partial x^2} - i\hbar\gamma(\ln \Psi - \langle \ln \Psi \rangle) + \frac{1}{2}m\omega^2(t)x^2\right]\Psi, \tag{3.23}$$

$\omega(t)$ being the time-dependent harmonic oscillator frequency. This equation gives the correct reduced frequency of this dissipative motion. It could also be used for a Brownian particle ($\omega = 0$) as an alternative nonlinear equation to the Kostin equation. In passing, it is worth noticing that Eq. (3.23) can be shown to be a special case of Eq. (4.96) from Chap. 4 if one sets $\kappa = \gamma/2$ (where κ characterizes the resolution of the continuous measurement and γ represents the friction coefficient). Chou has studied Eq. (3.23) in the complex plane [43]. In the following chapter, a quantum trajectory analysis of this equation will be carried out.

An extension to the nonlinear Schrödinger equation can be obtained by writing [44]

$$i\hbar \frac{\partial \Psi}{\partial t} = H\Psi + i\hbar D G(\Psi), \tag{3.24}$$

where $H = T + V$ and the nonlinear term is given by

$$G(\Psi) = \nabla^2 \Psi + \frac{|\nabla \Psi|^2}{|\Psi|^2} \Psi. \tag{3.25}$$

The continuity equation is then modified to a Fokker–Planck equation according to

$$\frac{\partial \rho}{\partial t} + \nabla \cdot \mathbf{J} = D \nabla^2 \rho, \tag{3.26}$$

where $\mathbf{J}$ is the usual quantum probability density current.

A phenomenological nonlinear wave equation with complex interaction was also proposed by Gisin [19, 20] to account for decaying states. This wave equation is

$$i\hbar \frac{\partial \Psi}{\partial t} = \left(1 - \frac{i\kappa}{2}\right) H\Psi + \frac{i\kappa}{2} \langle \Psi | H | \Psi \rangle \Psi, \tag{3.27}$$

where H is the usual Hamiltonian for the undamped system and κ is a dimensionless positive and real damping constant. This wave equation presents some advantages, such as: the norm is independent of time, it reduces to Schrödinger's equation when Ψ is an eigenstate of H, the rate of change of the energy expectation value is negative definite and the equation of motion for the damped harmonic oscillator is obtained in terms of $\langle x \rangle$.

Finally, it is worth mentioning that Razavy [8, 45], Wagner [46] and Schuch [47] have shown a connection between the Caldirola–Kanai Hamiltonian (see Chap. 2) and the log–nonlinear Schrödinger equation following Schrödinger's quantization procedure. This equation has also been derived by Skagerstam [48] and Yasue [49] and has found extensive use in many applications. The Kostin nonlinear logarithmic term has recently been suggested as an appropriate, practical bath functional in time-dependent density functional theory for open quantum systems with unitary propagation [50, 51].

3.3 The Schrödinger–Langevin–Bohm Equation

The purpose of this Section is simply to analyze Eq. (3.2) from a different theoretical (and conceptual) framework, the Bohmian mechanics [23, 52–55]. For the sake of simplicity, let us consider the dynamics of one particle in one dimension. If the wave function is written in polar form as usual

$$\Psi(x, t) = \phi(x, t) e^{iS(x,t)/\hbar} \tag{3.28}$$

where $\phi(x,t)$ and $S(x,t)$ are real valued functions and then is substituted in Eq. (3.2), the resulting Schrödinger–Langevin-Bohm (SLB) equation reads as

$$i\hbar\left[\frac{\partial\phi}{\partial t}+\frac{i}{\hbar}\frac{\partial S}{\partial t}\phi\right]=-\frac{\hbar^2}{2m}\left\{\left[\frac{\partial^2\phi}{\partial x^2}-\frac{\phi}{\hbar^2}\left(\frac{\partial S}{\partial x}\right)^2\right]+\frac{i}{\hbar}\left[2\frac{\partial S}{\partial x}\frac{\partial\phi}{\partial x}+\phi\frac{\partial^2 S}{\partial x^2}\right]\right\}$$
$$+\left[V(x)+V_r(x,t)+\gamma(S-\langle S\rangle)\right]\phi. \quad (3.29)$$

The quantum fluid dynamics description is better illustrated by expressing Eq. (3.29) in a different way, that is, by writing the real and imaginary parts separately. Thus, we have the continuity equation

$$\frac{\partial\rho}{\partial t}+\frac{\partial\rho v}{\partial x}=0 \quad (3.30)$$

where $\rho=\phi^2$ and the velocity is defined by $v=(\hbar/m)(\partial S/\partial x)$. The equivalent to the quantum Hamilton–Jacobi equation is now given by

$$\frac{\partial v}{\partial t}+v\frac{\partial v}{\partial x}+\gamma v=-\frac{1}{m}\frac{\partial(V+V_r+Q)}{\partial x}, \quad (3.31)$$

Q being the quantum potential defined in terms of ρ,

$$Q\equiv-\frac{\hbar^2}{2m}\frac{\nabla^2\rho^{1/2}}{\rho^{1/2}}=\frac{\hbar^2}{4m}\left[\frac{1}{2}\left(\frac{\nabla\rho}{\rho}\right)^2-\frac{\nabla^2\rho}{\rho}\right]. \quad (3.32)$$

In the following, some simple examples are analyzed under this hydrodynamical formulation when the random potential is neglected (dissipative case). Let us start by writing the probability density as a Gaussian function

$$\rho(x,t)=\frac{1}{\sqrt{2\pi\delta(t)^2}}e^{-(x-q(t))^2/2\delta(t)^2} \quad (3.33)$$

where $\delta(t)$ and $q(t)$ are the width and the center of mass of the wave packet, respectively. From Eq. (3.30), the velocity field turns out to be

$$v(x,t)=\frac{\dot{\delta(t)}}{\delta(t)}(x-q(t))+\dot{q}(t) \quad (3.34)$$

where dots on the variables mean the different orders in time derivation. The time integration of this velocity field is straightforward leading to the equation for the quantum trajectories

$$x(t) = q(t) + (x(0) - q(0))\frac{\delta(t)}{\delta(0)} \tag{3.35}$$

where the friction dependence of these trajectories is through the spreading of the Gaussian wave function in its time evolution and the velocity can be reexpressed as

$$v = \frac{\dot{\delta}}{\delta(0)}(x(0) - q(0)) + \dot{q}. \tag{3.36}$$

Equations (3.35) and (3.36) have the same dressing scheme than previously mentioned in Chaps. 1 and 2: a classical and quantum (nonlocal) contribution.

Consider now the interaction potential expanded in a Taylor series around q as

$$V(x,t) \approx V(q,t) + V'(q,t)(x-q) + \frac{1}{2}V''(q,t)(x-q)^2 + \cdots \tag{3.37}$$

where primes denote derivatives with respect to position and evaluated at $x = q$. By retaining up to quadratic terms in the interaction potential, substituting into Eq. (3.31) and separating terms proportional to $(x-q)^0$ and $(x-q)$ we have that

$$\ddot{q} + \gamma\dot{q} = -\frac{1}{m}V'(q,t) \tag{3.38}$$

and

$$\ddot{\delta} + \gamma\dot{\delta} + \omega^2(t)\delta = \frac{\hbar^2}{4m^2\delta^3} \tag{3.39}$$

where $\omega^2(t) = V''(q,t)/m$ is the frequency of the motion. Thus, the motion of q can be identified with that of the classical particle under the potential V and a linear friction. It should be noticed that Eq. (3.38) is just Ehrenfest's theorem for the potential (3.37) and distribution (3.33). On the other hand, the time evolution of the width is governed by the so-called damped (nonlinear) Pinney equation. This equation is well known to be solvable in terms of the linear time-dependent harmonic oscillator equation for the conservative case, but it only admits approximate solutions for weak damping and slowly varying non-vanishing frequencies $\omega(t)$ [52, 56]. In Chap. 2, Sect. 2.12, the differential equation governing the time evolution of the corresponding Gaussian width is simpler than here because we are in the linear regime of quantum mechanics.

For a free particle, $V = 0$, the dissipative motion given by Eq. (3.38) has a trivial classical solution

$$q(t) = q(0) + \frac{\dot{q}(0)}{\gamma}[1 - e^{-\gamma t}]. \tag{3.40}$$

On the contrary, Eq. (3.39) which is written now as

$$\ddot{\delta} + \gamma\dot{\delta} = \frac{\hbar^2}{4m^2\delta^3} \tag{3.41}$$

or, alternatively,

$$\frac{d}{dt}\left(\frac{\dot{\delta}^2}{2} + \frac{\hbar^2}{8m^2\delta^2}\right) = -\gamma\dot{\delta}^2 \leq 0 \tag{3.42}$$

has apparently no analytical solution. Indeed, for the conservative case, $\gamma = 0$,

$$\delta^2(t) = (\delta(0) + \dot{\delta}(0)t)^2 + \frac{\hbar^2 t^2}{4m^2\delta(0)^2} \tag{3.43}$$

which tells us that the width grows without bound with time (see Chap. 1). This is a problem if we want to use the unperturbed solution (for $\gamma = 0$) to obtain an approximate value of Eq. (3.41). In the overdamped regime, the second derivative or acceleration of the width can be neglected and by integrating the resulting first-order ordinary differential equation, the solution is given by

$$\delta(t) \simeq \left(\frac{\hbar^2 t}{m^2\gamma}\right)^{1/4} \tag{3.44}$$

at long times which is consistent with results found from nonlinear theories of quantum Brownian motion [34, 35]. Moreover, it is straightforward to show that

$$\frac{\ddot{\delta}}{\gamma\dot{\delta}} \simeq -\frac{3}{4\gamma t} \tag{3.45}$$

clearly justifying the omission of the acceleration term as long as $\gamma t >> 1$. In Ref. [52], a first order perturbative solution has been provided to be

$$\delta(t) = (t + c_1^4)^{1/4} + \frac{c_2 + (3/16)\ln(t + c_1^4)}{(t + c_1^4)^{3/4}} \tag{3.46}$$

where $c_1 > 0$ and c_2 are integrations constants, which are determined from the initial conditions

$$\delta(0) = c_1 + \frac{c_2}{c_1^3} + \frac{3\ln c_1}{4c_1^3} \tag{3.47}$$

and

$$\dot{\delta}(0) = \frac{3 - 12c_2 + 4c_1^4 - 9\ln c_1}{16c_1^7} \tag{3.48}$$

to be numerically solved only. A direct numerical integration of this equation should obviously be carried out for an exact evaluation of the time evolution of the width [36]. In any case, a more detail discussion below, when dealing the surface diffusion problem, is presented.

For the time-dependent harmonic oscillator case, $V = (1/2)m\omega^2(t)x^2$, the classical equation is

$$\ddot{q} + \gamma\dot{q} + \omega^2(t)q = 0 \tag{3.49}$$

and the differential equation (3.39) for the width is also valid for this case. In nonlinear dynamics, it is known that the integration of the damped Pinney equation is a real challenge. In fact, several methods have been proposed in the literature to look for approximate solutions within perturbation theory [56]. The perturbation procedure could be carried out to higher orders but the zero order solution is very often sufficient for most purposes. In any case, a direct numerical solution is also easy to be obtained.

On the other hand, by eliminating the frequency $\omega^2(t)$ between Eqs. (3.49) and (3.39), and after some manipulations, the quantity

$$I = \frac{1}{2}e^{2\gamma t}[(\alpha\dot{q} - q\dot{\alpha})^2 + (q/\alpha)^2] \tag{3.50}$$

is no longer an invariant (except in the trivial case $\gamma = 0$) with $\delta = (\hbar/m)\alpha^2$. Thus, Eqs. (3.49) and (3.39) constitute a non-Ermakov system (see Chap. 2).

Once the velocity is known after solving Eq. (3.30) for the probability density given by Eq. (3.33), the phase of the wave function is easily obtained by integrating over the position variable. The wave function is then written as

$$\Psi(x,t) = \frac{1}{(2\pi\delta^2)^{1/4}} e^{-(x-q)^2/4\delta^2} e^{\frac{im}{\hbar}[\frac{\dot{\delta}}{2\delta}(x-q)^2 + \dot{q}(x-q)]}. \tag{3.51}$$

An extra arbitrary function of time could have been added to S, but it has been disregarded since it has no effect on the probability density and the probability current. The Wigner function for a pure state defined by

$$f(x,p,t) = \frac{1}{2\pi\hbar}\int dy\, e^{ipy/\hbar}\, \Psi^*(x+\frac{y}{2},t)\Psi(x-\frac{y}{2},t) \tag{3.52}$$

has also a Gaussian shape and is positive defined when Eq. (3.51) is used

$$f(x,q,p,t) = \frac{1}{\hbar\pi} e^{-(x-q)^2/2\delta^2} e^{-\frac{2\delta^2}{\hbar^2}[p - \frac{m\dot{\delta}}{\delta}(x-q) - m\dot{q}]^2}. \tag{3.53}$$

3.4 The Ramsauer–Townsend Effect in a Dissipative Medium

The Ramsauer–Townsend (RT) effect is a quantum phenomenon involving the scattering of low-energy electrons by atoms of a noble gas such as Argon, Krypton, or Xenon [57–59]. The probability of collision between the electrons and gas atoms has a minimum value for electrons with a certain amount of kinetic energy (about 1 eV for Xenon gas). It was shown that when the incident electron wavelength is two times larger than the well width L, there is no reflection and the transmission is complete. In a certain sense, this behavior is just the opposite of the so-called quantum reflection [60, 61] where at very low collisional energies, the transmission can be totally suppressed. Quantum theory predicts the existence of the RT minimum by considering the atom as a finite square potential well [62–66]. Via the de Broglie–Bohm formalism, the RT effect is first analyzed when in the collisional process there is no energy dissipation. Subsequently, the dissipative dynamics of the RT effect is after studied by means of the Kostin equation for small dissipation $\hbar\gamma << (V + E)$ where γ, V and E are the coefficient of friction, the height of the potential and the energy of the particle, respectively.

Let us consider a stationary electronic flux with incident energy E, colliding with a potential well with height V and width L such as

$$V(x) = \begin{cases} 0 & , x \neq 0,\ L \\ -V & , 0 < x < L \end{cases} \tag{3.54}$$

which define the following regions

$$Incidence\ \ Region\ \ (1):\quad x < 0, \tag{3.55}$$

$$Scattering\ \ Region\ \ (2):\quad 0 < x < L, \tag{3.56}$$

$$Transmission\ \ Region\ \ (3):\quad x > L. \tag{3.57}$$

Since E > 0, the solution of Schrödinger's equation

$$i\,\hbar\,\frac{\partial\psi(x,\,t)}{\partial t} = -\,\frac{\hbar^2}{2\,m}\,\frac{\partial^2\,\psi(x,\,t)}{\partial x^2} + V(x)\,\psi(x,\,t)\,. \tag{3.58}$$

for these three mentioned above regions is given by

$$\psi_1(x,\,t) = (e^{i\,k\,x} + A\,e^{-\,i\,k\,x})\,e^{-\,i\,\omega\,t}\,, \tag{3.59}$$

$$\psi_2(x,\,t) = (C\,e^{i\,q\,x} + D\,e^{-\,i\,q\,x})\,e^{-\,i\,\omega\,t}\,, \tag{3.60}$$

$$\psi_3(x,\,t) = (B\,e^{i\,k\,x})\,e^{-\,i\,\omega\,t}\,, \tag{3.61}$$

where

$$k^2 = \frac{2\, m\, E}{\hbar^2}, \qquad q^2 = \frac{2\, m\, (E + V)}{\hbar^2}. \tag{3.62}$$

Thus, the reflection $\mid R \mid^2 = A\, A^*$ and the transmission $\mid T \mid^2 = B\, B^*$ coefficients are given by

$$\mid R \mid^2 = \frac{\left(\frac{k^2 - q^2}{2\, k\, q} \right)^2 sin^2\, (q\, L)}{1 + \left(\frac{k^2 - q^2}{2\, k\, q} \right)^2 sin^2\, (q\, L)}, \tag{3.63}$$

$$\mid T \mid^2 = \frac{1}{1 + \left(\frac{k^2 - q^2}{2\, k\, q} \right)^2 sin^2\, (q\, L)}. \tag{3.64}$$

Using Eqs. (3.63) and (3.64), a particular case can be considered, that is, when

$$q\, L = \pi\, . \tag{3.65}$$

Substitution of Eqs. (3.65) into (3.63) and (3.64) results

$$\mid T \mid^2 = 1, \qquad \mid R \mid^2 = 0, \tag{3.66}$$

which leads to the suppression of the reflection probability.

Let us now turn to the study of the RT effect when in the collision process there is an energy dissipation governed by the Kostin equation [62, 66]

$$i\, \hbar\, \frac{\partial \psi(x,\, t)}{\partial t} = -\frac{\hbar^2}{2\, m}\, \frac{\partial^2 \psi(x, t)}{\partial x^2} + \left[V(x,\, t) + \frac{\hbar\, \gamma}{2\, i} \ell n\, \frac{\psi(x,\, t)}{\psi^*(x,\, t)} \right] \psi(x,\, t), \tag{3.67}$$

where $\psi(x,\, t)$, $V(x, t)$ and γ represent, respectively, the wave function, the interaction potential and dissipation coefficient of our new system. Equation (3.67) will be studied now within the formalism of de Broglie–Bohm.

First, we write $\psi(x,\, t)$ as

$$\psi(x,\, t) = \Phi(x)\, exp \left[-\frac{i\, E}{\hbar\, \gamma}\, (1 - e^{-\gamma\, t}) \right], \tag{3.68}$$

which yields

$$i\,\hbar\,\frac{\partial\psi(x,\,t)}{\partial t} \;=\; E\,e^{-\gamma\,t}\,\psi(x,\,t)\;, \tag{3.69}$$

$$-\,\frac{\hbar^2}{2\,m}\,\frac{\partial^2\psi(x,\,t)}{\partial x^2} \;=\; -\,\frac{\hbar^2}{2\,m}\,\frac{\Phi''(x)}{\Phi(x)}\,\psi(x,\,t)\;, \tag{3.70}$$

$$\frac{\hbar\,\gamma}{2\,i}\,\ell n\,\frac{\psi(x,\,t)}{\psi^*(x,\,t)} \;=\; \frac{\hbar\,\gamma}{2\,i}\,\ell n\,\frac{\Phi(x)}{\Phi^*(x)} \;-\; E\,(1\;-\;e^{-\gamma\,t})\;. \tag{3.71}$$

Inserting Eqs. (3.69)–(3.71), into (3.67) and using Eq. (3.68) we obtain

$$\Phi''(x) + \left[\,q^2 \;-\; \frac{m\,\gamma}{i\,\hbar}\,\ell n\,\frac{\Phi(x)}{\Phi^*(x)}\,\right]\Phi(x) \;=\; 0\;. \tag{3.72}$$

Now, considering the Madelung transformation [67] (for convenience, $\hbar$ has been absorbed in the argument of the phase, $S(x)$)

$$\Phi(x) \;=\; \phi(x)\,e^{i\,S(x)}\;, \tag{3.73}$$

we see that Eq. (3.72) becomes

$$\phi'' \;+\; 2\,i\,\phi'\,S' \;-\; \phi\,(S')^2 \;+\; i\,\phi\,S'' \;+\; \left[q^2 \;-\; \frac{2\,m\,\gamma}{\hbar}\,S\right]\phi \;=\; 0\;. \tag{3.74}$$

Separating real and imaginary parts of the above expression we have,

$$\phi'' \;+\; \left(q^2 \;-\; \frac{2\,m\,\gamma}{\hbar}\,S\right)\phi \;=\; (S')^2\,\phi\;, \qquad 2\,\phi'\,S' \;+\; \phi\,S'' \;=\; 0. \tag{3.75}$$

Defining, as usual,

$$\rho(x) \;=\; \phi^2(x)\;, \tag{3.76}$$

and integrating Eq. (3.75) results

$$\frac{(S')'}{S'} \;=\; -\,\frac{2\,\phi'}{\phi} \;\;\rightarrow\;\; \int\frac{(S')'}{S'} \;=\; -\int\frac{2\,\phi'}{\phi} \;\;\rightarrow$$

$$\ell n\;S' \;=\; -\,2\,\ell n\;\phi \;+\; \ell n\;C \;=\; -\,\ell n\;\phi^2 \;+\; \ell n\;C \;=\; \ell n\,\frac{C}{\phi^2} \;\;\rightarrow$$

$$S'(x) \;=\; \frac{C}{\rho(x)}\;, \qquad S(x) \;=\; S(0) \;+\; C\int_0^x \frac{dx'}{\rho}\;. \tag{3.77}$$

Multiplying Eq. (3.75) by ϕ' and using Eqs. (3.76) and (3.77) we reach,

$$\phi'' \, \phi' \; + \; [q^2 \; - \; (S')^2] \, \phi \, \phi' \; = \; \frac{2 \, m \, \gamma}{\hbar} \, S \, \phi \, \phi' \quad \rightarrow$$

$$\frac{d}{dx} \left[\frac{1}{2} \, (\phi')^2 \; + \; \frac{1}{2} \, q^2 \, \phi^2 \; + \; \frac{1}{2} \, \frac{C^2}{\phi^2} \right] \; = \; \frac{2 \, m \, \gamma}{\hbar} \, S \, \phi \, \phi' \; . \tag{3.78}$$

Since Eq. (3.76) can be written as

$$\rho' \; = \; 2 \, \phi \, \phi' \; , \tag{3.79}$$

Equation (3.78) can be rewritten, using also Eq. (3.76), as

$$I'(x) \; = \; \frac{2 \, m \, \gamma}{\hbar} \, S(x) \, \rho'(x) \; , \qquad I(x) \; = \; \frac{[\rho(x)']^2}{4 \, \rho(x)} \; + \; q^2 \, \rho(x) \; + \; \frac{C^2}{\rho(x)} \; . \tag{3.80}$$

One can readily see that, taking Eq. (3.77), we obtain,

$$\frac{d}{dx} \, (S \, \rho) \; = \; S' \, \rho \; + \; S \, \rho' \; = \; S \, \rho' \; + \; C \; = \; S \, \rho' \; + \; \frac{d}{dx} \, (C \, x) \quad \rightarrow$$

$$S \, \rho' \; = \; \frac{d}{dx} \, (S \, \rho \; - \; C \, x) \; . \tag{3.81}$$

Substituting Eq. (3.81) into (3.80) we have

$$\frac{dI}{dx} \; = \; \frac{2 \, m \, \gamma}{\hbar} \, \frac{d}{dx} \, (S \, \rho \; - \; C \, x) \quad \rightarrow \quad \frac{d}{dx} \left[I \; - \; \frac{2 \, m \, \gamma}{\hbar} \, (S \, \rho \; - \; C \, x) \right] \; = \; 0 \quad \rightarrow$$

$$I \; - \; \frac{2 \, m \, \gamma}{\hbar} \, (S \, \rho \; - \; C \, x) \; = \; constant \; = \; I_0 \quad \rightarrow$$

$$I(x) \; = \; I_0 \; + \; \frac{2 \, m \, \gamma}{\hbar} \, [S(x) \, \rho(x) \; - \; C \, x] \tag{3.82}$$

$$I_o \; = \; \frac{[\rho'(x)]^2}{4 \, \rho(x)} \; + \; q^2 \, \rho(x) \; + \; \frac{C^2}{\rho(x)} \; -$$

$$- \, \frac{2 \, m \, \gamma}{\hbar} \left[\rho(x) \left(S(0) \; + \; C \int_0^x \frac{dx'}{\rho(x')} \right) \; - \; C \, x \right] . \tag{3.83}$$

By solving the differential Equations (3.80) using a variational-of-parameter technique one has

$$\rho(x) = \frac{1}{2\,q^2}\left[\, I(x) + \sqrt{I^2(x) - 4\,q^2\,C^2} \times cos\,(\,2\,q\,[x - \beta(x)]\,)\,\right], \tag{3.84}$$

where $\beta(x)$ is a variational unknown function. To determine $\beta(x)$ we take the derivative of Eq. (3.84), that is,

$$\rho'(x) = \frac{1}{2\,q^2}\left(\, I'(x) + \frac{I(x)\,I'(x)\,cos\left(\,2\,q\,[x - \beta(x)]\,\right)}{\sqrt{I^2(x) - 4\,q^2\,C^2}} - \right.$$

$$\left. - \sqrt{I^2(x) - 4\,q^2\,C^2}\; sin\left(\,2\,q\,[x - \beta(x)]\,\right) \times 2\,q\,[1 - \beta'(x)]\,\right), \tag{3.85}$$

where the following conditions must be obeyed

$$I'(x) + \frac{I(x)\,I'(x)\,cos\,[2\,\theta(x)]}{\sqrt{I^2(x) - 4\,q^2\,C^2}} + \sqrt{I^2(x) - 4\,q^2\,C^2} \times$$

$$\times\, 2\,q\,\beta'(x)\,sin\,[2\,\theta(x)] = 0\,, \qquad \theta(x) = q\,[x - \beta(x)]\,. \tag{3.86}$$

This implies that Eq. (3.85) is written as,

$$\rho'(x) = -\frac{\sqrt{I^2(x) - 4\,q^2\,C^2}}{q}\; sin\,[2\,\theta(x)]\,. \tag{3.87}$$

From Eqs. (3.80) and (3.87) we finally reach

$$\beta'(x) = \frac{m\,\gamma\,S(x)}{\hbar\,q^2}\left(1 + \frac{I(x)\,cos\,[2\,\theta(x)]}{\sqrt{I^2(x) - 4\,q^2\,C^2}}\right). \tag{3.88}$$

We study now the scattering of a stationary flux of particles with energy E and $k^2 = 2\,m\,E/\hbar^2$ by a potential well defined by Eq. (3.54). The particles flux, incident ($x < 0$) and transmitted ($x > L$), will be given by

$$\psi_I(x) = e^{i\,k\,x} + A\,e^{-i\,k\,x} = \phi(x)\,e^{i\,S(x)}\,, \tag{3.89}$$

$$\psi_T(x) = B\,e^{i\,k\,x} = \phi(x)\,e^{i\,S(x)}\,. \tag{3.90}$$

Since ψ and $\partial\psi/\partial x$ are assumed to be continuous at the boundaries of the potential well, we have that:
(a) At $x = 0$, Eq. (3.89) becomes

$$\psi_I(x=0) \quad \rightarrow \quad 1 \,+\, A \;=\; \phi(0)\; e^{i\; S(0)}\;, \tag{3.91}$$

$$\frac{\partial\psi_I}{\partial x}|_{x\,=\,0} \quad \rightarrow \quad 1 \,-\, A \;=\; \frac{e^{i\; S(0)}}{k}\; [-\, i\; \phi'(0) \,+\, \phi(0)\; S'(0)]\;. \tag{3.92}$$

Adding Eqs. (3.91) and (3.92), we obtain the following expression

$$2\,k = [cos\;\; S(0) + i\;\, sin\;\; S(0)]\Big(\; \phi(0)\;[k \;+\; S'(0)] - i\;\phi'(0)\;\Big),$$

which can be separated again in two parts, real and imaginary,

$$2\,k \;=\; cos\;\; S(0)\;\phi(0)\;[k \;+\; S'(0)] \;+\; \phi'(0)\;sin\;S(0)\;, \tag{3.93}$$

$$0 \;=\; sin\;\; S(0)\;\phi(0)\;[k \;+\; S'(0)] \;-\; \phi'(0)\;cos\;\; S(0). \tag{3.94}$$

Multiplying Eq. (3.93) by $sin\;\; S(0)$ and Eq. (3.94) by $cos\;\; S(0)$ and subtracting the expressions, results

$$2\;k\;sin\;\; S(0) \;=\; \phi'(0)\;. \tag{3.95}$$

On the other hand, multiplying Eq. (3.93) by $cos\;\; S(0)$ and Eq. (3.94) by $sin\;\; S(0)$ and adding the expressions, we have that

$$2\;k\;cos\;\; S(0) \;=\; \phi(0)\;[k \;+\; S'(0)]\;. \tag{3.96}$$

Squaring and adding Eqs. (3.95) and (3.96), the following expression is found

$$4\,k^2 \;=\; [\phi'(0)]^2 \;+\; \phi^2(0)\;[k \;+\; S'(0)]^2. \tag{3.97}$$

(b) At $x = L$, Eq. (3.90) leads to

$$\psi_T(x=L) \quad \rightarrow \quad B\; e^{i\;k\;L} \;=\; \phi(L)\; e^{i\;\; S(L)}\;, \tag{3.98}$$

$$\frac{\partial\psi_T}{\partial x}|_{x\,=\,L} \quad \rightarrow \tag{3.99}$$

$$B\; e^{i\;k\;L} \;=\; \frac{1}{k}\; e^{i\;\, S(L)} \times [-\, i\;\phi'(L) \;+\; S'(L)\;\phi(L)]\;. \tag{3.100}$$

Taking the real and imaginary parts of Eqs. (3.98) and (3.100) and using Eq. (3.77)

$$\phi(L) = \frac{1}{k} [- i \, \phi'(L) + S'(L) \, \phi(L)] \rightarrow \phi(L) = \frac{1}{k} S'(L) \, \phi(L) \rightarrow$$

$$S'(L) = k \,, \qquad \rho(L) = \frac{C}{k} \,, \tag{3.101}$$

$$\phi'(L) = 0 \,, \qquad \rho'(L) = 0 \,. \tag{3.102}$$

Subtracting Eqs. (3.91) and (3.92) and taking into account Eqs. (3.76), (3.77), (3.79), (3.95) and (3.96), we find

$$2 \, A = e^{i \, S(0)} \left(\phi(0) \left[1 - \frac{S'(0)}{k} \right] + \frac{i}{k} \, \phi'(0) \right) \rightarrow$$

$$A = \frac{2 \, i \, [k \, \rho(0) - C] - \rho'(0)}{2 \, i \, [k \, \rho(0) + C] + \rho'(0)} \,. \tag{3.103}$$

Taking the above expression, let us calculate the reflection ($| \, R \, |^2$) and transmission ($| \, T \, |^2$) coefficients

$$| \, R \, |^2 = A \, A^* = \frac{4 \, [k \, \rho(0) - C]^2 + [\rho'(0)]^2}{4 \, [k \, \rho(0) + C]^2 + [\rho'(0)]^2} \,, \tag{3.104}$$

$$| \, T \, |^2 = 1 - | \, R \, |^2 = \frac{4 \, k \, C}{\frac{[\rho'(0)]^2}{4 \, \rho(0)} + \frac{C^2}{\rho(0)} + k^2 \, \rho(0) + 2 \, k \, C} \,. \tag{3.105}$$

Substitution of $x = 0$ into Eq. (3.83), results

$$I_0 = \frac{[\rho'(0)]^2}{4 \, \rho(0)} + q^2 \, \rho(0) + \frac{C^2}{\rho(0)} - \frac{2 \, m \, \gamma}{\hbar} \, \rho(0) \, S(0) \,. \tag{3.106}$$

Now, substituting Eq. (3.106) into (3.105) the transmission coefficient is written as

$$| \, T \, |^2 = \frac{4 \, k \, C}{I_0 + [k^2 - q^2 + \frac{2 \, m \, \gamma}{\hbar} \, S(0)] \, \rho(0) + 2 \, k \, C} \,. \tag{3.107}$$

The above expression can be written in a different form. Indeed, considering Eqs. (3.76) and (3.101), and using Eqs. (3.98) and (3.100) we can write

$$| \, T \, |^2 = B \, B^* = \phi^2(L) = \rho(L) = \frac{C}{k} \,. \tag{3.108}$$

To obtain the final form for $| \, T \, |^2$ we need to determine the constant C. We find after long calculations [62]

$$|T|^2 = \frac{1}{1 + sin^2\,[q\,\beta(0)]\,\left(\frac{1-n^2}{2\,n}\right)^2\left[1 + \frac{m\,\gamma\,S(0)}{\hbar\,q^2}\left(\frac{n^2+1}{n^2-1}\right)\right]}, \tag{3.109}$$

and

$$|R|^2 = \frac{sin^2\,[q\,\beta(0)]\,\left(\frac{1-n^2}{2\,n}\right)^2\left[1 + \frac{m\,\gamma\,S(0)}{\hbar\,q^2}\left(\frac{n^2+1}{n^2-1}\right)\right]}{1 + sin^2\,[q\,\beta(0)]\,\left(\frac{1-n^2}{2\,n}\right)^2\left[1 + \frac{m\,\gamma\,S(0)}{\hbar\,q^2}\left(\frac{n^2+1}{n^2-1}\right)\right]} \tag{3.110}$$

where $n \equiv q/k$. Note that Eqs. (3.63) and (3.64) are obtained by setting $\gamma = 0$ into Eqs. (3.109) and (3.110). After substantial calculations we find that

$$\beta(0) \sim L\left(1 - \frac{\gamma\hbar}{V}\left[\frac{kL}{2} + S(0)\right]\right) \tag{3.111}$$

and are in conditions to write the final expression to explain the RT effect. Indeed, as was shown previously, this effect is characterized (see Eq. (3.65)) by

$$q\,L = \pi\,. \tag{3.112}$$

With the help of Eqs. (3.112) and (3.111), we have

$$q\,\beta(0) \sim \pi \quad \rightarrow \quad 2\,q\,\beta(0) \sim 2\,\pi\,. \tag{3.113}$$

Equations (3.86), (3.87) and (3.113) lead us to

$$\rho'(0) = 0\,, \qquad sin\,[q\,\beta(0)] = 0\,. \tag{3.114}$$

Equations (3.79), (3.95) and (3.114) yield

$$2\,k\,sin\,S(0) = \phi'(0) = \frac{\rho'(0)}{2\,\phi(0)} = 0 \quad \rightarrow \quad S(0) = 0\,. \tag{3.115}$$

Finally, Eqs. (3.109) and (3.114) show that the coefficients of transmission and reflection are given by

$$|T|^2 = 1 \text{ and } |R|^2 = 0, \tag{3.116}$$

in agreement with Eq. (3.66), which characterizes the RT effect. On the other hand, $|T|^2 = min$ and $|R|^2 = max$, provided that $q\,\beta(0) \rightarrow \pi/2$ where $\beta(0) \approx L\left(1 - \frac{m\gamma kL}{q^2-k^2}\right)$ (Fig. 3.2).

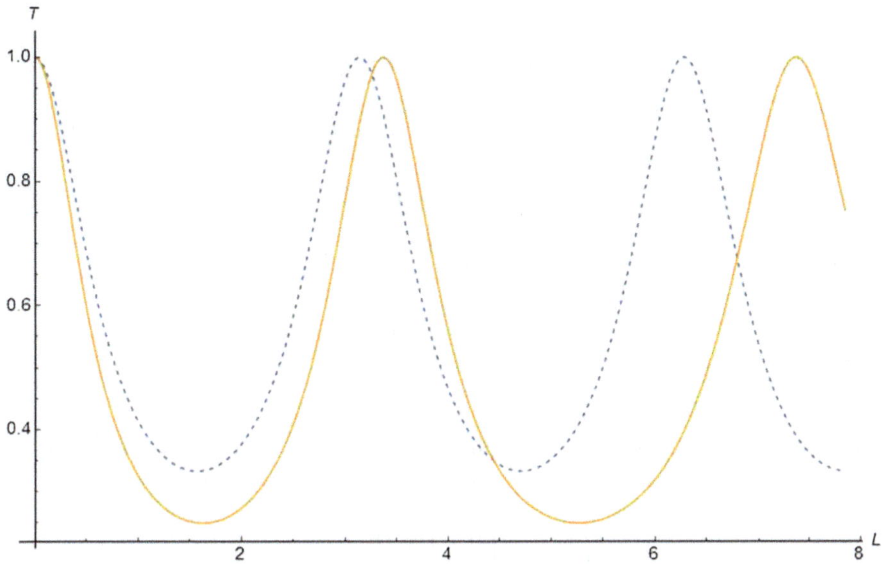

Fig. 3.2 Resonances in Scattering from a square potential: Transmission rate versus width of well with small friction $\gamma << (q^2 - k^2)/mkL$ (*solid line*) and without friction (*dashed line*). Note that the resonances for the dissipative case display a varying periodicity

3.5 Tunneling Through a Dissipative Barrier

Dissipative quantum tunneling is a very important and active field [2]. On the other hand, we have shown in Sect. 2.3 through Bohmian mechanics how more general boundary conditions follow from the continuity of mass, momentum and energy densities. With these new boundary conditions, a novel approach to tunneling through sharp-edged, dissipative potential barriers is presented. The usual boundary conditions are not only more general but the assumption that the wave function and its derivative need to be continuous at a boundary are physically incorrect for dissipative systems. To this end, let us consider the dynamics of a quantum particle in the tunneling region described by the Kostin equation (3.67), expressed the wave function in a polar form $\psi = \phi \exp(iS/\hbar)$ [67]. The proper boundary conditions where the potential undergoes a finite jump read

$$\rho_1(0) = \rho_2(0), \tag{3.117}$$

$$\rho_2(L) = \rho_3(L), \tag{3.118}$$

$$\rho_1{}'(0) = \rho_2{}'(0), \tag{3.119}$$

$$\rho_2{}'(L) = \rho_3{}'(L), \tag{3.120}$$

$$\rho_1(0)v_1(0) = \rho_2(0)v_2(0), \tag{3.121}$$

$$\rho_2(L)v_2(L) = \rho_3(L)v_3(L), \tag{3.122}$$

$$\left(\frac{\partial S_1}{\partial t}\right)_0 = \left(\frac{\partial S_2}{\partial t} + \gamma S_2\right)_0, \tag{3.123}$$

$$\left(\frac{\partial S_2}{\partial t} + \gamma S_2\right)_L = \left(\frac{\partial S_3}{\partial t}\right)_L, \tag{3.124}$$

which show the discontinuity in the phase of the wave function at the boundary (γ represents the friction coefficient as in Sect. 2.3). The potential barrier expressed as

$$V(x) = \begin{cases} 0 \ , x \neq 0, \ L \\ V \ , 0 \ < x \ < \ L \end{cases} \tag{3.125}$$

defines the following regions

$$Incidence \ \ Region \ \ (1): \quad x \ < \ 0, \tag{3.126}$$

$$Scattering \ \ Region \ \ (2): \quad 0 \ < \ x \ < \ L, \tag{3.127}$$

$$Transmission \ \ Region \ \ (3): \quad x \ > \ L. \tag{3.128}$$

Following Sects. 2.3 and 3.4, we find the wave functions for regions 1, 2 and 3 to be (for convenience, in the different interaction regions, $\hbar$ has also been absorbed in the arguments of the corresponding phases, $S_1(x)$, $S_2(x)$ and $S_3(x)$, respectively)

$$\psi_1(x, \ t) \ = \ \sqrt{\rho_1} \ e^{i \ S_1} \ =$$

$$= \sqrt{1 \ + \ a^2 \ + \ 2 \ a \ cos \ (2 \ k_1 \ x \ - \ \alpha)} \ \times exp\left[i \left(- \ \omega_1 \ t \ + \right.\right.$$

$$\left.\left. + \ \frac{\alpha}{2} \ + \ tan^{-1} \left[\frac{1 \ - \ a}{1 \ + \ a} tan \ \left(k_1 \ x \ - \ \frac{\alpha}{2}\right)\right]\right)\right], \tag{3.129}$$

$$\psi_2(x) \ = \ \sqrt{\rho_2} \ e^{i \ S_2} \ =$$

$$= \left(\frac{1}{\bar{Q}(x)} \left[c^2 \ exp\left(2 \int_o^x \bar{Q} \ dx\right) \ + \ d^2 \ exp\left(- \ 2 \int_o^x \bar{Q} \ dx\right) + \right.\right.$$

$$+ 2\, c\, d\, cos\,(\gamma\ -\ \delta)\Big]\Big)^{1/2} \times exp\left[i \left(\frac{\gamma\ +\ \delta}{2} + \right.\right.$$

$$+ tan^{-1} \left[\frac{c\ exp(\int_o^x \bar{Q}\, dx)\ -\ d\ exp(-\ \int_o^x \bar{Q}\, dx)}{c\ exp(\int_o^x \bar{Q}\, dx)\ +\ d\ exp(-\ \int_o^x \bar{Q}\, dx)} \times \right.$$

$$\left.\left.\left. \times\ tan\left(\frac{\gamma\ -\ \delta}{2}\right)\right]\right)\right] \tag{3.130}$$

$$\psi_3(x,\ t)\ =\ \sqrt{\rho_3}\ e^{i\ S_3}\ =$$

$$=\ \sqrt{b^2}\ exp\ [i\ (-\ \omega_3\ t\ +\ k_3\ x\ +\ \beta)]\ . \tag{3.131}$$

where

$$k_1^2 = \frac{2mE_1}{\hbar^2} \tag{3.132}$$

$$\bar{Q}^2(x)\ =\ \frac{2\ m}{\hbar^2}\ [V\ -\ E_1\ +\ \hbar\ \gamma\ \Delta\ S_2(x)] \tag{3.133}$$

$$k_3^2 = \frac{2mE_3}{\hbar^2}, \tag{3.134}$$

$$\omega_1 = \frac{E_1}{\hbar}, \tag{3.135}$$

$$\omega_3 = \frac{E_3}{\hbar}, \tag{3.136}$$

$$\Delta S_2(0)\ = 0, \tag{3.137}$$

$$\Delta S_2(L)\ =\ -\frac{(\omega_3 - \omega_1)}{\gamma}. \tag{3.138}$$

Thus, E_1,E_2 and E_3 are the energies of the particle in the incident, scattering and transmission regions, Eq. (3.125), and related to the wave numbers k_1, $\bar{Q}$ and k_3, respectively (as expressed above).

After lengthy calculations, we find the tunneling probability to be [62]

$$T^{-\,2} = b^{-\,2}\ =\ \frac{1}{D\ +\ 2\ B\ sinh^2\ \alpha}\left(A^2\ +\ 2\ A\ B\ sinh^2\ \alpha + \right.$$

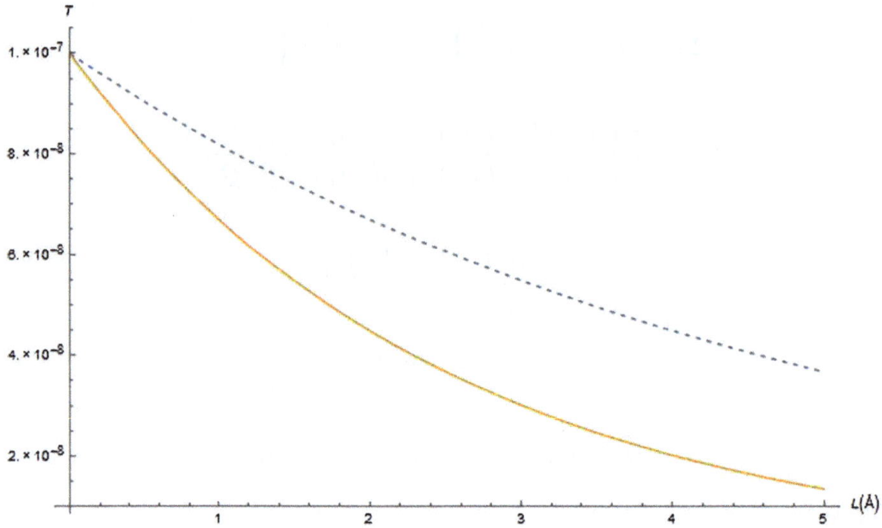

Fig. 3.3 Tunneling through a square potential versus width of well with friction (*solid line*) and without friction (*dashed line*): An electron with kinetic energy $E = 1$ eV tunneling through a barrier with $V = 10$ eV, width $L = 0.5$ nm and friction $\hbar\gamma\Delta S = 2$ eV

$$+ \ 4\, C \ sinh^2\, \alpha \left[1 \ + \ sinh^2\, \alpha \left(1 \ + \ \frac{k_1^2}{\bar{Q}^2(0)} \right) \right] \Bigg) \tag{3.139}$$

where

$$A \ = \ \frac{1}{2} \left[\ \frac{k_3}{k_1} \ + \ \frac{\bar{Q}(L)}{\bar{Q}(0)} \ \right], \quad B \ = \ \frac{1}{2} \ \frac{\bar{Q}^2(L) \ + \ k_3^2}{\bar{Q}(L) \ \bar{Q}(0)} \ , \tag{3.140}$$

$$C \ = \ \left[\frac{\bar{Q}^2(L) \ + \ k_3^2}{4 \ \bar{Q}(L) \ k_1} \right]^2, \quad D \ = \ \frac{\bar{Q}(L)}{\bar{Q}(0)} \ \text{ and } \alpha = \int_0^L \bar{Q}(x)\ dx. \tag{3.141}$$

Finally, by setting $\gamma = 0$, we recover the well-known result

$$T^{-\,2} = b^{-\,2} \ = \ 1 \ + \ \frac{(q^2 \ + \ k^2)^2}{4 \ k^2 \ q^2} \ sinh^2 \ (\bar{q} \ L) \tag{3.142}$$

where $k^2 = k_1^2 = k_3^2 = \frac{2mE}{\hbar^2}$ and $q^2(x) \ = \ \frac{2\,m}{\hbar^2} \ [V \ - \ E]$ (Fig. 3.3).

3.6 The Plasma Fluid Formulation Within Bohmian Mechanics

In the analysis of correspondence between classical and quantum dynamics, one can find a striking analogy between the quantum mechanics of a point particle and the dynamics of a fluid which has been pursued by a number of authors [68–73]. As shown in Chap. 1, Bohmian mechanics clearly offers an intuitive and appealing understanding of the quantum world in a classical-like fashion by means of trajectories evolving throughout configuration space. It provides an explanation of quantum phenomena in terms of point particles guided by wave functions. As has also been previously reported, Madelung [74] was the first to transform the Schrödinger equation for a particle into two fluid-dynamical equations: a continuity equation and an Euler-type equation. This description involves the density $\rho = \Psi^*\Psi$ and the velocity field v as primary quantities. Thus, the fluid dynamicist can gather experience of its effects by translating some of the elementary situations of the quantum theory into their corresponding fluid mechanical statements and vice versa. This fluid formulation was subsequently refined by de Broglie through his pilot wave theory and further developed by Bohm [75, 76]. The wave function (which satisfies Schrödinger's equation) is an objective field, not just as some ghost wave of information where the particle rides along on the wave at some position $\mathbf{r}(t)$.

In order to illustrate this point, we draw striking similarities with the classical plasma fluid theory describing the motion of charged particles (electrons) in a neutralized (ion) background subject to an external electric field by writing in the three dimensional configuration space [77–79]

$$\frac{\partial \rho(\mathbf{r}, t)}{\partial t} + \nabla \cdot (\rho(\mathbf{r}, t)\mathbf{v}(\mathbf{r}, t)) = 0 \tag{3.143}$$

and

$$\begin{aligned} \frac{\partial \mathbf{v}(\mathbf{r}, t)}{\partial t} &+ \mathbf{v}(\mathbf{r}, t) \cdot \nabla \mathbf{v}(\mathbf{r}, t) + \gamma \mathbf{v}(\mathbf{r}, t) \\ &+ \frac{k_B T}{m} \frac{\nabla \rho(\mathbf{r}, t)}{\rho(\mathbf{r}, t)} = -\frac{e\mathbf{E}(\mathbf{r}, t)}{m}, \end{aligned} \tag{3.144}$$

where m, e, ρ, $\mathbf{v}(\mathbf{r}, t)$ and $(k_B T/m)^{1/2}$, γ are the particle (electron) mass, charge, density, mean velocity, thermal speed and collision frequency, respectively. In fact, this collision frequency can also be seen as a collisional friction. $\mathbf{E}(\mathbf{r}, t)$ is the electric field and the second part of Eq. (3.144) refers to the Lorentz force.

To obtain the quantum fluid dynamical counterpart of the classical problem described above, we can translate the information into a form that is easier to comprehend by means again from the continuity equation given by (3.143) and

$$\frac{\partial \mathbf{v}(\mathbf{r},t)}{\partial t} + \mathbf{v}(\mathbf{r},t)\cdot\nabla\mathbf{v}(\mathbf{r},t) + \gamma\mathbf{v}(\mathbf{r},t) + \frac{kT}{m}\frac{\nabla\rho(\mathbf{r},t)}{\rho(\mathbf{r},t)} = -\frac{e\mathbf{E}(\mathbf{r},t)}{m} - \nabla\left(\frac{Q}{m}\right), \tag{3.145}$$

where as usual $\rho(\mathbf{r},t) = |\phi(\mathbf{r},t)|^2$ is the quantum fluid-particle density, $\mathbf{v}(\mathbf{r},t) = (\hbar/m)\nabla S(\mathbf{r},t)$ is the quantum fluid-particle velocity, and Q is the quantum potential which is responsible for all the non-local effects predicted by the theory. Equation (3.145) can be interpreted as the Hamilton–Jacobi equation for this new system, with the additional term proportional to $\hbar^2$ defining $Q(\mathbf{r},t)$.

By expressing, as usual, the wave function in the polar form [67] as in Eq. (3.28), the corresponding Schrödinger equation can be obtained

$$i\hbar\frac{\partial\Psi(\mathbf{r},t)}{\partial t} = -\frac{\hbar^2}{2m}\nabla^2\Psi(\mathbf{r},t) + \left(\frac{\hbar\gamma}{2i}\ln\frac{\Psi(\mathbf{r},t)}{\Psi^*(\mathbf{r},t)} + kT\ln|\Psi(\mathbf{r},t)|^2\right)\Psi(\mathbf{r},t) + e\mathbf{r}\cdot\mathbf{E}(\mathbf{r},t)\Psi(\mathbf{r},t). \tag{3.146}$$

which is the corresponding SL equation for this dissipative problem. The wave function $\Psi(\mathbf{r},t)$ is that of wave mechanics, but conceived in the tradition of Maxwell and Einstein, as an objective field [80]. The particle rides along on the wave at some position $\mathbf{r}(t)$ with velocity $\dot{\mathbf{r}}(t)$ with the help of the guiding equation

$$\dot{\mathbf{r}}(t) = \frac{\hbar}{m}Im(\nabla\ln\Psi(\mathbf{r},t)), \tag{3.147}$$

which constitutes the simplest first-order evolution equation for the trajectories of the particle – the Bohmian trajectories of an evolving particle of the wave packet ensemble with an initial position $\mathbf{r}_0$ – that is, compatible with the Galilean (and time-reversal) covariance of the Schrödinger evolution.

Interestingly enough, Eq. (3.144) corresponds to the classical standard Langenvin equation

$$\ddot{\mathbf{r}} + \gamma\dot{\mathbf{r}} = -e\mathbf{E}(\mathbf{r},t)/m \tag{3.148}$$

except for the nonlinear term $\mathbf{v}\cdot\nabla\mathbf{v}$ (which is called the inertial term in hydrodynamics) and the pressure term $p = (k_B T/m)\rho$. It only attains the form of Eq. (3.148) if, for instance, $\mathbf{v}$ and its derivatives are small and we consider the cold-plasma approximation. Furthermore, the analogy between Eq. (3.144) and the nonlinear SL Eq. (3.146) renders strong support for the exploration of manifold solutions, based on the well developed mathematical problem solving methods of nonlinear differential equations in the literature.

The term $\mu \equiv Q/m$ had been loosely referred to as a quantum pressure term (see also Chap. 1). This term is a misnomer [73] because: (i) the dimensionality is incorrect and it would be better regarded as a quantum (chemical) potential per unit mass, (ii) the word pressure suggests a phenomenon that depends only on the local thermodynamics state (here fixed ρ) and (iii) the presence of the derivatives in

$$\nabla_i \mu = -(\nabla_j \sigma_{ij})/\rho \tag{3.149}$$

where

$$\sigma_{ij} = (\hbar^2 \rho/4m^2)\nabla_i \nabla_j \ln \rho, \tag{3.150}$$

with $i, j = x, y, z$, shows that neighbouring points are involved in its definition. Thus, σ_{ij} is a properly dimensioned contender for the (anisotropic) quantum stress tensor, instead. Even more, the term $(k_B T/m)\rho$ has been proposed by Bialynicki-Birula and Mycielski [18]: a logarithmic Schrödinger-like equation with nonlinearity $b \ln |\Psi|^2$ which possesses soliton-like solutions of Gaussian form (see Sect. 2.11). The term $(\hbar\gamma/2i)\ln(\Psi/\Psi^*)$ in Eq. (3.146) represents quantum dissipation where γ stands for the friction coefficient (collisional friction). Caldeira and Leggett [81] have given support and justification for the use of nonlinear wave functions for the description of non-conservative systems, based on their conclusion that damping tends to destroy interference effects (decoherence) of two Gaussian wavepackets in a harmonic potential. Therefore, the protocol presented above through Bohmian mechanics can be used as a clue to a deeper understanding of nonlinear wave mechanics as well as to encourage usage of the well developed mathematical problem by solving methods of nonlinear differential equations to gather new insights and design some physical models.

Finally, it worth stressing that electron transport studies in nanostructure devices have also been described by the quantum hydrodynamic equations [82, 83].

3.7 The Schrödinger Equation for an Extended Electron via Bohmian Mechanics

About a century ago, Lorentz [84, 85] and Abraham [86] argued that when an electron is accelerated, there are additional forces acting due to the electron's own electromagnetic field. However, the so-called Lorentz–Abraham equation for a point-charge electron

$$m\frac{dV}{dt} = \frac{2e^2}{3c^3}\frac{d^2V}{dt^2} + F_{ext} \tag{3.151}$$

was found to be unsatisfactory because, for $F_{ext} = 0$, it admits runaway solutions. These solutions clearly violate the law of inertia.

Since the seminal works of Lorentz and Abraham, inumerous papers and textbooks have given great consideration to the proper equation of motion of an electron

[87–97]. The problematic runaway solutions were circumvented by Sommerfeld [91] and Page [92] by going to an extended model. In the nonrelativistic case of a sphere with uniform surface charge, such an electron obeys in good approximation the difference-differential equation [93–95]

$$m\frac{dV}{dt} = \frac{e^2}{3L^2c}\left[V(t-2L/c) - V(t)\right] + F_{ext}. \tag{3.152}$$

This extended model is finite and causal if the electron size L is larger than the classical electron radius $r_e = e^2/\mathrm{mc}^2$. In the following, discussion here is limited to the sphere with uniform surface charge since the case of a volume charge is considerably more complicated and adds nothing to the understanding of the problem.

The dynamics of charges is a key example of the importance of obeying the validity limits of a physical theory. If classical equations can no longer be trusted at distances of the order of (or below) the Compton wavelength $\lambda = h/mc$, c being the speed of light, what is the Schrödinger equation that can replace Eq. (3.152). Within quantum electrodynamics (QED), it has not been able to derive an equation of motion and it is unclear whether QED can actually produce an equation of motion at all. A simple solution to this problem in the nonrelativistic regime is proposed here.

A new quantum mechanical wave equation describing the dynamics of an extended electron is derived via Bohmian mechanics. The solution to this equation is found through a wave packet approach which establishes a direct correlation between a classical variable with a quantum variable describing the dynamics of the center of mass and the width of the electron wave packet. It is shown that the new Schrödinger equation is free of any runaway solutions or any acausal responses. Besides, this approach provides a comparatively clearer picture than the modern time quantum approach carried out by Moniz and Sharp [95]. They derived an infinite order differential equation, i.e., an infinite series of derivatives that apparently can not be summed.

As we know, the Bohmian framework ascribes a particle motion via the de Broglie guidance condition (in one dimension)

$$\frac{dx}{dt} = v(x,t)|_{x=x(t)} = \left.\frac{1}{m}\frac{\partial S}{\partial x}\right|_{x=x(t)} \tag{3.153}$$

where v represents the particle velocity and S is the phase of the wave function Ψ written in polar form, [67] Eq. (3.28). In this formalism, Schrödinger's equation can be recast as in Eqs. (3.30) and (3.31). Within this framework, a quantum extension to the Sommerfeld-Page equation (3.152) for an electron sphere with uniform surface charge in the absence of external forces can be accomplished by writing

$$\frac{\partial v}{\partial t} + v\frac{\partial v}{\partial x} = \frac{e^2}{3mL^2c}\left[v(t-2L/c) - v(t)\right] + \frac{1}{m}\frac{\partial}{\partial x}\left(\frac{\hbar^2}{2m\sqrt{\rho}}\frac{\partial^2\sqrt{\rho}}{\partial x^2}\right). \tag{3.154}$$

Then, Eqs. (3.28) and (3.154) yield

$$i\hbar\frac{\partial\Psi(x,t)}{\partial t} = -\frac{\hbar^2}{2m}\frac{\partial^2\Psi(x,t)}{\partial x^2} + \left\{\frac{i\hbar e^2}{6mL^2c}\ln\left(\frac{\Psi(x,t-2L/c)\Psi^*(x,t)}{\Psi^*(x,t-2L/c)\Psi(x,t)}\right)\right\}\Psi(x,t). \tag{3.155}$$

In order to find the most general Gaussian wave packet solution to Eq. (3.155), the usual ansatz is made according to Eq. (3.33). First, Eq. (3.33) is substituted into (3.30) and integrated the result being Eq. (3.34) where the constant of integration must be zero since ρ and $\rho\,(\partial S/\partial x)$ vanish for $|x| \to \infty$. In fact, any well-behaved function of $(x - q)$ multiplied by ρ clearly vanishes as $|x| \to \infty$. Then, substitution of Eqs. (3.33) and (3.145) into (3.154) yields

$$\frac{m}{2a(t)}\left\{\ddot{\delta}(t) - \frac{e^2}{3mL^2c}\left[\dot{\delta}(t-2L/c) - \dot{\delta}(t)\right] - \frac{\hbar^2}{4m^2\delta^3(t)}\right\}[x-q(t)]^2 +$$
$$m\left\{\ddot{q}(t) - \frac{e^2}{3mL^2c}\left[\dot{q}(t-2L/c) - \dot{q}(t)\right]\right\}[x-q(t)] = 0. \tag{3.156}$$

This polynomial equation is satisfied once the coefficients of $[x - q(t)]$ and $[x - q(t)]^2$ are set equal to zero, namely,

$$\ddot{q}(t) - \frac{e^2}{3mL^2c}\left[\dot{q}(t-2L/c) - \dot{q}(t)\right] = 0 \tag{3.157}$$

and

$$\ddot{\delta}(t) - \frac{e^2}{3mL^2c}\left[\dot{\delta}(t-2L/c) - \dot{\delta}(t)\right] = \frac{\hbar^2}{4m^2\delta^3(t)}. \tag{3.158}$$

The wave packet dynamics is now completely determined by Eqs. (3.157) and (3.158). The first equation is the Sommerfeld-Page equation (describing here the time evolution of center of the wave packet) which does not have runaway solutions [95]. The second equation describing the time evolution of the width of the wave packet can be considered as a new result. This equation is free of any runaway solutions or any acausal response due of the restrictive term on the right hand side: physically this means that for $t > 2e^2/3\mathrm{mc}^3$ (which is the time required for light to traverse the extended electron) this term settles down the dynamics of the wave packet.

To sum up, this new formulation to describe the nonrelativistic quantum dynamics of an extended electron provides then a comparatively clearer picture than other formulations using elaborative manipulation of infinite series of operators. The approach here is also reasonable because the electron is smeared out due to the uncertainty principle and has the appropriate feature of a wave packet. In fact, there exists as yet no proper formulation in QED (and may not be possible within QED as a perturbation theory). The only calculation by QED was reported by Low [98], who has not been able to derive an equation of motion and it is unclear whether QED can actually produce an equation of motion at all. As far we know, this treatment presents the only nonrelativistic Schrödinger equation available to describe the problem. Further, Eq. (3.155) may be used to investigate quantum tunneling through potential barriers in light of the work developed by Denef et al. on the so-called classical tunneling [99] and described previously.

Finally, Eqs. (3.30) and (3.31) have also been proposed to be macroscopic equations for many superconducting particles in the same state [100]. In this case, the external potential is derived from the Lorentz force when an electric and magnetic field are considered.

3.8 Surface Diffusion

The main purpose of spectroscopic experiments involving a probe and a system at thermal equilibrium with a reservoir (or thermal bath) consists of measuring the system response under the perturbation caused by the probe. As established by the so-called fluctuation-dissipation theorem, the intrinsic properties of matter can be extracted by analyzing the response which many times can be described by first order perturbation theory and determined by the spectrum of the spontaneous fluctuations of the reservoir. Within this linear approximation, the probability per time unit that the full system formed by the probe and reservoir changes from the initial state to the final state is given by the Fermi Golden rule. Within the Born approximation in scattering theory, the nature of the scattered particles as well as the details of the system-probe interaction potential are largely irrelevant, this essentially reducing the scattering problem to a typical statistical mechanics problem [101]. The linear response function of a system consisting of interacting particles, also known as the dynamic structure factor, can then be related to the spontaneous-fluctuation spectrum.

As mentioned in the Introduction, the system–plus–bath approach is perhaps the most successful and useful way to deal with stochastic dynamics, since it starts from the physical system of interest and its environment (or bath) forming all together a conservative system. For open quantum systems, it seems to be the most natural approach. Within this system–plus–bath approach, the corresponding dynamics is commonly described by a total Hamiltonian which is split up into three different parts

$$H = H_S + H_B + H_{SB}, \tag{3.159}$$

where H_S and H_B correspond to the free evolution of the system and the reservoir or bath, respectively, and H_{SB} describes the interaction between them. In this total Hamiltonian, the system usually consists of a few degrees of freedom, while the environment is formed by a huge number of them (even infinity). Moreover, it is reasonable to assume that the coupling between them is a linear function of the bath coordinates. This property of linearity is very convenient, since it is then very easy to eliminate the bath coordinates in an exact way.

Let us consider briefly the Caldeira-Leggett Hamiltonian model [102, 103]. The role of initial conditions has also been largely discussed in this model since some problems when solving the Langevin equation can appear [104, 105]. This model has been applied (and it is still applying) to physical problems with linear dissipation. In particular, we are going to focus here on atom surface diffusion and the vibrational relaxation of a particle adsorbed (or adsorbate or adparticle) on a surface [106, 107]. Within this context, space–time correlation functions play a key role, since they are used to describe the decay of spontaneous thermal fluctuations at surfaces, this being central to the study of transport phenomena. These functions are defined as the thermodynamic average of the product of two dynamical variables, each one expressing the instantaneous deviation from its corresponding equilibrium value at particular points on the surface and time. A complete description of the particle dynamics in a many–body system is then reached when the behavior of the corresponding correlation functions over the entire wavenumber range (small and long wavelengths) is studied. This range splits into different characteristic regions, each one associated with a different set of properties of the system. Space–time correlation functions can also be used to describe the linear response of a fluid under a weak, external perturbation.

In the quasielastic He atom scattering (QHAS) technique, time–of–flight measurements are converted to energy transfer spectra, from which a wide energy range can be spanned and several peaks are observed. The prominent peak around the zero energy transfer, namely the quasi–elastic peak (Q–peak), gives information about the adsorbate diffusion process. Additional weaker peaks at low energy transfers around the Q–peak are also observed. These peaks are attributed to the parallel (to the surface) frustrated translational motion of the adsorbates (T–mode) and also to the excitations of surface phonons —at positive energy transfers we have creation processes, and annihilation ones at negative energy transfers. The measurable quantity experimentally is the so–called dynamic structure factor or scattering law which gives the lineshapes of all those elementary processes; in particular, those corresponding to the Q and T peaks. The dynamic structure factor provides information about the dynamics and structure of the adsorbates through particle distribution functions and these ones are related to the nature of the adsorbate–substrate and adsorbate–adsorbate interactions.

In QAHS experiments, one is usually interested in the differential reflection coefficient, which can be expressed as

$$\frac{d^2\mathcal{R}(\Delta\mathbf{K},\omega)}{d\Omega d\omega} = n_d F S(\Delta\mathbf{K},\omega) \tag{3.160}$$

in analogy to scattering of slow neutrons by crystals and liquids [108, 109]. This magnitude gives the probability that the probe (He) atoms scattered from the diffusing collective (spread out along the surface) reach a certain solid angle Ω with an energy exchange $\hbar\omega = E_f - E_i$ and a parallel (to the surface) momentum transfer

$\Delta\mathbf{K} = \mathbf{K}_f - \mathbf{K}_i$. In Eq. (3.160), n_d is the (diffusing) surface concentration of adparticles; F is the atomic form factor, which depends on the interaction potential between the probe atoms in the beam and the adparticles on the surface; and $S(\Delta\mathbf{K}, \omega)$ is the dynamic structure factor, which gives the lineshapes of the Q and T peaks — other peaks can also be present, such as the inelastic ones related to surface phonon excitations— and provides a complete information about the dynamics and structure of the adsorbates through particle distribution functions. Experimental information about long distance correlations is obtained from the scattering law when considering small values of $\Delta\mathbf{K}$, while information on long time correlations is available at small energy transfers, $\hbar\omega$.

Surface diffusion is a dynamical problem that can be tackled by means of classical mechanics for heavy adsorbates or adparticles. For light adsorbates, quantum mechanics has to be applied and their positions are replaced by position operator. Thus, let us consider an ensemble of classical/quantum interacting particles on a surface. Their distribution functions are described by means of the so–called van Hove or time–dependent pair correlation function $G(\mathbf{R}, t)$ [108]. This function is related to the dynamic structure factor as

$$S(\Delta\mathbf{K}, \omega) = \iint G(\mathbf{R}, t) e^{i(\Delta\mathbf{K}\cdot\mathbf{R} - \omega t)}\, d\mathbf{R}\, dt. \tag{3.161}$$

Given an adparticle at the origin at some arbitrary initial time, $G(\mathbf{R}, t)$ represents the average probability for finding a particle (the same or another one) at the surface position $\mathbf{R} = (x, y)$ at a time t. Note that this function is a generalization of the well–known pair distribution function $g(\mathbf{R})$ from statistical mechanics [110], since it provides information about the interacting particle dynamics.

Depending on whether correlations of an adparticle with itself or with another one are considered, a distinction can be made between the self correlation function, $G_s(\mathbf{R}, t)$, and the distinct correlation function, $G_d(\mathbf{R}, t)$. The full pair correlation function can then be expressed as

$$G(\mathbf{R}, t) = G_s(\mathbf{R}, t) + G_d(\mathbf{R}, t). \tag{3.162}$$

According to its definition, $G_s(\mathbf{R}, t)$ is peaked at $t = 0$ and approaches zero as time increases because the adparticle loses correlation with itself. On the other hand, at $t = 0$, $G_d(\mathbf{R}, 0) \equiv g(\mathbf{R})$ gives the static pair correlation function (the standard pair distribution function), while it approaches the mean surface number density σ of diffusing particles as $t \to \infty$. Accordingly, Eq. (3.162) can be split up as

$$G(\mathbf{R}, 0) = \delta(\mathbf{R}) + g(\mathbf{R}) \tag{3.163}$$

at $t = 0$, and expressed as

$$G(\mathbf{R}, t) \approx \sigma \tag{3.164}$$

for a homogeneous system with $\|\mathbf{R}\| \to \infty$ and/or $t \to \infty$. At low adparticle concentrations (coverage, $\theta \ll 1$), when interactions among adsorbates can be neglected because they are far apart from each other, the main contribution to (3.162) is G_s (particle–particle correlations are negligible and $G_d \approx 0$). On the contrary, for high coverages, it is expected that G_d presents a significant contribution to (3.162).

Within this theoretical framework, the dynamic structure factor is better written as [108]

$$S(\Delta\mathbf{K}, \omega) = \int e^{-i\omega t} I(\Delta\mathbf{K}, t)\, dt, \tag{3.165}$$

where

$$I(\Delta\mathbf{K}, t) \equiv \frac{1}{N} \left\langle \sum_{j,j'}^{N} e^{-i\Delta\mathbf{K}\cdot\mathbf{R}_j(0)} e^{i\Delta\mathbf{K}\cdot\mathbf{R}_{j'}(t)} \right\rangle \tag{3.166}$$

is the so-called intermediate scattering function —note that this function is the space Fourier transform of $G(\mathbf{R}, t)$. In Eq. (3.166) the brackets denote an ensemble average and $\mathbf{R}_j(t)$ the trajectory of the j adparticle on the surface. This function can be split into two sums, distinct (I_d) and self (I_s), if the crossing terms are taken into account or not, respectively. Following neutron scattering language, the corresponding Fourier transforms of I and I_s give what is called the coherent scattering law, $S(\Delta\mathbf{K}, \omega)$, and incoherent scattering law $S_s(\Delta\mathbf{K}, \omega)$, respectively. In QHAS experiments, and with interacting adsorbates, coherent scattering is always obtained. The corresponding theoretical interpretation of that scattering is usually carried out in terms of the Vineyard convolution approximation [109] where the distinct correlation function G_d is expressed as a convolution of the self correlation function G_s. This approximation is known to fail at small distances where the surface lattice becomes important. Whereas in neutron scattering many attempts to improve the convolution approximation have been developed, in the QHAS context very little effort has been devoted to this goal.

In order to go a step further into the dynamics, a Hamiltonian has to be specified. In surface diffusion, the full system+bath Hamiltonian is usually written as [106, 107]

$$\begin{aligned} H = {} & \frac{p_x^2}{2m} + \frac{p_y^2}{2m} + V(x, y) \\ & + \frac{1}{2} \sum_{j=1}^{N} \left[\frac{p_{x_j}^2}{m_j} + m_j \omega_{x_j}^2 \left(x_j - \frac{c_{x_j}}{m_j \omega_{x_j}^2} x \right)^2 \right] \\ & + \frac{1}{2} \sum_{j=1}^{N} \left[\frac{p_{y_j}^2}{m_j} + m_j \omega_{y_j}^2 \left(y_j - \frac{c_{y_j}}{m_j \omega_{y_j}^2} y \right)^2 \right], \end{aligned} \tag{3.167}$$

where (p_x, p_y) and (x, y) are the adparticle momenta and positions with mass m; and (p_{x_i}, x_i) and (p_{y_i}, y_i) with $i = 1, \ldots, N$ are the momenta and positions of the bath oscillators (phonons), with mass and frequency given by m_i and ω_i, respectively.

Phonons with polarization along the z–direction are not considered. The Hamiltonian was originally considered by Magalinskii and Caldeira and Leggett, who used it for weak and strong dissipation (a general discussion about the Hamiltonian (3.167) can be found in [2]). In surface diffusion, $V(x, y)$ is in general a periodic function describing the surface corrugation at zero temperature. The harmonic frequencies of the bath modes and the coupling coefficients are expressed in terms of spectral densities, defined as

$$J_i(\omega) = \frac{\pi}{2} \sum_{j=1}^{N} \frac{c_{i_j}^2}{m_j \omega_{i_j}^2} \left[\delta(\omega - \omega_{i_j}) \right]. \tag{3.168}$$

with $i = x, y$. These densities enable the passage to a continuum model.

In the Heisenberg picture, the time evolution of the position operators is given by the generalized Langevin (GL) equation for each system coordinate

$$m\ddot{x}(t) + m \int_0^t \gamma_x(t - t')\, \dot{x}(t')\, dt' + \frac{\partial V(x, y)}{\partial x} = \xi_x(t), \tag{3.169a}$$

$$m\ddot{y}(t) + m \int_0^t \gamma_y(t - t')\, \dot{y}(t')\, dt' + \frac{\partial V(x, y)}{\partial y} = \xi_y(t), \tag{3.169b}$$

where the associated friction functions are defined through the cosine Fourier transform of the spectral densities,

$$\gamma_i(t) = \frac{2}{\pi m} \int_0^\infty \frac{J_i(\omega)}{\omega} \cos \omega t \; d\omega, \tag{3.170}$$

with $i = x, y$. The nonhomogeneity of (3.169) represents a fluctuating or random force ξ (when introducing the SL equation in Sect. 3.2, this force was labeled by F_r) for each degree of freedom which depend on the initial position of the system and initial positions and momenta of the oscillators of each bath [2]

$$\xi_x(t) = -\sum_j c_{x_j} \left\{ \left[x_j(0) + \frac{c_{x_j}(0)}{m_j \omega_{x_j}^2} x(0) \right] \cos(\omega_{x_j} t) + \frac{p_{x_j}(0)}{m_j \omega_{x_j}} \sin(\omega_{x_j} t) \right\}. \tag{3.171}$$

and

$$\xi_y(t) = -\sum_j c_{y_j} \left\{ \left[y_j(0) + \frac{c_{y_j}(0)}{m_j \omega_{y_j}^2} y(0) \right] \cos(\omega_{y_j} t) + \frac{p_{y_j}(0)}{m_j \omega_{y_j}} \sin(\omega_{y_j} t) \right\}. \tag{3.172}$$

For each cartesian component of the noise, it can be easily shown that its equilibrium (canonical ensemble) expectation value with respect to the heat bath including the corresponding bilinear coupling to the system vanishes. On the contrary, the noise autocorrelation function (each cartesian component) is a complex quantity because in general it does not commute at different times. In the classical limit $\hbar \to 0$, each noise correlation reduces to $mk_B T \gamma_i(t)$, with $i = x, y$. For Ohmic friction, $\gamma_i(t) = 2\gamma_i \delta(t)$, where γ_i is a constant and $\delta(t)$ is Dirac's δ–function. Under the assumption of Ohmic friction, it can be shown that noise in this model is white. The paradigm of this type of noise is the *Gaussian white noise*. Dealing with large systems (the surface seen as a thermal bath) where the number of collisions between substrate and adsorbate is very high, one of the fundamental theorems of the theory of probability, namely the central limit theorem, ensures that the fluctuations of the bath will be Gaussian distributed. Diffusion can then be described by a Brownian-type motion involving a continuous Gaussian stochastic process. In virtue of the fluctuation–dissipation theorem, such fluctuations can be related to the friction coming mainly from surface phonons: the phonon friction. Electronic friction due to low-lying electron–hole pair excitations is usually neglected in most of cases. Moreover, quantum mechanically [2], for Ohmic friction the imaginary part of each noise function is a step function and its real part goes with $\mathrm{csch}^2(\pi t/\hbar\beta)$. Thus, at zero surface temperature, the noise is still correlated even for long time (it decays as t^{-2}) in contrast to the classical case. These facts give rise to important differences with respect to the classical case such as, for example, the noise and the system coordinates are correlated instead of being zero. In particular, in classical mechanics, the random force can be neglected at zero temperature, but quantum-mechanically this force is always present due to the zero point motion.

In order to simplify this theoretical treatment, only classical noise will be considered, though keeping in mind that the quantum results will be only valid for not too low surface temperatures. Moreover, the motion of only one adsorbate is considered within the so-called single adsorbate approximation since, for very low coverage, adparticles are considered non-interacting. Thus, if Ohmic friction is assumed, Eqs. (3.169) reduce to two coupled standard Langevin equations[1] (Markovian approximation),

$$m\ddot{x}(t) + m\gamma_x(t)\,\dot{x}(t) + \frac{\partial V(x,y)}{\partial x} = \xi_x(t), \tag{3.173a}$$

$$m\ddot{y}(t) + m\gamma_y(t)\,\dot{y}(t) + \frac{\partial V(x,y)}{\partial y} = \xi_y(t). \tag{3.173b}$$

In Fig. 3.4, one classical stochastic trajectory driven by a Gaussian white noise is plotted, which simulates a realization of a two–dimensional Brownian motion. In order to obtain information about the diffusion process, a swarm of these trajectories (i.e., a sampling over many Brownian realizations) should be considered.

---

[1]The δ–function counts only one half when the integration is carried out from zero to infinity.

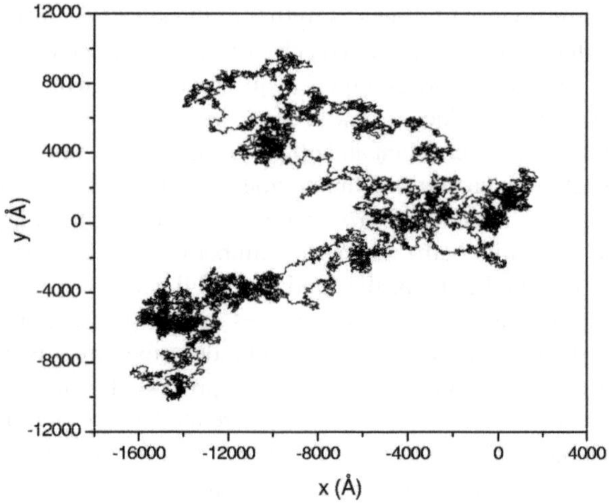

Fig. 3.4 As an example of Brownian motion driven by a Gaussian white noise, the classical stochastic trajectory pursued by a Na atom is displayed here at $T = 300\,\text{K}$, though on a flat surface ($V = 0$). The friction constant is $\gamma = 0.5\ \text{ps}^{-1}$ and the evolution is up to $t = 20,000\,\text{ps}$

When dealing with higher coverages, adsorbate–adsorbate interactions can no longer be neglected. In this case, the adsorbate–surface interaction can still be described as before, but pairwise potential functions accounting for the adsorbate–adsorbate interactions are usually introduced into Langevin molecular dynamics simulations [111]. Moreover, these simulations always result in a relatively high computational cost due to the time spent by the codes in the evaluation of the forces among particles. This problem is even worse when working with long–range interactions, since *a priori* they imply that one should consider a relatively large number of particles in order to get a good numerical simulation. An alternative approach is to consider a purely stochastic description for these interactions [112–115]. This description, which is denominated the interacting single adsorbate approximation, is a fully Langevin approach based on the theory of spectral–line collisional broadening developed by Van Vleck and Weisskopf [116] and the elementary kinetic theory of gases [110]. In this approach, the motion of a single adsorbate is modelled by a series of random pulses within a Markovian regime (i.e., pulses of relatively short duration in comparison with the system relaxation); the pulses simulate the collisions with other adsorbates. In particular, we describe these adsorbate–adsorbate collisions by means of a white shot noise as a limiting case of a colored shot noise [7]. For a good simulation of a diffusion process, one has to consider very long times in comparison to the timescales associated with the friction caused by the surface or with the typical vibrational frequencies observed when the adsorbates keep moving inside a surface well. This means that there will be a considerably large number of collisions during the time elapsed in the propagation, and therefore that, at some point, the past history of the adsorbate could be irrelevant. This memory loss is a signature of a Markovian dynamical regime, where adsorbates have reached what is called the statistical limit. Otherwise, for timescales relatively short, the interaction is not Markovian and it is very important to take into account the effects of the interaction on the particle and its dynamics (memory effects). The diffusion of a single adsorbate is thus modeled by a series of random pulses within a Markovian regime (i.e., pulses of relatively short duration in comparison with the system relaxation) simulating collisions between adsorbates. In particular, we describe these adsorbate–adsorbate collisions by means of a white shot noise [7] as a limiting case of colored shot noise. In this way, a typical molecular dynamics simulation problem involving N adsorbates is substituted by the dynamics of a single adsorbate, where the action of the remaining $N - 1$ adparticles is replaced by a random force given by the white shot noise. Furthermore, the surface coverage can be related to a collisional friction providing the average number of collision per unit time, γ_c [114, 115], where the probability of observing a given number of collisions, after a certain time, follows a Poisson distribution. In brief, the adsorbate is subject to a two white noises, one coming from the substrate and the other from the surrounding adsorbates; both noises being uncorrelated. Thus, the total friction coefficient η for the interacting single adsorbate model is a sum of two friction coefficients, $\eta = \gamma + \gamma_c$ and the total noise is given by $\xi = \xi_G + \xi_S$ (G comes from Gaussian and S from shot) for each degree of freedom (x, y).

In this interacting single adsorbate approximation, and within a classical framework, the distinction between self and distinct time–dependent pair correlation function does not exist and Eqs. (3.161) and (3.165) still hold. The intermediate scattering function reads now as

$$I(\Delta\mathbf{K}, t) \equiv \langle e^{-i\Delta\mathbf{K}\cdot[\mathbf{R}(t)-\mathbf{R}(0)]}\rangle = \langle e^{-i\Delta\mathbf{K}\cdot\int_0^t \mathbf{v}(t')\,dt'}\rangle \tag{3.174}$$

In order to get some analytical results and therefore a guide for the interpretation of the numerical Langevin simulations, the intermediate scattering function can be expressed as a second order cumulant expansion in $\Delta\mathbf{K}$,

$$I(\Delta\mathbf{K}, t) \approx e^{-\Delta K^2 \int_0^t (t-t')\mathcal{C}_{\Delta\mathbf{K}}(t')dt'}, \tag{3.175}$$

where

$$\mathcal{C}_{\Delta\mathbf{K}}(\tau) \equiv \langle v_{\Delta\mathbf{K}}(0)\, v_{\Delta\mathbf{K}}(\tau)\rangle \equiv \lim_{\mathcal{T}\to\infty} \frac{1}{\mathcal{T}} \int_0^{\mathcal{T}} v_{\Delta\mathbf{K}}(t)\, v_{\Delta\mathbf{K}}(t+\tau)\, dt \tag{3.176}$$

is the *autocorrelation function* of the velocity projected onto the direction of the parallel momentum transfer (whose length is $\Delta K \equiv \|\Delta\mathbf{K}\|$). Only differences between two times, τ, are considered because this function is *stationary*. This is the so–called *Gaussian approximation* [110], which is exact when the velocity correlations at more than two different times are negligible, thus allowing to replace the average acting over the exponential function by an average acting over its argument. These approximation results of much help in the interpretation of the numerical results as well as in getting an insight into the underlying dynamics.

The decay of $\mathcal{C}_{\Delta\mathbf{K}}(\tau)$ allows to define a characteristic time, the so-called correlation time,

$$\tilde{\tau} \equiv \frac{1}{\langle v_0^2 \rangle} \int_0^\infty \mathcal{C}_{\Delta\mathbf{K}}(\tau)\, d\tau, \tag{3.177}$$

where $\sqrt{\langle v_0^2 \rangle} = \sqrt{k_B T/m}$ is the average thermal velocity in one dimension—though the dimensionality is two, note that $\mathcal{C}_{\Delta\mathbf{K}}$ is defined along a particular direction (that given by $\Delta\mathbf{K}$) and therefore $\mathcal{C}_{\Delta\mathbf{K}}(0) \equiv k_B T/m$, i.e., the thermal velocity in one direction. The correlation time is related to the lineshape broadening in the sense that it provides a timescale for the decay of the intermediate scattering function and therefore information about the width of the Q–peak in the dynamic structure factor.

On the contrary, if we are dealing with quantum position operators, Eq. (3.174) is not longer valid due to the fact that the position operators at two different times do not commute. However, it is possible to factorize the intermediate scattering function in two factors, one of them being precisely that given by Eq. (3.174) [107].

In order to better understand the shape and broadening of the Q and T peaks within the SLB equation formalism, two simple cases are considered. In the first case, the simplest illustration for the Q-peak is to assume a Brownian motion of the adsorbates with $V(x, y) = 0$; that is, the diffusion of adsorbates on a flat surface. And the second one, for the T-peak, a good model is that provided by the harmonic oscillator motion in a potential well of a periodic corrugated surface.

3.8.1 The Bohmian–Brownian Motion

Apart from considering the interaction potential zero, one also can assume the diffusion of adsorbates on low corrugated surfaces where the energy provided by the surface temperature can easily surmount the corresponding energy barrier. Thus, the adparticle motion can be regarged as quasi–free since it is not ruled by a potential, but only influenced by stochastic forces. It can be shown [115])

$$\mathcal{C}(\tau) = \frac{k_B T}{m} e^{-\eta\tau}. \tag{3.178}$$

Unless the relaxation of the collisional effects can be of relevance, these effects and those caused by the surface should be indistinguishable and therefore (3.178) would describe accurately the loss of correlation, with $\tilde{\tau} = \eta^{-1}$.

The expression for the intermediate scattering function that results from (3.178) is

$$I(\Delta K, t) = \exp\left[-\chi^2 \left(e^{-\eta t} + \eta t - 1\right)\right], \tag{3.179}$$

where the so–called shape parameter χ is defined as

$$\chi^2 \equiv \frac{\langle v_0^2 \rangle \Delta K^2}{\eta^2} = \frac{D \Delta K^2}{\eta}. \tag{3.180}$$

From this relation we can extract both the mean free path, $\bar{l} \equiv \tilde{\tau}\sqrt{\langle v_0^2 \rangle}$, and the self–diffusion coefficient D which is also given by Einstein's relation

$$D = \lim_{t \to \infty} \int_0^t C(t')dt' \\ = \frac{k_B T}{m\eta}. \tag{3.181}$$

This ensures that the adparticle velocity distribution becomes Maxwellian asymptotically. The same result is reached from the mean squared displacement (MSD)

$$D = \lim_{t \to \infty} \frac{1}{4t} < (\mathbf{R}(t) - \mathbf{R}(0))^2 > . \tag{3.182}$$

When the coverage increases, the collisions among adsorbates are also expected to increase, and so λ and therefore η. As can be easily shown, (3.179) displays a Gaussian decay at short times that does not depend on the particular value of η, while at longer times it decays exponentially as ηt. Thus, two dynamical regimes can be clearly distinguished depending on the value of ηt. For $\eta t << 1$, collision events are rare and the adparticle shows an almost free motion with relatively long mean free paths. This is the ballistic or free-diffusion regime characterized by the MSD

$$\langle q^2(t) \rangle \sim \frac{k_B T}{m} t^2. \tag{3.183}$$

On the other hand, for $\eta t << 1$, there is no free diffusion since the effects of the stochastic force are dominant. This is the diffusive regime where the MSD is linear with time according to

$$\langle q^2(t) \rangle \sim \frac{2k_B T}{m\eta} t = 2Dt \tag{3.184}$$

which is an alternative way to express Einstein's law.

The type of decay is important concerning the width and shape of the dynamic structure factor which is analytically obtained from the time Fourier transform of Eq. (3.179),

$$S(\Delta K, \omega) = \frac{e^{\chi^2}}{\pi} \sum_{n=0}^{\infty} \frac{(-1)^n \chi^{2n}}{n!} \frac{(\chi^2 + n)\eta}{\omega^2 + [(\chi^2 + n)\eta]^2} . \tag{3.185}$$

As can be seen in the high friction limit, this function becomes a Lorentzian function which its full width at half maximum (FWHM) is $\Gamma = 2\eta\chi^2$. Thus, D is experimentally extracted from this width and following Eq. (3.180). Furthermore, Γ is approaching zero as η increases (narrowing effect). This in sharp contrast to what one could expect - as the frequency between successive collisions increases one would expect that the line shape gets broader (effect of the pressure in the spectral line shapes of gases [116]). The physical reason for this effect is relatively simple: as η increases the particle's mean free path decreases and therefore correlations are lost more slowly. In the limit case where friction is such that the particle remains in the same place, the space-time correlation function becomes a δ, the intermediate scattering function remains equal to one and the dynamic structure factor will also consist of a δ–function at $\omega = 0$. Conversely, in the low friction limit, the line shape is given by a Gaussian function whose width is $\Gamma = 2\sqrt{2\ln 2}\sqrt{k_B T/m}\,\Delta K$, which does not depend on η. This is precisely the case of a two dimensional free gas. This gradual change of the line shapes as a function of the friction and/or the parallel momentum transfer leading to a change of the shape parameter is known as the motional narrowing effect. Remember that in this formalism the friction is related to the coverage. Thus, at higher coverages a narrowing effect is predicted for a flat surface [115].[2]

In the quantum regime within the Heisenberg representation, the standard quantum Langevin equation has the following formal solution

$$\mathbf{R}(t) = \mathbf{R}(0) + \frac{\mathbf{P}(0)}{m\eta}\Phi(\eta t) + \frac{1}{m\eta}\int_0^t \Phi(\eta t - \eta t')\xi(t')dt' \tag{3.186}$$

where $\mathbf{P}(0)$ is the initial adparticle momentum operator and $\Phi(\eta t) = 1 - exp(-\eta t)$ and ξ is the sum of the two noises (ξ_G and ξ_S) for (x, y). As mentioned above, the operators $\mathbf{R}(t)$ and $\mathbf{R}(0)$ do not commute each other, leading to a much more involved diffusion dynamics according to Eq. (3.174); for example,the velocity autocorrelation function is complex with several decaying constants.

In Fig. 3.5, the real and imaginary parts of Eq. (3.174) for Na diffusion on a flat surface are plotted at two different surface temperatures, 50 and 100 K, and two coverages, 0.028 and 0.18 [107]. As can be clearly seen, the real part of $I(t)$ decreases faster with temperature and slower with coverage. On the other hand, the imaginary part of $I(t)$ displays maxima between 4–6% of the corresponding real part, depending on the temperature. It starts linearly with time and, after passing through a maximum, decays smoothly to zero. The corresponding quasielastic line shapes (around the zero energy exchange) will then display narrowing with the coverage and broadening with the surface temperature. This behavior could be experimentally confirmed for those systems where the diffusion barrier is smaller than the thermal energy $k_B T$. For light particles, the imaginary part is expected to be much more important keeping the same shape. The scattering law for the Q-peak is no longer a simple sum of Lorentzian functions [107]. However, quantum effects are mainly important at low surface temperatures, except if the adorbate is very light.

[2] At finite coverages, one usually distinguishes between two diffusion coefficients: the tracer and the collective diffusion constants [117]. The first one refers to the self–diffusion process and focuses on the motion of a single adsorbate. On the contrary, the second is related to the collective motion of all adsorbates which is governed by Fick's law. In any case, a Kubo–Green formula relates both diffusion coefficients with the velocity autocorrelation function of a single adsorbate or with the corresponding of the velocity of the center of mass, respectively.

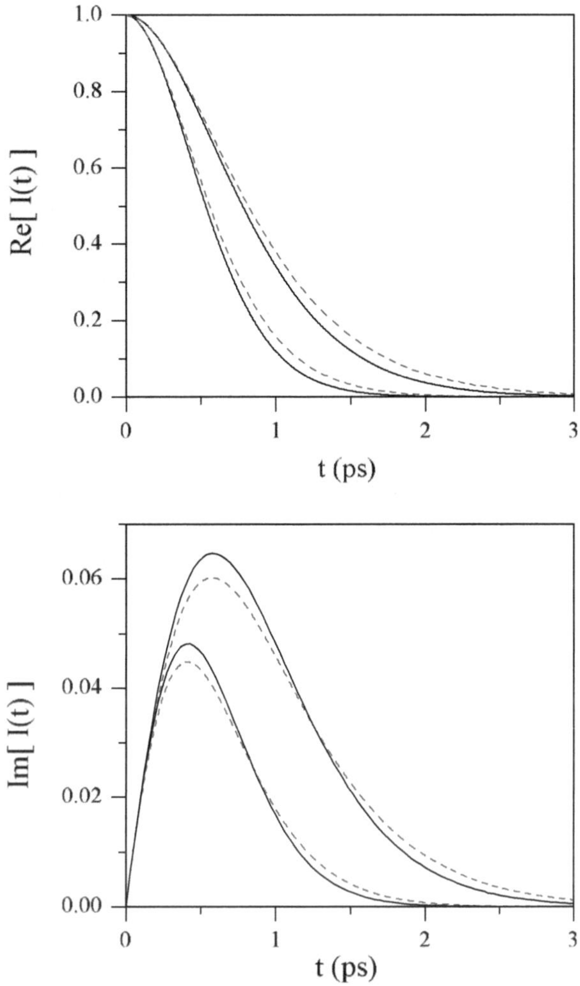

Fig. 3.5 Quantum intermediate scattering function (3.174) for Na diffusion on a flat surface at 50 and 100 K: **a** real part and **b** imaginary part. Two coverages are considered: $\theta = 0.028$ (*black solid line*) and $\theta = 0.18$ (*red dashed line*)

Once the simplest description of the surface diffusion (Brownian motion regime) is introduced, the next step is to see how the SLB formalism is applied to this very fundamental quantum process. In Sect. 3.3, we have discussed the free regime problem $V = 0$ with no random force. For simplicity and without loss of generality, we are going to deal with a one dimensional quantum surface diffusion (light adparticles). By starting from Eq. (3.29), and assuming a Gaussian probability density given by Eq. (3.33) for an adsorbate, the quantum trajectories are expressed by Eq. (3.38), which comes from a direct integration of Eq. (3.36) in time, and Eq. (3.39) ruling the time evolution of the wave packet width. Under the presence of noise, the quantum trajectory has the following expression

$$x(t) = q(0) + \frac{\dot{q}(0)}{\eta}\Phi(\eta t) + \frac{1}{m\eta}\int_0^t \Phi(\eta t - \eta t')\xi(t')dt' + (x(0) - q(0))\frac{\delta(t)}{\delta(0)} \tag{3.187}$$

where the classical trajectory $q(t)$ has been rewritten from Eq. (3.186).

In this diffusion problem, a critical function is the velocity autocorrelation function $\mathcal{C}_{\Delta\mathbf{K}}(\tau)$ which is defined along with the direction given by $\Delta\mathbf{K}$. For a flat surface, diffusion is isotropic (however, for a corrugated surface, that direction can be determined by a symmetry direction of the surface). After Eq. (3.36), we have that the quantum autocorrelation function is

$$\mathcal{C}_{\Delta\mathbf{K}}(t) \equiv \langle v(0)v(t)\rangle = \langle \dot{q}(0)\dot{q}(t)\rangle + \langle (x(0) - q(0))^2\rangle \frac{\dot{\delta}(0)}{\delta(0)^2}\dot{\delta}(t) \tag{3.188}$$

since $\langle \dot{q}(0)(x(0) - q(0))\rangle = 0$ and $\langle (x(0) - q(0))\dot{q}(t)\rangle = 0$, keeping the same structure than the quantum velocity, that is, a classical contribution given by the classical autocorrelation function and a quantum contribution governed by the time derivative of the Gaussian wave packet width (dressing scheme). Interestingly enough, when the initial velocity of spreading of the wave packet is zero, the quantum contribution disappears. In the long time limit, the classical contribution is simply given by Eq. (3.178). In this limit, the overdamped regime is established and therefore the acceleration term of the differential equation (3.41) for this width can be neglected. Thus, Eq. (3.188) is finally expressed

$$\langle v(0)v(t)\rangle = \frac{k_B T}{m} e^{-\eta\tau} + \langle (x(0) - q(0))^2\rangle \frac{\dot{\delta}(0)}{\delta(0)^2} \frac{\hbar^2}{4m\eta\delta(t)^3} \tag{3.189}$$

where $\delta(t)$ is asymptotically given by Eq. (3.44). Remember that the time dependence of $\delta(t)^{-3}$ comes from the quantum potential. Several important consequences are deduced from this time behavior. First, the line shape of the Q-peak is no longer a Lorentzian function; it is modified by the quantum contribution which goes with $t^{-3/4}$ according to Eqs. (3.175) and (3.165). Second, the relation of Einstein for the diffusion constant is also no longer valid. In fact, from Eq. (3.182),

$$< (x(t) - x(0))^2 > \propto t^\alpha \tag{3.190}$$

with $\alpha = 1/2$, indicating an anomalous diffusion process [118, 119]. In general, when the MSD is not linear in time, Einstein's relation is violated and the diffusion process can be described by a subdiffusion regime ($\alpha < 1$) or a superdiffusion regime ($\alpha > 1$), that is, slower or faster than ordinary Brownian motion [120]. Superdiffusion is originated by anomalous long jumps of a random walker while diffusion can be associated with unusually long waiting times between successive walks. A useful model to describe these processes is the so-called continuous time random walk. And third, this way is still completely unexplored and new studies need to be carried out for a better understanding of this quantum anomalous diffusion. An open question is if for very light adparticles, surface diffusion can exhibit this anomalous behavior.

3.8.2 The Harmonic Oscillator Motion

The harmonic potential model is an appropriate working model to understand the bound motion inside the wells of a corrugated surface and, therefore, to also understand the behavior associated with the T-mode (observed through the so-called T-peak). This mode comes from the oscillating behavior undergone by the adparticle when the diffusional motion is temporarily frustrated.

In contrast with the case of a dynamics where the interaction potential with the surface does not play a relevant role, we can devise a particle fully trapped within a harmonic potential well. The velocity autocorrelation function can be expressed [107, 115] as

$$\mathcal{C}(\tau) = \frac{k_B T}{m} e^{-\eta\tau/2} \left(\cos \bar{\omega}\tau - \frac{\eta}{2\bar{\omega}} \sin \bar{\omega}\tau \right), \tag{3.191}$$

where

$$\bar{\omega} \equiv \sqrt{\omega_0^2 - \frac{\eta^2}{4}} \tag{3.192}$$

and ω_0 is the harmonic frequency. Equation (3.191) can also be recast as

$$\mathcal{C}(\tau) = \frac{k_B T}{m} \frac{\omega_0}{\bar{\omega}} e^{-\eta\tau/2} \cos(\bar{\omega}\tau + \delta), \tag{3.193}$$

with

$$\delta \equiv (\tan)^{-1} \left(\frac{\eta/2}{\bar{\omega}} \right). \tag{3.194}$$

Note that Eq. (3.178) can be easily recovered after some algebra in the limit $\omega_0 \to 0$ from either Eq. (3.191) or (3.193). For anharmonic potentials, ω_0 represents the corresponding approximate harmonic frequency.

The only information about the structure of the lattice is found in the shape parameter through ΔK. When large parallel momentum transfers are under consideration, both the periodicity and structure of the surface have to be taken into account. Consequently, the shape parameter should be changed for different lattices. The simplest model including the periodicity of the surface is that developed by Chudley and Elliott [121] who proposed a master equation for the pair–distribution function in space and time assuming instantaneous discrete jumps on a two–dimensional Bravais lattice. Very recently, a generalized shape parameter has been proposed to be [122]

$$\chi_l(\Delta\mathbf{K}) \equiv \sqrt{\frac{\Gamma_\nu(\Delta\mathbf{K})}{2\gamma}}, \tag{3.195}$$

where

$$\Gamma_\nu(\Delta\mathbf{K}) = \nu \sum_{\mathbf{j}} P_{\mathbf{j}} \left[1 - \cos(\mathbf{j} \cdot \Delta\mathbf{K})\right], \tag{3.196}$$

ν being the total jump rate out of an adsorption site, and $P_{\mathbf{j}}$ the relative probability that a jump with a displacement vector $\mathbf{j}$ occurs.

Introducing now Eq. (3.193) into (3.175), this leads to the following expression of the intermediate scattering function

$$I(\Delta K, t) = \exp\left\{-\frac{\chi_l^2\eta^2}{\bar{\omega}\omega_0}\left[\cos\delta - e^{-\eta t/2}\cos(\bar{\omega}t - \delta)\right]\right\}. \tag{3.197}$$

The argument of this function displays an oscillatory behavior around a certain value with the amplitude of the oscillations being exponentially damped. This translates into an also decreasing behavior of the intermediate scattering function, which also displays oscillations around the asymptotic value. This means that after relaxation the intermediate scattering function has not fully decayed to zero unlike the free–potential case. It can be easily shown that in the limit $\omega_0 \to 0$, Eq. (3.197) approaches the result given by Eq. (3.179).

From (3.197) it is now straightforward to derive an expression for the dynamic scattering factor (the time Fourier transform of the corresponding intermediate scattering function), which results

$$S(\Delta K, \omega) \propto \sum_{m,n=0}^{\infty} \frac{(-1)^{m+n}}{m!\,n!} \chi_l^{2(m+n)} \frac{(2\chi_l^2 + m + n)\eta/2}{[\omega - (m-n)\bar{\omega}]^2 + [(2\chi_l^2 + m + n)\eta/2]^2}. \tag{3.198}$$

For a harmonic potential, there is no diffusion and, therefore, the line shapes (or scattering law) are only observed when $m \neq n$ consisting of Lorentzian functions related to vibrational excitations (creation and annihilation events of the T-mode). These Lorentzians are characterized by a width given by $\Gamma = (2\chi_l^2 + m + n)\eta/2$, which increases as η becomes larger. This broadening (proportional to η) undergone by the dynamic structure factor is thus contrary to the narrowing effect observed for a flat surface. It can be then assigned to the confined or bound motion displayed by the adparticle ensemble when trapped inside the potential wells. Hence, in order to detect broadening of the lineshapes in surface diffusion experiments, adparticles must spend some time confined inside potential wells, since it is induced by the presence of temporary vibrational motions.

On the other hand, the corresponding quantum trajectory in one-dimension has the following expression

$$
\begin{aligned}
x(t) = {} & \left[q(0) cos(\bar{\omega} t) + \left(\frac{\dot{q}(0)}{m\bar{\omega}} + \frac{\gamma}{2\bar{\omega}} q(0) \right) sin(\bar{\omega} t) \right] e^{-\eta t/2} \\
& + \int_0^{\infty} \chi(t - t') \xi(t') dt' \\
& + (x(0) - q(0)) \frac{\delta(t)}{\delta(0)}
\end{aligned} \tag{3.199}
$$

where the standard classical trajectory $q(t)$ is written in terms of the response function

$$
\chi(t) = \Theta(t) \frac{1}{m\bar{\omega}} e^{-\eta t/2} sin(\bar{\omega} t) \tag{3.200}
$$

with $\Theta(t)$ being the Heaviside function to ensures causality and $\delta(t)$ the width of the Gaussian wave packet which is governed by Eq. (3.39).

As said above, the velocity autocorrelation function has also two contributions, classical and quantum. Unfortunately, analytical expressions for this function as well as for the T-peak are no longer possible.

3.9 The Generalized Schrödinger–Langevin–Bohm Equation

As mentioned previously in the Introduction, in the system-plus-bath model, the dynamics is carried out in a conservative scenario. The so–called Caldeira-Leggett Hamiltonian [102] is the starting point leading to the generalized Langevin equation (GLE). One of the key issues is the interaction term which by construction is linear in the bath coordinates. For example, in one dimensional systems, the dependence on the system variable is through a function $f(x)$ which usually is separable and linear (linear dissipation). This scenario is known as a state–independent dissipation and can be seen as a measurement of particle's position by a reservoir in von Neumann's sense [2]. This function also appears in an additional term in the total Hamiltonian in order to avoid the renormalization of the interaction potential. In the Markovian regime with Ohmic (constant) friction, this GLE becomes the standard Langevin equation for a Brownian particle where noise is additive. As we have seen along this Chapter, Kostin [16, 17] established the link between this standard Langevin equation with the Schrödinger equation. The resulting logarithmic nonlinear equation is quite different from others existing in the literature such as the so–called stochastic Schrödinger equation and the Linblad equation for the density matrix [3].

For a nonlinear function $f(x)$, the open quantum system displays a state dependent dissipation process and the corresponding GLE exhibits multiplicative noise [33, 123]. Typical examples of nonlinear functions take place, for example, in rotational tunneling systems [124], quasi-particle tunneling in Josephson systems [125], in the Langevin canonical formulation of chiral two level systems [126], in atom surface scattering [127] and so on. Within the Markovian regime, the standard Langevin equation with multiplicative noise is reached. The main purpose of this Section is to derive a generalization of the SLB equation for nonlinear dissipation.

Without loss of generality, a one dimensional problem is again considered. As mentioned at the beginning of this Chapter and in the previous section, for open systems, it is usual to split the total Hamiltonian into three parts including system, bath and mutual coupling after Eq. (3.159). Let us consider explicitly the different Hamiltonians

$$H_S = \frac{p^2}{2m} + V(x) \tag{3.201}$$

which stands for the Hamiltonian of the isolated system in presence of a force field given by the potential $V(x)$;

$$H_B = \frac{1}{2}\sum_i \left(\frac{p_i^2}{m_i} + m_i \omega_i^2 x_i^2 \right) \tag{3.202}$$

is the Hamiltonian for the bath, which acts as a reservoir, and can be represented as an infinite set of harmonic oscillators; and

$$H_{SB} = \sum_i \left[\frac{f^2(x) d_i^2}{m_i \omega_i^2} - 2 d_i f(x) x_i \right] \tag{3.203}$$

expressing the interaction term between the isolated system and the bath, d_i being appropriate coupling constants. The function $f(x)$ is, in general, a nonlinear function of the system variable x. The term with the squared of $f(x)$ gives the so–called counter term introduced to compensate the renormalization of the potential. Following the standard procedure where the bath degrees of freedom are eliminated, the equation of motion for the corresponding system dynamics in the Heisenberg picture of quantum mechanics is given by the GLE [102]

$$\begin{aligned} f'[x(t)]\,\xi(t) &= m\ddot{x}(t) + V'(x) \\ &\quad + m f'[x(t)] \int_0^t dt' \gamma(t-t') f'\left[x(t')\right] \dot{x}(t') \end{aligned} \tag{3.204}$$

where the time–dependent friction (memory kernel) is given by

$$\gamma(t) = \frac{1}{m} \sum_i \frac{d_i^2}{m_i \omega_i^2} \cos(\omega_i t) \tag{3.205}$$

and the external force (noise term) is expressed as (written in Sect. 3.2 as F_r)

$$
\xi(t) = -\sum_i d_i \left[\left(x_i(0) + \frac{d_i}{m_i \omega_i^2} f(0) \right) \cos(\omega_i t) \right]
- d_i \sum_i \left[\frac{p_i(0)}{m_i \omega_i} \sin(w_i t) \right]. \tag{3.206}
$$

In the Markovian regime, the memory kernel is a δ–function in time, giving place to an Ohmic dissipation with a time–independent friction. Within this regime and for a nonlinear coupling, the corresponding standard Langevin equation reads as

$$
m\ddot{x}(t) + m\gamma \left[f'(x) \right]^2 \dot{x}(t) + V'(x) = f' \left[x(t) \right] \xi(t). \tag{3.207}
$$

Notice that the random force is multiplied by the derivative of the function $f(x)$ giving place to a stochastic process with multiplicative noise. When the system–bath coupling is linear, that is, for $f(x) = x$, the standard Langevin equation for additive noise (that is, when the noise term is not multiplied by any system function) with Ohmic friction is recovered

$$
m\ddot{x}(t) + m\gamma\dot{x}(t) + V'(x) = \xi(t). \tag{3.208}
$$

Following the Kostin procedure [16], the generalized SL equation (that is, for any $f(x)$) can be obtained by writing first the Schrödinger equation as

$$
i\hbar \frac{\partial \Psi}{\partial t} = \left[\frac{1}{2m} \left(-i\hbar \frac{\partial}{\partial x} \right)^2 + V(x) + V_D + V_r \right] \Psi, \tag{3.209}
$$

where the random potential V_r is given by

$$
V_r(x, t) = -f(x)\xi(t). \tag{3.210}
$$

and V_D is the damping potential, Eq. (3.4). On the other hand, the quantum mechanical current is defined as

$$
J = \frac{1}{2m} \left[\Psi^* \left(-i\hbar \frac{\partial}{\partial x} \right) \Psi + \Psi \left(-i\hbar \frac{\partial}{\partial x} \right) \Psi^* \right] \tag{3.211}
$$

Then, from $\partial \Psi / \partial t$ and $\partial \Psi^* / \partial t$ and Eq. (3.209) we have that

$$
\frac{d}{dt} \langle x \rangle = \int J dx
$$
$$
\frac{d^2}{dt^2} \langle x \rangle = \int \frac{\partial J}{\partial t} dx. \tag{3.212}
$$

where < . > is the expectation value of a given operator. Now, by performing the same type of averaging into Eq. (3.207)

$$m\langle\ddot{x}(t)\rangle + m\gamma\langle\left[f'(x)\right]^2 \dot{x}(t)\rangle + \langle V'(x)\rangle = \langle f'(x)\xi(t)\rangle, \tag{3.213}$$

and comparing it with Eq. (3.209) and using Eq. (3.212), we can identify terms leading to

$$\int \Psi^* \left(-\frac{\partial V_D}{\partial x}\right) \Psi dx = \gamma m \int f'(x)^2 J dx. \tag{3.214}$$

Thus, the damping potential is a functional of the wave function and can be expressed as

$$V_D\left[\Psi, \Psi^*, f\right] = -m\gamma \int \frac{\tilde{J}}{\Psi\Psi^*} dx, \tag{3.215}$$

where the new quantum mechanical current is now

$$\tilde{J} \equiv f'(x)^2 J \tag{3.216}$$

which is coupling–dependent. Finally, the corresponding generalized SL equation can then be expressed as [54]

$$i\hbar\frac{\partial\Psi}{\partial t} = \left[\frac{1}{2m}\left(-i\hbar\frac{\partial}{\partial x}\right)^2 + V(x) - m\gamma\int\frac{\tilde{J}}{\Psi\Psi^*}dx - f(x)\xi(t) - W(t)\right]\Psi. \tag{3.217}$$

where $W(t) = \langle V_D\rangle$ arises from the requirement that the integration of Eq. (3.215) with respect to x must be equal to the expectation values of the kinetic and potential energies through the total Hamiltonian. As mentioned previously, this term can be removed from Eq. (3.217) by introducing the transformation of the wave function given by Eq. (3.6). We also note that when the coupling function f is assumed to be linear in the system variable, Eq. (3.217) reduces to the standard Kostin equation.

Equation (3.217) can also be written in terms of the quantum trajectory formulation by expressing the wave function in polar form [67] as in Eq. (3.28). Doing this, Eq. (3.209) can be split into a system of two coupled equations,

$$\frac{\partial S}{\partial t} = -\frac{1}{2m}\left(\frac{\partial S}{\partial x}\right)^2 - (V + V_d + V_r + Q) \tag{3.218}$$

$$\frac{\partial\phi^2}{\partial t} = -\frac{1}{m}\frac{\partial S}{\partial x}\frac{\partial\phi^2}{\partial x} - \frac{\phi^2}{m}\frac{\partial^2 S}{\partial x^2}. \tag{3.219}$$

Moreover, the current density is expressed in this formalism as

$$J = \frac{\Psi\Psi^*}{m}\frac{\partial S}{\partial x}. \tag{3.220}$$

By differentiating Eq. (3.218) with respect the position, the time evolution of $p(x, t)$ is given by

$$\frac{\partial p}{\partial t} = -\frac{p}{m}\frac{\partial p}{\partial x} - \frac{\partial}{\partial x}\left(V + V_d + V_r + Q\right). \tag{3.221}$$

According to Eq. (3.207), which can be interpreted in terms of the Lagrangian framework of hydrodynamics, the corresponding quantum Newton–Langevin equation including the dissipative and random sources can be expressed as

$$\begin{aligned}\frac{\partial p}{\partial t} = &-\frac{p}{m}\frac{\partial p}{\partial x} - \frac{\partial}{\partial x}\left(V + Q\right)\\ &- \gamma f'(x)^2 p - f'(x)\xi(t).\end{aligned} \tag{3.222}$$

Then, by integrating Eq. (3.222) with respect to x, we obtain [54]

$$\begin{aligned}-\frac{\partial S}{\partial t} = &\frac{p^2}{2m} + V + Q + \gamma\int f'(x)^2 p dx\\ &+ f(x)\xi(t) + C(t),\end{aligned} \tag{3.223}$$

where $C(t)$ is an arbitrary time function resulting from the space integration to be specified later on. This equation gives the evolution of the wave function phase in presence of damping which corresponds to a generalized Caldeira–Leggett coupling.

Furthermore, the partial integration in Eq. (3.223) leads to

$$\int \left(\frac{df}{dx}\right)^2 p dx = \left(\frac{df}{dx}\right)^2 S - 2\int S\frac{df}{dx}\frac{d^2 f}{dx^2}dx, \tag{3.224}$$

which gives place naturally to the coupling–dependent phase

$$\tilde{S} \equiv \left(\frac{df}{dx}\right)^2 S - 2\int S\frac{df}{dx}\frac{d^2 f}{dx^2}dx. \tag{3.225}$$

which can be straightforwardly expressed as

$$\tilde{S} = m\int \frac{\tilde{J}}{\Psi\Psi^*}dx \tag{3.226}$$

with

$$V_D\left[\Psi, \Psi^*, f\right] = -\gamma\tilde{S}. \tag{3.227}$$

Importantly, Eq. (3.227) includes, as a special case, the Bohmian version of the Kostin equation when the coupling is linear. In this case, $\tilde{J} \rightarrow J$ and $V_D[\Psi, \Psi^*, f] \rightarrow V_D[\Psi, \Psi^*, x] = -\gamma S$, which is the dissipative potential expressed within the Bohmian formalism [8, 53, 128].

The constant of integration $C(t)$ can be defined in such a way that the overall phase of the wave function should not affect its evolution. This requirement is satisfied when $C(t) = -\gamma\langle\tilde{S}\rangle$. Therefore, $C(t) \equiv W(t)$ and for a general nonlinear coupling, the generalized SLB equation can be rewritten in terms of the modified quantum action, $\tilde{S}$, as

$$i\hbar\frac{\partial\Psi}{\partial t} = \left[\frac{1}{2m}\left(-i\hbar\frac{\partial}{\partial x}\right)^2 + V(x) - \gamma\left(\tilde{S} - \langle\tilde{S}\rangle\right) - f(x)\xi(t)\right]\Psi. \tag{3.228}$$

Thus, the corresponding generalized Hamilton–Jacobi equation for the nonlinear dissipation is given by Eq. (3.218) with V_D given by Eq. (3.227) and V_r by Eq. (3.210). Notice, however, that the continuity equation is the same expressed by Eq. (3.219).

A similar theoretical analysis could be carried out with the nonlinear Eq. (3.22) where the continuity equation has an extra term, a source or sink term.

References

1. Percival, I.: Quantum State Diffusion. Cambridge University Press, Cambridge (1998)
2. Weiss, U.: Quantum Dissipative Systems. World Scientific, Singapore (1999)
3. Breuer, H.-P., Petruccione, F.: The Theory of Open Quantum Systems. Oxford University Press, Oxford (2002)
4. Sanz, A.S., Miret-Artés, S.: A Trajectory Description of Quantum Processes. I. Fundamentals. Lecture Notes in Physics, vol. 850, pp. 1–299 (2012)
5. Sanz, A.S., Miret-Artés, S.: A Trajectory Description of Quantum Processes. II. Applications. Lecture Notes in Physics, vol. 831, pp. 1–333 (2014)
6. Accardi, L., Lu, Y.G., Volovich, I.: Quantum Theory and its Stochastic Limit. Springer, Berlin (2002)
7. Gardiner, C.W., Zoller, P.: Quantum Noise. Springer, Berlin (2004)
8. Razavy, M.: Classical and Quantum Dissipative Systems. Imperial College Press, London (2005)
9. Ford, G.W., O'Conell, R.F.: There is no quantum regression theorem. Phys. Rev. Lett. **77**, 798–801 (1996)
10. Kohen, D., Tannor, D.J.: Phase space approach to dissipative molecular dynamics. Adv. Chem. Phys. **111**, 219–398 (2000)
11. Oxtoby, D.W.: Dephasing of molecular vibrations in liquids. Adv. Chem. Phys. **40**, 1–48 (1979)
12. Levine, A.M., Shapiro, M., Pollak, E.: Hamiltonian theory for vibrational dephasing rates of small molecules in liquids. J. Chem. Phys. **88**, 1959–1966 (1988)
13. Bader, J.S., Berne, B.J., Pollak, E., Hänggi, P.: The energy relaxation of a nonlinear oscillator coupled to a linear bath. J. Chem. Phys. **104**, 1111–1119 (1996)
14. Bader, J.S., Berne, B.J.: Quantum and classical relaxation rates from classical simulations. J. Chem. Phys. **100**, 8359–8366 (1994)

15. Egorov, S.A., Berne, B.J.: Vibrational energy relaxation in the condensed phases: quantum vs classical bath for multiphonon processes. J. Chem. Phys. **107**, 6050–6061 (1997)
16. Kostin, M.D.: On the Schrödinger-Langevin equation. J. Chem. Phys. **57**, 3589–3591 (1972)
17. Kostin, M.D.: Friction and dissipative phenomena in quantum mechanics. J. Stat. Phys. **12**, 145–151 (1975)
18. Bialynicki-Birula, I., Mycielski, J.: Nonlinear wave mechanics. Ann. Phys. **100**, 62–93 (1976)
19. Gisin, N.: A simple nonlinear dissipative quantum evolution equation. J. Phys. A **14**, 2259–2267 (1981)
20. Gisin, N.: Microscopic derivation of a class of non-linear dissipative Schrödinger-like equations. Phys. A **111**, 364–370 (1982)
21. Nassar, A.B.: Derivation of a generalized nonlinear Schrödinger-Langevin equation. Phys. Lett. A **109**, 1–3 (1985)
22. Nassar, A.B.: Ermakov and non-Ermakov systems in quantum dissipative models. J. Math. Phys. **27**, 755–758 (1986)
23. Nassar, A.B.: Time-dependent invariant associated with nonlinear Schrödinger-Langevin equation. J. Math. Phys. **27**, 2949–2952 (1986)
24. Nassar, A.B.: New method for the solution of the logarithmic nonlinear Schrödinger equation via stochastic mechanics. Phys. Rev. A **33**, 3502–3505 (1986)
25. Khrennikov, A.: Nonlinear Schrödinger equations from prequantum classical statistical field theory. Phys. Lett. A **357**, 171–176 (2006)
26. Weinberg, S.: Testing quantum mechanics. Ann. Phys. **194**, 336–386 (1989)
27. Polchinski, J.: Weinberg's nonlinear quantum mechanics and the Einstein–Podolsky–Rosen paradox. Phys. Rev. Lett. **66**, 397–400 (1991)
28. Bassalo, J.M.F., da Silva, D.G., Nassar, A.B., Cattani, M.S.D.: The Feynman's propagators for non-linear Schrödinger equations. J. Adv. Math. Appl. **1**, 1–30 (2012)
29. Senitzky, I.R.: Dissipation in quantum mechanics. The harmonic oscillator. Phys. Rev. **119**, 670–679 (1960)
30. Ford, G.W., Kac, M., Mazur, P.: Statistical mechanics of assemblies of coupled oscillators. J. Math. Phys. **6**, 504–515 (1965)
31. Katz, R., Gossiaux, P.B.: The Schrödinger–Langevin equation with and without thermal fluctuations. Ann. Phys. (NY) **368**, 267–295 (2016)
32. Messer, J.: Friction in quantum mechanics. Acta Phys. Austriaca **50**, 75–91 (1979)
33. Hänggi, P., Jung, P.: Colored noise in dynamical systems. Adv. Chem. Phys. **LXXXIX**, 239-326 (1995)
34. Tsekov, R., Vayssilov, G.N.: Quantum Brownian motion and classical diffusion. Chem. Phys. Lett. **195**, 423–426 (1992)
35. Tsekov, R.: Nonlinear theory of quantum Brownian motion. Int. J. Theor. Phys. **48**, 85–94 (2009)
36. Tsekov, R.: Thermo-quantum diffusion. Int. J. Theor. Phys. **48**, 630–636 (2009)
37. Hasse, R.W.: On the quantum mechanical treatment of dissipative systems. J. Math. Phys. **16**, 2005–2011 (1975)
38. Nassar, A.B.: Irreversible quantum hydrodynamics from kinetic theory. Ann. Phys. **210**, 137–177 (1991)
39. Schuch, D.: Nonunitary connection between explicitly time-dependent and nonlinear approaches for the description of dissipative quantum systems. Phys. Rev. A **55**, 935–940 (1997)
40. Schuch, D., Chung, K.M., Hartmann, H.: Nonlinear Schrödinger equation-type field equation for the description of dissipative systems. I. Derivation of the nonlinear field equation and one-dimensional example. J. Math. Phys. **24**, 1652–1660 (1983)
41. Schuch, D., Chung, K.M., Hartmann, H.: Nonlinear Schrödinger equation-type field equation for the description of dissipative systems. III. Frictionally damped free motion as an example for an aperiodic motion. J. Math. Phys. **24**, 3086–3092 (1984)
42. Risken, H.: The Fokker-Planck Equation. Springer, Berlin (1984)
43. Chou, C.-C.: Dissipative quantum trajectories in complex space: damped harmonic oscillator. Ann. Phys. (NY) **373**, 325–345 (2016)

44. Doebner, H.-D., Goldin, G.A.: On a general nonlinear Schrödinger equation admitting diffusion currents. Phys. Lett. A **162**, 397–401 (1992)
45. Razavy, M.: Quantization of dissipative systems. Z. Phys. B **26**, 201–206 (1977)
46. Wagner, H.-J.: Schrödinger quantization and variational principles in dissipative quantum theory. Z. Phys. B **95**, 261–273 (1994)
47. Schuch, D.: Effective description of the dissipative interaction between simple and model-system and their environment. Int. J. Quantum Chem. **72**, 537–547 (1999)
48. Skagerstam, B.K.: Stochastic mechanics and dissipative forces. J. Math. Phys. **18**, 308–311 (1977)
49. Yasue, K.: Quantum mechanics of nonconservative systems. Ann. Phys. (NY) **114**, 479–496 (1978)
50. Tsekov, R.: Dissipative time dependent density functional theory. Int. J. Theor. Phys. **48**, 2660–2664 (2009)
51. Yuen-Zhou, J., Tempel, D.G., Rodriguez-Rosario, C.A., Aspuru-Guzik, A.: Time-dependent density functional theory for open quantum systems with unitary propagation. Phys. Rev. Lett. **104**, 043001(1–4) (2010)
52. Haas, F., Bassalo, J.M.F., da Silva, D.G., Nassar, A.B., Cattani, M.: Time-dependent Gaussian solution for the Kostin equation around classical trajectories. Int. J. Theor. Phys. **52**, 88–95 (2013)
53. Garashchuk, S., Dixit, V., Gu, B., Mazzuca, J.: The Schrödinger equation with friction from the quantum trajectory perspective. J. Chem. Phys. **138**, 054107(1–7) (2013)
54. Bargueño, P., Miret-Artés, S.: The generalized Schrödinger–Langevin equation. Ann. Phys. (NY) **346**, 59–65 (2014)
55. Chou, C.-C.: Trajectory approach to the Schrödinger–Langevin equation with linear dissipation for ground states. Ann. Phys. (NY) **362**, 57–73 (2015)
56. Haas, F.: The damped Pinney equation and its applications to dissipative quantum mechanics. Phys. Scr. **81**, 025004(1–7) (2010)
57. Ramsauer, C.: Über den Wirkungsquerschnitt der Gasmoleküle gegenüber langsamen Elektronen. Annalen der Physik **396**, 513–540 (1921)
58. Townsend, J.S., Bailey, M.A.: The motion of electrons in argon. Phil. Mag. **43**, 593–600 (1922)
59. Ramsauer, C., Kollath, R.: Über den Wirkungsquerschnitt der Edelgasmoleküle gegenüber Elektronen. Annalen der Physik **395**, 536–564 (1929)
60. Côté, R., Heller, E. J., Dalgarno, A.: Quantum suppression of cold atom collisions. Phys. Rev. A **53**, 234–241 (1996)
61. Zhao, B.S., Meijer, G., Schöllkopf, W.: Quantum reflection of He_2 several nanometers above a grating surface. Science **331**, 892–894 (2011)
62. Bassalo, J.M.F., Cattani, M.S.D., Alencar, P.T.S., Nassar, A.B.: Tópicos da Mecânica Quântica de de Broglie-Bohm (2010). Capítulo 5: http://publica-sbi.if.usp.br/PDFs/pd1655.pdf
63. Kukolich, S.G.: Demonstration of the Ramsauer–Townsend effect in a Xenon Thyratron. Am. J. Phys. **36**, 701–703 (1968)
64. Mott, N.F., Massey, H.S.W.: The Theory of Atomic Collisions. Clarendon Press, Oxford (1971)
65. Brode, R.B.: The quantitative study of the collision of electrons with atoms. Rev. Mod. Phys. **5**, 257–279 (1933)
66. Nassar, A.B.: Effect of dissipation on scatterring and tunneling through sharp-edged potential barriers. DFUFPA, preprint (1998)
67. Sometimes, it is convenient to express the wave function in either polar form: $\psi = \phi \exp iS/\hbar$ or $\psi = \phi \exp iS$, where the velocity $v = \frac{1}{m}\frac{\partial S}{\partial x}$ or $v = \frac{\hbar}{m}\frac{\partial S}{\partial x}$, respectively
68. Guerra, F.: Structural aspects of stochastic mechanics and stochastic field theory, Phys. Rep. **77**, 263–312 (1981). (The same volume **77** is entirely devoted to New Stochastic Methods in Physics, edited by C. DeWitt-Morette, K.D. Elworthy)
69. Ghosh, S.K., Deb, B.M.: Quantum fluid dynamics with a relativistic density-functional framework. J. Phys. A Math. Gen. **17**, 2463–2473 (1984)
70. Putterman, S.J., Roberts, P.: Classical non-linear waves in dispersive nonlocal media, and the theory of superfluidity. Phys. A **117**, 369–397 (1983)

71. Takabayasi, T.: Vortex, spin and triad for quantum mechanics of spinning particle. I: general theory. Prog. Theor. Phys. **70**, 1–17 (1983)
72. Bohm, D., Hiley, B.J.: Measurement understood through the quantum approach. Found. Phys. **14**, 255–274 (1984)
73. Nassar, A.B.: Fluid formulation of a generalised Schrödinger–Langevin equation. J. Phys. A Math. Gen. **18**, L509–L511 (1985)
74. Madelung, E.: Quantentheorie in hydrodynamischer form. Z. Phys. **40**, 322–326 (1926)
75. Bohm, D.: A suggested interpretation of the quantum theory in terms of "hidden" variables. I. Phys. Rev. **85**, 166–179 (1952)
76. Bohm, D.: A suggested interpretation of the quantum theory in terms of "hidden" variables. II. Phys. Rev. **85**, 180–193 (1952)
77. Krall, N.A., Trivelpiece, A.W.: Principles of Plasma Physics. McGraw-Hill, New York (1973)
78. Tanenbaum, B.S.: Plasma Physics. McGraw-Hill, New York (1967)
79. Mengoli, M., Molinari, V.G., Pizzo, F.: Nonlinear electron oscillation driven by an external wave in a plasma. Il Nuovo Cimento B **76**, 130–138 (1983)
80. Bell, J.S.: Speakable and Unspeakable in Quantum Mechanics. Cambridge University Press, Cambridge (1987)
81. Caldeira, A.O., Leggett, A.J.: Influence of damping on quantum interference: an exactly soluble model. Phys. Rev. A **31**, 1059–1066 (1985)
82. Grubin, H.L., Kreskovsky, J.P.: Quantum moment balance equations and resonant tunneling structures. Solid-State Electron. **32**, 1071–1075 (1989)
83. Grubin, H.L., Kreskovsky, J.P., Grovindan, T.R., Ferry, D.K.: Uses of the quantum potential in modelling hot-carrier semiconductos devices. Semicond. Sci. Technol. **9**, 855–858 (1994)
84. Lorentz, H.A.: La theórie eléctromagnetique de Maxwell et son application aux corps mouvemants. Arch. Neérl. Sci. Exactes Nat. **25**, 363–552 (1892)
85. Lorentz, H.A.: The Theory of Electrons, Teubner, Leipzig (1909). (2nd edn, 1916)
86. Abraham, M.: Theorie der Elektrizitat. Vol II: Elektromagnetische Theorie der Strahlung, Teubner, Leipzig (1905)
87. Rohrlich, F.: The dynamics of a charged sphere and the electron. Am. J. Phys. **65**, 1051–1056 (1997)
88. Rohrlich, F.: The self-force and radiation reaction. Am. J. Phys. **68**, 1109–1112 (2000)
89. Rohrlich, F.: Classical Charged Particles. Addison-Wesley, Reading, MA (1965)
90. Griffiths, D.: Introduction to Electrodynamics. Prentice Hall, New York (2004)
91. Sommerfeld, A.: Simplified deduction of the field and the forces of an electron moving in any given way. Proc. R. Acad. Amst. **7**, 346–367 (1905))
92. Page, L.: Is a moving mass retarded by the reaction of its own radiation? Phys. Rev. **11**, 376–400 (1918)
93. Kim, K.-J., Sessler, A.M.: 8th Workshop on Advanced Acceleration Concepts. Baltimore, MD (1998)
94. Jackson, J.D.: Classical Electrodynamics. Wiley, New York (1998)
95. Moniz, E.J., Sharp, D.H.: Radiation reaction in nonrelativistic quantum electrodynamics. Phys. Rev. D **15**, 2850–2865 (1977)
96. Levine, H., Moniz, E.J., Sharp, D.H.: Motion of extended charges in classical electrodynamics. Am. J. Phys. **45**, 75–78 (1977)
97. Cook, R.J.: Radiation reaction revisited. Am. J. Phys. **52**, 894–895 (1984)
98. Low, F.: Run-Away Electrons in Relativistic Spin 1/2 Quantum Electrodynamics. Preprint MIT-CTP-2522 (1997)
99. Denef, F., Raeymaekers, J., Studer, U.M., Troost, W.: Classical tunneling as a consequence of radiation reaction forces. Phys. Rev. E **56**, 3624–3627 (1997)
100. Feynman, R.P.: Statistical Mechanics. Perseus Books, Cambridge (1998)
101. Lovesey, S.W.: Theory of Neutron Scattering from Condensed Matter. Clarendon, Oxford (1984)
102. Caldeira, A.O., Leggett, A.J.: Influence of dissipation on quantum tunneling in macroscopic systems. Phys. Rev. Lett. **46**, 211–214 (1981)

103. Caldeira, A.O., Leggett, A.J.: Path integral approach to quantum Brownian motion. Phys. A **121**, 587–616 (1983)
104. Ford, G.W., Kac, M.: On the quantum Langevin equation. J. Stat. Phys. **46**, 803–810 (1987)
105. Sánchez-Cañizares, J., Sols, F.: Translational symmetry and microscopic preparation in oscillator models of quantum dissipation. Phys. A **122**, 181–193 (1994)
106. Miret-Artés, S., Pollak, E.: The dynamics of activated surface diffusion. J. Phys. Condens. Matter **17**, S4133–S4150 (2005)
107. Martínez-Casado, R., Sanz, A.S., Vega, J.L., Rojas-Lorenzo, G., Miret-Artés, S.: Linear response theory of activated surface diffusion with interacting adsorbates. Chem. Phys. **370**, 180–193 (2010)
108. Van Hove, L.V.: Correlations in space and time and Born approximation scattering in systems of interacting particles. Phys. Rev. **95**, 249–262 (1954)
109. Vineyard, G.H.: Scattering of slow neutrons by a liquid. Phys. Rev. **110**, 999–1010 (1958)
110. McQuarrie, D.A.: Statistical Mechanics. Harper and Row, New York (1976)
111. Ellis, J., Graham, A.P., Hofmann, F., Toennies, J.P.: Coverage dependence of the microscopic diffusion of Na atoms on the Cu(001) surface: a combined helium atom scattering experiment and molecular dynamics study. Phys. Rev. B **63**, 195408(1–9) (2001)
112. Martínez–Casado, R., Vega, J.L., Sanz, A.S., Miret–Artés, S.: Surface diffusion and low vibrational motion with interacting adsorbates: a shot noise description. Phys. Rev. E **75**, 051128(1–12) (2007)
113. Martínez–Casado, R., Vega, J.L., Sanz, A.S., Miret–Artés, S.: Line shape broadening in surface diffusion of interacting adsorbates with quasielastic He atom scattering. Phys. Rev. Lett. **98**, 216102(1–4) (2007)
114. Martínez–Casado, R., Vega, J.L., Sanz, A.S., Miret–Artés, S.: Stochastic theory of lineshape broadening in quasielastic He atoms scattering with interacting adsorbates. Phys. Rev. B **77**, 115414(1–5) (2008)
115. Martínez–Casado, R., Vega, J.L., Sanz, A.S., Miret–Artés, S.: Quasielastic He atom scattering from surfaces: a stochastic description of the dynamics of interacting adsobates. J. Phys.: Condens. Matter **19**, 305002(1–27) (2007)
116. van Vleck, J.H., Weisskopf, V.F.: On the shape of collision-broadened lines. Rev. Mod. Phys. **17**, 227–235 (1945)
117. Gomer, R.: Diffusion of adsorbates on metal surface. Rep. Prog. Phys. **53**, 917–1002 (1990)
118. Wolfgang, P., Baschnagel, J.: Stochastic Processes. From Physics to Finance. Springer, Berlin (1999)
119. Metzler, R., Klafter, J.: The random walk's guide to anomalous diffusion: a fractional dynamics approach. Phys. Rep. **339**, 1–77 (2000)
120. Vega, J.L., Guantes, R., Miret-Artés, S.: Chaos and transport properties of adatoms on solid surfaces. J. Phys. Condens. Matter **14**, 6193–6232 (2002)
121. Chudley, C.T., Elliott, R.J.: Neutron scattering from a liquid on a jump diffusion model. Proc. Phys. Soc. **77**, 353–361 (1961)
122. Martínez–Casado, R., Vega, J.L., Sanz, A.S., Miret–Artés, S.: A generalized Chudley-Elliott vibration-jump model in activated atom surface diffusion. J. Chem. Phys. **126**, 194711(1–5) (2007)
123. Sancho, J.M., San Miguel, M., Katz, S.L., Gunton, J.D.: Analytical and numerical studies of multiplicative noise. Phys. Rev. A **26**, 1589–1609 (1982)
124. Stevens, K.H.: The theory of planar methyl rotation. J. Phys. C **16**, 5765–5772 (1983)
125. Ambegaokar, V., Eckern, U.: Quantum dynamics of tunneling between superconductors: microscopic theory rederived via an oscillator model. Z. Physik B **69**, 399–407 (1987)
126. Dorta-Urra, A., Peñate-Rodriguez, H.C., Bargueño, P., Rojas-Lorenzo, G., Miret-Artés, S.: Dissipative geometric phase and decoherence in parity-violating chiral molecules. J. Chem. Phys. **136**, 174505(1–6) (2012)
127. Miret-Artés, S., Pollak, E.: Classical theory of atom-surface scattering: the rainbow effect. Surf. Sci. Rep. **67**, 161–200 (2012)
128. Nassar, A.B., Miret-Artés, S.: Dividing line between quantum and classical trajectories in a measurement problem: Bohmian time constant. Phys. Rev. Lett. **111**, 150401(1–5) (2014)

Chapter 4
Continuous Quantum Measurements in the Bohmian Framework

Abstract Following a phenomenological approach for continuous measurement, a general theory for decoherence in the framework of restricted path integrals (RPI) has been proposed by Mensky. The corresponding propagator is modified according to the information provided by the measurement through the so-called quantum corridors, which correspond to different readouts of the measurement. The measured system is also considered in this theory as an open system since the back reaction of the environment, considered as a measuring apparatus, is taken into account implicitly. In this chapter, the RPI formulation is briefly presented and discussed since our approach is inspired from Mensky's procedure. The linear time dependent Schrödinger equation derived from this formulation is reached for a non-Hermitian (complex) effective Hamiltonian which takes into account the measurement readout. The RPI formulation is equivalent to a master equation, a special case of the Linblad equation, and to a stochastic Schrödinger equation. Within the Kostin framework, a logarithmic nonlinear Schrödinger equation is proposed and discussed by extending Mensky's approach to be analyzed in terms of Bohmian trajectories, the so-called SLB equation for continuous measurement. When a time-dependent Gaussian shape is assumed for the probability density, soliton-like solutions are first analyzed in order to establish the dividing line between the quantum and classical trajectories, leading to the concept of Bohmian time. Afterwards, the solutions of the Schrödinger–Langevin–Bohm (SLB) equation for continuous measurement are analyzed in terms of mathematical stability for three simple cases, that is, the free particle and the linear and harmonic potentials. In particular, the continuous position measurement is seen that it is no longer governed by a standard continuity equation. The sign of the extra source/sink term appearing in this new equation is critical for the stability of the solutions. Two types of general solutions are found: (i) the stationary wave packet considered becomes unstable (that is, this solution is not a physically acceptable solution for that process since, as time evolves, the width of the corresponding wave packet becomes more delocalized) and (ii) the probability density becomes more and more localized at asymptotic times (the stationary wave packet is then an attractor of the corresponding width dynamics); in terms of trajectories, this analysis indicates that the Bohmian trajectories are not approaching the corresponding

A.B. Nassar and S. Miret-Artés, *Bohmian Mechanics, Open Quantum Systems and Continuous Measurements*, DOI 10.1007/978-3-319-53653-8_4

classical ones. These conclusions are also corroborated by carrying out an analysis of the entropy. Finally, it is also briefly discussed the question of coupling classical variables (apparatus) to quantum ones (system) in terms of Lyapunov exponents.

4.1 Introduction

It is well known that the path integral formulation is equivalent to the Schrödinger picture of quantum mechanics. Thus, the restricted path integral (RPI) formalism due to Mensky [1, 2] is also directly related to a linear Schrödinger equation but with complex Hamiltonian and the measuring process can then be described in terms of wave functions. A straightforward extension to a master equation in terms of the density matrix can also be carried out. Doing this, the *non-selective* description of the measurement is reached when the concrete readout is not known. In other words, the summation or integration is carried out over the set of alternatives readouts. Furthermore, dealing with open systems, it is natural to ask oneself if the RPI formulation is related to a stochastic (nonlinear) Schrödinger equation [3, 4].

Inspired by Mensky's work, Nassar has proposed that the evolution of the wave function of the quantum system under continuous measurement can also be described in terms of a logarithmic nonlinear Schrödinger equation [5]. This equation clearly provides a good starting point for a hydrodynamical description of the measuring process in terms of quantum trajectories. We would like to show that one of the fundamental aspects of Bohmian mechanics is its ability to tackle more intuitively the quantum measurement problem. Following the same theoretical scheme, Nassar and Miret-Artés [6] have also proposed that the time evolution of the wave function of a quantum dissipative system under continuous measurement can also be described in terms of a more general nonlinear Schrödinger equation. The logarithmic nonlinearity for the dissipative part is the same one proposed by Kostin when considering open quantum systems (see Chap. 3) [7]. Recently, this nonlinear Schrödinger equation has been investigated by Zander et al. [8].

As pointed out by Bell [9], the lack of clarity in regards to where the transition between the classical and quantum regimes is located is one aspect of the measurement problem. This problem represents one of the most important conceptual difficulties in quantum mechanics. A solution to the challenge posed by Bell on the dividing line between the quantum and classical regimes in a measurement problem is proposed here within the Bohmian context. Via the quantum trajectory method, a solution to this equation reveals a novel result: it displays a time threshold which establishes the dividing line between the quantum and classical trajectories. It is shown that continuous measurements and damping not only disturb the particle but compel the system to converge in time to a Newtonian regime without any assumption of collapse. In particular, it is shown that damping tends to suppress further quantum effects on a time scale shorter than the relaxation time of the system following also a gradual decoherence process. The solution to this equation is found through a wave packet approach which establishes a direct correlation between a classical variable with a quantum variable describing the dynamics of the center of mass and the width of the wave packet. The approach presented here gives a comparatively clearer picture than approaches using restricted path integrals and master equation.

For these goals, this chapter is organized as follows. In Sect. 4.2, the so-called characteristic function of a quantum measurement which is very convenient for the RPI formulation is described. In Sect. 4.3, since our approach is inspired from Mensky's procedure, a very short introduction to path integrals and restricted path integrals is presented In Sect. 4.4, some equations derived from the RPI formalism are presented and discussed, such as the linear time dependent Schrödinger equation for a non-Hermitian (complex) effective Hamiltonian which takes into account the measurement readout, a master equation which is a special case of the so-called Linblad equation, and a stochastic Schrödinger equation. In Sect. 4.5, a logarithmic nonlinear Schrödinger equation for continuous measurement is proposed and discussed by extending Mensky's approach within the Bohmian formalism which is equivalent to the one presented in Chap. 3, Eq. (3.22). In Sect. 4.6, this nonlinear equation is again extended within the Kostin framework to be analyzed in terms of Bohmian trajectories, resulting the Schrödinger–Langevin–Bohm equation for continuous measurement. Some important consequences are also analyzed when considering Eq. (3.22) instead. In Sect. 4.7, it is established the dividing line between the quantum and classical trajectories, leading to the concept of Bohmian time. Section 4.8 is devoted to the analysis of the solutions of the Schrödinger–Langevin–Bohm in terms of mathematical stability for three simple cases, that is, the free particle and the linear and harmonic potentials. When a time-dependent Gaussian shape is assumed for the probability density, the continuous quantum measurement process (in particular, the continuous position measurement) is no longer governed by a standard continuity equation. The extra source/sink term is mainly dependent on the sign (positive/negative) of such a parameter. Two types of general solutions are found: (i) the stationary wave packet considered becomes unstable (that is, this solution is not a physically acceptable solution for that process since, as time evolves, the width of the corresponding wave packet becomes more delocalized) and (ii) the probability density becomes more and more localized at asymptotic times (the stationary wave packet is then an attractor of the corresponding width dynamics). In terms of stochastic trajectories, this analysis indicates that the corresponding Bohmian trajectories are not approaching the corresponding classical ones. These conclusions are also corroborated by carrying out an analysis of the entropy. Finally, Sect. 4.9 sets out the question of coupling classical variables (apparatus) to quantum ones (system) in terms of Lyapunov exponents.

4.2 Measurement and the Characteristic Function

As is well known, the Heisenberg uncertainty principle (HUP) has a double interpretation. First, it concerns the relationship between two observables in a given quantum state or, in general, between the so-called conjugate observables. And, second, when dealing with *instantaneous* quantum measurements, it asserts that it is impossible to measure simultaneously two conjugate observables with arbitrary precision. The most general expression of this principle is given by

$$\Delta A \Delta B \geq \frac{1}{2}\langle\Psi|[A,B]|\Psi\rangle \tag{4.1}$$

where A and B are two Hermitian operators. When these two operators are canonically conjugate to each other, or their commutation relation is the same as for the coordinate and momentum, the usual expression is obtained.[1] The squared of the uncertainty or quantum variance of an operator A is defined as

$$\Delta A^2 = \langle\Psi|A^2|\Psi\rangle - \langle\Psi|A|\Psi\rangle^2. \tag{4.2}$$

Following Mensky [2], if the uncertainties ΔA and ΔB characterize the state before the measurement, and δA and δB refer to the state after the measurement, and DA is the measurement error or resolution, then

$$\delta B \geq \frac{1}{2}\frac{\hbar}{DA} \tag{4.3}$$

for two canonically conjugate observables. If, on the contrary, $[A, B] = iC$, then

$$\delta B \geq \frac{1}{2}\frac{|\langle C\rangle|}{DA} \tag{4.4}$$

where the mean value of C is over the final state, that is, after the measurement. These expressions are no longer valid when continuous quantum measurements are carried out. For example, when monitoring the position up to a certain error, information about the momentum can not be expressed as dictated by the Heisenberg principle.

It is very well known that when a quantum system is measured, its corresponding dynamics is unavoidably affected. Following the standard theory due to von Neumann about quantum measurements, an idealized measurement is described by the reduction postulate or collapse of the wave function. The conversion of a superposition of states to the mixture of the same states is called decoherence. Interaction between the system and its environment or measuring apparatus leads to entanglement or quantum correlation. When the measurement is carried out, the measured system decoheres. This measurement in standard quantum mechanics is described in terms of projection operators, a complete set of orthogonal projectors.

In general, when dealing with composite systems (for example, $S = S_1 + S_2$), the corresponding Hilbert space is the tensor product space of the Hilbert spaces describing its subsystems ($\mathcal{H} = \mathcal{H}_1 \otimes \mathcal{H}_2$). The corresponding density matrix when both subsystems are uncorrelated takes also the form of a tensor product, $\rho = \rho_1 \otimes \rho_2$,

[1] As $[q, p] = i\hbar$, then $\Delta q \Delta p \geq \frac{1}{2}\hbar$ and, therefore, $\Delta A \Delta B \geq \frac{1}{2}\hbar$.

and the corresponding expectation value of any tensor product of operators pertaining to the subsystems factorizes. The reduced density matrix of one of the subsystems is obtained by tracing over the second Hilbert space, $\rho_1 = \mathrm{Tr}_2\rho$. This is called a coarse grained description since very often we are not interested in all details of one of the subsystems, usually the more extended one which is considered as the environment. A state of $\mathcal{H}$ is a product state if and only if the reduced density matrices ρ_1 and ρ_2 describe pure states. On the other hand, if a state of $\mathcal{H}$ can not be written as a tensor product of states of the subsystems is said to be *entangled*.

The physical process leading to the transition from the density matrix (describing the pure state $|\psi\rangle$) $\rho = |\psi\rangle\langle\psi|$ to the density matrix of a mixed state where all the coherences are zero (or when phase relations are lost) is an alternative way to describe *decoherence*. The off-diagonal terms are typically decaying as an exponential function where the decoherence time is inversely proportional to the squared difference between the measured values of the corresponding observable. In other words, decoherence appears when quantum interference is destroyed or suppressed. In a certain sense, it could be said that decoherence leads to the emergence of a classical world in quantum mechanics. Environment induced decoherence [10] is omnipresent and, in general, it is a short time phenomenon. When measuring, the dynamics of the system tends to be decoherent. In the weak coupling approximation between the system and environment, the trace operation carried out to obtain the reduced density matrix is key to the existence of any master equation. However, this operation is questionable when the system and environment are entangled at all times, including the initial state. The role of the initial conditions has also be widely discussed in the corresponding literature.

In a measurement, the physical mechanism leading to decoherence consists of the interaction of the measured system with its environment (for example, a measuring device). As mentioned above, this interaction also leads to their entanglement or quantum correlation of the two systems. Thus, the information about the system is somehow recorded in the state of the environment. This process can also be described by the reduced density matrix of the measured system (when tracing over the degrees of freedom of the device). Doing this, it can easily be shown (starting from a product state of wave functions), that one more property is needed for decoherence, the orthogonality of the measuring device wave functions before and after the measurement. This condition is satisfied for macroscopically distinct states and the decoherence is irreversible. When dealing with microscopic environments where such a condition is not satisfied, the decoherence process is reversible. An alternative way of seeing this is by considering the environment split into two parts, one microscopic part directly interacting with the system, usually called a *meter* or (*pointer*), and this, in turn, interacts with a macroscopic part usually named *reservoir*. Thus, decoherence induced by entanglement through a unitary evolution of both the system and meter is practically reversible. When the reservoir is also present, decoherence may be considered reversible, in principle, but it is irreversible from a practical point of view. However, the *many-worlds* interpretation by Everett is an attempt to overcome irreversibility completely. It seems to be impossible to verify it or to be testable; in other words, to be falsified.

When one wants to obtain information on the state of the system by means of the measurement on the meter, we speak about the indirect (or *ancilla*) measurement [11, 12]. As said before, in this composite system, correlations between the measured system and the meter are built up during their interaction. The time-dependent interaction Hamiltonian is assumed to vanish outside the time interval $[0, \tau]$. If the time interaction is sufficiently small to suppose that only the interaction Hamiltonian is acting, the unitary time evolution operator takes a very simple form in terms of the integrated coupling strength G (given by the integration of the time-dependent coupling $g(t)$ in that time interval). For example, if the position operator Q, canonically conjugate to the momentum operator P of the meter, is coupled to an observable A the interaction Hamiltonian is given by

$$H_I(t) = g(t)AQ \tag{4.5}$$

the evolution operator being then written as

$$U = \exp(-iGAQ). \tag{4.6}$$

It is a straightforward exercise to show that the variance of the inferred eigenvalue of A (through one the measurement outcomes of P) is the sum of the variance of A and the variance of P in the initial state divided by G^2. Thus, the measurement is the more accurate the lower the momentum uncertainty of the meter and the stronger the coupling between the system and meter.

In standard textbooks, the measurement is introduced through the projection procedure due to von Neumann (the so-called von Neumann's reduction postulate). In this ideal or instantaneous measurement, the system is projected onto one of the possible eigenstates of a given observable. In a given basis set of the observable A, the state of the system is expressed as

$$|\psi\rangle = \sum_m c_m |m\rangle. \tag{4.7}$$

The probability that the system is projected onto $|m\rangle$ is $|c_m|^2$. In other words, the transition from the state given by Eq. (4.7), before the measurement, to one of its eigenstates, after the measurement, is the simplest form of the state reduction or wave function collapse. This collapse is considered to be an irreversible process. In this sense, it is claimed that this type of measurement is complete. If $P_m = |m\rangle\langle m|$ is the projection operator onto the eigenstate $|m\rangle$ and the initial state of the system is $\rho = |\psi\rangle\langle\psi|$, the m possible outcome of the final state is given by

$$\rho_f = |m\rangle\langle m| = \frac{P_m \rho P_m}{\mathrm{Tr}\, P_m \rho P_m} \tag{4.8}$$

with probability $P(m) = \mathrm{Tr}(P_m \rho P_m) = |c_m|^2$ and $\mathrm{Tr}\rho_f = 1$. This is the well-known von Neumann–Lüders projection postulate. Under the collapse condition, one may ask for the density matrix after the measurement to have a particular measurement outcome or alternative (*selective measurement*) or, for one reason or another, one does not register the outcomes or alternatives (*Nonselective measurement*). In this case, we have to sum over all possible outcomes.

The von Neumann–Lüders postulate can be generalized to a set of operators Ω_m with the condition $\sum_{m=1}^{M} \Omega_m^\dagger \Omega_m = I$, I being the identity operator. The different outcomes can be again written as

$$\rho_f = \frac{\Omega_m \rho \Omega_m^\dagger}{\mathrm{Tr}(\Omega_m \rho \Omega_m^\dagger)} \tag{4.9}$$

with $P(m) = \mathrm{Tr}(\Omega_m \rho \Omega_m^\dagger) = \mathrm{Tr}(\Omega_m^\dagger \Omega_m \rho)$, being the probability of having the mth outcome. These generalized measurements are referred to as *positive- operator-valued measure*. The measurements described by such operators are minimally disturbing. Furthermore, the positive operator $F = \sum_{m=a}^{b} \Omega_m^\dagger \Omega_m$ determines the probability for obtaining an outcome lying in the range of real numbers $[a, b]$. This positive operator is also called the *effect* and the operator $\Omega_m \rho \Omega_m^\dagger$ is called an *operation*. A measurement is a quantum non-demolition (QND) measurement for the observable A if the operators Ω_m describing the change of the quantum system induced by the measurement commute with A. This condition is always fulfilled for an ideal measurement of the observable A. Moreover, in a composite system, if A is a constant of motion under the free evolution of a quantum system, a general sufficient condition for a QND measurement takes place when the commutator of A and the interaction Hamiltonian is zero. The more general forms for operations and effects are established by the so-called representation theorem for quantum operations. The reader interested in more details about the properties of such operators are addressed to Refs. [10–17]

The Ω operators could be thought to be a weighted (for example, a Gaussian function; the weight function is also called a *characteristic function* of the measurement) sum of projectors onto the eigenstates $|m\rangle$, each one peaked around a different value of the observable such as

$$\Omega_m = \frac{1}{\bar{N}} \sum_n e^{-(n-m)^2/2\sigma_0^2} |n\rangle\langle n| \tag{4.10}$$

where $\bar{N}$ is a normalization constant chosen to fulfill the normalization condition for $\Omega_m^\dagger \Omega_m$ previously established, the mean value is given by m and the variance by σ_0^2. After the measurement, with an outcome m, the state of the system is given by

$$\rho_f = \frac{\Omega_m \rho \Omega_m^\dagger}{\mathrm{Tr}(\Omega_m \rho \Omega_m^\dagger)} = \frac{1}{\bar{N}} \sum_n e^{-(n-m)^2/2\sigma_0^2} |n\rangle\langle n| \tag{4.11}$$

that is, the final state is peaked around the eigenvalue m with a width given by the variance σ_0^2. The width σ_0 of this function corresponds to the resolution of the measurement; sometimes, we also speak about the uncertainty of the observable. For large variances, we have *fuzzy, soft or unsharp measurements* (sometimes, the adjective *weak* is also used but we are going to reserve this adjective for the weak measurement due to Aharonov et al. [18]); conversely, *strong, projecting or sharp measurements* are obtained for small variances. Measurement devices have in general a finite resolution, given by the variance in the example above. If the eigenvalues of a given observable are not too closely spaced, one can measure such an observable with a very small uncertainty, provided the experimental resolution is sufficiently large. However, if such a resolution is not high enough to distinguish neighbouring eigenvalues, our measure is quite approximate [11, 12]. The renewed interest in coupling classical systems to quantum ones has been revived by a number of authors [2, 19, 20] who have examined continuous quantum measurements. Nowadays, continuous or repeated (prolonged in time) measurements of a quantum system is a very active field mainly for two reasons. First, it is a direct way to look into the intrinsic features of quantum theory and second, such measurements are gaining step by step a predominant role in typical experimental setups [2]. In this context, one of the most well known illustrations is the so-called quantum Zeno effect [21], as well as the anti-Zeno effect [22], where the decaying of an unstable system can be delayed or accelerated, respectively. This process is carried out in a series of projecting (sharp, hard or strong) measurements. In fact, the Zeno effect is a nice illustration of the reduction or collapse postulate of the wave function due to von Neumann. Each observation collapses the wave function, resetting the clock and allowing to delay, in principle, indefinitely or freezing a given decaying process. Clearly, this series of measurements has to be carried out under certain conditions.

A continuous measurement is performed when information is continually extracted from the system, by reducing such an information with the duration of the measurement. This time is divided into a sequence of intervals of length Δ where a fuzzy measurement is carried out. If, for example, the observable we want to measure is the position of a particle, X, which displays a continuous spectrum of eigenvalues x with $\langle x|x'\rangle = \delta(x-x')$, one can consider a fuzzy measurement in each time interval Δt described by the operators [23]

$$\Omega(\alpha) = \left(\frac{1}{\pi\sigma_0^2\Delta t}\right)^{1/4} \int_{-\infty}^{\infty} e^{-(x-\alpha)^2/2\sigma_0^2\Delta t} |x\rangle\langle x| dx \tag{4.12}$$

where each operator is a Gaussian-weighted sum of projectors onto the eigenstates of X peaked at α, which is a real number. In other words, the continuum of measurement outcomes is labelled by α. The probability density $P(\alpha)$ of the outcome α, when Δt is small, is then given by

$$P(\alpha) = \mathrm{Tr}(\Omega(\alpha)^\dagger\Omega(\alpha)|\psi\rangle\langle\psi|) = \frac{1}{\sigma_0\sqrt{\pi\Delta t}} \int_{-\infty}^{\infty} e^{-(x-\alpha)^2/\sigma_0^2\Delta t} |\psi(x)|^2 dx \tag{4.13}$$

and the mean value by

$$\langle X\rangle = \int_{-\infty}^{\infty} x|\psi(x)|^2 dx = \int_{-\infty}^{\infty} \alpha P(\alpha) d\alpha = \langle\alpha\rangle \tag{4.14}$$

For Δt sufficiently small, the Gaussian function is much broader than $|\psi(x)|^2$, which could be replaced by a delta function centered at the expected position of X. Thus,

$$P(\alpha) = \frac{1}{\sigma_0\sqrt{\pi\Delta t}} e^{-(\alpha-\langle\alpha\rangle)^2/\sigma_0^2\Delta t}. \tag{4.15}$$

Now, following Ref. [23], the probability density given by Eq. (4.15) can be transformed in the corresponding of a Wigner process if

$$\alpha - \langle\alpha\rangle = \frac{\sigma_0^2 \Delta W}{\sqrt{2}} \tag{4.16}$$

where ΔW denotes the small displacement of a Brownian particle [24].

A different approach is by considering a fuzzy measurement which results from the generalization of von Neumann's projector theory. For example, when measuring the position of a particle, a certain finite resolution should be considered. The simplest way of describing this fact is by means of an indicator or characteristic function of the measurement. Thus, by considering the interval $[x_i - \Delta x, x_i + \Delta x]$ where x_i is the center of the interval and Δx the finite resolution, the projector P_i is defined as a function to be unity inside the points of the above interval, and zero otherwise. Then, the projector simply multiplies the wave function by one inside that interval

$$P_i\Psi(q) = \begin{cases} \Psi(q) \ , if \ |q - q_i| < \Delta a \\ 0 \qquad , otherwise. \end{cases} \tag{4.17}$$

Obviously, this characteristic function which is discrete and rectangular can be improved by considering a more general scheme [2]. Let us define a set of non-orthogonal characteristic functions $\{R_i\}$ with $\sum_i R_i^\dagger R_i = 1$ such that

$$|\Psi_i\rangle = R_i|\Psi\rangle \tag{4.18}$$

with

$$P_i = ||\Psi_i||^2 = \langle\Psi|R_i^\dagger R_i|\Psi\rangle \tag{4.19}$$

and

$$\sum_i P_i = 1. \tag{4.20}$$

If these characteristic functions are assumed to be continuous functions (for example, a Gaussian function), the index i is replaced by a continuous index x_c in such a way that the continuous set of operators is now given by $\{R_{x_c}\}$ with $\int dx_c R^\dagger_{x_c} R_{x_c} = 1$ and

$$|\Psi_{x_c}\rangle = R_{x_c}|\Psi_{x_c}\rangle \tag{4.21}$$

with

$$P_{x_c} = ||\Psi_{x_c}||^2 = \langle\Psi|R^\dagger_{x_c} R_{x_c}|\Psi\rangle \tag{4.22}$$

and

$$\int dx_c P_{x_c} = 1. \tag{4.23}$$

In the literature, the set of positive operators $\{R^\dagger_{x_c} R_{x_c}\}$ represents a positive-operator-valued measure on the set $\{x_c\}$ of all possible values of such a parameter.[2] For a Gaussian function, we have

$$R_{x_c}\Psi(x) = R(x - x_c)\Psi(x) \tag{4.24}$$

where x_c is the center of the Gaussian and its width gives the resolution of the measurement (or fuzziness of the measurement). The measurements described by such operators are called *minimally disturbing* since only the amplitude of the wave function is rescaled, but not its phase. When the characteristic function includes a phase of the momentum, a transition of the linear momentum is accompanied with the measurement.

The characteristic function can be generalized to any observable. Thus, if the observable A has a spectrum $\{a_i\}$, the wave function in the representation of A is written as

$$|\Psi\rangle = \sum_i c_i|a_i\rangle \tag{4.25}$$

and

$$R_{a_c}|\Psi\rangle = \sum_i R(a_i - a_c)c_i|a_i\rangle. \tag{4.26}$$

The next interesting step is to consider the effect of repeated fuzzy measurement and free evolution in the intervals between the measurements at each instant. Decoherence is settled down after a measurement has been carried out. Obviously, in the limit of many, many measurements during a finite time interval, we have a continuous measurement with the corresponding gradual decoherence. Each fuzzy measurement has a poor resolution leading to a poor information about the observable to be measured. However, the repetition of this type of measurement improves the resolution of the eigenvalues of the corresponding observable by projecting more and more the state of the system onto one of the eigenvectors. Thus, with a long

---

[2]A Hermitian operator is called positive, $R \geq 0$, if for an arbitrary state $|\phi\rangle$ then $\langle\phi|R|\phi\rangle \geq 0$.

series of fuzzy measurements, a model for gradual decoherence is provided leading to von Neumann's projection. The sequence thus defined by the measurements can be seen as a stochastic process [2].

4.3 A Short Introduction to Restricted Path Integrals

When dealing with classical systems, each individual trajectory or path forms a complete description of its time evolution. However, for quantum systems, only the sum or integration over all paths according to certain rules describes its corresponding evolution.

The time evolution of a closed system in terms of its wave function is described by the so-called propagator (or evolution operator), starting at t_0, according to

$$\Psi(t) = U(t, t_0)\psi(t_0) \tag{4.27}$$

which is equivalent to solve the time-dependent Schrödinger equation. For a time-independent Hamiltonian, this operator is defined as

$$U(t, t_0) = \theta(t - t_0) \exp\left[-\frac{iH(t - t_0)}{\hbar}\right] \tag{4.28}$$

where θ is the step function (1 for positive argument and 0, otherwise). In one-dimension, its matrix elements are given by

$$U(q'', t''; q', t') = \langle q''|U(t'', t')|q'\rangle \tag{4.29}$$

with $q(t') = q'$ and $q(t'') = q''$. The evolution of the wave function is then given by

$$\Psi(q'', t'') = \int dq' U(q'', t''; q', t')\Psi(q', t'). \tag{4.30}$$

The combined effect of all possible paths is defined by means of a propagator, which acts according to *Huygens' principle*. That is, following an optical analogy, the (quantum) wave function at the time t'' is formed by the interference of a series of secondary wavelets starting from a wave function at a previous time t'.

As is well known, this propagator can also be expressed as an integral over all paths connecting the initial and final points [1, 2, 25, 26], the so-called Feynman path integral, which in the one dimensional configuration space, is written as

$$U(q'',t'';q',t') = \int d\,[q]\, e^{iS[q]/\hbar}, \tag{4.31}$$

where the integral is understood over all paths of the coordinate q fulfilling the condition

$$[q] = \{q(t)|t' \leq t \leq t''\},$$

and the action S of the corresponding system is given by

$$S\,[q] = \int_{t'}^{t''} L(q,\dot{q},t)dt \tag{4.32}$$

$L(q,\dot{q},t)$ being its Lagrangian, which is related to the Hamiltonian as

$$H(p,q,t) = p\dot{q} - L(q,\dot{q},t), \tag{4.33}$$

with $p = \partial L/\partial \dot{q}$. In the phase space, the path integral for U is written as

$$U(q'',t'';q',t') = \int d\,[p] \int d\,[q]\; e^{\frac{i}{\hbar}\int_{t'}^{t''}(p\dot{q}-H(p,q,t))dt}, \tag{4.34}$$

where the paths of the momentum are defined as

$$[p] = \{p(t)|t' \leq t \leq t''\},\ p(t') = p',\ p(t'') = p''.$$

Equation (4.31) can be written in a more explicit way by means of the so-called discretization. This procedure consists of replacing the continuous paths by continuous piecewise linear curves or broken lines. The nodes of the broken line lie on the path and are at the time and space positions $q_j = q(t_j)$ and $q_{j+1} = q(t_{j+1})$ with $t_j = j\Delta t$. Thus, the integration over continuous path $d\,[q]$ is replaced by integration over all possible node positions

$$\prod_{j=1}^{N-1} dq_j$$

leading to the correct value of the integral when the time interval between nodes of the broken lines tends to zero, i.e., $\Delta t = (t'' - t')/N \to 0$ or $N \to \infty$. For example, for a particle of mass m with a Lagrangian of the form

$$L(q, \dot{q}, t) = \frac{1}{2} m \dot{q}^2 - V(q, t) \tag{4.35}$$

the propagator given by Eq. (4.31) is finally expressed as

$$U(q'', t''; q', t') = \lim_{N \to \infty} \left(\frac{m}{2\pi i \hbar \Delta t} \right)^{1/2} \int \prod_{j=1}^{N-1} \left(\frac{m}{2\pi i \hbar \Delta t} \right)^{1/2} dq_j$$
$$. \exp \left(\frac{i}{\hbar} \sum_{j=1}^{N} \left[\frac{1}{2} m \frac{(q_j - q_{j-1})^2}{\Delta t} - V(q_j, t_j) \Delta t \right] \right) \tag{4.36}$$

where a normalization factor has been introduced.

For Eq. (4.34), the measure $d\,[q]\,d\,[p]$ is discretized as

$$\prod_{j=1}^{N-1} dq_j \prod_{j=1}^{N} \frac{dp_j}{2\pi\hbar}$$

where piecewise constant curves are used for paths $[p]$. The resulting expression for the propagator is finally given by

$$U(q'', t''; q', t') = \lim_{N \to \infty} \int \prod_{j=1}^{N-1} dq_j \prod_{j=1}^{N} \frac{dp_j}{2\pi\hbar}$$
$$. \exp \left(\frac{i}{\hbar} \sum_{j=1}^{N} \left[p_j (q_j - q_{j-1}) - \left(\frac{p_j^2}{2m} - V(q_j, t_j) \right) \Delta t \right] \right). \tag{4.37}$$

For an n-dimensional system, where q and p are n-vectors, the generalization of the propagator is straightforward by taking into account the corresponding measures [1, 2, 25, 26].

The next step is to describe sequential measurements of, for example, the position of the system, in order to reach the continuous quantum measurement. If the output of the measurement is q', with an error equal to Δq, by simulating the device by a rectangular characteristic function,[3] we could write this operation by a set of operators $\{R_{q'}\}$ proportional to projectors as

$$R_{q'} = \frac{1}{\sqrt{2\Delta q}} P_{q'}, \tag{4.38}$$

where $P_{q'}$ is defined as in Eq. (4.17)

---

[3]Real devices correspond to smoother characteristic functions such as a Gaussian one.

$$P_{q'}\Psi(q) = \begin{cases} \Psi(q) \ , if \ |q - q'| < \Delta q \\ 0 \qquad , otherwise \end{cases} , \tag{4.39}$$

with the probability density

$$P(q') = \langle \Psi | R_{q'}^2 | \Psi \rangle, \tag{4.40}$$

the normalization relation being

$$\int_{-\infty}^{+\infty} P(q')dq' = 1. \tag{4.41}$$

Usually, the sequence of measurements is carried out with intervals of free evolution of the system between them. Thus, after $N - 1$ measurements with outputs $\{q_1, q_2, \ldots, q_{N-1}\}$ during the period T, we have that

$$\Psi(N) = U_{q_1, q_2, \ldots, q_{N-1}} \Psi(0) \tag{4.42}$$

where

$$U_{q_1, q_2, \ldots, q_{N-1}} = U(t_N, t_{N-1}) R_{q_{N-1}} \cdots U(t_2, t_1) R_{q_1} U(t_1, t_0) \tag{4.43}$$

and the probability density of the outputs of the sequential measurement is

$$P(q_1, q_2, \ldots, q_{N-1}) = P(q_1)P(q_2) \cdots P(q_{N-1}) = \langle \Psi(N) | \Psi(N) \rangle \tag{4.44}$$

due to the unitarity of the free evolution. Finally, the normalization condition is

$$\int \cdots \int P(q_1, q_2, \ldots, q_{N-1}) dq_1 dq_2 \cdots dq_{N-1} = 1. \tag{4.45}$$

In the limit $N \to \infty$, $\Delta t = T/N \to 0$, the continuous measurement process in the finite time interval T is reached. As seen in Fig. 4.1, the sequence of outputs determines in this limit a quantum corridor or band of width $2\Delta q$. It is clear why path integrals can be used to calculate characteristics of continuous measurements.

The limit of the operator $U_{q_1, q_2, \ldots, q_{N-1}}$ is the propagator for a system under continuous quantum measurement. This propagator can also be expressed in this limit as

$$U(q'', t''; q', t') = \int dq_1 \int dq_2 \cdots \int dq_{N-1} U(q'', t''; q_{N-1}, t_{N-1}) \cdots U(q_2, t_2; q_1, t_1) U(q_1, t_1; q', t') \tag{4.46}$$

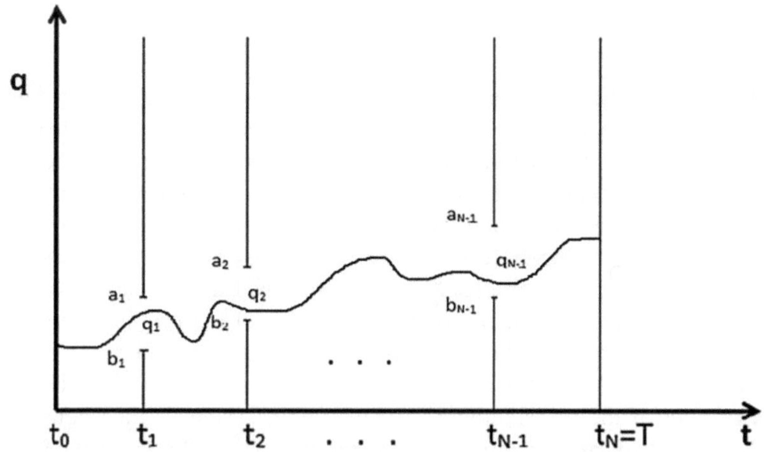

Fig. 4.1 The discretization procedure of both space and time required to evaluate the propagator (4.31) is plotted which is analogous to carrying out a Riemannian integral

which satisfies the Chapman–Kolmogorov condition

$$U(q'', t''; q, t) = \int dq' U(q'', t''; q', t') U(q', t'; q, t). \tag{4.47}$$

As shown by Kac [26, 27], the connection between path integration and Brownian motion is so close that they are nearly indistinguishable. The Brownian motion leads us to the Wiener integral which is a *bona fide* integral on the space of paths, this being defined from a Wiener measure. The space on which the proposed measure is imposed is that of the continuous functions $[q]$ where $q(0) = 0$ and $q(t) = q$. The σ-algebra is built by forming countable unions and intersections of *cylinder sets* (in our case, quantum corridors) defined as follows

$$K[a_1, b_1, t_1; a_2, b_2, t_2; ...; a_n, b_n, t_n] = \{q(.)|a_j < q(t_j) < b_j, j = 1, \ldots, N\}$$

and the measure is the probability that a particle starting at $q(0) = 0$ is between a_1 and b_1 at time t_1, is between a_2 and b_2 at $t_2, \ldots,$ and is between a_N and b_N at time t_N with $a_j < b_j$ and $t_j < t_{j+1}$, that is,

$$\int_{a_1}^{b_1} dq_1 ... \int_{a_N}^{b_N} dq_N G(q_1, t_1) G(q_2 - q_1, t_2 - t_1) ... G(q_N - q_{N-1}, t_N - t_{N-1})$$

where the G-functions are Gaussian functions.

4.3.1 Quantum Corridors

Opposite to strong measurements, fuzzy (unsharp, soft or weak) continuous measurements have also been proposed for monitoring quantum transitions leading to a new and deeper insight of the measuring process. In this chapter, when discussing about weak measurements is to be understood as a fuzzy one, not to be confused with the weak measurement due to Aharonov et al. discussed in Chap. 1. For fuzzy continuous measurements, the theory technically developed by Mensky provides a general phenomenological theory of gradual decoherence in the framework of restricted path integrals (RPI) [2]. This notion of restricted path integral was originally proposed by Feynman [25], the propagator being modified according to the information provided by the measurement through the so-called *quantum corridor*, which corresponds to a given readout of the measurement. The measured system is then considered in the RPI theory of measurements as an open system since the interaction with the environment is taken into account implicitly. In general, interaction of the system with its environment may be interpreted as measuring the system, the corresponding information being recorded in the environment. A phenomenological approach is sometimes preferred since no model for such interaction or measurement is chosen and the rather cumbersome procedure of considering first explicitly the environment and afterwards excluding it turns out to be unnecessary. In this theory, no special postulate is introduced in order to take into account the measuring process, it is just derived from the Feynman formulation of quantum mechanics.

In the path integral formalism, one does not deal with quantum states but with alternatives or paths and it results a quite natural framework to describe continuous quantum measurements. The analogue of the HUP under these conditions is the so called action uncertainty principle (AUP) [2]. In particular, the RPI formalism turns out to be very convenient for such a goal. This formalism takes into account the measuring process without explicit reference to the apparatus or environment, the back influence of the environment being considered implicitly. It makes the emphasis on the dynamical role of information issued from the continuous measurement as treated as an open quantum system. Thus, in the coordinate representation, the evolution operator or propagator is written as

$$U_{[a]}(q'', q') = \int_{I_{[a]}} d[q] e^{iS[q]/\hbar} \tag{4.48}$$

where the integral is restricted to the set $I_{[a]}$ of paths $[q] = \{q(t)|t' \leq t \leq t''\}$ (with $q'' = q(t'')$ and $q' = q(t')$) representing the measurement readout $[a]$ of a given observable A. A similar expression can be written in the phase space. This measurement readout may be expressed by the function or curve $[a] = \{a(t)|0 \leq t \leq T\}$ during the measuring period of time T and characterized by an error or resolution $[\Delta a] = \{\Delta a(t)|0 \leq t \leq T\}$, establishing a quantum corridor between $[a - \Delta a, a + \Delta a]$ (see Fig. 4.2). In this way, $|q(t) - a(t)| \leq \Delta a$. In principle, this resolution could also be a time dependent function. This evolution operator $U_{[a]}$ can

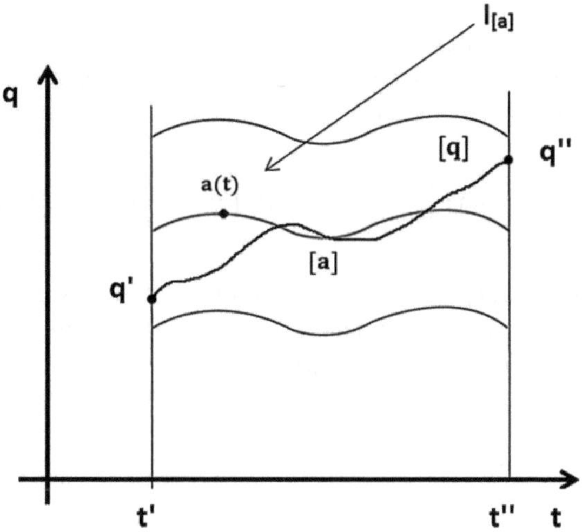

Fig. 4.2 Restriction of a path integral to the quantum corridor $I_{[a]}$ which corresponds to the readout $[a]$ when monitoring the position. The quantum corridor is defined by $[a - \Delta a, a + \Delta a]$

be interpreted as an amplitude for the measurement to give the readout $[a]$. The most probable measurement output always corresponds to the classical trajectory $[q]_{class}$. However, due to the quantum effects, outputs may differ with a large probability. The variance we are dealing with in the HUP is with respect to the classical output. The simplest (and weakest) way to express this AUP is

$$\delta S \geq \hbar, \tag{4.49}$$

which means that a continuous measurement produces information such that the uncertainty in the action is not less than the quantum of action. Thus, the most probable output corresponds to the classical output $[a]_{class}$ and to the minimum of the action. However, outputs different from the classical one with larger values of the action are also probable but different from the minimum action by a term of the order of the quantum action, $\hbar$. Therefore, the variance of the action δS should be not less than $\hbar$.

The AUP is an inequality allowing one to specify the subset of the measurement readout $[a]$ having comparatively high probabilities. When monitoring a coordinate, this principle (in a stronger form) is given by the condition [2]

$$max\left|\int_0^T \frac{\delta S[q]}{\delta q(t)}\Delta q(t)dt\right| \leq \hbar \tag{4.50}$$

where the functional derivative is taken at $[q] = [a]$ and the maximum should be taken for all $[\Delta q]$ such that $[a + \Delta q] \in I_{[a]}$ (condition for the measurement readout $[a]$ to be highly probable). When monitoring an observable $A(t) = A(p(t), q(t), t)$ such as $|A(t) - a(t)| \leq \Delta a(t)$, then the AUP is expressed as [2]

$$max|\sum_i \int_0^T dt \left[\Delta p_i \left(\dot{q}_i - \frac{\partial H}{\partial p_i} \right) - \Delta q_i \left(\dot{p}_i + \frac{\partial H}{\partial q_i} \right) \right] | \leq \hbar \tag{4.51}$$

where $H(p, q, t)$ is the Hamiltonian of the system with components q_i and p_i for the coordinate and momentum, respectively. Again, the maximum is taken for $|\Delta A(t)| = |A(p'(t), q'(t), t) - A(p(t), q(t), t)| \leq \Delta a(t)$. If we write an effective Hamiltonian as $H_e(p, q, t) = H(p, q, t) - \delta F(t) A(p, q, t)$, then the AUP can be reexpressed as

$$\int_0^T |\delta F(t)| \Delta a(t) \leq \hbar \tag{4.52}$$

the function $\delta F(t)$ being the deviation of the measured system from the classical behavior. This can be seen as a fictitious force describing the measurement process on the system of interest. The readout then follows a Langevin-type equation where $\delta F(t)$ plays the role of noise.

4.3.2 *Some Simple Applications of the RPI Formalism*

When a continuous measurement is performed in the interval $[t', t'']$, the information of this measurement is then expressed by the positive-valued functional $0 \leq \omega_{[a]} \leq 1$, the measurement output being $[a]$. Thus, paths $[q]$ for which this functional is close to unity are probable. Then, the so-called measurement amplitude is defined by

$$U^{[a]}(q'', q') = \langle q' | U^{[a]} | q'' \rangle = \int_{q'}^{q''} d\,[q]\, \omega_{[a]}[q] e^{iS[q]/\hbar}, \tag{4.53}$$

which its modulus squared roughly estimates the probability of the corresponding output or readout, through the probability density $P_{[a]} = |U^{[a]}|^2$. For the classical output, this probability is maximum. Furthermore, we have also the following inequality

$$|q(t) - a(t)| \leq \Delta a \tag{4.54}$$

where the output is given by the curve $[a] = \{a(t) | t' \leq t \leq t''\}$, the quantum corridor being defined by $[a - \Delta a, a + \Delta a]$ and symbolized by $I_{[a]}$.

If the $\omega_{[a]}$ functional is considered to be Gaussian as described by the following expression

$$\omega_{[a]}[q] = \exp\left(-\frac{2}{T\Delta a^2}\int_0^T ((q(t) - a(t))^2 dt\right) \tag{4.55}$$

the position monitoring may be called measurement of a path. The error Δa is usually taken to be a constant. Let us consider now a driven harmonic oscillator of mass m [1]. The Lagrangian is given by

$$L = \frac{1}{2}m\dot{q}^2 - \frac{1}{2}m\omega^2 q^2 + qF(t) \tag{4.56}$$

where ω is its frequency and $F(t)$ the driving force. The measurement amplitude is then written as

$$U_{[a]}(q'', q') = e^{-\frac{2}{T\Delta a^2}\int_0^T a(t)^2 dt}\int_{q'}^{q''} d[q]e^{i\tilde{S}[q]/\hbar} \tag{4.57}$$

since an effective action and effective Lagrangian can be defined as

$$\tilde{S}[q] = \int_0^T \tilde{L}(q, \dot{q}, t) \tag{4.58}$$

and

$$\tilde{L} = \frac{1}{2}m\dot{q}^2 - \frac{1}{2}m\tilde{\omega}^2 q^2 + q\tilde{F}(t), \tag{4.59}$$

respectively. This also leads to an effective driven harmonic oscillator with complex frequency

$$\tilde{\omega}^2 = \omega^2 - i\nu^2 \tag{4.60}$$

and new driving force

$$\tilde{F}(t) = F(t) - im\nu^2 a(t). \tag{4.61}$$

The imaginary part of the complex frequency is governed by the error of the measurement,

$$\nu^2 = \frac{4\hbar}{mT\Delta a^2} \tag{4.62}$$

showing that a very precise measurement leads to very high contributions of the imaginary parts of the frequency and driving force and, therefore, to a strong influence on the measured system.

When spectral measurements are considered, the theoretical procedure is different [1]. The path of an oscillator is clearly periodic and it can be expanded in a Fourier series as

$$q(t) = \sum_{n=1}^{\infty} q_n sin\frac{n\pi}{T}t \tag{4.63}$$

in the time interval $[0, T]$. The continuous measurement is carried out for the q_n components with the errors Δa_n for the frequencies $\Omega_n = n\pi/T$. If a given device is effective in the range $I = [\Omega_a, \Omega_b]$, only those frequencies Ω_n inside such an interval with the corresponding errors are considered a spectral measurement. The weight functional can then be written for one of the n components as

$$\omega_{a_n,\Delta a_n}[q] = \exp\left(-\frac{(q_n - a_n)^2}{\Delta a_n^2}\right). \tag{4.64}$$

and for all components inside that interval of frequencies as a product of exponential functions as written above. The measure for the measurement amplitude is given by

$$d[q] = \prod_{n=1}^{\infty} dq_n. \tag{4.65}$$

A detailed account of this problem can be found in Mensky's book [1],

An interesting and illustrative result is also when we want to measure the momentum of a free particle. It can be shown [1] that the variance of the linear momentum at the instant T after continuous measurement is given by

$$\Delta p = \sqrt{\Delta p_0^2 + \frac{2\hbar^2}{\Delta a^2}} \tag{4.66}$$

which shows that it does not change with time because of the momentum conservation. When there is no measurement, $\Delta a = \infty$. However, when a measurement is carried out by monitoring the position, the variance is higher due to the error of the measurement. In a certain sense, the particle undergoes scattering as a result of the continuous measurement. The second sum inside the square root can be easily understood from the uncertainty principle.

Finally, another aspect of the measurement process is worth commenting. The variance of the outputs or readouts of a precise measurement can be considered to be a quantum measurement noise. For a certain level of precision, this noise becomes an inconvenient. For a class of continuous measurements, this fact is not valid and they have been called quantum non-demolition (QND) leading to no absolute restriction on observability. In other words, the idea of the QND measurement is to measure a variable such that the unavoidable disturbance of the complementary one does not disturb the evolution of the chosen variable. This is clearly seen when measuring the linear momentum of a free particle since it is a conserved property, even disturbing the position. In general, a given variable is quantum nondemolition (as the linear momentum for a free particle) when the commutation relation of this observable at two different times is zero. Mensky has clearly shown the radical difference between the linear momentum and the velocity of a particle when considering this quantum property.

4.4 Some Equations Derived from the Restricted Path Integral

4.4.1 The Schrödinger Equation

Using Mensky's notation, his quantum corridor approach can be outlined as follows. If the system of interest undergoes a continuously prolonged in time measurement (time interval T) and, therefore, is considered as being open, the propagator can then be expressed by restricted path integrals. When monitoring an observable $A = A(p, q, t)$ by continuous measurement, the measurement readout may be given by the curve

$$[a] = \{a(t)|0 \leq t \leq T\}, \tag{4.67}$$

characterizing the values of this observable in different time moments (see Fig. 4.2 where the position is measured). However, due to the finite resolution of the measurement, Δa, the actual value of A at t may differ from $a(t)$. In this sense, a quantum corridor can be defined by including all curves or paths which differ from $[a]$ by not more than Δa. In general, the width of the corridor also depends on time performing the measurement, Δa_T. Thus, the restricted path integral representing the monitoring of A has to be taken over paths in the corridor around $[a]$.

If the squared average deflection

$$\left\langle (A-a)^2 \right\rangle_T = \frac{1}{T} \int_0^T [A(t) - a(t)]^2 dt \tag{4.68}$$

is seen as a measure of the deviation of the observable $A(t) \equiv A(p(t), q(t), t)$ from the readout $a(t)$, then the most natural weight functional describing the measurement may be written in the Gaussian form as

$$\omega_{[a]}[p, q] = \exp\left[-\kappa \int_0^T [A(t) - a(t)]^2 dt \right]. \tag{4.69}$$

In general, $\omega_{[a]}$ is a positive-valued functional fulfilling $0 \leq \omega_{[a]} \leq 1$. The constant κ characterizes the resolution of the measurement and may be expressed in terms of the "measurement error" Δa_T which is achieved during the period T of the measurement

$$\kappa = \frac{1}{T \Delta a_T^2}, \tag{4.70}$$

where Δa_T decreases if its duration T increases. In general, the κ parameter can also be a time dependent function. For small values of κ, the measurement is weak or fuzzy; on the contrary, for large values, a sharp measurement is settled down.

The measurement amplitude (partial evolution operator) in the phase space representation is then given by

$$U_T^{[a]}(q'', q') = \int_0^T d[p]d[q] \exp\left\{ \frac{i}{\hbar} \int_0^T (p\dot{q} - H)dt - \kappa \int_0^T [A(t) - a(t)]^2 dt \right\}, \tag{4.71}$$

where $q'' = q(t''), q' = q(t')$ and H the corresponding Hamiltonian. This integration is carried out over the subset of paths compatible with the $[a]$ readout. However, the time evolution of a continuously measuring system (open system) is described, in general, by a set of partial evolution operators $\{U_T^{[a]}\}$. Each partial evolution operator is not unitary because it describes only one curve of the total evolution. Therefore, a summation or integration over all possible alternatives or measurement readouts gives rise to what is known *non-selective* description of the measurement (that is, without fixing a certain measurement readout) and it is given only in terms of the density matrix,

$$\rho_T = \int d[a]\rho_T^{[a]} = \int d[a]U_T^{[a]}\rho_0(U_T^{[a]})^\dagger, \tag{4.72}$$

with $d[a] = \prod_{t=0}^T da(t)$. Obviously, the generalized unitary condition or normalization of the RPI is

$$\int d[a](U_T^{[a]})^\dagger U_T^{[a]} = 1, \tag{4.73}$$

and the probability density of the readout $[a]$ is given by

$$P([a]) = Tr(U_T^{[a]}\rho_0(U_T^{[a]})^\dagger). \tag{4.74}$$

Equation (4.71) has the form of a conventional Feynman path integral but with the non-Hermitian (complex) effective Hamiltonian

$$H_{[a]}(p, q, t) = H(p, q, t) - i\kappa\hbar[A(p, q, t) - a(p, q, t)]^2. \tag{4.75}$$

Now, instead of calculating a restricted path integral, one may solve the corresponding (linear) Schrödinger equation with the effective Hamiltonian given by

$$\frac{\partial}{\partial t}\Psi(q, t) = -\frac{i}{\hbar}H_{[a]}\Psi(q, t) = \left(-\frac{i}{\hbar}H - \kappa[A - a(t)]^2\right)\Psi(q, t). \tag{4.76}$$

If this equation is solved for the initial wave function Ψ_0 describing the initial state of the measured system, then the solution Ψ_T in the final time moment represents the state of the system after the measurement, under the condition that the measurement readout is $[a]$.

A special case is when considering the monitoring of the energy where $A \equiv H_0$ is the free evolution of the system for a total Hamiltonian given by $H = H_0 + V$, V being the interaction term responsible for transitions between stationary states. Similarly, $[a] \equiv [E]$ and $[E] = \{E(t), 0 \leq t \leq T\}$. The fuzziness or resolution of the measurement is better characterized by

$$\Delta_T = \frac{1}{\kappa T} \tag{4.77}$$

than by κ since it takes also into account the duration T (the corresponding resolution clearly decreases when increasing the duration of the measurement). The weight functional is then expressed as

$$\omega_{[E]} = \exp\left(-\frac{\langle (H_0 - E)^2 \rangle_T}{\Delta E_T^2}\right) \tag{4.78}$$

with

$$\langle (H_0 - E)^2 \rangle_T = \frac{1}{T}\int_0^T (H_0(q, p, t) - E(t))^2 dt \tag{4.79}$$

and the repetition of this fuzzy measurement improves the sharpness of the measurement. This analysis allows one to deal with the monitoring of a transition between two stationary states. If the characteristic difference between energy levels is ΔE, the T duration can then be controlled in order to monitor a given energy transition (when $\Delta E_T \ll \Delta E$). The corresponding linear Schrödinger equation is now expressed as

$$\frac{\partial}{\partial t}\Psi(q, t) = \left(-\frac{i}{\hbar}H - \kappa[H_0 - E(t)]^2\right)\Psi(q, t). \tag{4.80}$$

Interestingly enough, each partial evolution operator describes only one channel of the evolution. The probability of a single curve $[a]$ ($P[a] = \langle \Psi(T)|\Psi(T)\rangle$) is zero since, strictly speaking, the pure state evolution of the corresponding Schrödinger equation, Eq. (4.76), is not possible. Finite probabilities are only adscribed to a bundle $\mathcal{A}$ of curves $[a]$

$$P_{[a]\in\mathcal{A}} = \int_{\mathcal{A}} P[a]d[a]. \tag{4.81}$$

Thus, the state of the system has to be represented by a density matrix and its time evolution is written as follows

$$\rho_T = \int_{\mathcal{A}} d[a]\rho_T^{[a]} = \int_{\mathcal{A}} d[a]U_T^{[a]}\rho_0(U_T^{[a]})^{\dagger}. \tag{4.82}$$

When the bundle $\mathcal{A}$ is quite narrow and the initial state is pure, the final state may be considered to be pure to a good approximation, being represented by Eq. (4.76).

Several weight functionals $\omega_{[a]}$ can be envisaged leading to different generalizations of the concept of monitoring an observable. Thus, for example, we can write

$$\omega_{[a]}[p,q] = \exp\left[-\kappa \int_0^T W(A,a)dt\right], \tag{4.83}$$

where $W(A,a)$ is an arbitrary complex-valued function of two multidimensional variables A and a, when simultaneous continuous measurements of several observables with different resolutions are carried out. The corresponding effective Hamiltonian has then the expression

$$H_{[a]} = H - i\kappa\hbar W(A, a(t)). \tag{4.84}$$

As a final remark of Mensky's approach it is worth mentioning that this theoretical framework is always considered within a linear theory where the Schrödinger equation is valid.

4.4.2 The Master Equation

The description of the RPI in terms of the density matrix opens the door to link this formalism with the master equation theory. If the weight functional is chosen to be a Gaussian function, it simplifies enormously this correspondence. Mensky [2] proved that the non-selective RPI description of the measurement is

$$\begin{aligned}
\rho_t(q',q'') &= \int d[a] \langle q'|U_t^{[a]}\rho_0(U_t^{[a]})^\dagger|q''\rangle \\
&= \int d[p']d[q']d[p'']d[q'']\exp\left\{\frac{i}{\hbar}\int_0^t (p'\dot{q}' - H(p',q',t))dt\right\} \\
&\quad .\exp\left\{-\frac{i}{\hbar}\int_0^t (p''\dot{q}'' - H(p'',q'',t))dt\right\} \\
&\quad .\exp\left\{-\frac{\kappa}{2}\int_0^t (A(p',q',t) - A(p'',q'',t))^2 dt\right\} \rho_0(q_0',q_0'').
\end{aligned} \tag{4.85}$$

In order to derive a master equation, within the system-plus-device framework, the density matrix of the system may be obtained by tracing out degrees of freedom of the device (or apparatus). The simplest master (phenomenological) equation is given by

$$\dot{\rho} = -\frac{i}{\hbar}[H, \rho] - \frac{\kappa}{2}[A, [A, \rho]] \tag{4.86}$$

for monitoring of the observable $A(p, q)$. This equation is a special case of the so-called Linblad equation which was derived without using any particular model and under the assumption of the Markovian character of the process. The first term results from the dynamical properties of the free system and the second term provides the decay of the off-diagonal matrix elements. It is a straightforward exercise to prove that Eq. (4.85) satisfies the above master equation.

When dealing with Hermitian and positive operators representing the fuzziness or the characteristic function of the measurement, the modulus of the wave function is scaled, but not its phase. This kind of measurements carried out by such operators is minimally disturbing. Mensky also showed the equivalence between the corresponding RPI density matrix and a much more involved master equation.

Finally, from a practical point of view, the assumption that the measuring device resolves time with absolute accuracy is an approximation (Markovian approximation). In this sense, the weight functional used in the RPI formalism has to be changed and no differential equation in time exits for the resulting density matrix due to non-local character in time [2]. The replacement of the value of an observable $A(t)$ by its time coarse graining $\langle A \rangle_t$ has to be carried out (non-Markovian approach).

4.4.3 The Stochastic Schrödinger Equation

Dealing with open systems, it is natural to ask oneself if the RPI formulation is related to a stochastic Schrödinger equation. With appropriate change of variables, it has been shown by several authors [3, 4] that the resulting equation is written as

$$d|\Phi\rangle = \left[-\frac{i}{\hbar} H - \kappa (A - c)^2 \right] |\Phi\rangle dt + \sqrt{2\kappa}(A - c)|\Phi\rangle dW \tag{4.87}$$

with $a = c + (\xi/\sqrt{2\kappa})$, $\Phi = exp[(1/2) \int_0^t dt \xi^2]\Psi(0)$ and $dW = \xi dt$. W is a Wiener stochastic process with the property $dW^2 = dt$. To have the norm of the wave function conserved, it should occur that $c = \langle \Phi | A | \Phi \rangle$. Therefore, this stochastic differential equation is of Itô type and non-linear. Different stochastic equations have also been proposed in the literature for the description of continuous quantum measurements [28–30].

4.4.4 *The Bohmian Equations*

The Bohmian formalism could obviously be applied to Eq. (4.76) in order to obtain the counterpart of this linear Schrödinger equation in terms of the quantum action coming from the phase of the wave function. Thus, by writing the wave function in one dimension in the usual form as

$$\Psi(x,t) = \rho^{1/2}(x,t)e^{iS(x,t)/\hbar}, \tag{4.88}$$

and after introducing Eq. (4.88) into the time-dependent Schrödinger Eq. (4.76), the following two real, coupled partial differential equations are obtained from the resulting imaginary and real parts, respectively

$$\frac{\partial \rho}{\partial t} + \frac{\partial(\rho v)}{\partial x} = -2\kappa(x-q)^2\rho, \tag{4.89a}$$

$$\dot{v} + \frac{1}{m}\frac{\partial V_{\text{eff}}}{\partial x} + vv' = 0, \tag{4.89b}$$

where the effective potential is again

$$V_{\text{eff}}(x,t) = V(x) + Q(x,t). \tag{4.90}$$

Q being the standard quantum potential and v' is the first derivative of the velocity with respect to x. Notice that Eq. (4.89a) is no longer the standard continuity equation since the source term $(-2\kappa(x-q)^2\rho)$ is added due to the continuous quantum measurement process. Furthermore, Eqs. (4.76) and (4.89) do not exhibit important properties such as separability of noninteracting systems, normalization, etc.

4.5 The Nonlinear Schrödinger Equation for Continuous Measurement

By considering only position measurements, monitoring a quantum particle by the measure of the deviation of the actual path $x(t)$, from the classical trajectory $q(t)$, may be given by the negative of the logarithmic of a weight functional $w_{[x]}$ as follows

$$[x(t) - q(t)]^2 \propto - \ln w_{[x]}. \tag{4.91}$$

Now, the back influence of the selective measuring device onto the measured system can be explicitly taken into account by the non-Hermitian effective Hamiltonian

$$H_{[x]}(x, t) = H(x, t) - i\hbar\kappa \ln w_{[x]}, \tag{4.92}$$

where κ measures the strength of the interaction between the measuring apparatus and the measured system and the choice of the weight functional depends on the class of measurements under consideration. Nassar have proposed that the evolution of the wave function of the quantum system $\Psi(x, t)$ under continuous measurement can be described in terms of the logarithmic nonlinear Schrödinger equation [5]

$$i\hbar\frac{\partial\Psi(x,t)}{\partial t} = \left(H(x,t) - i\hbar\kappa\left[\ln|\Psi(x,t)|^2 - \left\langle\ln|\Psi(x,t)|^2\right\rangle\right]\right)\Psi(x,t). \tag{4.93}$$

which is identical to Eq. (3.22) of Chap. 3. This generalization can be better understood from the following arguments. If we start from a minimum-uncertainty wave packet solution around the measurement readout or record $q(t)$ written as

$$|\Psi(x, t)| = [2\pi\delta^2(t)]^{-1/4} \exp\left(-\frac{[x(t) - q(t)]^2}{4\delta^2(t)}\right) \tag{4.94}$$

where $\delta(t)$ is the total width of the Gaussian wave packet, then $\ln|\Psi(x, t)|^2 \propto [x(t) - q(t)]^2$ by considering that continuous position measurements produce and maintain localization as a necessary result of the information it provides. This minimum-uncertainty wave packet solution is further supported by recent, alternative stochastic approaches [31] which have demonstrated that individual trajectories remain minimum-uncertainty localized wave packets for all times, the localization being stronger the smaller $\hbar$ becomes.

The last term of Eq. (4.93) arises from the requirement that the integration of this equation, with respect to the variable x, under the condition that for a particle the expectation value of the energy $< E(t) >$ must be equal to the expectation values of the kinetic and potential energies where

$$< E(t) > \equiv \int_{-\infty}^{+\infty} \Psi^*(x, t)E(t)\Psi(x, t)dx. \tag{4.95}$$

as seen in the Schrödinger–Langevin or Kostin equation discussed in Chap. 3. Furthermore, Eq. (4.93) has several unique properties easy to prove. First, it guarantees the separability of noninteracting subsystems. Other nonlinear modifications can introduce an interaction between two subsystems even when there are no real forces acting between them. Second, the stationary states can always be normalized. For other nonlinearities, stationary solutions have their norms fully determined and after multiplication by a constant they cease to satisfy the equation. Third, this equation possesses simple analytical solutions in a number of dimensions, especially non-spreading wave packet solutions. A remarkable feature of this equation is the existence of exact soliton-like solutions of Gaussian shape. Hefter [32] has given physical grounds for the use of this logarithmic nonlinear equation by applying it to nuclear physics and obtaining qualitative and quantitative positive results. He argues that this type of equation can be applied to extended objects such as nucleons and alpha particles. Fourth, it is fundamentally different from the equation proposed by Bialynicki-Birula and Mycielsky due to the imaginary coefficient in front of the logarithmic terms and the last term $\langle \ln |\Psi(x,t)|^2 \rangle$ [33]. Fifth, as mentioned in Chap. 3, the irreversible dynamics in this case due to the continuous measurement leads to fulfil a Fokker–Planck equation. And sixth, it provides a good starting point for a hydrodynamical description of the measuring process in terms of quantum trajectories. In fact, we want to show that one of the fundamental aspects of Bohmian mechanics is its ability to tackle more intuitively the quantum measurement problem. As we know, the wave function plays a dual role in the Bohmian framework; it determines the probability of the actual location of the particle and choreographs its motion. As pointed out by Bell [9], in physics the only observations we must consider are position observations - a definite outcome in an individual measurement is determined by the relevant position variable associated with the apparatus. It is a great merit of the Bohmian picture to force us to consider this fact.

4.6 The Schrödinger–Langevin–Bohm Equation for Continuous Quantum Measurement

For simplicity, let us consider again a one-dimensional problem. The time evolution of the wave function $\Psi(x,t)$ of a quantum stochastic system under continuous measurement can also be described in terms of a more general nonlinear Schrödinger equation. This equation has to combine two types of logarithmic nonlinearities [6]: (i) For the description of a system under continuous measurement, Eq. (4.93) is used with the logarithmic nonlinear term $-i\hbar\kappa \ln |\Psi|^2$ and (ii) For the description of quantum stochastic processes, Eq. (3.2) is also used but with the logarithmic nonlinear term $(i\gamma\hbar/2) \ln(\Psi/\Psi^*)$. Thus, by combining both nonlinearities, the resulting equation now reads

$$i\hbar \frac{\partial \Psi(x,t)}{\partial t} = \left[H(x,t) + i\hbar \left(W_\kappa(x,t) + W_\gamma(x,t)\right)\right] \Psi(x,t), \quad (4.96)$$

where

$$H(x,t) = -\frac{\hbar^2}{2m}\nabla^2 + V(x) + V_r(x,t) \quad (4.97)$$

V_r being the random potential (see Chap. 3),

$$W_\kappa(x,t) = -\kappa \left[\ln |\Psi(x,t)|^2 - \langle \ln |\Psi(x,t)|^2 \rangle\right], \quad (4.98)$$

and

$$W_\gamma(x,t) = \frac{\gamma}{2}\left[\ln \frac{\Psi(x,t)}{\Psi^*(x,t)} - \left\langle \ln \frac{\Psi(x,t)}{\Psi^*(x,t)} \right\rangle\right], \quad (4.99)$$

and where, as we know, the coefficient κ characterizes the resolution of the continuous measurement and the coefficient γ represents the friction coefficient. The terms in $< . >$ arise from the requirement that the integration of Eq. (4.96) with respect to the variable x must be equal to the expectation values of the kinetic and potential energies through the Hamiltonian H, see Eq. (4.95). Equation (4.96) has the same unique properties than Eq. (4.93) mentioned before.

Equation (4.96) can now be further developed via the method of quantum trajectories. In order to simplify the resulting equations and discussion, let us consider $V = 0$ and $V_r = 0$ (or zero temperature, no thermal fluctuations or noise is involved) and whence only a dissipative free dynamics is present. To this end, the wave function is again expressed in the polar form as in Eq. (4.88) After substitution of this equation into Eq. (4.96), we obtain the Schrödinger–Langevin–Bohm (SLB) equation

$$i\hbar\left[\frac{\partial \phi}{\partial t} + \frac{i}{\hbar}\frac{\partial S}{\partial t}\phi\right] =$$

$$= -\frac{\hbar^2}{2m}\left\{\left[\frac{\partial^2 \phi}{\partial x^2} - \frac{\phi}{\hbar^2}\left(\frac{\partial S}{\partial x}\right)^2\right] + \frac{i}{\hbar}\left[2\frac{\partial S}{\partial x}\frac{\partial \phi}{\partial x} + \phi\frac{\partial^2 S}{\partial x^2}\right]\right\}$$

$$- i\hbar\kappa\left[\ln \phi^2 - < \ln \phi^2 >\right]\phi + \gamma\left[S - \langle S \rangle\right]\phi. \quad (4.100)$$

As we have carried out in similar contexts, Eq. (4.100) can be separated into real and imaginary parts. By defining the quantum hydrodynamical density ρ, velocity v and quantum potential Q respectively as

$$\rho(x,t) = \phi^2(x,t), \tag{4.101}$$

$$v = \frac{1}{m}\frac{\partial S}{\partial x}, \tag{4.102}$$

$$Q = -\frac{\hbar^2}{2m\phi}\frac{\partial^2 \phi}{\partial x^2}, \tag{4.103}$$

we have the corresponding splitting of the SLB equation into

$$\frac{\partial v}{\partial t} + v\frac{\partial v}{\partial x} + \gamma v = -\frac{1}{m}\frac{\partial Q}{\partial x} \tag{4.104}$$

$$\frac{\partial \rho}{\partial t} + \frac{\partial}{\partial x}(\rho v) + 2\kappa\left[\ln \rho - \langle \ln \rho \rangle\right]\rho = 0. \tag{4.105}$$

Equation (4.104) is again an Euler-type equation describing trajectories of a fluid particle, with momentum $p = mv$. On the other hand, Eq. (4.105) describes the evolution of the quantum fluid density ρ but it is no longer a continuity equation since a source term is added due to the presence of the continuous measurement. This density is interpreted as the probability density of a particle being actually present within a specific region. Such a particle follows a definite space-time trajectory that is determined by its wave function through an equation of motion in accordance with the initial position, formulated in a way that is consistent with the Schrödinger time evolution. An essential and unique feature of the quantum potential is that the force arising from it is unlike a mechanical force of a wave pushing on a particle with a pressure proportional to the wave intensity. By assuming that the wave packet is initially centered at $x = 0$ and $\rho(x,0) = \left[2\pi\delta^2(0)\right]^{-1/2}\exp\left[-x^2/2\delta^2(0)\right]$ and ρ vanishes for $|x| \to \infty$ at any time, it can be written as Eq. (4.94) Now, Eq. (4.94) can be readily used to demonstrate that

$$\int_{-\infty}^{+\infty} \left([x - q(t)]^2\right) \rho(x, t) dx = \delta^2(t). \tag{4.106}$$

Substitution of Eq. (4.94) into (4.105) yields

$$\frac{\partial \rho}{\partial t} = \left[-\frac{\dot{\delta}}{\delta} + \frac{(x-q)}{\delta^2}\dot{q} + \frac{1}{\delta^3}(x-q)^2\dot{\delta} \right] \rho, \tag{4.107}$$

and

$$\frac{\partial(\rho v)}{\partial x} = \left(\frac{\dot{\delta}}{\delta} - \kappa\right)\rho + \left[\left(\frac{\dot{\delta}}{\delta} - \kappa\right)(x-q) + \dot{q}\right]\left(-\frac{(x-q)}{\delta^2}\right)\rho, \tag{4.108}$$

which implies that

$$v(x, t) = \left(\frac{\dot{\delta}}{\delta} - \kappa\right)(x - q) + q \tag{4.109}$$

according to the standard dressing scheme.

Analogously, substitution of Eq. (4.109) into (4.104) yields

$$\left(\ddot{\delta}(t) + (\gamma - 2\kappa)\dot{\delta}(t) + (\kappa^2 - \kappa\gamma)\delta(t) - \frac{\hbar^2}{4m^2\delta^3(t)}\right)(x-q)^1 \\ +\delta(\ddot{q}(t) + \gamma\dot{q})(x-q)^0 = 0, \tag{4.110}$$

which implies that

$$\ddot{\delta}(t) + (\gamma - 2\kappa)\dot{\delta}(t) + (\kappa^2 - \kappa\gamma)\delta(t) = \frac{\hbar^2}{4m^2\delta^3(t)} \tag{4.111}$$

and

$$\ddot{q}(t) + \gamma \dot{q} = 0. \tag{4.112}$$

Equations (4.111) and (4.112) show that a continuous measurement of a quantum dissipative wave packet gives specific features to its evolution: the appearance of distinct classical and quantum elements, respectively. This measurement consists of monitoring the position of the quantum dissipative system and the result is the measured classical trajectory $q(t)$ for t within a quantum uncertainty $\delta(t)$. If the random potential is included (temperature different from zero), the above derivation changes slightly the resulting equations. Thus, in Eq. (4.104) the extra random force has to be added and Eq. (4.112) becomes the standard Langevin equation with the noise term. When $\kappa = 0$, the same coupled partial differential equations for the amplitude and phase of the wave function, as seen in Chap. 3, are recovered. Furthermore, when $\gamma = 0$, the corresponding Bohmian equations for continuous measurements are finally obtained.

As far as the Schuch–Chung–Hartmann equation (3.23) is concerned, notice that if we set

$$\kappa = \frac{\gamma}{2} \tag{4.113}$$

in Eq. (4.111), we obtain such an equation and results as a special case, such that

$$\ddot{\delta}(t) - \frac{\gamma^2}{4}\delta(t) = \frac{\hbar^2}{4m^2\delta^3(t)} \tag{4.114}$$

or for the harmonic oscillator

$$\ddot{\delta}(t) - \left(\omega^2 - \frac{\gamma^2}{4}\right)\delta(t) = \frac{\hbar^2}{4m^2\delta^3(t)} \tag{4.115}$$

with the corrected reduced frequency for the dissipative harmonic oscillator, as well as

$$\ddot{q}(t) + \gamma\dot{q} + \omega^2 q(t) = 0. \tag{4.116}$$

Interestingly enough, a new starting point is the following differential equation

$$i\hbar\frac{\partial\Psi(x,t)}{\partial t} = \left[H(x,t) + i\hbar\left(W_\kappa(x,t) + W_\gamma(x,t)\right)\right]\Psi(x,t), \tag{4.117}$$

where W_κ is defined as in Eq. (4.98) but the dissipative operator is now given by

$$W_\gamma(x,t) = -\gamma\left[\ln|\Psi(x,t)|^2 - \langle\ln|\Psi(x,t)|^2\rangle\right], \tag{4.118}$$

according to Eq. (3.22). Now, following the same theoretical procedure as before, the final important equations are

$$\frac{\partial v}{\partial t} + v\frac{\partial v}{\partial x} = -\frac{1}{m}\frac{\partial Q}{\partial x} \tag{4.119}$$

and

$$\frac{\partial \rho}{\partial t} + \frac{\partial}{\partial x}(\rho v) + 2(\kappa + \gamma)\left[\ln\rho - \langle\ln\rho\rangle\right]\rho = 0. \tag{4.120}$$

Again, starting from the Gaussian shape for the density given by Eq. (4.94), the differential equations fulfilled by the width and center of the Gaussian wave packet are, respectively,

$$\ddot{\delta}(t) - 2(\kappa + \gamma)\dot{\delta}(t) + (\kappa + \gamma)^2\delta(t) = \frac{\hbar^2}{4m^2\delta^3(t)} \tag{4.121}$$

and

$$\ddot{q}(t) = 0 \tag{4.122}$$

and the velocity is written as

$$v(x, t) = \left(\frac{\dot{\delta}}{\delta} - (\kappa + \gamma)\right)(x - q) + \dot{q}\,. \tag{4.123}$$

As can be seen from Eq. (4.122), the classical dissipative motion is not recovered. The dissipative operator can not have the same nonlinearity term than that of the continuous measurement; in other words, as expected, the two mechanisms (measurement and friction) have to be different source.

The same theoretical analysis carried out here for Gaussian wave packets could be extended to Airy packets introduced in Sect. 2.7 and given by Eq. (2.149),

$$\psi(x, t) = Ai^2\left[\frac{B}{\hbar^{2/3}}\left(x - \frac{B^3t^2}{4m^2}\right)\right]\exp\left[i\left(\frac{B^3t}{2m\hbar}\right)\left(x - \frac{B^3t^2}{6m^2}\right)\right]. \tag{4.124}$$

where Ai is the Airy function, m is the mass of the particle and B an arbitrary constant. As we know, Airy wave packets continue to propagate without spreading even when a spatially uniform and time-varying force is acting.

4.7 Bohmian Time: Dividing Line Between Quantum and Classical Trajectories

As pointed out by Bell [9], the lack of clarity in regards to where the transition between the classical and quantum regimes is located is one aspect of the measurement problem. This problem represents one of the most important conceptual difficulties in quantum mechanics. Consequently, this topic of research has gained considerable interest in the last decades [2, 5]. The presence of a classical apparatus considerably affects the behavior of the observed quantum system and measurements typically fail to have outcomes of the sort the theory was created to explain.

The solution to a challenge posed by Bell on the dividing line between the quantum and classical regimes in a measurement problem is proposed here from the theory developed in the previous Section. It is shown that continuous measurements and damping not only disturb the particle but compel the system to converge in time to a Newtonian regime without any assumption of collapse. While the width of the wave packet may reach a stationary regime, its quantum trajectories converge exponentially in time to classical trajectories. In particular, it is shown that damping tends to suppress further quantum effects on a time scale shorter than the relaxation time of the system.

The associated Bohmian trajectories of an evolving i^{th} particle of the ensemble with an initial position x_{oi} can be calculated by first substituting

$$\dot{x}_i(t) = v_i(x, t) \tag{4.125}$$

into Eq. (4.109) to obtain (dressing scheme)

$$x_i(t) = q(t) + (x_{0i} - q_{0i})\frac{\delta(t)}{\delta_0}e^{-\kappa t}, \tag{4.126}$$

where $\delta(0) = \delta_0$ is the initial width. The position of the center of mass of the wave packet (the classical trajectory) is represented by $q(t)$ and its initial value q_{0i}, while x_{0i} is the initial position of the ith individual particle in the Gaussian ensemble corresponding to the wave function given by Eq. (4.88). Now, Eq. (4.111) admits analytic Gaussian-shaped soliton-like solution or gausson where the width is constant and, therefore, $\dot{\delta} = \ddot{\delta} = 0$. For $\kappa \neq 0$, a stationary regime can be reached and the width of the wave packet can be related to the resolution of measurement as follows

$$\kappa = \frac{\gamma}{2} \pm \sqrt{\frac{\gamma^2}{4} + \frac{\hbar^2}{4m^2\delta_0^4}}, \tag{4.127}$$

which means that if an initially free wave packet is kept under a certain continuous measurement, its width may not spread in time. The two basic existing decoherence mechanisms are put on equal footing. The contribution $\hbar/2m\delta_0$ is exactly the same associated with the spreading velocity of a free Gaussian wave packet [34]. Equation (4.127) displays a very similar form to that found for the renormalized frequency of a damped harmonic oscillator (see below) [35]. It is worth stressing that when the friction mechanism is added, the resolution of the apparatus is changed, showing the intertwining role played by both mechanisms when Eq. (4.127) holds.

The main characteristic of the transition from quantum to classical trajectories can then be defined as the Bohmian time constant as $\tau_B \equiv \kappa^{-1}$ and Eq. (4.126) can be further simplified to

$$x_i(t) = q(t) + x_{0i}e^{-t/\tau_B}. \tag{4.128}$$

It follows from Eq. (4.128) that if $x_{0i} = 0$, then the particle follows the Newtonian trajectory at any time. If, however, x_{0i} is positive, then the particles distributed in the right half of the initial ensemble are accelerated whereas the particles distributed in the left half of the initial ensemble are decelerated. Nevertheless, there is only a temporary asymmetry in the Bohmian velocities between any two symmetric particles since the rate of the asymmetry diminishes with time. After a short time, the distance in position space shifted by the particles initially lying at positive and negative $x_{0i}'s$ converges to a constant value. Thus, continuous measurements not only disturb the particle but compel it to eventually converge to a classical position leading to the violation of the noncrossing rule of quantum trajectories. It is also noticeable that damping tends to suppress further quantum effects on a time scale shorter than the relaxation time of the system. For a small friction coefficient ($\gamma < \hbar/m\delta_0^2$), the Bohmian time constant can be expressed as

$$\tau_B \simeq \frac{2m\delta_0^2}{\hbar}\left(1 - \frac{\nu m\delta_0^2}{\hbar}\right). \tag{4.129}$$

Further, from Eqs. (4.103) and (4.126) we have that the quantum force is given by

$$F_{qu} = -\frac{\partial Q}{\partial x} = -\frac{\partial}{\partial x}\left[-\frac{\hbar^2}{8m\delta_0^4}(x-q)^2 + \frac{\hbar^2}{4m\delta_0^2}\right] = \frac{\hbar^2}{4m\delta_0^4}x_{0i}e^{-t/\tau_B}. \tag{4.130}$$

Thus, the convergence of the quantum particle trajectories to classical trajectories is due to the influence of the measuring apparatus and friction through the quantum force. From Eq. (4.94), we note that the expectation value of the quantum force vanishes at all times since $-1/m\langle \partial Q/\partial x\rangle = [\hbar^2/(2m\delta^2)]\langle x - q(t)\rangle = 0$. The quantum force is directly proportional to the initial position of the ith particle and decays exponentially in time (it drops 63% of its initial value after a time constant τ_B). Likewise, the quantum position $x_i(t)$ - the initial position of the ith individual particle in the Gaussian ensemble - approaches its classical value. In this way, friction and continuous observation of a wave packet may lead to a gradual freezing (gradual decoherence) of the quantum features of the particle. This is translated in a gradual violation of the noncrossing rule of quantum trajectories.

Finally, if the initial wave packet width for an electron is taken to be equal to 2.8×10^{-15} m (the approximate size of an electron [36]) and the coefficient of friction is made very small ($\gamma \ll \frac{\hbar}{m\delta_0^2}$), the Bohmian time constant is found to have an upper limit

$$\tau_B \simeq 10^{-26}\ \text{s}. \tag{4.131}$$

This result provides an answer to a challenge posed by Bell [9, 37] on the lack of clarity about the line between the quantum and classical regimes in a measurement problem: The Bohmian time constant above may establish that dividing line.[4]

4.8 Some Simple Applications

As seen previously, the logarithmic nonlinear Schrödinger equation proposed for continuous measurement in a dissipative environment for a one-dimensional problem is given by Eq. (4.96). The corresponding Hamiltonian and the nonlinear operators involving the action of the continuous measurement and the stochastic part are expressed by means of Eqs. (4.97)–(4.99), respectively. In Fig. 4.3, a schematic view of such a measurement is displayed. Following the work by Zander et al. [8], the parameter characterizing the continuous measurement κ is going to be considered a free real (mathematical) parameter and noted as χ. Doing this, the new nonlinear differential equations to be analyzed is

$$i\hbar\frac{\partial\Psi(x,t)}{\partial t} = \left[H(x,t) + i\hbar\left(W_\chi(x,t) + W_\gamma(x,t)\right)\right]\Psi(x,t), \tag{4.132}$$

where

$$W_\chi(x,t) = \chi\left[\ln|\Psi(x,t)|^2 - \left\langle\ln|\Psi(x,t)|^2\right\rangle\right], \tag{4.133}$$

Some of the interesting properties of this equation have already been previously mentioned; in particular, the breakdown of the superposition principle in this theoretical framework. Two important aspects are also worth commenting. First, along

[4]One of the most interesting uses of an RC circuit is the electronic pacemaker, which can make a stopped heart start beating again. The electrodes are implanted in the heart, and the circuit contains a capacitor and resistor. The pulsing rate depends on the time constant $\tau = RC$, which regulates the frequency of the heartbeat.

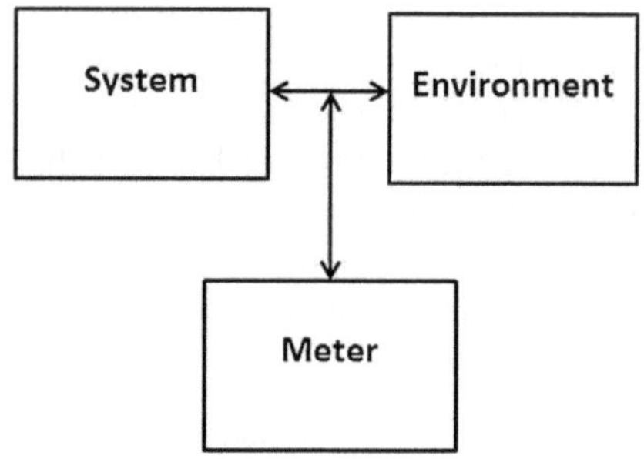

Fig. 4.3 Schematic view of the continuous measurement process on a system surrounded by an environment

this monograph (in particular, see Chap. 3), it has been pointed out several times that the environment also plays the role of a measuring device. However, it is clear from Eq. (4.132) that the sources of the two nonlinearities are drastically different; the one coming from the continuous measurement goes with the amplitude of the wave function and the one coming from the environment goes with the phase of the wave function. The second aspect concerns the interaction potential. For example, when the particle is considered free, these two nonlinearities play the role of a certain interaction potential (imaginary potential) and the corresponding dynamics is no longer "free", apart from the random potential arising from thermal fluctuations of the environment (monitoring the position of the Brownian particle).

On the other hand, the Bohmian description of the measuring process under the presence of an environment is easily reached when the wave function is, as usual, expressed in the polar form as in Eq. (4.100) and substituted in Eq. (4.132). This leads to the splitting of the Langevin–Schrödinger–Bohm equations into two coupled equations,

$$\frac{\partial v}{\partial t} + v\frac{\partial v}{\partial x} + \gamma v = -\frac{1}{m}\frac{\partial (V + V_r + Q)}{\partial x} \tag{4.134}$$

$$\frac{\partial \rho}{\partial t} + \frac{\partial}{\partial x}(\rho v) = 2\chi \left[\ln \rho - \langle \ln \rho \rangle\right] \rho. \tag{4.135}$$

which rule the time evolutions for the velocity and the probability density, respectively. Equation (4.134) is again an Euler-type equation describing trajectories of a fluid particle, with momentum $p = mv$. The two parameters governing this dynamics, χ and γ, that is, the two basic existing decoherence mechanisms here, are put on equal footing, leading to an intertwining or entanglement among them.

In the Bohmian formalism, Eq. (4.135) describes the evolution of the quantum fluid density ρ with a source/sink term $R(x,t) = 2\chi[\ln\rho - \langle\ln\rho\rangle]\rho$, the continuity equation being no longer valid. This fact has very important consequences when analyzing the relationship between the classical and quantum trajectories and the stability of the stationary solutions. Even more, when the divergence or spatial derivative term is neglected and only the source term is taken into account in Eq. (4.135), the corresponding solution of the probability density is given by [8]

$$\rho(x,t) = \frac{\exp[e^{2\chi t}\ln\rho(x,0)]}{\int_{-\infty}^{+\infty} dx' \exp[e^{2\chi t}\ln\rho(x',0)]} \tag{4.136}$$

which is properly normalized for all times t. For an initial Gaussian distribution given by $\rho(x,0) = (1/\sqrt{2\pi})\exp(-x^2/2)$, Eq. (4.136) becomes

$$\rho(x,t) = \frac{1}{\sqrt{2\pi e^{-2\chi t}}}\exp[-x^2/2e^{-2\chi t}] \tag{4.137}$$

and then

$$R(x,t) = 2\chi\left[\frac{1}{2} - \frac{x^2}{2e^{-2\chi t}}\right]\rho(x,t). \tag{4.138}$$

It is clear from the above equations, and what follows below, that the sign of the χ parameter plays a very important role; the $R(x,t)$-term can be negative (or sink term) or positive (or source term). Furthermore, when $\chi < 0$, the width of the Gaussian function increases with time whereas, when $\chi > 0$, the corresponding width decreases leading to a localization of the wave packet due to the measurement process of the particle position. Thus, Eq. (4.135) provides us the time evolution of the probability density resulting from the corresponding evolution of the density current ρv and the source/sink term, depending on the sign of χ. It should be noted that the normalization of ρ is then conserved globally, not locally, since it is not governed by a continuity equation.

The evaluation of the entropy $\mathcal{S}[\rho]$ and its time evolution is also a useful tool to characterize the degree of localization exhibited by ρ due to the measurement process. From

$$\mathcal{S}[\rho] = -\int \rho\ln\rho \tag{4.139}$$

its time evolution is governed by [8]

$$\begin{aligned}\dot{\mathcal{S}} &= -\int \frac{\partial\rho}{\partial t}\ln\rho\, dx \\ &= \left[\int_{-\infty}^{\infty}\frac{\partial v}{\partial x}\rho\, dx\right] - 2\chi\int_{-\infty}^{\infty}\rho\,(\ln\rho - <\ln\rho>)^2\, dx.\end{aligned} \tag{4.140}$$

where Eq. (4.135) has been used and an integration by parts has been carried out by considering that the probability density goes to zero fast enough in the integration limits.

4.8.1 Stability of Quantum Stochastic Trajectories

Following the same theoretical scheme used in previous chapters, a good starting point to analyze the stability of the solutions to Eqs. (4.134) and (4.135) is by considering the interaction potential $V(x)$ as a Taylor series expansion around the classical trajectory $q(t)$ up to second order

$$V(x) \approx V(q) + V'(q)(x-q) + \frac{1}{2}V''(q)(x-q)^2 + \cdots \tag{4.141}$$

as well as its first derivative,

$$V'(x) \approx V'(q) + V''(q)(x-q) + \cdots \tag{4.142}$$

where primes denote derivatives with respect to position and evaluated at the center of the wave packet which is time-dependent, $q(t)$, and $\omega^2(t) = V''(q)/m$ provides the frequency of the harmonic motion up to second order expansion. Thus, these derivatives are only functions of time through $q(t)$. Within the spatial region where the wave function is appreciably different from zero, this local approximation is assumed to be valid [38].

For a fluid particle, ρ is interpreted as the probability density of a particle being actually present within a specific region. This particle follows a definite space-time trajectory that is determined by the time evolution of the wave function through an equation of motion in accordance with the initial position. As usual assumed in this monograph, a time dependent Gaussian ansatz given by Eq. (4.107) is considered. Then, it can be readily shown that

$$\int_{-\infty}^{+\infty} \left([x-q(t)]^2\right) \rho(x,t)dx = \langle (x-q(t))^2 \rangle = \delta^2(t). \tag{4.143}$$

and the nonlocal character of the dynamical equations is given by the first spatial derivative of the quantum potential Q appearing in Eq. (4.134). For this Gaussian ansatz, this term contributes as

$$-\frac{1}{m}\frac{\partial Q}{\partial x} = -\frac{\hbar^2}{4m^2\delta^4}, \tag{4.144}$$

which is only a time-dependent function.

From Eqs. (4.107) and (4.135), the velocity field is then governed by the time differential equation but in terms now of the parameter characterizing the continuous measurement

$$v(x,t) = \left(\frac{\dot{\delta}}{\delta} + \chi\right)(x-q) + \dot{q}\,. \tag{4.145}$$

Analogously, substitution of Eq. (4.145) into (4.134) yields

$$\left(\ddot{\delta}(t) + (\gamma + 2\chi)\dot{\delta}(t) + (\chi^2 + \chi\gamma)\delta(t) - \frac{\hbar^2}{4m^2\delta^3(t)}\right)(x-q) \\ +\delta(t)\left(\ddot{q} + \gamma\dot{q} + \frac{1}{m}\frac{\partial V_r}{\partial x}\right) = -\frac{\delta(t)}{m}\frac{\partial V}{\partial x}, \tag{4.146}$$

which implies, according to Eqs. (4.141) and (4.142), that

$$\ddot{\delta}(t) + (\gamma + 2\chi)\dot{\delta}(t) + (\chi^2 + \chi\gamma)\delta(t) - \frac{\hbar^2}{4m^2\delta^3(t)} + \frac{\delta(t)}{m}V''(q) = 0 \tag{4.147}$$

and

$$\ddot{q} + \gamma\dot{q} + \frac{1}{m}\frac{\partial V_r}{\partial x} + \frac{1}{m}V'(q) = 0. \tag{4.148}$$

where, as mentioned above, the spatial derivatives are evaluated at the center of the wave packet, $q(t)$. Equation (4.147) gives the time differential equation for the width of the Gaussian wave packet and Eq. (4.148) is the standard Langevin equation fulfilled by the center of the wave packet with $V_r(x,t) = -xF_r(t)$, $F_r(t)$ being the random force or noise. It is worth stressing that both dynamical equations are not coupled each other and their time evolutions are totally independent except that the force field has to be evaluated at the points of the classical trajectory followed by the center of the wave packet. Moreover, the Langevin equation is not affected by the parameter characterizing the continuous measurement. As seen in Chap. 3, when $\chi = 0$, the same partial differential equations for δ and q are recovered.

As clearly showed by Zander et al. [8], Eq. (4.147) can also be reexpressed by assuming the following Gaussian wave function

$$\Psi(x,t) = e^{a(t)x^2 + b(t)x + c(t)} \tag{4.149}$$

where the time dependent functions a, b and c are complex functions. From the resulting six first-order coupled differential equations for the corresponding real and imaginary parts, only the two equations governing the evolution of the complex function $a(t)$ do not involve the b and c coefficients. In particular, with the initial conditions for the real part of a, $a_R(0)$ and $\dot{a}_R(0)$, we have that

$$a_R(t) = -\frac{1}{4\delta^2} \tag{4.150}$$

$a_R(t)$ being the real part of $a(t)$. Furthermore, Eq. (4.147) can be formally written as [8]

$$\ddot{\delta}(t) = -(2\chi + \gamma)\dot{\delta}(t) - \frac{\partial U}{\partial \delta} \tag{4.151}$$

representing an equation of motion of a particle of unit mass and coordinate δ moving in a one-dimensional effective potential given by

$$U = \frac{1}{2}(\chi^2 + \chi\gamma)\delta^2 + \frac{\hbar^2}{8m^2\delta^2} + \frac{\delta^2}{2m}V''(q), \tag{4.152}$$

the second spatial derivative being evaluated at $x = q$. This is a good starting point to analyze the stability properties of the wave packet width and, therefore, of the corresponding quantum trajectories.

After Eq. (4.145), it is easy to show that quantum trajectories are governed by

$$x(t) = q(t) + (x(0) - q(0))\frac{\delta(t)}{\delta(0)}e^{\chi t}, \tag{4.153}$$

displaying clearly the relationship with classical trajectories, $q(t)$ (dressing scheme). In principle, the gradual decoherence process due to the continuous measurement of the particle position is easily established. At asymptotic times, the exponential term, with negative χ values, is approaching zero and the decoherence should be complete; in other words, the quantum trajectory is approaching the classical one with time. Thus, in this asymptotic regime, the noncrossing rule of quantum trajectories is no longer valid. However, we are going to see that this simply analysis is not always right when studying the stability properties of the solutions due to the source/sink term in Eq. (4.135) which is not, strictly speaking, a continuity equation. Moreover, the velocity field (4.145) is then written as

$$v(t) = \dot{q}(t) + \left(\frac{\dot{\delta}(t)}{\delta(t)} + \chi\right)(x(0) - q(0))\frac{\delta(t)}{\delta(0)}e^{\chi t}, \tag{4.154}$$

showing again that the quantum velocity has clearly two independent contributions, one coming from the velocity of the center of the wave packet and the other one from the velocity of its spreading. The important difference is now the presence of the χ parameter characterizing the continuous measurement process.

4.8.2 *Free Gaussian Wave Packet*

4.8.2.1 Stationary Solutions

Apart from the discussion of soliton-like solutions seen in Sect. 4.8, a second case worth discussing is when the width of the Gaussian function reaches a stationary value, δ_s, after a certain time. When no interaction potential ($V = 0$) is present, the time differential equations governing such a free motion are issued from Eqs. (4.147) and (4.148) to be

$$\ddot{\delta}(t) + (\gamma + 2\chi)\dot{\delta}(t) + (\chi^2 + \chi\gamma)\delta(t) - \frac{\hbar^2}{4m^2\delta^3(t)} = 0 \tag{4.155}$$

and

$$\ddot{q} + \gamma\dot{q} + \frac{1}{m}\frac{\partial V_r}{\partial x} = 0. \tag{4.156}$$

The stability of this solution can be analyzed from Eqs. (4.151) and (4.152) with the effective potential written as

$$U = \frac{1}{2}(\chi^2 + \chi\gamma)\delta^2 + \frac{\hbar^2}{8m^2\delta^2}. \tag{4.157}$$

In this way, this potential depends on time only through δ. The stability is then obtained when it displays a local minimum, that is, when $U'(\delta_s) = 0$ and $U''(\delta_s) > 0$. Then, a dissipative drag force requires, after Eq. (4.151), that

$$\chi > -\frac{\gamma}{2} \tag{4.158}$$

and the condition for a stationary point is

$$\chi^2 + \chi\gamma - \frac{\hbar^2}{4m^2\delta_s^4} = 0 \tag{4.159}$$

which implies that

$$\delta_s^4 = \frac{\hbar^2}{4m^2(\chi^2 + \chi\gamma)} \tag{4.160}$$

showing the close relationship between the stationary point and the measurement resolution. The second derivative is positive at the stationary point δ_s when

$$\chi(\chi + \gamma) > 0 \tag{4.161}$$

which together with condition (4.158) implies that

$$\chi > 0 \tag{4.162}$$

and, therefore, the stationary point is necessarily a minimum. These facts can be verified by numerically solving Eq. (4.155). Thus, we have that $\chi = \kappa > 0$.

The positive value of χ or κ leads to very interesting properties concerning the behavior of the probability density and entropy, after Eqs. (4.139) and (4.140), as well as the Bohmian trajectories given by Eq. (4.153). As already mentioned when discussing the $R(x,t)$-term, it becomes positive (source) and tends to localize ρ with time whereas the current density has exactly the opposite behavior resulting that the Bohmian trajectories exhibit an exponential deviation from the classical trajectories. Due to the fact the source term is the dominating one, a stable solution or a stationary wave packet is reached. This observation is consistent with a continuous quantum measurement process. As pointed out by Zander et al. [8], this process could be describing the collapse of the wave function (localization). Thus, for the Gaussian ansatz given by Eq. (4.107), the entropy becomes

$$\mathcal{S}(\delta) = \frac{1}{2} + \ln \sqrt{2\pi\delta(t)^2} \tag{4.163}$$

and

$$\dot{\mathcal{S}} = \frac{\dot{\delta}}{\delta} + \kappa - 2\kappa \int \rho \left(\ln \rho - \langle \rho \rangle\right)^2 dx \tag{4.164}$$

and the localization effect contributes to the decreasing of the entropy. The time derivative of the entropy can be seen as a balance between two contributions, one of them being local, $(\dot{\delta}/\delta)+\kappa$, with no definite sign, and the other one being nonlocal and formed by the integral term with a definite sign. A stable solution means that it is an attractor of the probability density, being a more localized solution as time progresses. A characteristic time can then be defined as the time makes $\delta(t)$ to reach the stationary point δ_s. Following the mechanical analogy of Ref. [8], from the mechanical energy defined as $\epsilon(t) = \dot{\delta}^2/2 + U(\delta)$, the stationary state is estimated to be reached at time $\tau = 1/2(\gamma + 2\kappa)$. Finally, the opposite behavior of χ (negative values) makes the solution unstable and can not be considered an acceptable description of such a measurement. In other words, the very appealing property of the Bohmian trajectories approaching the classical ones is lost.

4.8.2.2 Monitoring the Bohmian–Brownian Particle

One of the paradigmatic examples where the interaction potential is negligible or constant (we are going to assume that $V(x) = 0$) is the Brownian motion. The diffusion process of an adsorbate on a flat surface, at a given surface temperature, and interacting with other adparticles, can be carried out under the presence of a continuous monitoring of the position, the Bohmian–Brownian particle. In Eq. (4.148), the random forces or noises arising from the thermal fluctuations of the surface and the collision among particles are acting on the adparticle. This problem has already been discussed in Chap. 1, Sect. 1.3.2., from a different perspective, that is, within a linear theoretical framework and, therefore, without any continuous quantum measurement. In the previous subsection, emphasis has been only put on the stability of the solutions in terms of the χ parameter because the friction parameter γ is always a positive real number. As discussed in Chap. 3, and along this subsection, when dealing with surface diffusion, we are going to assume a simple model where the interaction of the adparticle with other adparticles is described by a collisional friction, λ, coming from a shot noise. Thus, the total friction is given by $\eta = \gamma + \lambda$ [35, 39]. So far no discussion about the classical solution $q(t)$ has been carried out.

Consider an adparticle of mass m initially placed at a given position on a flat surface. This particle can be represented by a Gaussian wave packet,

$$\Psi(x,0) = (2\pi\delta^2(0))^{-1/4} e^{-(x-q(t))^2/4\delta^2(0)}, \tag{4.165}$$

with initial width, $\delta(0)$. The adparticle is also assumed to be initially in equilibrium with the reservoir or heat bath (surface) at a temperature T, but weakly coupled to the environment, so that dissipation can be neglected. The corresponding standard Langevin equation given by Eq. (4.148) can be conveniently written for Ohmic (constant) friction as

$$\ddot{q}(t) = -\eta\dot{q}(t) + \Delta F_r(t). \tag{4.166}$$

where the mass has been absorbed in the noise or random force term (per mass unit). Two independent noise terms are supposed to well describe this simple stochastic dynamics. Lattice vibrational effects due to the surface temperature are usually represented by a Gaussian (G) white noise and collisions among adsorbates by a shot (S) white noise. These two random forces affect importantly the wave packet dynamics of the adsorbate, $F_r = F_r^G + F_r^S$. It is customarily to write the total noise fluctuations acting on the adparticle as $\Delta F_r(t) = F_r(t) - \langle F_r \rangle$, its autocorrelation function at two different times being described by a Dirac δ-function, that is, $\langle \Delta F_r(t)\Delta F_r(t')\rangle = (2\eta k_B T/m)\delta(t-t')$, k_B being the Boltzmann constant [35, 39]. The solution of Eq. (4.166) is readily shown to be (see Chaps. 2 and 3)

$$q(t) = q(0) + \frac{v(0)}{\eta}\Phi(\eta t) + \frac{1}{\eta}\int_0^t \Phi(\eta t - \eta t')\Delta F_r(t')dt' \tag{4.167}$$

where $\Phi(\eta t) = 1 - \exp(-\eta t)$. Now, after Eq. (4.153), the quantum stochastic trajectories with continuous measurement $\chi = \kappa$ are then given by

$$x(t) = q(0) + \frac{v(0)}{\eta}\Phi(\eta t) + \frac{1}{\eta}\int_0^t \Phi(\eta t - \eta t')\Delta F_r(t')dt' + (x(0) - q(0))\frac{\delta(t)}{\delta(0)}e^{\kappa t}, \tag{4.168}$$

where the requirements to have stable stationary solutions for κ are the same as previously mentioned.

As discussed in Chap. 3, in this diffusion problem, the velocity autocorrelation function $C_{\Delta\mathbf{K}}(\tau)$, which is defined along with the direction of observation given by $\Delta\mathbf{K}$, provides us information about the diffusion coefficient. It should be recalled that $\Delta\mathbf{K}$ is the parallel momentum transfer of the atoms monitoring the position of the adsorbates. For a flat surface, diffusion is isotropic and, after Eq. (4.154), we have that the corresponding quantum velocity autocorrelation function is

$$\langle v(0)v(t)\rangle = \langle \dot{q}(0)\dot{q}(t)\rangle + \langle (x(0) - q(0))^2\rangle \left(\frac{\dot{\delta}(t)}{\delta(t)} + \kappa\right)\left(\frac{\dot{\delta}(0)}{\delta(0)} + \kappa\right)\frac{\delta(t)}{\delta(0)}e^{\kappa t} \tag{4.169}$$

since $\langle \dot{q}(0)(x(0) - q(0))\rangle = 0$ and $\langle (x(0) - q(0))\dot{q}(t)\rangle = 0$, keeping a similar structure than the quantum velocity, that is, a classical contribution given by the classical autocorrelation function and a quantum contribution governed by the time derivative of the Gaussian wave packet width as well as terms involving the κ-parameter. Interestingly enough, when the initial velocity of spreading of the wave packet is zero, the quantum contribution does not disappear due precisely to those terms. In the long time limit, the classical contribution is the same given in Chap. 3. In this limit, the overdamped regime is also established and hence the acceleration term of the differential Eq. (4.155) for this width can again be neglected. However, the corresponding time evolution is marked by its stationary value reached after a certain time.

4.8.3 Linear Potential

As we have studied in previous chapters, the linear potential plays an important role when a gravitational or electric field is present. As in Chap. 2, Sect. 2.12.3, let us assume that $V(x) = -max$ with $a > 0$. In this case, the Taylor expansions given by Eqs. (4.141) and (4.142) are only retained up to first order with $V'(q) = -ma$, which corresponds to a constant force. With this linear potential, the time differential equations governing such a motion are given by Eq. (4.155) and

$$\ddot{q} + \gamma\dot{q} + a + \frac{1}{m}\frac{\partial V_r}{\partial x} = 0. \tag{4.170}$$

The stability of the corresponding stationary solutions can again be analyzed from Eqs. (4.151) and (4.157) since the effective potential has only the dependence on the second spatial derivative of the interaction potential. Therefore, the stability conditions are the same as previously derived for the free particle motion; that is, the effective potential displays a minimum at δ_s when $\chi = \kappa > 0$. Again, these conditions can easily be verified from a numerical evaluation. The stationary state means that the wave packet finally reaches a constant width.

Concerning the analysis of trajectories, we start again from the resulting standard Langevin equation issued from Eq. (4.148) for Ohmic friction

$$\ddot{q}(t) = -\gamma\dot{q}(t) - a + \Delta F_r(t). \tag{4.171}$$

where the mass has been again absorbed in the noise or random force term (per mass unit), which is assumed to be a Gaussian white noise. The solution of Eq. (4.171) is readily shown to be (see Chap. 2, Sect. 2.12.2)

$$q(t) = q(0) + \frac{a}{\gamma}t + \frac{1}{\gamma}\Phi(\gamma t)\left(v(0) - \frac{a}{\gamma}\right) + \frac{1}{m\gamma}\int_0^t \Phi(\gamma t - \gamma t')\Delta F_r(t')dt' \tag{4.172}$$

where $\Phi(\gamma t)$ has been defined above. Again, after Eq. (4.153), the quantum stochastic trajectories with continuous measurement are then given by

$$x(t) = q(t) + (x(0) - q(0))\frac{\delta(t)}{\delta(0)}e^{\kappa t}, \tag{4.173}$$

with $q(t)$ given by Eq. (4.172). It is clear that, as expected, the behavior of the quantum dynamics with time is drastically different from that obtained in Chaps. 2 and 3

4.8.4 Quadratic Potential

When a harmonic oscillator of frequency ω is subject to dissipation, noise and continuous measurement, the Taylor expansions given by Eqs. (4.141) and (4.142) are fully considered for the potential $V(x) = m\omega^2/2$ with $V'(q) = m\omega^2 q$ and $V''(q) = m\omega^2$ (as in Chap. 2, Sect. 2.12.4). In this case, the time differential equations governing such a damped harmonic motion are given by Eq. (4.147) rewritten as

$$\ddot{\delta}(t) + (\gamma + 2\chi)\dot{\delta}(t) + (\chi^2 + \chi\gamma + \omega^2)\delta(t) - \frac{\hbar^2}{4m^2\delta^3(t)} = 0 \tag{4.174}$$

and

$$\ddot{q} + \gamma \dot{q} + \frac{1}{m}\frac{\partial V_r}{\partial x} + \omega^2 q = 0. \tag{4.175}$$

The stability of the corresponding stationary solutions can again be analyzed from Eqs. (4.151) and (4.152) where the effective potential includes the dependence on the second spatial derivative of the interaction potential. Therefore, the stability conditions, $U'(\delta_s) = 0$ and $U''(\delta_s) > 0$, are also ruled by such a derivative. The minimum of the effective potential is given by ($\chi = \kappa > 0$)

$$\delta_s^4 = \frac{\hbar^2}{4m^2(\kappa^2 + \kappa\gamma + \omega^2)}. \tag{4.176}$$

provided that

$$\kappa^2 + \kappa\gamma + \omega^2 > 0. \tag{4.177}$$

This inequality is valid whenever κ is a positive real number, which also ensures a stochastic dynamics as well as that the wave packet reaches a constant width δ_s or stationary motion.

If the frequency of the harmonic oscillator is a constant, ω_0, as previously discussed in Chap. 2, three regimes can be established: (i) $\omega_0 > \gamma/2$ (the oscillatory motion is that of underdamped oscillator and the radicand is negative), (ii) $\omega_0 < \gamma/2$ (the overdamped motion is present and the radicand is positive) and (iii) $\omega_0 = \gamma/2$ (the critical damped motion is established). It is clear that in Eq. (4.177) is always fulfilled. As discussed in Chap. 3, this type of motion can be responsible for the so-called T-peak in the dynamical structure factor observed when an adsorbate is oscillating between two potential barriers of a corrugated surface. For simplicity, we are going to focus on the regime where $\omega_0 > \gamma/2$. Concerning the analysis of trajectories, we start again from the resulting standard Langevin equation issued from Eq. (4.148) for Ohmic friction

$$\ddot{q}(t) = -\gamma\dot{q}(t) + \Delta F_r(t) - \omega_0^2 q. \tag{4.178}$$

which is the equation of motion of a damped harmonic oscillator and where the mass has been again absorbed in the noise or random force term (per mass unit). In analogy to Eqs. (3.199) and (3.200), the corresponding quantum trajectories are given by

$$\begin{aligned} x(t) = {} & \left[q(0)cos(\bar{\omega}t) + \left(\frac{v(0)}{\bar{\omega}} + \frac{\gamma}{2\bar{\omega}} q(0) \right) sin(\bar{\omega}t) \right] e^{-\gamma t/2} \\ & + \frac{1}{m\bar{\omega}} \int_0^\infty dt' e^{-\gamma t/2} sin(\bar{\omega}t) \Delta F(t') \\ & + (x(0) - q(0)) \frac{\delta(t)}{\delta(0)} e^{\kappa t} \end{aligned} \tag{4.179}$$

where

$$\bar{\omega} = \sqrt{\omega_0^2 - \frac{\gamma^2}{4}} \tag{4.180}$$

is the reduced harmonic frequency. This frequency is obtained from the classical motion as well as from the differential equation governing the time evolution of the width of the wave packet. This is important since the continuous measurement process is able to provide such a reduced frequency. Equation (4.115) describes such a motion resulting from the application of the Schuch–Chung–Hartmann equation obtained when $\kappa = \gamma/2$.

4.8.5 The Schuch–Chung–Hartmann Equation for Continuous Quantum Measurement

When dealing with the dissipative harmonic oscillator, the Schuch–Chung–Hartmann equation [40, 41] could be extended to continuous measurement as follows

$$i\hbar\frac{\partial\Psi(x,t)}{\partial t} = \left[-\frac{\hbar^2}{2m}\frac{\partial^2}{\partial x^2} + \frac{1}{2}m\omega^2(t)x^2 + i\hbar\left(W_\chi(x,t) + W_\gamma(x,t)\right)\right]\Psi(x,t), \tag{4.181}$$

the dissipative operator being now given by

$$W_\gamma(x,t) = -\gamma\left[\ln\Psi(x,t) - \langle\ln\Psi(x,t)\rangle\right], \tag{4.182}$$

and $W_\chi(x,t)$ by Eq. (4.133). Following the same theoretical procedure as before, the final new Bohmian equations are then

$$\frac{\partial v}{\partial t} + v\frac{\partial v}{\partial x} - \gamma v + \omega^2(t)x = -\frac{1}{m}\frac{\partial Q}{\partial x} \tag{4.183}$$

and

$$\frac{\partial\rho}{\partial t} + \frac{\partial}{\partial x}(\rho v) - (2\chi - \gamma)\left[\ln\rho - \langle\ln\rho\rangle\right]\rho = 0. \tag{4.184}$$

Then, by starting from the Gaussian shape for the density given by Eq. (4.107), the differential equations fulfilled by the width and center of the Gaussian wave packet are, respectively,

$$\ddot{\delta}(t) - (2\chi - \gamma)\dot{\delta}(t) + \left(\omega^2(t) + \left(\chi^2 - \frac{\gamma^2}{4}\right)\right)\delta(t) = \frac{\hbar^2}{4m^2\delta^3(t)} \tag{4.185}$$

and

$$\ddot{q}(t) - \gamma q + \omega^2(t)q = 0. \tag{4.186}$$

Then, the velocity is written as

$$v(x,t) = \left(\frac{\dot{\delta}}{\delta} + (\chi - \frac{\gamma}{2})\right)(x - q) + \dot{q}\,. \tag{4.187}$$

and the quantum trajectories as

$$x(t) = q(t) + (x_0 - q_0)\frac{\delta(t)}{\delta_0}e^{(\chi-\gamma/2)t} \tag{4.188}$$

with x_0, q_0 and δ_0 being the initial values of the quantum trajectory, the classical trajectory and the width of the Gaussian wave packet.

A similar stability analysis can be carried out for these new dynamical equations.

4.9 Classical and Quantum Lyapunov Exponents

In this Section, we show how the extremely irregular character of classical chaos can be reconciled with the smooth and wavelike nature of phenomena on the atomic scale. It is demonstrated that a wave packet under continuous quantum measurement displays both chaotic and non-chaotic features. The Lyapunov characteristic exponents for the trajectories of classical particle and the quantum wave packet center of mass are calculated and their chaoticities are demonstrated to be about the same. Nonetheless, the width of the wave packet exhibits a non-chaotic behavior and allows for the possibility to beat the standard quantum limit by means of transient, contractive states [42, 43]. In particular, the periodically driven Duffing oscillator, which has become a classic model for analysis of nonlinear phenomena [31, 44, 45], is studied, and its classical chaos is shown to crossover into the quantum regime. The theory presented below deals with some unresolved features posed by chaos and on the correspondence principle which is the main focus of many new experiments with excited atomic and molecular systems. These experiments can directly probe the realm of high quantum numbers or classically chaotic motion.

The question of coupling classical variables to quantum variables is intimately connected to the question of how certain variables become classical in the first place [46]. In reality, there are no fundamentally classical systems, only quantum systems that are effectively classical under certain conditions. One must start from the underlying quantum theory of the whole composite system and then derive the effective form of the classical theory. The starting point is to think of the classical particle as continuously monitoring the quantum particle position and responding to the measured value. To this end, consider a classical particle of mass M with position X in a nonlinear potential, the periodically driven Duffing oscillator, coupled to a quantum oscillator of frequency ω and mass m

$$M\ddot{X}(t) + BX^3(t) - AX(t) + \lambda q(t) = \Lambda \cos(\Omega t), \tag{4.189}$$

where $q(t)$ is associated with the measurement record of the quantum system and Ω is the driving frequency. The evolution of the wave function of the quantum system Ψ can be expressed at first in terms of the path-integral for the unnormalized wave function

$$\Psi(x', t') = \int \mathcal{D}[x(t)] \exp\left[\frac{i}{\hbar}\int_0^{t'} dt \left(\frac{1}{2}m\dot{x}^2(t) - \frac{1}{2}m\omega^2 x^2(t) - \lambda x(t)X(t)\right)\right] \times \exp\left(-\int_0^{t'} dt \frac{[x(t) - q(t)]^2}{4\sigma^2(t)}\right)\Psi(x_0, 0), \tag{4.190}$$

where the path integral is over paths $x(t)$ satisfying $x(0) = x_0$ and $x(t') = x'$. The quantity σ in the equation above represents the resolution of the effective measurement of the particle by the classical system, as indicated by previous works [2, 19, 20]and Sections. However, some differences here are worth mentioning. One is the time dependence of the quantity $\sigma(t)$: most importantly is the novelty that the general resolution of the measurement evolves according to a nonlinear differential equation. Another difference relates to the dimension of the quantity $\sigma(t)$: it should be considered only proportional to the actual position uncertainty in the measurement of the quantum particle. Thus, an explicit connection to a wave packet approach can be established by writing $\sigma^2(t) = \tau\delta^2(t)$, where δ and τ have dimensions of space and time, respectively. This point can be further elucidated by approximating the last term of Eq. (4.190) around an average time $\bar{t}$, i.e., $\sim \exp -[(x(\bar{t}) - q(\bar{t}))^2/4\delta(\bar{t})^2] \exp(-\bar{t}/\tau)$, where $\delta(t)$ clearly stands for the position uncertainty (width of the wave packet) and τ characterizes the time constant (relaxation time) of the measurement.

Now, the squared of the absolute value of Eq. (4.190) yields the probability density for different measurement outputs at different times and, from this equation, the associated Schrödinger equation describing the system undergoing continuous measurement can be written as

$$i\hbar\frac{\partial\Psi(x,t)}{\partial t} = -\frac{\hbar^2}{2m}\frac{\partial^2\Psi(x,t)}{\partial x^2} + \left(\frac{1}{2}m\omega^2x^2 + \lambda x X(t)\right)\Psi(x,t) - \frac{i\hbar}{4\tau}\left(\frac{[x-q(t)]^2}{\delta^2(t)} - 1\right)\Psi(x,t). \quad (4.191)$$

Next, a solution to this equation can be found by considering previous works [19, 31] which have shown that continuous position measurement produces and maintains localization in phase space as a necessary result of the information it provides. In addition to localizing the state, a continuous position measurement can also introduce noise in its evolution: the measured value $q(t)$ can be associated with a mean value $< x(t) >$ plus a noise-dependent component $\xi(t)$. Thus, in order to obtain a semiclassical protocol one must be in a regime in which the localization is relatively strong and the noise sufficiently weak. However, a protocol based on a complete hierarchy of stochastic equations associated with the average value of the position $< x(t) >$ makes it difficult to obtain an analytic solution to the problem [19].[5] The details of the variances and resulting noise strength permit only partial solutions based on varying $\hbar$ and steady state regimes. Therefore, a formalism that keeps the measurement record quantity $q(t)$ without dealing with the details of the variances can circumvent this difficult task and give a direct description of the evolution of the quantum system. This rationale entails a wave packet solution around the measurement record $q(t)$ according to Eq. (4.94). As we know, this minimum-uncertainty wave packet solution is further supported by recent, alternative stochastic approaches [31] which have demonstrated that individual quantum trajectories remain minimum-uncertainty localized wave packets for all times: the localization being stronger the smaller $\hbar$ becomes. Similar localization properties hold also for a variety of quantum trajectory methods [49–52] where the mean uncertainty product $M[\Delta x \Delta p]/\hbar$ remains close to 1 almost independent of $\hbar$, thus corroborating the minimum-uncertainty ansatz (4.94).

In the Bohmian framework, and using the standard procedure, the auxiliary functions of time $\delta(t)$ and $q(t)$ of the wave packet fulfill the following equations

$$\ddot{q}(t) + \omega^2 q(t) + \left(\frac{\lambda}{m}\right) X(t) = 0. \quad (4.192)$$

and

$$\ddot{\delta}(t) + \frac{1}{\tau}\dot{\delta}(t) + \left(\omega^2 + \frac{1}{4\tau^2}\right)\delta(t) = \frac{\hbar^2}{4m^2\delta^3(t)} \quad (4.193)$$

[5] If the noise is significant, thermal effects compete with quantum ones. To obtain the average probability solution for this stochastic process, use can be made of Chandrasekhar's convolution lemma: $\langle p(x,t)\rangle = \int_{-\infty}^{+\infty} p(x-\langle x\rangle, t) W(\langle x\rangle, t) d\langle x\rangle$, where $p(x,t) = |\psi(x,t)|^2$ and $W(\langle x\rangle, t)$ is the thermal probability of some value of $\langle x\rangle$ at time t for given initial conditions [47, 48].

Equations (4.192) and (4.193) show that a continuous measurement of a quantum oscillator gives specific features to its evolution: the appearance of distinct classical and quantum elements. This measurement consists of monitoring the position of the quantum system and the result is the measured path $q(t)$ for t within an uncertainty $\delta(t)$.

Equation (4.192) demonstrates the claim that continuous measurement can effectively obtain classical mechanics from quantum mechanics. The Lyapunov exponents that separate different time scales of motion can be established for both classical and quantum solutions as follows

$$\Lambda_{(cl,qu)} = \lim_{\substack{t\to\infty \\ \Delta(0)\to 0}} \left\{\ln\left[\Delta_{(cl,qu)}(t)/\Delta_{(cl,qu)}(0)\right]/t\right\}, \tag{4.194}$$

where

$$\Delta_{(cl,qu)}(t) = \left\{[(X^+, q^+) - (X^-, q^-)]^2 + [(\dot{X}^+, \dot{q}^+) - (\dot{X}^-, \dot{q}^-)]^2\right\}^{1/2} \tag{4.195}$$

represents the renormalized classical, quantum Euclidean distances of the trajectories in phase space, respectively. Equation (4.194) describes explicitly the asymptotic rate of exponential divergence of the classical and quantum trajectories evolving from two initially close initial conditions, respectively. It appears that, at least initially, the logarithmic divergence of trajectories with a very small perturbation in the initial conditions is roughly linear, indicating an exponential relationship. To find the exponent we need to find a line that fits the logarithm of the data. Thus, it is appropriate to use only the data up to the point where the difference is of order one. Although a perturbation causes exponential divergence locally, solutions near this initial condition are attracted to a strange attractor, which is a bounded set with zero area. Since this set is bounded, the divergence can not continue indefinitely. A regression on the data provides us a reasonable exponential function to model the divergence: for the classical case, $8 \times 10^{-7}e^{0.17(1)t}$ and for the quantum case $5 \times 10^{-7}e^{0.16(8)t}$. Thus, the behavior of a quantum wave packet center of mass and the monitoring classical coordinate are equally chaotic and the Lyapunov exponents for both cases is found to be

$$\Lambda_{qu} \simeq \Lambda_{cl} = 0.17. \tag{4.196}$$

On the other hand, Eq. (4.193) shows that the width of the wave packet exhibits a non-chaotic behavior. In this context, a solution to Eq. (4.193) for a free particle ($\omega = 0$) supports qualitatively Yuen's conclusions [42, 43] so far as showing the possibility to beat the standard quantum limit by means of transient, contractive states. Extensive deliberations on how to defend or beat the standard quantum limit for both discrete and continuous measurements of the position of a quantum particle can be found in the literature [53–59]. Accurate measurements of the position of a particle is of much interest in the context of gravitational-wave detection where questions have arisen as to whether there are fundamental quantum mechanical limits on detection sensitivity. The point here is that discrete or continuous measurements may introduce squeezing that affects subsequent measurements. Besides, the resolution squared $[\sigma^2(t) = \tau\delta^2(t)]$ of the measurement can reach a stationary regime, namely

$$\sigma_0^2 = \frac{\hbar\tau^2/m}{\left(1 + 4\omega^2\tau^2\right)^{1/2}}, \tag{4.197}$$

which indicates that localization can occur on a time scale which might be extremely short compared to the oscillator frequency ω. For the low-frequency limit $\omega\tau \ll 1$ (the free particle limit $\omega = 0$), this result reduces to $\sigma_0^2 = \hbar\tau^2/m$. On the other hand, for the high-frequency limit $\omega\tau \gg 1$, $\sigma_0^2 = \hbar\tau/2m\omega$. These results show that the resolution σ of the effective measurement increases as the characteristic time constant τ (relaxation time) increases.

We have thus established a direct correlation between a classical variable X with the measurement record of the quantum system q. Moreover, it is shown how the extremely irregular character of classical chaos can be reconciled with the smooth and wavelike nature of phenomena by demonstrating that a wave packet under continuous quantum measurement displays both chaotic and non-chaotic features. The Lyapunov characteristic exponents for the trajectories of classical particle and the quantum wave packet center of mass are calculated and their chaoticities are demonstrated to be about the same. On the other hand, the width of the wave packet exhibits a non-chaotic behavior and allows for the possibility to beat the standard quantum limit by means of transient, contractive states.

References

1. Mensky, M.B.: Continuous Quantum Measurements and Path Integrals. Springer-Science, Heidelberg (1993)
2. Mensky, M.B.: Quantum Measurements and Decoherence. Springer-Science, Heidelberg (2000)
3. Presilla, C., Onofrio, R., Tambini, U.: Measurement quantum mechanics and experiments on quantum Zeno effect. Ann. Phys. (NY) **248**, 95–121 (1996)
4. Albeverio, S., Kolokol'tsov, V.N., Smolyanov, O.G.: Continuous quantum measurement: Local and global approaches. Rev. Math. Phys. **9**, 907–920 (1997)
5. Nassar, A.B.: Quantum trajectories and the Bohm time constant. Ann. Phys. (NY) **331**, 317–322 (2013)
6. Nassar, A.B., Miret-Artés, S.: Dividing line between quantum and classical trajectories in a measurement problem: bohmian time constant. Phys. Rev. Lett. **111**, 150401-1,5 (2014)
7. Kostin, M.D.: On the Schrödinger-Langevin equation. J. Chem. Phys. **57**, 3589–3591 (1972)
8. Zander, C., Plastino, A.R., Díaz-Alonso, J.: Wave packet dynamics for a non-linear Schrödinger equation describing conti uous position measurements. Ann. Phys. **362**, 36–56 (2015)
9. Bell, J.S.: Speakable and Unspeakable in Quantum Mechanics. Cambridge University Press, Cambridge (1987)

10. Wheeler, J.A., Zurek, W.H.: Quantum Theory and Measurement. Princeton University Press, Princeton (1983)
11. Braginsky, V.B., Khalili, V.B., Ya, F.: Quantum Measurement. Cambridge University Press, Cambridge (1992)
12. Breuer, H.-P., Petruccione, F.: The Theory of Open Quantum Systems. Clarendon Press, Oxford (2006)
13. Davies, E.B.: Quantum Theory of Open Systems. Academic Press, London (1976)
14. Gottfried, K.: Quantum Mechanics. W. A. Benjamin, Reading (1974)
15. Ballentine, L.E.: Quantum Mechanics. A Modern Development. World Scientific, Singapore (1998)
16. Mittelstaedt, P.: The Interpretation of Quantum Mechanics and The Measurement Process. Cambridge University Press, Cambridge (1998)
17. Alicki, R., Lendi, K.: Quantum Dynamical Semigroups and Applications. Lecture Notes in Physics, vol. 717. Springer, Heidelberg (2007)
18. Aharonov, Y., Albert, D.Z., Vaidman, L.: How the resuslt of a measurement of a component of the spin of a spin-1/2 particle can turn out to be 100. Phys. Rev. Lett. **60**, 1351–1354 (1988)
19. Bhattacharya, T., Habib, S., Jacobs, K.: Continuous quantum measurement and the emergence of classical chaos. Phys. Rev. Lett. **85**, 4852–4855 (2000)
20. Diósi, L., Halliwell, J.J.: Coupling classical and quantum variables using continuous quantum measurement theory. Phys. Rev. Lett. **81**, 2846–2849 (1998)
21. Misra, B., Sudarshan, E.C.G.: The Zeno's paradox in quantum theory. J. Math. Phys. **18**, 756–763 (1977)
22. Kofman, A.G., Kurizki, G.: Acceleration of quantum decay processes by frequent observations. Nature **405**, 546–550 (2000)
23. Jacobs, K., Steck, D.A.: A straightforward introduction to continuous quantum measurement. Contemp. Phys. **47**, 279–303 (2006)
24. Gardiner, C.W.: Handbook of Stochastic Methods. Springer, Berlin (1983)
25. Feynman, R.P., Hibss, A.R.: Quantum Mechanics and Path Integrals. McGraw-Hill Book Company, New York (1965)
26. Schulman, L.S.: Techniques and Application of Path Integration. Wiley-Interscience, New York (1981)
27. Kac, M.: Probability and Related Topics in the Physical Science. Interscience, New York (1959)
28. Gisin, N.: Stochastic quantum dynamics and relativity. Helv. Phys. Acta **62**, 363–371 (1989)
29. Diosi, L.: Models for universal reduction of macroscopic quantum fluctuations. Phys. Rev. A **40**, 1165–1174 (1989)
30. Belavkin, V.P.: A new wave equation for a continuous nondemolition measurement. Phys. Lett. A **140**, 355–358 (1989)
31. Strunz, W.T., Diósi, L., Gisin, N., Yu, T.: Quantum trajectories for Brownian motion. Phys. Rev. Lett. **83**, 4909–4913 (1999)
32. Hefter, E.F.: Application of the nonlinear Schrödinger equation with a logarithmic inhomogeneous term to nuclear physics. Phys. Rev. A **32**, 1201–1204 (1985)
33. Bialynicki-Birula, I., Mycielski, I.: Nonlinear wave mechanics. Ann. Phys. (NY) **100**, 62–93 (1976)
34. Sanz, A.S., Miret–Artés, S.: A trajectory-based understanding of quantum interference. J. Phys. A **41**, 435303(1–23) (2008)
35. Sanz, A.S., Miret-Artés, S.: A trajectory description of quantum processes. I. Fundamentals. Lect. Notes Phys. **850**, 1–299 (2012)
36. Griffiths, D.J.: Introduction to Quantum Mechanics. Prentince-Hall, Englewood Cliffs (1995)
37. Bernstein, J.: More about Bohm's quantum. Am. J. Phys. **79**, 601–606 (2011)
38. Gutzwiller, M.C.: Chaos in Classical and Quantum Mechanics. Springer, New York (1990)
39. Sanz, A.S., Miret-Artés, S.: A trajectory description of quantum processes. II. Applications. Lect. Notes Phys. **831**, 1–333 (2014)
40. Schuch, D., Chung, K.M., Hartmann, H.: Nonlinear Schrödinger equation-type field equation for the description of dissipative systems. I. Derivation of the nonlinear field equation and one-dimensional example. J. Math. Phys. **24**, 1652–1660 (1983)

41. Schuch, D., Chung, K.M., Hartmann, H.: Nonlinear Schrödinger equation-type field equation for the description of dissipative systems. III. Frictionally damped free motion as an example for an aperiodic motion. J. Math. Phys. **24**, 3086–3092 (1984)
42. Yuen, H.P.: Contractive states and the standard quantum limit for monitoring free-mass positions. Phys. Rev. Lett. **51**, 719–722 (1983)
43. Yuen, H.P.: Yuen reponds. Phys. Rev. Lett. **52**, 1730–1730 (1984)
44. Habib, S., Jacobs, K., Mabuchi, H., Ryne, R., Shizume, K., Sundaram, B.: Quantum-classical transition in nonlinear dynamical systems. Phys. Rev. Lett. **88**, 040402-1,4 (2002)
45. Brun, T.A., Gisin, N., O'Mahony, P.F., Rigo, M.: From quantum trajectories to classical orbits. Phys. Lett. A **229**, 267–272 (1997)
46. Caron, L.A., Huard, D., Kröger, H., Melkonyan, G., Moriarty, K.J.M., Nadeau, L.P.: Is quantum chaos weaker than quantum chaos? Phys. Lett. A **322**, 60–66 (2004)
47. Chandrasekhar, S.: Stochastic problems in physics and astronomy. Rev. Mod. Phys. **15**, 1–89 (1943)
48. Nassar, A.B.: Logarithmic nonlinearities within the framework of stochastic mechanics. Phys. Rev. A **33**, 2134–2136 (1986)
49. Gisin, N., Percival, I.C.: Quantum state diffusion, localization and quantum dispersion entropy. J. Phys. A **26**, 2233–2244 (1993)
50. Schack, R., Brun, T., Percival, I.C.: Quantum state diffusion, localization and computation. J. Phys. A **28**, 5401–5414 (1995)
51. Strunz, W.T., Percival, I.C.: Classical mechanics from quantum state diffusion-a phase-space approach. J. Phys. A **31**, 1801–1814 (1998)
52. Nassar, A.B.: New quantum squeezed states for the time-dependent harmonic oscillator. J. Opt. B: Quantum Semiclass. Opt. **4**, S226–S228 (2002)
53. Mabuchi, H.: Standard quantum limits for broadband position measurement. Phys. Rev. A **58**, 123–127 (1998)
54. Ozawa, M.: Measurement breaking the standard quantum limit for free-mass position. Phys. Rev. Lett. **60**, 385–388 (1988)
55. Ni, W.T.: Quantum measurements and the standard quantum limit. Phys. Rev. A **33**, 2225–2229 (1986)
56. Caves, C.M.: Defense of the standard quantum limit for free-mass position. Phys. Rev. Lett. **54**, 2465–2468 (1985)
57. Lynch, R.: Repeated contractive-state position measurements and the standard quantum limit. Phys. Rev. Lett. **54**, 1599–1601 (1985)
58. Braginsky, V.B., Khalili, F.Ya.: Quantum Measumerent. Cambridge University Press, Cambridge (1992)
59. Braginsky, V.B., Manukin, A.B.: Measurement of Weak Forces in Physics Experiments, Nauka, Moscow (1974) [English translation edited by Douglass, D.H., University of Chicago, Chicago (1977)]

Epilogue

Throughout this monograph, our aim has been to introduce the reader to an alternative way to deal with open quantum systems and the continuous quantum measurement within the framework of nonrelativistic Bohmian mechanics. The time evolution of the open quantum system is carried out following the well-known Schrödinger–Langevin equation or Kostin equation and the measuring process treatment is inspired by the so-called restricted path integral due to Mensky. This analysis has been carried out by means of quantum (Bohmian) stochastic trajectories, not to be confused with the same terminology used in different contexts, namely, quantum optics and the stochastic Schrödinger equation. This way of considering the process of measuring as an open quantum process leads us to consider its corresponding dynamics as causal, nonlocal, nonlineal, nonunitary and stochastic and without any collapse of the wave function. Within this theoretical formalism, two contributions of very different character, but considered on equal footing through a given coupling scheme, are present; one coming from the interaction with the measuring device and, the second one, from the interaction with its surroundings. Entanglement among the system, environment and measuring apparatus results in decoherence, leading to the emergence of the classical world. In our opinion, the inflection point in this field is when the measurement problem is no longer considered as a philosophical problem but a quantum process, following the well known rules of quantum mechanics and beyond. When it is established as a process, the corresponding dynamics has somehow to be solved by means of computational/numerical methods.

The measuring process of position has been described in a hydrodynamics-like formulation. When a Gaussian shape for the probability density is assumed, two nonlinear coupled equations for the density and the corresponding flow velocity are obtained. One of the nice features issued from those coupled equations is that the corresponding quantum dynamics can be clearly splitted into two contributions. Thus, following the same dressing scheme obtained in the Bohmian formalism, where a quantum trajectory consists of a classical trajectory dressed by a nonlocal term, a quantum stochastic trajectory is also the result of a classical stochastic trajectory plus a quantum (nonlocal) contribution, consisting of the time evolution of the width of the Gaussian probability density multiplied by a critical exponential function governed

A.B. Nassar and S. Miret-Artés, *Bohmian Mechanics, Open Quantum Systems and Continuous Measurements*, DOI 10.1007/978-3-319-53653-8

by the characteristic parameter defining the continuous quantum measurement. For special cases, soliton-like solutions in a dissipative medium, this exponential function decreases with time and, at asymptotic times, the quantum trajectory is approaching the classical one, establishing a natural dividing line (according to Bell's proposal) for decoherence by means of a characteristic time, the so-called Bohmian time. However, in general, when the localization of the particle is settled down due to the measuring process, that is, the width of the Gaussian function reaches its stationary value, this exponential function is not a decaying function of time. In other words, the classical trajectory is no longer an attractor and the decoherence process follows a different route. Several simple examples have been developped to illustrate how this nonlinear and stochastic theory works; in particular, special emphasis has been put on surface diffusion where the Brownian motion is a paradigmatic case.

In our opinion, the increasing development of experimental setups combining weak and strong measurements and continuous measurements can lead to an important advance and use of classical-like objects in all branches of physics, chemistry, and biology. Thus, recent work claiming the direct measurement of the photon transverse wave function or the inference of average trajectories of single photons in a two-slit interferometer open new and exciting perspectives. This is the first step to observing or measuring wave functions and paths or trajectories of massive particles. In the near future, the combination of the measurement process in terms of weak values and stochasticity is going to be of paramount importance for a more complete and deeper understanding of quantum processes in terms of trajectories, beyond the standard meaning of being a purely classical concept. As mentioned in the Preface and illustrated here, it is clear that the enormous effort invested by many researchers in the measurement problem within the standard quantum mechanics outweighs by far that invested in Bohmian mechanics. In our opinion, the only way to overcome such a disparity starts at the universities, when teaching the first courses of quantum mechanics. This mechanics should be seen as an alternative and complementary way to describe the measuring process, at the same level than any other standard picture of quantum mechanics. We sincerely think that this approach will have a long and fruitful life in the future with really unexpected and promising results. As the reader will soon realize, there is plenty left to find out and develop in the near future, this monograph being a small and humble step towards a more general and complete theory of continuous quantum measurement within the Bohmian picture of quantum mechanics. In this regard, if this work has served to foster the use of this picture of quantum mechanics and to make the reader aware of the large variety of potentialities displayed by Bohmian theory for a better understanding of the continuous quantum measurement, all the efforts devoted to this monograph will have been worthwhile.

Index

A.B. Nassar and S. Miret-Artés, *Bohmian Mechanics, Open Quantum Systems and Continuous Measurements*, DOI 10.1007/978-3-319-53653-8

Water Law for the Twenty-First Century

National and International Aspects of Water Law Reform in India

Edited by Philippe Cullet, Alix Gowlland Gualtieri, Roopa Madhav and Usha Ramanathan

FNSNF
SWISS NATIONAL SCIENCE FOUNDATION

International Environmental Law Research Centre
ielrc·org

Routledge
Taylor & Francis Group
LONDON AND NEW YORK

First published 2010
by Routledge
2 Park Square, Milton Park, Abingdon, Oxon, OX14 4RN

Simultaneously published in the USA and Canada
by Routledge
270 Madison Avenue, New York, NY 10016

Routledge is an imprint of the Taylor & Francis Group, an informa business

Typeset in Baskerville by
RefineCatch Limited, Bungay, Suffolk
Printed and bound in Great Britain by
CPI Antony Rowe, Chippenham, Wiltshire

British Library Cataloguing in Publication Data
A catalogue record for this book is available from the British Library

Library of Congress Cataloging-in-Publication Data
Water law for the twenty-first century : national and international aspects of water law reform in India / edited by Philippe Cullet, Alix Gowlland Gualtieri, Roopa Madhav and Usha Ramanathan.
p. cm.
Includes bibliographical references.
1. Water—Law and legislation. 2. Water—Law and legislation—India. I. Cullet, Philippe. II. Gowlland Gualtieri, Alix. III. Madhav, Roopa. IV. Ramanathan, Usha.
K3496.W367 2009
346.04′691—dc22 2009016640

ISBN10: 0–415–47753–0 (hbk)
ISBN13: 978–0–415–47753–6 (hbk)

ISBN10: 0–203–86776–9 (ebk)
ISBN13: 978–0–203–86776–1 (ebk)

Table of contents

Acknowledgments

This book would not have been possible without the support of the Swiss National Science Foundation (SNF), which sponsored a three-year research partnership on water law reforms (www.ielrc.org/water) from 2006 to 2009, within which a significant part of the work for this book was undertaken. The editors and authors would like to thank the SNF for supporting the research as well as the workshops that helped develop the ideas presented here.

Delhi/Geneva/London, March 2009

List of abbreviations and acronyms

ADB	Asian Development Bank
ARWSP	Accelerated Rural Water Supply Programme
BCM	billion cubic metres
BIS	Bureau of Indian Standards
CAD	command area development
CADA	Command Area Development Authority
CBDR	Common but Differentiated Responsibility
CCA	Cultivable Command Area
CDD	Community-Driven Development
CESCR	Committee on Economic, Social and Cultural Rights
CPCB	Central Pollution Control Board
CPHEEO	Central Public Health and Environmental Engineering Organization
crore	ten million
CWC	Central Water Commission
DSB	Dispute Settlement Body
Dublin Conference	International Conference on Water and the Environment
FAO	Food and Agriculture Organization
GATS	General Agreement on Trade in Services
GATT	General Agreement on Tariffs and Trade
General Comment 15	Committee on Economic, Social and Cultural Rights, General Comment 15: The Right to Water (2002).
HRC	United Nations Human Rights Council
IBRD	International Bank for Reconstruction and Development
ICCPR	International Covenant on Civil and Political Rights
ICESCR	International Covenant on Economic, Social and Cultural Rights
ICJ	International Court of Justice

IDA	International Development Association
IFIs	international financial institutions
ILA	International Law Association
ILM	International Legal Materials
IMF	International Monetary Fund
IMT	irrigation management transfer
IWRM	integrated water resources management
lakh	hundred thousand
lpcd	litres per capita per day
MDBs	Multilateral Development Banks
MDGs	Millennium Development Goals
Mha	million hectare
MoA	Ministry of Agriculture
MoEF	Ministry of Environment and Forestry
MoRD	Ministry of Rural Development
MoWR	Union Ministry of Water Resources
MWRRA	Maharashtra Water Resources Regulatory Authority
NAFTA	North American Agreement in Free Trade
NWB	National Water Board
NWRC	National Water Resources Council
PIM	participatory irrigation management
PMU	Project Management Unit
PRIs	panchayati raj institutions
RGNDWM	Rajiv Gandhi National Drinking Water Mission
RWSS	Rural Water Supply and Sanitation
SPCB	State Pollution Control Boards
Swajal project	World Bank Uttar Pradesh Rural Water Supply and Environmental Project
TBT	Agreement on Technical Barriers to Trade
UN	United Nations
UNCED	United Nations Conference on Environment and Development, 1992
UNDP	United Nations Development Programme
UNECE	United Nations Economic Commission for Europe
UNESCO	United Nations Educational, Scientific and Cultural Organization
UNICEF	United Nations Children's Fund
UN Watercourses Convention	United Nations Convention on the Law of the Non-navigational Uses of International Watercourses, 1997
UT	Union Territory
VWSC	village water supply and sanitation committee

WHO	World Health Organization
WMO	World Meteorological Organization
WQAA	Water Quality Assessment Authority
WSR	water sector reform
WTO	World Trade Organization
WUA	water users association

Contributor biographies

Philippe Cullet is a Reader in Law at the School of Oriental and African Studies – University of London (SOAS) where he teaches law related to the environment, natural resources and intellectual property. He is also a Founding Programme Director of the International Environmental Law Research Centre (IELRC) and the Editor-in-Chief of the *Journal of Law, Environment and Development* (LEAD-journal.org). He studied law at the University of Geneva and King's College London (LLM). He received an MA in development studies from SOAS and his doctoral degree in international environmental law from Stanford Law School. He is the author of *Differential Treatment in International Environmental Law* (Ashgate, 2003), *Intellectual Property and Sustainable Development* (Butterworths, 2005) and *Water, Law and Development in the Context of Water Sector Reforms in India* (Oxford University Press, 2009), and the editor of *The Sardar Sarovar Dam Project: Selected Documents* (Ashgate, 2007).

Alix Gowlland Gualtieri is a Research Fellow at IELRC. She holds a law degree from the University of Geneva, Faculty of Law, and an LLM from the University of California, Berkeley School of Law. In 2003, she received her doctoral degree (DPhil) in international environmental law from the University of Oxford. She is currently working on water issues in the framework of IELRC's project on water sector reforms. Other research interests include trade and environment, accountability of international financial organizations, and human rights and environment. She has also worked at the Environment Department of the United Nations Economic Commission for Europe (UNECE) and the Secretariat to the Aarhus Convention on Access to Information, Public Participation in Decision-making and Access to Justice in Environmental Matters. She is co-editor of *Key Materials in International Environmental Law* (Aldershot: Ashgate, 2004) and *Water Law at the Crossroads – National and International Perspectives with Special Emphasis on India* (New Delhi: Cambridge University Press, 2009).

Adil Hasan Khan is reading for a Masters in International Studies (MIS) with specialization in international law at the Graduate Institute in International and Development Studies in Geneva. He holds a BA/LLB (Hons) from the National Law Institute University, Bhopal, India and a Post-graduate Diploma in Public International Law and Diplomacy from the Indian Academy of International Law and Diplomacy. He worked on the Indo-Swiss water law research partnership from 2006 to 2008. His areas of interest include international natural resources law and international legal theory and history.

Sujith Koonan worked on the Indo-Swiss water law research partnership from 2006 to 2009. He holds an LLM from Cochin University of Science and Technology in human rights and environmental law. Currently, he is pursuing an MPhil in International Law at the Centre for International Legal Studies of Jawaharlal Nehru University, New Delhi. His research interest includes human rights, environmental law and international economic law in relation to human rights and natural resources.

Roopa Madhav worked on the Indo-Swiss water law research partnership from 2006 to 2008. Her primary research interests include labour law, human rights and environmental law. She holds an LLM from New York University and a BA/LLB from the National Law School of India University, Bangalore. She has been a visiting faculty at the National Law School, has had extensive work experience with trade unions and was the President and Founder Member of the Alternative Law Forum, Bangalore.

Andrés Olleta is Law Researcher at IELRC and a PhD candidate at the international law section of the Graduate Institute of International and Development Studies (GIIDS), Geneva. He is a law graduate from the Universidad Nacional del Sur, Bahía Blanca, Argentina and a Masters in International Relations from the GIIDS. His PhD thesis is on the permeability of the processes of international norm creation to power and the challenges that international law-making poses to developing countries for the incorporation of their interests to new norms. He is a member of research groups in international law in both Argentina and Switzerland and is a consultant for the Competition Research for Economic Development project of the International Development Research Centre, Canada.

Usha Ramanathan is an internationally recognized expert on law and poverty. She studied law at Madras University, the University of Nagpur and received her PhD from Delhi University. She has taught in many universities around the world, including the Indian Law Institute in Delhi and the School of Oriental and African Studies (SOAS) in London where she is regular guest professor. Her research interests include human rights, displacement, torts and environment. She has published extensively in India and abroad. In particular, she has devoted her attention to a number of specific issues such as the Bhopal gas disaster, the Narmada valley dams or slum eviction in Delhi. She is a frequent adviser to non-governmental organizations and international organizations, in particular concerning human rights issues. She is also the South Asia Editor of the *Law, Environment and Development Journal* (LEAD Journal). She has written numerous articles, many of which can be found at www.ielrc.org/about_ramanathan.php and is co-editor of *Water Law at the Crossroads – National and*

International Perspectives with Special Emphasis on India (New Delhi: Cambridge University Press, 2009).

Priya Sangameswaran holds a PhD in economics from the University of Massachusetts Amherst. She is currently a Fellow in environmental studies in the Centre for Studies in Social Sciences, Calcutta. She works broadly at the intersection of developmental studies and environmental studies, drawing on heterodox economic theory, post-development theory and political ecology to understand how community-based efforts and questions of equity (particularly in the realm of water) have been conceptualized and practised. Currently, she is researching sector reforms in water in the state of Maharashtra in western India, and more specifically, how these changes link to larger debates about neoliberal development, rights, equity and commodification of resources.

David Takacs holds a PhD in Science & Technology Studies from Cornell University. He also holds an LLM from SOAS and a JD from the University of California, Hastings College of the Law. He is the author of *The Idea of Biodiversity* (Johns Hopkins University Press, 1996). He is currently working as a consultant on legal issues related to property rights and forest carbon for Conservation International and is also a Research Associate at IELRC.

Irina Zodrow worked on the Indo-Swiss water law research partnership in 2008. She holds an LLM from Melbourne University and a BA/LLB from the University of Tuebingen. She has extensive working experience as a lawyer in civil and common law systems and within the environment sector of the United Nations. Her primary research area is international environmental law, with a particular focus on water, environment and trade, agriculture and climate change.

I. Introduction

Philippe Cullet, Alix Gowlland Gualtieri and Roopa Madhav

This book is for the most part the outcome of the work carried out in the context of a three-year research partnership on water law reforms sponsored by the Swiss National Science Foundation (SNF). The research work was carried out from early 2006 to early 2009 by two teams of researchers based in Delhi and Geneva. The project was coordinated in Switzerland by the International Environmental Law Research Centre (IELRC).

The research project was initially conceived as an attempt to map out and analyze the evolution of water law since the early 1990s as there appeared to have been little sustained academic work on the legal aspects of the water sector in India since a landmark research project that gave rise to the publication of two books.[1] The underlying assumption was that the main need was to update existing research in the context of the gradual evolution of water law. It quickly became apparent that the work needed in the second part of the present decade was of an altogether broader scope and different nature. Indeed, what had been an area of law relatively untouched by significant changes, such as the development of environmental law and human rights, suddenly started a process of fundamental change that is still unravelling. A number of signs of these new developments were visible from the late 1990s with the adoption of some early reform laws and policy documents. Yet, it is over the past five years that most of the important law reforms have taken place in India.

One of the reasons for ongoing reforms coming up as a 'surprise' is that most of them are not part of a long-term process of evolution that progressively adapts and updates existing laws or adopts new laws in areas that are not yet regulated. Rather, the current wave of water law reforms is characterized by a 'big bang' approach, which brings in new laws that are built on a conceptual framework that is either different from or opposed to the existing basic principles of water law.

The task of the researchers working on this project was thus much broader than envisaged at the outset. It included in a first phase mapping and analyzing the changes to existing laws that had taken place over the past fifteen years. This

1 Chhatrapati Singh, *Water Rights and Principles of Water Resources Management* (Bombay: N.M. Tripathi, 1991) and Chhatrapati Singh (ed.) *Water Law in India* (New Delhi: Indian Law Institute, 1992).

provided the background for the analysis of the broad process of reform that is currently unfolding. At this juncture, it was realized that a broad distinction had to be made between water sector reforms and water law reforms. The former relate to a broad process of change that includes a policy dimension but no focus on water law, or focus only as a subsidiary process. The latter constitute the actual legal changes that are taking place and changing the face of water law. The link between the two is that water law reforms are for the most part directly influenced by the principles of water sector reforms. Since water sector reforms proceed in large part from an international consensus crafted in a variety of mostly non-UN contexts over the past two decades, our work has thus been largely divided into two main strands. On the one hand, we have analyzed water sector reforms as they arose at the international level because of the high level of congruence between the principles proposed at that level and the principles endorsed at the national level in a number of countries, including in India. On the other hand, we have focused on the actual legal reforms taking place in India and analyzed them both in their specific context and in the broader policy context of water sector reforms that informs them. This division of work is reflected in the present book which includes chapters looking at water sector reforms in general, chapters looking at water sector reforms and water law reforms in specific countries to identify some of the commonalities visible around the world and chapters focusing specifically on water law and policy reforms in India.

A. Water sector reforms

Since the late 1990s, and particularly over the past five years, the water sector has witnessed significant changes. These are generally known as 'water sector reforms', a process of reform based on a specific set of principles that has been proposed as valid both for developed and developing countries. The challenges that water sector reforms seek to address include diminishing per capita availability, increasing problems in water quality, efficiency in water allocation and use, and increasing competition for access to freshwater. The basic principle buttressing the reorganization of the water sector is the notion of integrated management of water resources. More specific changes suggested include institutional, technical, financial, legal and management reforms to support the new approach, as well as encouraging the greater participation of private actors. Water sector reforms are driven by the belief that addressing the challenges faced by the water sector across the globe by enabling efficient and sustainable water use, is crucial to avert future water crises.

Water sector reforms have on the whole tried to foster better management of water resources. Guided by policy prescriptions and technocratic solutions, the reforms are characterized largely by a rolling back of the state and increased private participation. One central principle guiding the reform process is that all uses of water should be viewed from the perspective of its economic value, as it enables a more 'efficient' and 'sustainable' use of water. The emphasis of this approach is to view water as an economic good. Flowing from this perspective, the reforms promote the principle of cost recovery, the setting up of water rights, user participation, decentralization and the redefinition of the role of the government. Since the public sector is seen as unable to deliver better management, water sector reforms propose comprehensive governance reforms, including the setting up of independent water regulatory authorities. Water sector reforms are also driven by international actors, such as international financial institutions (IFIs) – the Asian Development Bank (ADB), the World Bank and the International Monetary Fund (IMF) – as well as other international bodies such as the World Water Council and the Global Water Partnership.

The central organizing principle for water sector reforms is the concept of integrated water resources management (IWRM). The concept has been introduced and promoted in the water policies of a number of countries and is seen as a positive move away from a sectoral approach to institutions and decision making to a more integrated multi-sectoral approach to water management. A key component of this approach is greater stakeholder participation through decentralization to user groups and associations. IWRM also promotes the shift from a supply-side focus of the water sector to the demand-side management of water resources. Primarily, it seeks to improve access to water and improve environmental aspects of water resource use. Variations of the IWRM concept are now being experimented with across the world and this is manifested in changes to water policy, law and administrative structures. This book analyzes the rapidly evolving international and national legal and institutional framework as a primary component of the water sector reforms being undertaken in a number of countries.

Water sector reforms have been actively implemented in India in the last two decades. At a national level, water sector restructuring has included the introduction of a water policy, reforms of the legal and regulatory framework, institutional changes, increased farmers' participation in irrigation management, increased user participation in drinking water sector, increased water rates, the privatization of the urban water supply, regulation of groundwater and a policy shift to reduce the role of the state in the water sector. The process of reform can be said to have formally started in the late 1980s with the adoption of the first national water policy (revised in 2002). It has, since then, led to the introduction of a number of changes at the national level and in a number of states. This book analyzes the new policies and their wide-ranging impact on water resource use.

India has also been under intense pressure from IFIs to reform its water sector. In fact, the ADB and the World Bank have been key players in India, working behind the scenes in building consensus over water policy reform both at the Union level and in the states. They both promote the principles of water sector reforms they have also contributed towards developing.

B. Water law reforms

Water sector reforms constitute the general framework under which the legislation relevant to water is being reformed. Two main factors can be ascribed to the need for water law reforms in India, as in other countries, namely the growing awareness of economic and physical water scarcity, and the fact that existing legislation is often outdated. For instance, laws that link access to groundwater and control over land are neither socially equitable nor environmentally sustainable. However, in addressing existing problems, important tensions emerge between different perspectives. While some proposals aim to de-link water and land in order to ensure that water rights can eventually become tradable entitlements, others emphasize the need for groundwater to be given the status of a public trust in order to lead to more equitable and environmentally sustainable outcomes. These tensions are a common feature in water law reforms generally and reflect broader opposition between the conflicting views of water as a good or commodity on the one hand, and as a human right or public trust on the other.

Water law reforms have been introduced as a precursor to the water sector reforms themselves, and while based on the same principles, seek to find new ways to achieve their purposes. This is the case, for instance, with regard to the legislation making the setting up of water users associations (WUAs) compulsory throughout a state, or of the legislation creating independent water regulatory authorities whose mandate includes, among other things, the privatization of water services. The introduction of legislation as a way to introduce a set of reforms in the water context can be viewed as having two particularities in the Indian context. First, water-related laws that have been adopted during the past decade tend to be sectoral, in that they address either one specific issue (e.g., participatory irrigation management), one specific type of water (e.g. groundwater), or one specific part of the regulatory framework (e.g., institutional reorganization). Thus, fragmentation is one characteristic of the reforms, with the result that there is no integrated comprehensive legal framework setting out all the principles applicable to water issues in general.

Second, water law reforms are modelled on the principles of the water sector reforms described above, mainly because in most states, new legislation has been or is adopted in the context of projects grounded on these principles. The similarities in the laws introduced over the past decade in different states can thus be traced to the fact that a number of these laws are adopted in the context of water projects funded by the ADB or the World Bank. This results in a relative uniformity in water law reforms.

Water law reforms can be undertaken through the adoption of different types of instruments. Legislative assemblies have adopted a number of new acts over the past decade. These tend to focus on the same specific areas, with the greatest number concerned with participatory irrigation management followed by groundwater legislation and legislation for institutional reforms in the water sector. These reforms are adopted according to the formal procedures set out for the adoption of legislation at the state level. Beyond these easily identifiable legal instruments, there is a range of reforms that does not qualify as legislation, but is nevertheless very important. For instance, this is the case with regard to the provision of drinking water where there is little by way of legislation, but where changes to the policy framework have been a central part of ongoing reforms in the form of government resolutions and other subsidiary instruments. The adoption of the Swajaldhara Guidelines is a good illustration of this. A major reform initiative, the Guidelines were adopted in the form of an administrative decision of the Ministry of Rural Development, and can be viewed as the single most significant change in drinking water policy at least since the early 1970s.

This means that, interestingly, water law reforms are not limited to legislation, but involve a range of other instruments. This is the case with respect to government orders – those that can be used to take decisions that have widespread repercussions whether or not the measures derive from an existing law. These were used in the case of the transfer of operation and maintenance responsibilities under the guise of the principle of full cost recovery.

C. Book outline

This book is divided into three parts. Part 1 examines the international and comparative framework in which water law reforms are being implemented and which provide some of the basic concepts that buttress these reforms. Chapter II first maps out trends in the development of water law in different regions of the world to provide an overview of the kinds of changes that have been taking place and gives a general framework to the changes concerning India, which are analyzed in more detail later in the book. Chapter III then provides an introduction to the international water law context informing ongoing water law reforms at the national level. This chapter provides a reminder that existing international freshwater law largely fails to directly address most of the issues that are at the centre of ongoing reforms, thus ensuring that non-binding international water policy instruments play a more important role in shaping ongoing reforms than binding international water law. Chapter IV examines in more detail legal issues related to the commodification of water in the context of trade law, an issue that has not yet directly impacted reforms at the national level, but that will likely have increasing immediate relevance in the future. Chapter V closes this section with a chapter that specifically looks at the role of IFIs, in particular the World Bank, in the development of the international policy consensus around water sector reforms at international level and in their implementation at national level, in particular in the context of law conditionality linked to water sector projects.

Part II then turns to water law reforms as they have been implemented in India. Several distinct areas have been chosen to illustrate the nature of ongoing water law reforms. This is linked to the sectoral nature of the reforms that have addressed different issues in different parts of the water sector separately. Chapter VI starts by analyzing the context for water law reforms in India. This chapter reviews the present status of water law reforms. In explaining the context for the water sector and water law reforms, it seeks to briefly map the history of the water sector administration and water law. It points specifically at the paradigm shifts in the institutional and legal frameworks that are being adopted by ongoing reforms and examines these changes in a bid to understand and analyze the policies and principles that inform water law reforms in India. Chapter VII then goes on to examine institutional reforms for water. This chapter looks at new structures such as water regulatory authorities that are being established and reviews the changes, both minor and dramatic, being effected to the public sector concerned with water. Additionally, it examines the legal experience of privatization and public–private partnership in India, and the myriad legal concerns associated with the process. Chapter VIII focuses on the reforms taking place in the drinking water sector that derive their importance from the link between drinking water and the fundamental right to water. This chapter examines the specificities of drinking water reforms that, despite the direct links with the judicially sanctioned human right to water, are not based on any legislative instruments. The changes introduced for drinking water (i.e., the introduction of the principle of cost recovery) are not different from what has been proposed in other sectors, but the mode of delivery through executive decisions, such as the introduction of the Swajaldhara Guidelines, is noteworthy. Chapter IX then follows with a focus on groundwater legal reforms. These reforms deserve special mention because they are the only main changes proposed that do not directly follow the model of water sector reforms. This is due to the fact that ongoing reforms are based on a conceptual framework developed in 1970, which sought to restrict landowners' control over groundwater. This chapter examines the proposed model and critically analyzes it in the context of the need for groundwater legislation that takes into account the environmental and social aspects of its use, now the main source of water for all water uses. Chapter X examines the extensive changes introduced in irrigation law. For instance, the operation and maintenance of irrigation sub-systems have been vested with WUAs and water charges have been revised in many states. Reforms in the irrigation sector have led to the introduction of a new framework that enables the setting up of water users associations and the devolution of powers to them. Underpinning these changes is a perception of farmers as clients rather than beneficiaries. The policy shift in irrigation development has introduced the decentralized/empowerment model of reform. The principal thrust of the reform process has been a revamp of the institutional set-up by reducing the role of the state, the creation of water users associations, the setting up of water regulatory authorities and greater private sector participation in the irrigation sector. This chapter highlights that in India, though the empowerment model has been adopted, in practice, it is limited by many factors such as elite capture, caste dynamics and resistant bureaucracies.

Part III builds on the analysis of specific law and policy reforms in India and examines some of the broader issues that are directly related to water law reforms but not given prominence in existing reforms. Chapter XI focuses on the human right to water. It provides a broad analysis of the right in its national and international dimensions. The relative lack of specificity of the right, which is clearly recognized in India, but not further defined in any legal instrument, is thus examined through an analysis of different developments around the world, in particular at the international level. Chapter XII then moves on to examine the environmental aspects of water law reforms. The importance given to the environment in water sector reforms has not been matched by a similar emphasis on environmental law principles in water law reforms. This chapter provides a broad analysis of the role and place of environmental aspects in water law reforms, something which is relevant in India as well as in many other countries implementing water sector reforms. Chapter XIII analyzes the health dimension of water law. While health has been a central underlying rationale for water law, it is surprising that neither the existing nor reformed water law framework gives it the importance it deserves (e.g., in terms of binding water quality standards). This chapter emphasizes the need for giving water quality a much more prominent role in future water law reforms.

Part 1

Background and Historical Development

II. An overview of common trends in the water legislation of selected jurisdictions

Andrés Olleta

INTRODUCTION

Awareness of the growing scarcity of freshwater has led to significant changes in the domestic legal and regulatory frameworks for water management and water services provision in most countries. As water has become scarcer, states have intervened and used legal tools to protect the resource and to manage it wisely to achieve an economically efficient and environmentally sustainable management of water that contributes to offsetting the effects of its increasing scarcity of freshwater.

Current water law is the result of debates that have taken place on the international and domestic fronts. On the one hand, as the shortage and deterioration in quality of existing stocks of water have become a source of concern for all nations, a series of international events were convened with the purpose of agreeing water management principles that would serve as guidelines for local action. On the other hand, at the national level, states simultaneously acknowledged water management as a matter of national importance that deserved exhaustive regulation in domestic legislation. In recent years, therefore, many states have felt the need to either adopt new legal instruments for the protection and better management of this natural resource, or to improve existing water legislation by incorporating standards and principles that have emerged to meet the challenges that water scarcity has brought up.

For the past twenty years, domestic water legislation has aimed to deal with water comprehensively and in an environmentally-conscious way. The most recent legal instruments seek to regulate the integrated management of water; an exploitation of water that ensures that its different uses – domestic, agricultural, industrial, and recreational, among others – are made compatible, and that private rights are balanced against the public interest. Such a goal cannot be attained if priorities are not unmistakably set and regulated accordingly. This first test that domestic water regulation faces – to find a balance among the different uses of water – is joined by another significant predicament: since water regulation simultaneously pertains to human rights law, environmental law, administrative law and the law of property rights, contradictions between separate instruments – or at least hurdles in their concurrent application – can arise.

In this sense, the way national jurisdictions have faced the test of regulating the water sector is worth studying. Recent legislation specifically attempts to bring a balance among the divergent interests at stake, while stressing the importance of preserving the resource for future generations. An assessment of the latest developments in water legislation provides a broad picture in which to contextualize the water law reforms that are taking place in India. The topics covered in this chapter include access to water as a human right, state control over water resources and water rights, participation of users and private companies in the water sector, and the inclusion of environmental provisions in water legislation. They have been chosen, first, because they indicate the ways in which contemporary water law is changing. It is undeniable that they are part of an international consensus that has surfaced as a series of best practice policies to improve water management and water services provision. These policies have been applied with the support of international donors everywhere where governments face the problem of a growing scarcity of acceptable quality water and the economic, social and environmental impacts and issues that are created as a result. India is one among these countries. Secondly, these issues have been at the centre of recent water policy debates, in particular controversial issues such as the privatization of water services, the decentralization of water management and the emergence of markets for water rights. The polemic that surrounds them reflects the complexities and intricacies that characterize water regulation, to which India is no stranger.

The jurisdictions and the provisions that are referenced in each part of this chapter have been selected because they are good examples of existing trends. Section A tackles the insertion in domestic law of provisions that either consecrate the human right to water, or work towards that end. Section B dwells on issues related to the control of the state over the resource, and the distribution of water use permits to private persons. Section C analyzes the incorporation of users and civil society to water management and water services provision schemes. Section D reviews how national jurisdictions have implemented reforms that foresee the participation of private companies in water services delivery. Finally, Section E underlines the incorporation in domestic legislation of environmental provisions for the sustainable exploitation of water resources. The introduction of principles and guidelines for water management at the highest level of the domestic legal order – the constitution – will be referenced throughout these sections when appropriate, since it demonstrates the importance that water regulation has gained since the 1990s.

A. Water as a human right

An explicit legal consecration of access to water as an individual human right is still an oddity in national legislation. Yet, it is undeniable that, as water has become more scarce, provisions that either enshrine this right or impose on the state the obligation to guarantee access to water to the population have gradually emerged. Furthermore, in recent times, the judiciary has had an important role in upholding the existence and enforceability of the right to water by deriving it from other basic rights.

The legal consecration of access to water as a human right is of the utmost practical importance. Even though it does not necessarily entail automatic implementation, it cannot but work towards improving the access of the population to water services. As the Centre on Housing Rights and Evictions remarks, 'the establishment within constitutions of the right to water, and the corresponding series of State obligations to create the necessary legal, social and economic conditions, represent important legal foundations for a range of further actions towards ensuring this right'.[1]

The incorporation of the right to the domestic legal order can be the consequence of the direct applicability of international law instruments or be the product of a specific provision, either in the constitution or, more rarely, in hierarchically inferior legislation.

1. Right to water in international human rights conventions

Over time, access to water as a human right has often been mentioned in international documents of varied normative value.[2] However, it has been included in two binding instruments: the 1979 Convention on the Elimination of all Forms of Discrimination against Women[3] and the 1989 Convention on the Rights of the Child.[4] By becoming parties to these agreements, national governments have committed themselves to protecting and realizing the rights of women and children to water and developing and undertaking actions and policies oriented to such ends. They have also agreed to hold themselves accountable before the international community for the fulfilment of their obligations in the framework of both conventions.

However, since the conventions mentioned above only protect women and children, a right to water as an individual prerogative for all is still lacking in the international human rights system. The latest step in the ongoing legal process towards the consecration of the right to water internationally is a United Nations Human Rights Council resolution, which builds on the work of the Committee on Economic, Social and Cultural Rights (CESCR) and the Office of the High Commissioner for Human Rights and affirms that the Covenant on Economic, Social and Cultural Rights entails a series of obligations in relation to access to safe drinking water and sanitation.[5]

1 Centre on Housing Rights and Evictions, Legal Resources for the Right to Water: International and National Standards 45 (Geneva: COHRE, 2004).

2 See Chapter XI.

3 Convention on the Elimination of all Forms of Discrimination against Women, GA Res. 34/180, 18 December 1979, Art. 14(2)h.

4 Convention on the Rights of the Child, GA Res. 44/25, 20 November 1989, Art. 24(2)c.

5 Human Rights Council, Human Rights and Access to Safe Drinking Water and Sanitation, UN Doc. A/HRC/7/L.16 (2008).

The internationally recognized right to water may become part of the national legal order and therefore be able to have an impact at the domestic level. The right to water in the two conventions mentioned above – and the obligations implied in the Covenant – might actually enjoy the highest legal hierarchy domestically through their adoption as a complement to the bill of rights of national constitutions.[6] Alternatively, national constitutions may include an article broadly stating that international law in general – including of course ratified conventions – is part of domestic legislation.[7] In both cases, the consecration at the international level of the right to water would immediately make it both operative and enforceable at the domestic level.

2. Access to water in national legislation

During the past fifteen years, provisions that specifically address the right to water have started to surface in national constitutions. The trend sees national legislation switching from general guidelines to more assertive articles expressly naming water as a human right. However, instead of securing access to sufficient, clean water as an individual human right, most of the jurisdictions analyzed have instead underlined the duty of the state to provide citizens with water services.

The latter formulation is the same as can be found in most international declarations; it is a solution that is likely preferred by most countries due to the fact that they regard economic and social rights as inherently programmatic. Choosing the duty of the state approach is the result of understanding that the realization of economic and social rights is inherently progressive and that an assessment of their violation cannot be made on the basis of the same criteria in force for civil and political rights. Since the realization of economic and social rights is heavily dependent on the availability of resources, many countries have opted to regulate access to water as a basic need that has to be satisfied, instead of as a human right. In practice, the implications of each approach are different: access to water as a duty of the state means that the state is encouraged to do everything within its purview to achieve the aim of universal access to water services through the implementation of adequate policies and the execution of infrastructure projects. Proclaiming it as an individual human right, on the other hand, would additionally entail for the state a set of concrete obligations to assist and protect the poor that might be beyond its economic capabilities. Besides, it would make the state accountable before the international community for its failure to realize the right and have other consequences in terms of justiciability. The issue is then whether the alleged differences between civil and political rights, on the one hand, and economic and social rights, on the other, justify distinguishing the latter as programmatic. Additionally, this begs the question

6 See e.g., Constitution of Argentina, 1994, Art. 75.22.

7 See e.g., Constitution of Paraguay, 1992, Arts. 137 and 141.

whether in fundamental legal provisions it is better to aspire to the full realization of human rights or to keep the realization of these rights within a realistic framework.

Several African Constitutions have adopted the duty of the state approach. The Gambian Constitution affirms that the 'State shall endeavour to facilitate equal access to clean and safe water',[8] while the Ugandan Constitution indicates that the 'State shall endeavour to fulfil the fundamental rights of all Ugandans to social justice and economic development and shall, in particular, ensure that [. . .] all Ugandans enjoy rights and opportunities and access to [. . .] clean and safe water'.[9] In a similar vein, the Ethiopian Constitution provides that to 'the extent the country's resources permit, policies shall aim to provide all Ethiopians access to public health and education, clean water, housing, food and social security'.[10] The wording undoubtedly reflects a commitment of the state to strive for the achievement of that goal, treating access to water for all as an obligation of means and not of results. Venezuela, for instance, establishes that it is the obligation of the state – in fact of municipalities – to foster the improvement in general of the living conditions of the community, in areas such as potable water services, sewerage, irrigation canals and disposal of wastewater.[11] These constitutions reflect the CESCR's approach of access to water as an obligation of the state towards the population whose fulfilment depends on the state's economic and legal resources.[12]

In contrast to the examples above, a small number of constitutions have consecrated access to water as an individual human right. Section 27 of the South African Constitution opts for an unambiguous formulation declaring that 'everyone has the right to have access to [. . .] sufficient food and water'. However, this statement is immediately nuanced when it is affirmed that the state 'must take reasonable legislative and other measures, within its available resources, to achieve the progressive realization of each of these rights', thus making the realization of the right a matter of progressive work limited by economic and legal constraints. On the other side of the Atlantic, the constitutional reform approved by a referendum in 2004 in Uruguay followed a more radical path. Besides declaring water a resource essential to life, Article 47 of the Uruguayan Constitution now proclaims that access to drinking water and sewerage services 'is a basic human right'; affirms that the provision of water services 'will have to be done placing reasons of social order before those of economic order'; and indicates that such services should be provided by state legal persons, instead of private corporations.[13] Other constitutional texts in the American continent are not as clear about water

8 Constitution of the Gambia, 1997, Art. 216(4). An almost identical formulation can be found in the Constitution of Zambia, 1996, Art. 112.

9 Constitution of Uganda, 1995, Art. 14.

10 Constitution of Ethiopia, 1995, Art. 90.

11 Constitution of the Bolivarian Republic of Venezuela, 1999, Art. 178.

12 See e.g., Constitution of Colombia, 2005, Art. 366.

13 See below, Section 4 of this chapter.

being a human right as the Uruguayan provision. The 1998 Constitution of Ecuador comes close, since its Article 23(20) includes the human right to a quality of life that comprises access to potable water, among other elements.[14] However, most provisions related to water at the constitutional level do not go any further than placing water stocks and their management under the control of the state and including environmental principles for the sustainable exploitation of natural resources in general.[15]

In national laws, the majority of water laws are not as direct as the constitutional provisions commented on above. Access to water is not treated as an enforceable individual right, but rather as a fundamental principle for the interpretation and application of the text of the law.[16] Conversely, a number of water acts and codes merely indicate that the provision of water services to the population should observe principles such as universality, non-discrimination, equality, sustainability, affordability and others, which have been mentioned in the CESCR's General Comment No. 15 as part of the normative content of the right to water.[17]

There are also laws that include the right as an operative provision. For instance, the use of water for domestic purposes (i.e., for personal drinking and watering livestock) has been deemed as a right of the population in Zambia. Article 8 of its Water Act, 1997 states that 'any person shall have right to the primary use of public water' and given that Article 2 defines primary use as 'the use of water for domestic purposes and the support of animal life', this provision ultimately indicates Zambia's legal acceptance of the right to water. In addition, in some cases a formal consecration of the right to water has been upheld by the judiciary where the constitution lacks such a provision. In Argentina, the right to water had been construed through case law as forming part of the constitutional right to a healthy environment.[18] The regulatory framework for the company operating water services in the city of Buenos Aires now mentions the right directly.[19] The same situation can be observed in France where the right is upheld in the French Environmental Code,[20] in Belgium where the Water Code provides that each person has the right to have access to potable water, in quality and quantity enough to satisfy their elementary, domestic and health needs,[21] and in

14 See also Art, 42 of the same Constitution, which commends to the state the realization of the right to health through, among others, the provision of water services.

15 Constitution of Guatemala, 1985, Art. 127, Constitution of Mexico, 1995, Art. 27 and Constitution of Malawi, 1994, Art. 13.d.iii.

16 Namibian Water Resources Act, 2004, Arts. 3a and 3c.

17 Bolivian Water and Sewerage Act, 1999, Arts. 5 and 22.

18 Constitution of Argentina, 1994, Art. 41 and the decision of the Civil and Commercial Court of the 8th Nomination of the province of Cordoba, Argentina, in 'Marchisio y otros s/Amparo' (Expte. No 500003/36), 19 October 2004.

19 Law 26.221, 2007, Art. 58.

20 French Environmental Code, Art. 210–1, as modified by Law 2006–1772, 2006.

21 Code de l'Eau of the Wallon Region in Belgium, 2004, Art. 1.3.

Pakistan, where the 2006 Draft for the National Drinking Water Policy declares that 'access to safe water supply is a basic human right'.[22]

A remarkable feature related to the incorporation of the right to water in national legislation is the priority given, in more or less explicit terms, to the domestic use of water over other uses.[23] It is not necessary to request an official authorization for such use, which comprises mainly drinking and personal hygiene, but can also include other uses deemed to be domestic such as fire-fighting.[24] General Comment No. 15 underlined that this priority is a legal basis for the right to water,[25] since national policies adopted to implement it at local level can only contribute to the satisfaction of the water needs of the population. A common article found in several water instruments provides that any person can take water free of charge for domestic purposes as long as they take it lawfully and only if they do not cause any damage to third parties by their acts and do not prevent others from exercising the same right.[26] Nevertheless, individuals are usually not entitled to construct any works for facilitating the catchment of water for domestic purposes, unless a specific permit has been granted to that effect.[27]

Some water codes have chosen the approach often adopted at the constitutional level, which is to handle the issue of access to water as a goal towards which the authorities in the sector must work. Such is the case of the Nigerian Water Act, 2004, which commands the Secretary of the Water Sector to make proper provision for 'the adequate supplies of suitable water for the watering of animals, irrigation, agricultural purposes, domestic, and non-domestic use'.[28] Other than that, the fulfilment of the water needs of the population can also make a marginal appearance in domestic regulation as one of the objectives of water sector management plans, which are policy documents elaborated by and addressed to the water sector authorities and agencies, explaining the national strategy for the integral, sustainable and equitable exploitation of water resources. In Swaziland, for example, the water sector management plan is called to 'secure sufficient water resources for domestic, agricultural and industrial needs'.[29]

22 National Drinking Water Policy, Draft 001, 2006, Section D.4.1.

23 Water Resources Management Proclamation of Ethiopia, 2000, Art. 7.1; Water Act of Kenya, 2002, Art. 32.2; Water Law of Cape Verde, 1984, Art. 59.1; Water Code of the Province of Buenos Aires, Argentina, 1999, Art. 2.c; Framework Law on Water Services of Honduras, 2003, Art. 3; Law of National Waters of Mexico, 1992, Art. 14*bis*5.XXII; Law on Waters of Bulgaria, 1999, Art. 50.4; Water Act of the Czech Republic, 2001, Section 29.1; Water Law of Spain, 2001, Art. 60.3; Water Law of China, 2002, Art. 21; among others.

24 South African National Water Act, 1998, Art. 4.1.

25 Committee on Economic, Social and Cultural Rights, General Comment No. 15: The Right to Water (Arts. 11 and 12 of the International Covenant on Economic, Social and Cultural Rights), UN Doc. E/C.12/2002/11 (2002), para. 6.

26 Water Act of Nigeria, 2004, Art. 2.a.i; Water Utilization (Control and Regulation) Act of Tanzania, 2002, Art. 10; Water Act of Newfoundland, Canada, 2002, Section 4; Water Law of China, 2002, Art. 48.

27 Water Utilization (Control and Regulation) Act of Tanzania, 2002, Art. 10.

28 Nigerian Water Act, 2004, Art. 5(a)(i).

29 Water Act of Swaziland, 2003, Art. 10.6.b.

3. Recognition of the right to water by domestic judiciary organs

As mentioned above, the right to water might be in force domestically thanks to the work of local judges that have construed it from other rights, such as the right to life, food, health and to a healthy environment. An early precedent in this sense was a 1981 decision of the Supreme Court of India, in which it was stated that '[the] right to life guaranteed in any civilized society implies the right to water, decent environment, and the right to live with human dignity and all that goes along with it'.[30]

The right to a healthy environment, in particular, has become a common feature in modern constitutions ever since the UN Conference on Environment and Development and has been regarded as the closest that countries have come to a universal recognition of the right to water in domestic legislation.[31] The right to a healthy environment strives for general conditions related to the quality of life of the individuals and 'may provide a legal basis for the improvement of water quality, for example through the prevention of pollution and the provision of adequate sanitation services'.[32] Access to water services and to a minimal provision of water for domestic purposes has been seen as an indivisible part of the right. This has been the case in Belgium, where it was stated that 'the right of each person to a minimum delivery of potable water [. . .] flows from Article 23 of the Constitution', which recognizes the right to the protection of a healthy environment.[33]

In summary, the human right to water may be in force domestically under several guises. It might be embedded in the national legal order through international legal instruments to which the country is a party to, or might have been adopted independently, under terms clearer or not than those of international conventions. Access to water services for all might be set as a goal of a plan for action, a guideline for water management or a principle for the interpretation of water legislation. It might be mentioned as an obligation of the officials working in related sectors or be construed as an aim of domestic water policy. Access to water services for all is increasingly being recognized as a constitutional right and its normative contents and practical implications in national laws, administrative policies and other operative provisions are being further clarified and developed. The role of the judiciary in this process has been crucial, for judges have construed the existence and enforceability of the right to water from other rights

30 *Francis Coralie Mullin v. Administrator, Union Territory of Delhi* 2 SCR 516 (1981).

31 Art. 41 of the Constitution of Argentina, Art. 27 of Benin's, Art. 55 of Bulgaria's, Art. 79 of Colombia's, Art. 23 of Ecuador's, Art. 24 of Greece's, Art. 35 of Korea's, Art. 37 of Moldova's, Art. 72 of Slovenia's, Art. 45 of Spain's, Art. 56 of Turkey's, among others.

32 COHRE, note 1 above, 45.

33 Decision of the Belgian Constitutional Court in its Arrêt No 36/98, 1 April 1998, Commune de Wemmel, *Moniteur Belge*, 24 April 1998, para. B.4.3. For more on the right to water in Belgium, see Chiara Armeni, The Right to Water in Belgium (Geneva: International Environmental Law Research Centre, Briefing Paper 2008-02, 2008) available at www.ielrc.org/content/f0802.pdf.

whose status in the domestic legal order is not disputed. As a result of this process, the recognition of water as a human right, either legally or jurisprudentially, is actually more common than initially thought. This trend is expected to continue as long as the problem of lack of access to water services is not solved.

B. Ownership, trusteeship of and control over water

The first step in regulating water and, in particular, the use of water in any given jurisdiction, is to define the legal nature of water resources. This is required to subsequently determine responsibilities for water management and to delineate the rights and obligations of the state and users. It is curious that the worldwide concern about freshwater availability has led to a progressive homogenization of water law. Previous legal distinctions made between groundwater and surface water are disappearing, as it is revealed that both kinds of water are interrelated in the hydrological cycle and therefore a different legal treatment for each becomes unjustified. Simultaneously, whether states belong to the common law or civil law legal tradition, water stress has made them reorient the basic principles of water law towards strictly limiting or eradicating private ownership of water resources.[34] As a consequence, they have increased the control of the state over water and put in place a system headed by governmental authorities to distribute water rights to private persons who apply to use water beyond the satisfaction of their domestic needs.

The application of the doctrines of public domain – in countries of civil law tradition – and of public trusteeship – in those of common law tradition – over water resources heralded a renewed and strengthened role for the state in water management. It made possible state control over water by putting limits to extraction of water by private persons from surface and underground sources and thus by restricting private appropriation of the resource. New legal and regulatory instruments progressively abandoned land-based approaches to water rights, in the belief that such traditional approaches did not only cause problems among users during times of scarcity, but also threatened the effective management of water resources since those rights did not fully take into account environmental considerations.

The declaration of water as part of the public domain of the state, in countries of civil law tradition, ultimately seeks to prevent individuals from obtaining rights to use water through acquisitive prescription or any other way outside the official system adopted for issuing water rights. Countries of common law tradition declare the state public trustee of water resources to emphasize the role of the state as guarantor of the preservation and availability of water for present and future generations. South Africa, for instance, mentions that role expressly by stating that the resource belongs to all people and naming the national

34 Water Act of Zimbabwe, 1998, Section 4; Water Resources Act of Nigeria, 1993, Section 1; Water Act of the Czech Republic, 2001, Section 3.1.

government the public trustee of the nation's water resources.[35] The state thus becomes the sole allocative authority on water and controls its public and private uses. The doctrine determines that the state becomes responsible before the population for the availability, distribution, quality and sustainable exploitation of water.[36]

Neither doctrine entails that water belongs to the state in terms of ownership. They rather stress that, since water is primordially destined to satisfy essential basic requirements of the population, the state has the ultimate responsibility in managing it. This mission can only be accomplished if the resource is primarily controlled by the public authority. Under these doctrines, only the government can decide who uses water and under which conditions, beyond the legally recognized authorization to private persons to take water from any surface or underground source, for strictly domestic uses. Water management is thus capable of being taken on by either the central government, the provinces or even local communities and rural water users associations (WUAs), should the management have been decentralized. In each jurisdiction, water legislation then foresees the extent of the participation of users, civil society and the private sector in the general scheme adopted.

In theory, a right to use water at the community or national level can be acquired by the effect of law (*ministerio legis*), by appropriation (through prior use), or after filing an application to obtain a water permit.[37] The first of these scenarios occurs in a number of countries of civil law tradition that declare that water resources belong to the public domain of the state but still recognize private ownership over water to the owner of the portion of land where it is found. In effect, national legislation in these countries follows the Roman principle according to which surface or underground water acquires the legal nature of the portion of land where it is found, as long as it does not flow to neighbouring portions of land.[38] Where this happens, water is automatically considered to be within the public domain.[39] Consequently, in some countries, the private ownership of rainwater falling over private land or of water obtained from wells or natural fountains located within private land is still allowed, as long as it does not flow outside of the portion of land owned by the private person. Ivory Coast, for instance, authorizes the artificial accumulation of rainwater as long as it remains

35 National Water Act of South Africa, 1998, Preamble and Art. 3. *See also* Alix Gowlland Gualtieri, 'South Africa's Water Law and Policy Framework: Implications for the Right to Water', in Philippe Cullet, Alix Gowlland Gualtieri, Roopa Madhav and Usha Ramanathan (eds.), *Water Law at the Crossroads – National and International Perspectives with Special Emphasis on India* 389 (New Delhi: Cambridge University Press, 2009).

36 See below Section 4 on the privatization of water services and the regulatory role of the state.

37 Antoinette Hildering, *International Law, Sustainable Development and Water Management* 96 (Delft: Eburon Academic Publishers, 2004).

38 See, e.g., Water Code of Brazil, 1975, Arts. 2, 6, 7, 8.

39 This is the case in the Law on Waters of Bulgaria, 1999, Art. 24; the Water Law of Costa Rica, 1999, Art. 4; and the Water Law of Spain, 2001, Art. 10.

within the limits of the private property where it fell on, and its use is compatible with the laws of the country.[40] However, as it is the case for any other private property, everyone is expected to use the resource rationally and not to abuse their rights. In case of severe scarcity, water owned by a private person can be declared public through the usual legal mechanisms, its management being accordingly transferred to the state as long as it is necessary to palliate the emergency.

The second way through which the individuals can in theory acquire the right to use water beyond their domestic needs is the legal institution known as acquisitive prescription. That is, the legal acknowledgment of a right to use a source of water after the continuous, uninterrupted use of it throughout a certain period of time. Most states seem to have abandoned this mechanism. In civil law countries, it is a logical corollary of declaring water within the public domain of the state. In common law countries, it has been expressly stated.[41]

A third option to acquire a water right is a system of state-issued water permits and licences that grant to individuals the use of water beyond their domestic needs.[42] A majority of states have opted for this mechanism. Recent water legislation has in effect assimilated the regime of groundwater to the one that was in force for flowing surface water. In the words of Caponera, new legal instruments purport 'to limit individual and exclusive rights of ownership in favour of a centralized administrative control over groundwater through the introduction of a formal separation between the two concepts of "ownership" and "right to use" '.[43] The water rights system adopted by each state has its own particularities, which are mainly determined by the availability of water in the country. Still, these systems follow common patterns, especially those related to the procedure for obtaining the licence or permit and in the sanctions imposed on users who abuse it or violate the terms under which the right was granted.

The right to use water authorized by the government may be in the form of an easement, implying that the right is transmitted with the ownership of the property benefiting from it. However, some countries have opted for the exact opposite, detaching the right from the land and introducing the free trade of water rights among private persons. The case of Chile is emblematic, since water right transactions are not regulated in that country. The 1980 Constitution of Chile provides that 'the rights of private citizens over waters, recognized or constituted in conformity with the law, shall grant proprietorship to the owners thereof', while its 1981 Water Code provides that the licence is an entitlement owned by its proprietor, who is authorized to use and dispose of it.[44] These provisions allow for the free trade among private persons of the licences granted by the government

40 Ivory Coast – Water Code, 1998, Arts. 22 and 23. Water Law of Spain, 2001, Art. 54; Water Code of Chile, 1981, Arts. 10; and Project for a Comprehensive Water Law of Colombia, 2005, Art. 16.

41 Water Resources Act of Newfoundland, Canada, 2002, s. 9(1).

42 Water Act of the state of Victoria, Australia, 2005, ss. 7 and 8.

43 Dante A. Caponera, *Principles of Water Law and Administration: National and International* 249 (Rotterdam: A.A. Balkema, 1992).

44 Constitution of Chile, Art. 19(24) and Chilean Water Code, 1981, Art. 6, respectively.

and for the emergence of local water markets along watercourses. These markets are meant to allocate water efficiently among alternative economic uses for water. However, it did not take long before private companies providing water services became the main buyers of these rights from farmers.[45]

The procedure for obtaining water use permits is led by a state agency that is usually created by legislation. The new organ is in fact trusted with many other functions related to the management of water resources, such as undertaking inventories of national stocks, monitoring their evolution and providing for their conservation.[46] To this end, the state agency is usually in charge of delineating a national water sector management strategy.[47] In addition to administrative duties in the delivery of permits for water use, these organs also have an important, though not exclusive, role in setting the fees that private persons are required to pay for the licences or concessions contracted, in collecting the proceeds and in allocating them according to the priorities set in the law or water sector management plan. In Namibia, for example, the issue of water pricing is in the hands of a separate board.[48] In certain countries, community-based water management initiatives have delegated pricing to rural WUAs.

The procedure for obtaining a permit is usually open to public opposition. It must be given adequate publicity to allow others to raise their objections against applications under review by the authorities.[49] Permits can be used as long as the hydrological situation – as monitored by the state – is optimal for them to remain in force. However, reasons of public interest, or the need to balance the interests of a right holder with those of another legitimate user might force the suspension, revocation or modification of a permit, with compensation being paid to the holder, if required.[50] There is no compensation if the terms under which the water use authorizations were granted are violated. In this situation, the suspension or revocation of the right can be joined by other measures, such as a fine to the permit holder and prosecution for offences or even crimes.[51] Criminal proceedings are only undertaken for grave offences, such as endangering public health by large-scale pollution of water resources.[52] The growing space that water legislation dedicates to the regulation of sanctions and penalties for the misuse, wastage and pollution of water can also be read as a reflection of the economic value that countries have assigned to it.[53]

45 For an assessment of the benefits that this system has brought and the controversies that it has sparked, see Mónica Ríos Brehm and Jorge Quiroz, The Market for Water Rights in Chile (Washington: World Bank, World Bank Technical Paper No. 285, 1995).

46 Water Law of Mongolia, 1995, Arts. 3 to 7.

47 Water Act of Swaziland, 2003, ss 8a and 10(2).

48 Water Resources Act of Namibia, 2004, s. 87.

49 Federal Water Act of Germany, 2007, Part I, Art. 9, and Water Act of Kenya, 2002, s. 29(4) to (6).

50 Water Resources Management Proclamation of Ethiopia, 2000, Art. 9(1).

51 Water Law of Romania, 1996, Arts. 56 to 58 for an illustration of cases in which water rights can be either withdrawn or suspended. Chapter VI (Arts. 86 to 106) elaborates on infringements in the water management field and the ensuing penalties.

52 *Ibid.*, Art. 92.

53 *Ibid.*, Art. 80.

In sum, as a result of the awareness of the finitude of water resources and of the strategic value that their own stocks have acquired, most countries have increased the control of the state over water through the application of the doctrines of public domain and public trusteeship, and have put systems in place for delivering water use permits to private persons. The permits can either be personal, a right attached to the ownership of a portion of land, or a tradable right in certain few jurisdictions,[54] but in all cases it is the state in its capacity of responsible manager that decides which option is best suited to attain the objectives set by itself in its water management plans. Naturally, the reinvigorated role of the state does not forsake private sector participation; it only entails a more active presence in water management that is explained by water becoming a matter of national security, and it underlines that the state has the ultimate responsibility for an efficient and sustainable management of the resource and for monitoring the quality and extent of water services. As public trustee of water, the state has the power and duty to correct any defects or shortcomings in the overall management, something which is included in some water legislation. This is the case of Ivory Coast's Water Code, which affirms that the main role of the state is to ensure 'the management of water resources preserving the quality of the sources, preventing wastage and ensuring availability thereof'.[55] It further affirms that the state guarantees 'the provision of drinking water; the protection, conservation and integrated management of water resources; the satisfaction of other needs' as well as 'the development and protection of infrastructure and hydraulic works; the prevention and remedy of water diseases. It monitors and protects the state of water through its competent agents'.[56]

C. Participation in water management and water services operation

There are three concrete ways in which people participate in water management and water services operation. These are:

(a) representation in decision-making organs;
(b) access to judicial or quasi-judicial remedies in order to defend their interests as stakeholders and/or users of water services; and
(c) policies of democratization of water management that delegate rural water management and service provision to small communities.

These are examined in turn, after some remarks on transparency.

54 Besides the case of Chile, other jurisdictions that have developed water rights markets include South Africa, Mexico and certain provinces in Spain and Australia. Contrary to the Chilean case, trade in these jurisdictions is regulated.

55 Ivory Coast – Water Code, 1998, Art. 59 (author's translation).

56 Ivory Coast – Water Code, 1998, Art. 60 (author's translation).

1. Transparency in domestic legislation

Transparency is a premise to enable users and civil society to participate in the water sector. It is also a means of holding public officials accountable and of fighting corruption, as it provides information to the media and to the public and allows citizens to review budgets and financial statements, as well as to participate in discussions of laws, rules and decisions. A number of countries have included openness to society and transparency among the fundamental principles that guide their water sector policies,[57] although in many cases this is nothing but an implementation of the individual right to information, which is commonly recognized in modern democratic societies.[58] Domestic legislation usually comprises provisions adapting the right to information to the field of natural resources – the water sector in particular. The importance of the interaction between the population and public authorities in environmental issues and management of natural resources has been underlined by the UNECE Convention on Access to Information, Public Participation in Decision-making and Access to Justice in Environmental Matters.[59] This Convention, though regional in scope, is regarded as the most accomplished incarnation of Principle 10 of the Rio Declaration, which stresses the need for the participation of citizens in environmental issues and for access to information held by public authorities on the environment.

Water acts and codes oscillate between general provisions giving access to information and transparency as part of public participation in the management of the resource, and provisions that put that right into practice. Of the former category, Article 14*bis*(5)(XIX) of the Mexican Water Law is an apt example. It provides that one of the principles of the national water policy is the right of society and its institutions to information on the existence, availability and needs of water and on related matters that are indispensable for the management of the resource. Sub-paragraph (XX) then links access to information with the concrete participation of the population, by adding that informed and responsible participation of the society is the basis for a better management of water resources and, in particular, their conservation. In effect, sharing information and promoting education is a prerequisite for a participative management of natural resources that is effective in protecting them and using them in a sustainable way.[60]

With regard to operative provisions, most of them establish the obligation of public authorities to inform people of the status of water stocks and, especially, on their quality. In this sense, the obligation to inform on the quantity and character-

57 Water Resources Act of Namibia, 2004, Art. 3.f.

58 The right usually includes access to any information of public interest. See, e.g., Constitution of Romania, Art. 31.

59 Convention on Access to Information, Public Participation in Decision-Making and Access to Justice in Environmental Matters, Aarhus, 25 June 1998, UN Doc. ECE/CEP/43.

60 Federal Law on the Protection of Waters of Switzerland, 1991, Art. 50, and Project for a Comprehensive Water Law of Colombia, 2005, Art. 93.

istics of effluents spilled in streams is a common feature.[61] Article 13(III) of the French Water Law, 1992 states that information on the quality of water must be accessible to users and the government must make it public through appropriate means. Water instruments that are oriented towards the regulation of the basic services of drinking water and sanitation deal with the specific right to information of users of these services, in order to enable them to exercise their rights as such. For instance, Article 25(2) of the Framework Law for the Potable Water and Sewerage Services of Honduras affirms that information on the provision of the service, the tariffs regime and the plans for the development of the service network is indispensable for protecting their interests and rights as users.

2. *Representation in decision-making organs*

Turning now to the concrete ways in which the population has a role in water management and water services provision in domestic legal frameworks, the first of them would be through their representation in decision-making organs. Above all, public participation is possible through the delineation of frameworks that allow users to take part in the conception and execution of projects and policies that could have an incidence on their rights. Article 1 of Ivory Coast's Water Code, 1998 defines the principle of information and participation as granting everybody the right to be informed about the state of water resources and to participate in any procedures prior to making decisions likely to have a negative effect on them.[62] Many instruments have specific operative provisions about the public contribution to decision making by ensuring that users have a seat in the appropriate bodies or, at least, have public participation as a principle informing state action.[63] Legislation has provided for the creation of WUAs as fora where water users can agree on common positions to be then defended in the plenary meetings of the water management organs.[64] The representation of users has been assured, particularly, in the composition of the water authority that is often created to oversee the management of the resource.[65] The representative is usually a senior official within the WUA. In Honduras, the delegate is elected in an assembly gathering the representatives from different WUAs.[66]

In entities hierarchically placed below the national authority, the presence of

61 Water Law of Spain, 2001, Art.15.

62 See also Water Resources Act of Namibia, 2004, Art. 3.j.

63 Water Law of Spain, 2001, Art. 14.1.

64 Water Resources Management Proclamation of Ethiopia, 2000, Art. 27; Water Resources Management Act of Namibia, 2004, Art. 16; Water Act of Swaziland, 2003, Art. 76; Water Statute of Uganda, 1995, Art. 50; Law on Waters of Bulgaria, 1999, Art. 88; Water Code of the Province of Buenos Aires, Argentina, 1999, Art. 127.

65 Water Law of Morocco, 1995, Arts. 14 and 21; Water Act of Swaziland, 2003, Art. 21; Water Law of Spain, 2001, Art. 18.1.b.

66 Honduran Framework for the Drinking Water and Sanitation Law, 2003, Art. 7.7.

users is also valued. Legislation ensures that a certain minimum number of users are necessary to have quorum, or at least that a designated number of seats inside the organ are compulsorily accorded to users.[67] The designation of delegates to regional and federal boards and the formation of local management councils and basin committees, in which users and the public in general, to a greater or lesser extent, are given representation and thus the opportunity to express their views, is intended to efficiently address their concerns in water management.[68] The National Water Plan, in the design of which these organs are involved, must also in certain jurisdictions receive input from water users and the general public. Article 13 paragraph 4 of the 2005 Colombian Project for a Comprehensive Water Law states this without ambiguity and also provides that minorities such as native peoples of the country are also consulted. In Bulgaria, river basin management plans are discussed by the public in sessions specially convened by the Director of the River Basin Committee.[69] In Zimbabwe, public participation in the local management plans are assured by giving that task to catchment councils formed by water users.[70]

3. Participation of users through the defence of their rights before judicial and quasi-judicial organs

A second way in which users participate in water management and water services provision is by using in this context their right of access to justice. In effect, domestic water legislation regularly features the right of private persons to initiate or join judicial, arbitral and/or administrative proceedings by claiming that their interests are at stake due to public or private initiatives related to water. Certain water acts and codes create an administrative organ exercising judicial functions that specifically settle these disputes and dedicate a full section to this matter.[71] For example, citizens can oppose a certain water use licence or concession, whether in writing or in public audiences open to that effect.[72] This is a corollary of the civil right to justice, and provides civil society institutions with legal personality with the possibility of intervening as well. The distribution of licences and concessions for water use is usually put into motion by disseminating information on the applications received and currently under examination by the government, so that those that are opposed to a certain use of water can manifest their opposition, sometimes without needing to demonstrate that they hold a concrete interest in the matter.[73]

67 See, e.g., French Water Law, 1992, Art. 5.

68 Water Act of Zimbabwe, 2003, Art. 20(b), which enables the Minister of the sector to fix a number of water users to represent their group within the water catchment council.

69 Law on Waters of Bulgaria, 1999, Art. 162.

70 Water Act of Zimbabwe, 2003, Art. 21(1)a.

71 Water Resources Act of Namibia, 2004, Part XVIII (Arts. 118 to 123).

72 Federal Water Act of Germany, 2002, Art. 9; Water Code of the Province of Buenos Aires, Argentina, 1999, Art. 41; and Water Act of Kenya, 1995, Art. 57(6) and (7).

73 Project for a Comprehensive Water Law of Colombia, 2005, Art. 91.

In those jurisdictions where water services have been privatized, the legal and regulatory frameworks for the services have usually contemplated a role for users in the supervision of the performance of the private company. The users that are against a certain measure or practice taken or adopted by the private company have the opportunity to defend their rights before administrative or judicial organs, either individually, through associations for the defence of their rights as users of a service, or with the assistance of an ombudsman. The private company is obliged to facilitate this role by sharing with its clients information on issues such as tariffs and investment projects directly or through the regulatory authority.

4. Democratization of water management

The participation of the population in the water sector has grown exponentially with the introduction of the democratization of water management in rural areas and the privatization of national utilities operating water services in urban areas. The latter – including its public participation aspect – will be dealt with in the next section. Here, before tackling the democratization phenomenon, some mention should be made of decentralization.

The decentralization of water management and of the water services provision precedes both the democratization of the former and the privatization of the latter. Decentralization is a policy supported by international financial institutions for reducing public expenditure, by shrinking the burden that water management and its services entail for the state. Thus, where water management is concentrated in the hands of national authorities at the federal level, functions are then delegated to provincial and/or local entities. Where water services are delivered by a national public utility, countries are encouraged to downsize and restructure it in smaller units under the jurisdiction of the provinces or cities. Through decentralization it is affirmed that resources are saved and water management and the operation of water services increase in efficiency. As stated above, the decentralization of water services is commonly seen as the first step towards privatization in urban areas and democratization in rural areas, where private companies are not interested in investing. These two processes have been encouraged and supported widely by different international organizations. For instance, the UN Secretariat has stated that:

> [There] is now a growing body of experience showing the important role that the private sector can play in the generation of financial resources and in the delivery of services. There is also a growing understanding of the fact that peri-urban and rural communities can play an important role in the generation of financial resources and in the operation and management of local water supply and sanitation systems.[74]

74 UN Secretariat, Report of the Secretary General on the Progress Made in Providing Safe Water Supply and Sanitation for All During the First Half of the 1990s 11, para. 34, UN Doc. A/50/213-E/1995/87 (1995).

Democratization consists of empowering rural communities to manage their immediate sources of water and even operate their own water services – sometimes with local private sector participation – as a way to allegedly achieve universal and equal access to water services in rural areas. The term democratization of water management has evidently been chosen for its positive connotation. In theory, it involves public participation in decision making and control, while applying values of social justice, equity, sustainability, transparency and solidarity between communities. The democratization process – it is affirmed – would increase sustainable management while respecting local uses and customs. This strategy was first noticeable in irrigation water schemes, but is nowadays a common feature in the management of water resources in areas under stress and the provision of water services in rural areas in developing countries.[75] The trend of vesting the competences of the national water authority on WUAs managing irrigation water and local associations for the self-provision of water services has intensified with the introduction of initiatives funded by international financial institutions such as Community-Driven Development (CDD) and Participatory Irrigation Management (PIM). These projects have actually delegated the operation of water services and the construction of the necessary infrastructure to local groups, and have accordingly brought along changes to the water legal and regulatory frameworks of a number of developing countries.[76] Under CDD schemes, communities are actually in charge of choosing the Rural Water Supply and Sanitation (RWSS) model that they deem necessary for their particular circumstances, of implementing the project with the help of NGOs and technicians, of contributing financially to building infrastructure[77] and finally, of managing the water services delivery themselves.[78]

As the FAO has recently underlined, the increased participation of users in water management and in the provision of water services can be seen as a policy 'aimed at building consensus and support for difficult water allocation and management decisions, and at sharing government responsibilities with concerned users'.[79] The National Policy for Safe Water Supply and Sanitation of Bangladesh illustrates this point by stating that one of the principles of its strategy is the 'participation of users in planning, development, operation and maintenance through local government and community based organizations of the

75 World Bank/Bank-Netherlands Water Partnership, Rural Water Supply and Sanitation Toolkit for Multisector Projects (Washington: World Bank, Working Paper, 2004), available at http://go.worldbank.org/7C95VQ9TV0.

76 Concerning India, see Chapter VII concerning regulatory authorities and Chapter X concerning irrigation. For an early version of the application of these ideas, see the scheme of village committees in the 1988 Water Supply Regulations of Tonga.

77 Some countries that have foreseen such roles for citizens include Uganda, Water Statute, 1995, Art. 55.3; Zambia, Water Act, 1997, Art. 43; Zimbabwe, Water Act, 2003, Art. 24.8.

78 For a description of the alleged benefits of CDD applied to the water sector, see www.worldbank.org/cdd.

79 Food and Agricultural Organization, Law and Sustainable Development since Rio – Legal Trends in Agriculture and Natural Resource Management 157 (Rome: FAO, *Legislative Studies* 73, 2002).

stakeholders'.[80] However, the empowerment of small, rural communities could also have controversial derivations. Under the guise of public participation, local communities are often stifled by obligations that are actually the state's. The fact that users are even asked to contribute financially to develop infrastructure raises questions about whether those unable to pay will have access to water. The policy further provides percentages of the infrastructure costs for rural water supply that must be borne by users, and, in addition, states that 'user communities will be responsible for the operation and maintenance of water supply facilities and shall bear its total costs'.[81] The rights that public participation at this level is supposed to entail must be carefully balanced with the obligations that are bestowed upon the public, or democratization risks being discredited as a policy that truly serves the users' interests. India is one of the countries that has incorporated the democratization of water resources management into its water legislation, though this is mostly noticeable in rural water management for irrigation.[82]

D. Privatization of urban water services

The privatization of urban water services is a trend which has spread quickly and consistently around the world, particularly throughout the 1990s. The participation of the private sector in the provision of water services is a policy that international financial institutions strongly recommend in developing countries, heralding it as the solution to meet the investment needs that the public sector cannot afford. As a result of the privatization wave and of the concentration of the water business in a handful of multinational enterprises, domestic legislation gradually addressed the issue, incorporating provisions related to the insertion of private companies as a third party in water services operations. Uruguay, the Netherlands and Nicaragua are among the few jurisdictions that have taken a stand against privatization.[83] In Nicaragua, access to water is a right that must be assured to the people by the government. Water services must always remain the responsibility of the state and cannot be privatized.[84] However, other jurisdictions not only do not discard privatization, but actually embrace it as a legitimate option for achieving optimal provision of services, particularly those related to the costly infrastructure development needed in urban areas. In this sense, one of the policy principles of the National Policy for water services in Bangladesh argues that:

80 Bangladesh, National Policy for Safe Water Supply and Sanitation, 1998, Section 4(b).

81 *Ibid.*, Section 8(1).

82 On the role of WUAs in rural water management in India, see Chapter X.

83 Constitution of Uruguay after the 2004 referendum, Art. 47, Senate of the Dutch Parliament (2003), 28 339: Revision of the Law on Water Services (Property of Water Services Companies), prohibiting the provision of essential water services by societies other than state-owned or controlled and General Law on Water of Nicaragua, 2007.

84 *Ibid.* Art. 4.

> [Many] functions of the water supply and sanitation sector can be undertaken by private organizations. This will promote increased service coverage and thereby lessen the burden on the government. It is necessary to strengthen an administratively and financially enabling environment for the private sector to participate and contribute to sector development. Involvement of the private sector is essential to establish a closer relationship between the quality of services of the sector and its financial viability.[85]

The rationale behind provisions refusing privatization appears to be the belief that only the state is able to grant access to water to all. Activists against the privatization of water services claim that private operators will hinder the realization of the human right to water, as they could refuse access to those that are unable to pay.[86] Those that participate in the private sector argue that the state can use private investment to ensure that the network for drinking water and sanitation is maintained and expanded, and still be able to intervene in a privatized system of service delivery through subsidies and other mechanisms of the kind that ensure that the poorest people are connected to a supply and that water remains affordable for all. Private–public partnerships have been put forward and the importance of a sound legal and regulatory framework has been stressed by those that insist that the private sector has a vital role to play in water services provision.[87]

Domestic water legislation does not normally indicate the position taken on this issue. There may be indications in parliamentary debates, but a better idea of the degree of acceptance of privatization in a certain jurisdiction should be evident after studying its law on public contracts and its corresponding parliamentary debates. As this chapter focuses on water legislation, the instruments addressing the issue of privatization remain few. The issue is often mentioned incidentally, through a casual acknowledgment of the fact that service provision might be in the hands of private operators. The majority of the articles referring to privatization call for the state to evaluate the relationship between provider and user, to supervise the quality of the service and to respect national and international standards. The provision of water is deemed an essential public service and the state remains responsible for monitoring the performance of private providers and for taking appropriate and opportune measures when it is not satisfactory. This explains why there is no contradiction in stating that control over water is assumed by the state, while the delivery of water services and the infrastructure related to it can be ceded to private companies through public contracts signed between them and the state. The change of role for the state from provider to regulator should not, in effect, minimize its overall responsibility for realizing the

85 National Policy for Safe Water Supply and Sanitation of Bangladesh, 1998, Section 7.

86 Public Citizen, Water Privatization Fiascos: Broken Promises and Social Turmoil (Washington D.C.: Public Citizen, 2003).

87 World Bank, Water Resources Sector Strategy: Strategic Directions for World Bank Engagement (Washington, D.C.: World Bank, 2004).

right to water. Its first duty should be the delineation of an optimal regulatory framework and the creation of a regulatory agency in charge of monitoring the performance of private companies and of working as a link between them, the users and the state. This is not surprising, given that high-profile failures in the privatization processes have been attributed to deficiencies in the regulatory frameworks – mainly in privatization contracts – and to the poor institutional quality of host countries.[88]

The challenge in domestic water law has been to reconcile the public interest with a number of safeguards that the private operators of water services demand in order to be in charge of water services. The specifications aimed at achieving this balance are left for administrative instruments and the privatization contract itself, but provisions at the legal level already set a series of guidelines, particularly in what the responsibilities of the state are concerned. The Namibian Water Act mentions, for example, the provision of water services by third parties and the regulatory role of the State over the privatized service, signalling that:

> (1) The Minister must ensure that all Namibians are provided with an affordable and a reliable water supply that is adequate for basic human needs.
> (2) For the purposes of ensuring the adequacy, affordability and reliability of water supply, the Minister must:
>
> (a) develop reliable standards of performance and facilities that are applicable to any person who supplies water for domestic, commercial, industrial or agricultural use;
> (b) periodically review the performance of every person who supplies water for domestic, commercial, industrial or agricultural use to evaluate compliance with the standards of performance developed pursuant to sub-paragraph (a);
> (c) take corrective action with regard to any water supplier who fails to meet a standard of performance, which may include:
>
> (i) the secondment of managerial or technical personnel;
> (ii) the repair or replacement of water supply works; or
> (iii) the amendment, suspension or cancellation of any relevant licence to abstract and use water; and
>
> (d) establish an administrative mechanism that will enable water users to be heard regarding the adequacy and reliability of their water supply.[89]

This provision makes clear that the state has the power to revoke contracts in case of breach of the standards of performance contained therein, which are namely set to create an adequate, affordable and reliable service. Part XV of the same

88 Ioannis N. Kessides, 'The Challenges of Infrastructure Privatisation', 1 *CESifo DICE Report* 19 (2005).
89 Water Resources Management Act of Namibia, 2004, Art. 26.

instrument is dedicated to the creation of a Water Regulatory Board to 'exercise control over water and effluent pricing in the water supply and effluent discharge industries and to provide quality independent assessment of water pricing proposals by water utilities and suppliers'.[90]

In rural areas, on the other hand, where multinational companies are not interested in operating due to the unfavourable rate of costs and revenues, the privatization of water services has taken another guise. The privatization of water services in this context does not have to do with service delivery by big companies through concessions and other types of privatization contracts, but with the delegation of tasks to users themselves. Water services in villages are increasingly being managed by cooperatives, local committees and other forms of user associations, a trend that has been reinforced through the World Bank's new approach to RWSS and CDD. Accordingly, a number of water acts and codes have addressed this trend as well. Article 17 of Honduras' Framework Law on Water Services, for example, states that water services are under municipal authority and that the local users association should have priority in being granted the authorization for the operation of water services; however, it allows private companies to subsidiarily be in charge of the provision of drinking water and sewerage, as long as the majority of the beneficiaries choose that option through their vote. This provision indicates that local private firms can still participate in water service provision in small communities, and it is commendable that users have the opportunity of democratically choosing whether or not to privatize in the proper sense of the term.

E. Environmental protection of water resources

Provisions related to the protection and conservation of water resources are a common feature of domestic legislation of the sector. Whether through a special chapter or in the form of separate articles along the body of the text, water acts and codes have all incorporated environmental principles, particularly since the 1992 United Nations Conference on Environment and Development and the adoption of the Rio Declaration that included principles such as sustainable development, prevention and polluter-pays. For instance, Colombia's 2005 Project for a Comprehensive Water Law includes in a list of twelve basic principles for the integrated management of the resource, three that involve environmental aspects: sustainable development (No. 3); polluter-pays (No. 10); and conservation and preservation of the resource (No. 11).[91]

The concern for the preservation of national water stocks appears in the initial articles of water acts and codes, especially in those that set guidelines for action for the state agencies in charge of water management.[92] Included or not among

90 Water Resources Management Act of Namibia, 2004, Art. 86.

91 Project for a Comprehensive Water Law of Colombia, 2005, Art. 2.

92 National Water Act of South Africa, 1998, Preamble and Art. 2(h); Law on Potable Water and Sewage Services of Bolivia, 1999, Art. 5(g); Water Resources Management Act of Namibia, 2004, Art. 3(g) and 3.1.

the principles ruling water management, the law may additionally dictate that those principles aiming at the preservation and ecological use of water are included as well in water management sector plans.[93] Whatever legislative technique is preferred, what is noteworthy is the universal urge for the rational and environmentally-friendly use of water, and for its sustainable exploitation.

This concern is also noticeable in that the procedure for obtaining a licence or concession for the use of water requires that those interested present an environmental impact assessment. The protection of the resource is also granted by establishing sanctions for unlawful uses of water or for attacks against its integrity.[94] Water waste, pollution, misuse or use in violation of the terms of the licence or concession granted to private persons, all represent grounds for the application of punitive measures of diverse severity.

Another legislative measure related to environmental goals that has been taken in recent water acts and codes has been the creation of water reserves. An ecological reserve under South African law means 'the quantity and quality required [. . .] to protect aquatic ecosystems in order to secure ecologically sustainable development and use of the relevant water resource'.[95] National water resource and catchment management strategies must be drawn taking into account the formation of the reserves and ensuring that the issue of licences and concessions to private persons do not prejudice them. This is explicit, for example, in the amendments to the Water Act of Victoria, Australia, 1989. Section 5 states that water rights are not conferred, licences are not issued, and works and activities are not approved 'unless regard is had to the need to maintain the environmental water reserve in accordance with the environmental water reserve objective'.[96] Article 27 of the Namibian Water Resources Act, 2004 provides for a similar reservoir for environmental purposes, among them the protection of biodiversity and the maintenance of essential ecosystem functions. Other countries with analogous provisions include Spain, Jamaica, Vanuatu and Mexico.[97]

93 This is the solution preferred by Swaziland (Water Act, 2003, Art. 10.6(d) and Spain (Water Law, 2001, Art. 40.1).

94 Interestingly, the 1988 Water Supply Regulations of Tonga and the 1997 Water Act of Zambia only address pollution in the chapter dedicated to offences (Art. 9 and Art. 55, respectively).

95 South African National Water Act, 1998, Art. 1(1)xviii. Part Three is dedicated to the regulation of the 'reserve'.

96 See in general Water (Resource Management) Act of Victoria, Australia, 2005, Part Two; in particular, ss. 4.A and 4.B.

97 Water Law of Spain, 2001, Art. 42(d); Water Law of Jamaica, 1995, Art. 32 under which the reserve is made 'for a public purpose', an expression comprising environmental conservation; Water Resources Management Act of Vanuatu, 2002, Division 2 'Water Supply Conservation and Development', s. 9; and Law of National Waters of Mexico, 1992, Arts. 38 to 43.

CONCLUSION

Recent legislation aims to avoid legal overlaps that are the product of the accumulation along the years of different instruments regulating specific aspects of water management. Such has been the case in Morocco, where legal instruments of different kinds – some of which dated from as far back as 1914 – had been regulating different water-related issues up to the enactment of the current water legislation in 1995.[98] They also purport to update obsolete water legislation through the incorporation of the outcomes of international debates that have lately focused on water.

As such, these new domestic instruments provide us with a glimpse of ongoing changes in water policy around the world, oriented towards an integrated and sustainable management of the resource. Article 2 of the French Water Law summarizes in a way the goal of current water legislation: an integrated, balanced and environmentally-friendly management of water. It reads:

> This balanced management aims at granting: the safeguarding of water ecosystems and wetlands; [. . .] protection against any type of pollution and restoration of the quality of surface, underground and maritime waters within the limits of the territory; the development and conservation of water resources; the valuation of water as an economic resource and its distribution as such, in order to satisfy or to reconcile, during the various uses made of it, the requirements of salubrity, public health, civil security and supply of drinking water to the population; the conservation and the free flow of water, and the protection against floods; agriculture, fishing, energy production, transport, tourism, leisure such as water sports, as well as any other legal human activity.[99]

The awareness of the importance of water and of the implications of its increasing global scarcity has led to modifications to the legal and regulatory frameworks for water management and water services provision. On the one side, the role of the state seems to have been reinforced with the incorporation of new obligations related to the gradual realization of the human right to water, the prioritization of the use of water for personal and domestic purposes and the need to put effective schemes for the protection and conservation of water resources in place. It is evident that the state cannot attain these goals without first assuming a more active control over the resource, abolishing or limiting private ownership of water resources within domestic water legislation, and forming a strong agency charged with coordinating the multiple uses of water and delivering water use permits to private persons. The review of water legislation shows that states have assumed

98 *See* the 'exposé de motifs' of the Water Law of Morocco, Law 10–95, *Bulletin Officiel* No 4325, 20 September 1995, 626–662.

99 Water Law of France, 1992, Art. 2 (author's translation).

this mission. They have also engaged in detailed water planning, in many cases organizing the territory in catchment areas, and have implemented a battery of solutions adapted to their own water context, in an attempt to optimize the exploitation of water resources and prevent wastage.

On the other side, the state has seemingly taken a step back once it opened the door to private participation in service provision and user participation in water management. At least that would be the logical conclusion after assessing the combination of privatization initiatives for urban water supply since the late 1980s and of the more recent CDD/PIM projects for RWSS. This may appear to go against the strengthening of the role of the state as trustee of water resources but, in fact, the incorporation of private companies and of village committees in urban and rural water services and in rural water management does not alleviate the duty of the state for water management and, above all, the duty to provide people with access to basic services. On the contrary, domestic water legislation has often emphasized and regulated the new role of the state as supervisor of the performance of private companies in charge of urban water services and of users' organizations that have been put in charge of their water resources management and services provision.

The success of water reforms will naturally depend on the quality of the legal and regulatory design, the institutional capacity with which these initiatives are put into practice, and the transparency measures that are taken to keep actors informed, thus enabling them to participate in the overall scheme of water management and water services provision. An optimal management of water resources and an efficient delivery of water services are only possible if clear rules are enacted, efficient institutions put in place, stakeholders given the chance to participate in the sector and the exploitation of the resource done in a way that is respectful to the environment. In general, current domestic water legislation is a reflection of those broad ideas; however, the challenge for the sector is to come up with good policies along those lines, accommodate them to local specificities, defend them against extraneous interferences and not fail at the implementation level.

III. International aspects of water law reforms

Irina Zodrow

INTRODUCTION

Water is the precondition for human life. As a shared resource it is at the core of human interdependence, both within and between countries. Thus, ensuring access to clean and sufficient water for all, including through comprehensive laws and policies regulating the use and management of water, must be a priority of any government authority involved in the development or implementation of water management systems. The world has never been further away from universal access to water. According to the World Health Organization (WHO) and the United Nations Children's Fund (UNICEF), in 2004 over one billion people globally were without access to improved water supply (which amounts to one human being out of six).[1] Though this number dropped to 884 million by 2006,[2] the challenge remains grave. In 2004, there were over four billion cases of diarrhoea which caused 1.8 million deaths, mostly among children under five.[3] While the need to address this situation has been widely recognized, the attempts of the international community to solve the problem have not gone far enough. Emblematic of this, the water-related Millennium Development Goals (MDG) do not contribute to the realization of access to water for everyone, but only aim to halve the number of people without access to water by 2025.[4]

The reason for the global lack of access to water is not necessarily the growing water scarcity that has recently attracted increasing international attention. The 2003 United Nations water report also mentions mismanagement, global warming, population growth and political inertia.[5] In fact, as stated by the United

1 World Health Organization (WHO) and United Nations Children's Fund Joint Monitoring Programme (JMP), Progress in Drinking-water and Sanitation: Special Focus on Sanitation 23 (New York/Geneva: UNICEF/WHO, 2008).

2 *Ibid.*

3 WHO, Water, Sanitation and Hygiene links to Health – Facts and Figures 1 (WHO, Geneva 2004).

4 UN General Assembly Resolution 55/2, United Nations Millennium Declaration, UN Doc. A/RES/55/2, 2000, para. 19.

5 United Nations, *Water for People – Water for Life* (Paris: UNESCO, 2003).

Nations Development Programme (UNDP), '[a]bsolute scarcity is the exception, not the rule. Most countries have enough water to meet household, industrial, agricultural and environmental needs'.[6] One of the problems that countries face in addressing access to water is the lack of a comprehensive international legal framework, which could guide national decision-makers in the development and implementation of equitable, fair and sustainable water law reforms. However, there is no lack of existing laws and regulations. On the contrary, international water law is made up of a broad range of laws and principles, some of which have existed since Mesopotamian times. Yet, though extensive, these mostly transboundary instruments are generally limited in geographical scope and/or in content, or have not entered into force. Thus, despite the large number of existing legal instruments, the international legal system is still fragmented and not comprehensive. In the absence of a sound legal framework water projects are often determined by purely economic reasons, leading to frequent failure and increasing numbers of people without access to water. As Fitzmaurice points out:

> The example of the Sardar Sarovar Project[7] precisely illustrates the problems that can arise in the absence of a coherent body of universal principles relating to water (including human rights elements, which featured so prominently in this project) – principles that are integral in the management of water resources at both the national and international levels.[8]

The purpose of this chapter is to examine the existing international laws and principles applicable to national authorities when developing water law and water sector reforms. Though Fitzmaurice notes that 'not all the principles applicable at the international level have the same usefulness at the national level of water management',[9] recent practice in water law reforms has shown that many relevant international law principles are neglected. The chapter will focus on non-transboundary issues arising in water law reforms, in particular access to water for all parts of the population and fair, equitable and sustainable use among different national users.

Part A gives an overview of international hard and soft freshwater law

6 United Nations Development Programme, *Human Development Report 2006 – Beyond Scarcity: Power, Poverty and the Global Water Crisis* 133 (New York: UNDP, 2006).

7 The Sardar Sarovar Project involves mostly domestic law issues, the main international element in this case being the involvement of the World Bank. Jacob Werksman, 'Greening Bretton Woods', *in* Philippe Sands ed., *Greening International Law* (London: Earthscan, 1993 73–74); and Philippe Cullet, 'Human Rights and Displacement: The Indian Supreme Court Decision of Sardar Sarovar in International Perspective', 50 *Int'l & Comp. L.Q.* 973 (2001).

8 Malgosia Fitzmaurice, 'General Principles Governing the Cooperation between States in Relation to Non-Navigational Uses of International Watercourses', 14 *Yb. Int'l Envtl. L.* 3, 12–13 (2005).

9 *Ibid.*

instruments and their applicability to water law reforms. Part B examines the different dimensions of freshwater that influence water law reforms.

A. Overview of International Freshwater Law

1. The UN Watercourses Convention

For centuries, international freshwater management has occupied states and international organizations through the creation of numerous international, regional and bilateral water agreements and declarations, which are applicable primarily on the transboundary level. The period between the early 1970s and today marks perhaps the most important period in the development of international freshwater law. Already in 1966, the International Law Association (ILA) had made a first attempt to develop an international freshwater agreement on water management in adopting the Helsinki Rules on the Uses of the Waters of International Rivers (Helsinki Rules).[10] Until recently, these played a key role in the codification and definition of international freshwater law, as they have been used as a foundation for customary law, frequently serving as a template and guidance for regional and international agreements on freshwater resource management.[11]

The UN International Law Commission initiated the first global project on the codification of the law of non-navigational uses of international watercourses in 1970. This process led to the adoption of the UN Convention on Non-Navigational Uses of International Watercourses (UN Watercourses Convention) in 1997.[12] Though not yet in force, parties have begun to apply and adjust the Convention,[13] which has also been used as a template for a number of regional laws on shared water issues (e.g., for water management in the Nile Basin).[14] The Convention has been frequently criticized. Some authors pointed out that the Convention does not, or only partly, develop customary international law.[15]

10 International Law Association (ILA), The Helsinki Rules on the Uses of the Waters of International Rivers, *in* Report of the Fifty-Second Conference (Helsinki) (London: International Law Association, 1966).

11 Joerg Barandat, 'Wasserrecht – Keine trockene Juristerei', 17(1) *Vierteljahresschrift fuer Sicherheit und Frieden* 37 (1999).

12 Convention on the Law of the Non-navigational Uses of International Watercourses, New York, 21 May 1997, UN Doc. A/51/869.

13 Stephen McCaffrey, *The Law of International Watercourses* 361 (Oxford: Oxford University Press, 2007).

14 Mohammed Abdo, 'The Relevance and Contribution of the Watercourses Convention toward Resolving the Problems of the Nile Basin' (2005), available at www.dundee.ac.uk/cepmlp/journal/html/Vol15/Vol15_8.pdf.

15 Joseph W. Dellapenna, 'The Berlin Rules on Water Resources: The New Paradigm for International Water Law', *in* Randall Graham (ed.) The *World Environmental and Water Resource Congress 2006: Examining the Confluence of Environmental and Water Concerns* (Omaha: American Society of Civil Engineers, 2006), available at www.ualg.pt/5cigpa/comunicacoes/Berlin%20Rules%20Summary.doc and McCaffrey, note 13 above, 376.

According to others, the Convention does little to clarify the uncertainty surrounding the general principles concerning cooperation,[16] and will have no positive effects as it neither furthers sustainable water use nor regulates the discretionary powers of watercourse states.[17] However, others note the positive influence of the Convention on the development of peaceful and cooperative regional water agreements,[18] and its assistance in promoting and implementing purposes and principles set out in Articles 1 and 2 of the UN Charter.[19]

Though focusing on transboundary waters only, the UN Watercourses Convention is often seen as the most important current global freshwater agreement. By enshrining three main principles of international freshwater law, namely the principles of equitable and reasonable use,[20] the principle of no significant harm[21] and the principle of information exchange and notification,[22] it reflects the concept of fair, equitable and cooperative transboundary uses of international watercourses. Moreover, Article 10(2) requires that vital human needs should be given supremacy in cases of conflict, a concept which, according to McCaffrey, 'should be informed by the human right to water.'[23] However, Fitzmaurice points out that this provision 'does not indicate that vital human needs should override other uses in case of non-conflicting uses.'[24] It must also be noted that though the Convention calls for the protection and preservation of the environment,[25] it does not refer to fundamental environmental principles such as precaution and common but differentiated responsibilities (CBDR), nor does it offer a real perspective of sustainable and equitable management and development of water resources.[26] This lack of a broader human rights and environmental perspective limits the applicability of the UN Watercourses Convention to water law reforms.

16 See Fitzmaurice, note 8 above, 22.

17 Ellen Hey, 'The Watercourses Convention : To What Extent does it Provide a Basis for Regulating Uses of International Watercourses?', 7 *Rev. of European Community & Intl Envtl L.* 297–298 (1998).

18 Jorg Barandat and Aytul Kaplan, 'International Waterlaw: Regulations for Cooperation and the Discussion of the International Water Convention', *in* Waltina Scheumann and Manuel Schiffler (eds.) *Water in the Middle East* 361 (Berlin: Springer, 1998).

19 John R. Crook and Stephen McCaffrey, 'The United Nations Starts Work on a Watercourses Convention', 91 *American J. Intl L.* 374, 377 (1997). See also McCaffrey, note 113 above, 377 and Attila Tanzi and Maurizio Arcari, *The United Nations Convention on the Law of International Watercourses* (The Hague: Kluwer Law International, 2001).

20 Convention on the Law of the Non-navigational Uses of International Watercourses, New York, 21 May 1997, UN Doc. A/51/869, Art 5.

21 *Ibid.* Art. 7.

22 *Ibid.* Arts. 12 and 13.

23 McCaffrey, note 13 above, 369.

24 Fitzmaurice, note 8 above, 32.

25 Convention on the Law of the Non-navigational Uses of International Watercourses, New York, 21 May 1997, UN Doc. A/51/869, Art. 20.

26 On the environmental aspects of the UN Watercourses Convention, see Chapter XII in this volume.

2. *The Helsinki Convention and its Protocols*

Originally only a regional agreement, the 1992 UNECE Convention on the Protection and Use of Transboundary Watercourses and International Lakes (Helsinki Convention)[27] was opened for accession for all countries in 2003,[28] although to date there have been no signatures by non-regional states. Based on the principles of precaution, polluter-pays and inter-generational equity, it obliges all state parties to prevent, control and reduce water pollution, monitor transboundary water conditions, cooperate and exchange information.[29] In addition, State parties are mandated to enter into bilateral or multilateral agreements, or other arrangements, to establish joint bodies, and to hold consultations.[30]

Two protocols have been adopted under the Convention. The 1999 Protocol on Water and Health focuses specifically on attaining an adequate supply of safe drinking water and adequate sanitation for all, and to protect water used as a source of drinking water.[31] Focusing on water-related health issues, it embodies a range of international principles such as equity, no harm, public participation, sustainability, precaution, polluter-pays and prevention. The 2003 Protocol on Civil Liability and Compensation for Damage Caused by the Transboundary Effects of Industrial Accidents on Transboundary Waters, which is not yet in force, fills a gap regarding the compensation of environmental damages caused in neighbouring countries, ensuring access to judicial proceedings and remedies for all people affected by environmental damage from utilization of shared watercourses.[32] Like the UN Watercourses Convention, the limited scope and content of all three instruments as well as the lack of a sustainable approach and differential treatment of water users or providers prevent them to provide a basis for comprehensive water management.

3. *Soft law instruments*

Soft law instruments play a central role in international freshwater law, as most of its principles are not enshrined in binding multilateral agreements, but rather result from developments in state practice. In particular the activities of international legal associations, international organizations and international

27 Convention on the Protection and Use of Transboundary Watercourses and International Lakes, Helsinki, 17 March 1992, UN Doc. ENWA/R.53.

28 Amendment to Articles 25 and 26 of the UNECE Convention on the Protection and Use of Transboundary Watercourses and International Lakes, UN Doc. ECE/MP.WAT/14 (12 January 2004).

29 Convention on the Protection and Use of Transboundary Watercourses and International Lakes, Helsinki, 17 March 1992, UN Doc. ENWA/R.53, Arts. 2, 3, 4 and 6.

30 *Ibid.* Arts 9, 10 and 11.

31 Protocol on Water and Health to the 1992 Convention on the Protection and Use of Transboundary Watercourses and International Lakes, London, 17 June 1999, UN Doc. MP.WAT/AC.1/1999/1.

32 Protocol on Civil Liability and Compensation for Damage Caused by the Transboundary Effects of Industrial Accidents on Transboundary Waters, UN Doc. MP/WAT/2000/1 (2003).

courts and tribunals have created a bulk of soft law instruments and principles, some of which are crucial for current water law reforms.

The Dublin Principles are often seen as one of the most influential soft law instrument and a trendsetter for the modern approach to water management.[33] They were adopted as part of the Dublin Statement at the 1992 International Conference on Water and the Environment in Dublin, Ireland, as a follow-up to the 1977 Mar del Plata Action Plan, and have since been enshrined in many contemporary water agreements.[34] Aiming at the equitable use of water resources, the Dublin Principles enshrine the need for a sustainable approach to water as a human right, participation and integrated water resources management (IWRM), women's involvement and the economic value of water. Thus, they attempt to establish a comprehensive approach to water management, taking into account the different dimensions of water. However, the comprehensive approach has been jeopardized by the fact that the economic focus is often seen as perhaps the most influential aspect of the Dublin Principles. This led to a policy reversal from the perspective that water cannot be appropriated, to the introduction of water rights and the possibility of trading water entitlements.[35] In fact, the particular relevance of the Dublin Principles for water law reforms is based on this new approach as it has served as the basis for both the 1993 and 2003 water strategies of the World Bank and most IFI-led reforms. As critics point out, the fundamental demand of the Dublin Principles for participation has fallen by the wayside and the central role of women has been reduced to a subordinate clause.[36]

In 2001, the Bonn Ministerial Declaration was adopted by Ministers of the International Conference in Bonn to 'increase water security and to achieve sustainable management of water resources' through equitable and sustainable use and protection of the world's water resources.[37] Whilst referring to Agenda 21 and building on the 2000 Ministerial Declaration of The Hague on Water Security in the 21st Century, the Bonn Declaration requests governments, private business, non-governmental organizations and all other stakeholders to base

33 Dublin Statement on Water and Sustainable Development, International Conference on Water and the Environment, Dublin, 31 January 1992. See Chapter XII.

34 For example, Agenda 21, Report of the UNCED, Rio de Janeiro, 3–14 June 1992, UN Doc. A/CONF.151/26/Rev.1 (Vol. 1, Annex II), Convention on the Law of the Non-navigational Uses of International Watercourses, New York, 21 May 1997, UN Doc. A/51/869, and International Law Association, The Berlin Rules on Water Resources, *in* Report of the Seventy First Conference (London: ILA, 2004) at international level as well as several national laws in Germany, England and China. Miguel Solanes and Fernando Gonzales-Villarreal, The Dublin Principles as Reflected in a Comparative Assessment of Institutional and Legal Arrangements for Integrated Water Resources Management 30-24 (Stockholm: Global Water Partnership/Swedish International, Background paper for the Technical Advisory Committee of the Global Water Partnership, 1999) available at www.gwpforum.org/gwp/library/Tac3.pdf.

35 Solanes & Gonzales, note 34 above, 30–42.

36 Uwe Hoering and Ann-Kathrin Schneider, King Customer? The World Bank's 'New' Water Policy and its Implementation in Sri Lanka and India 8 (Brot fuer die Welt, 2004).

37 Bonn Ministerial Declaration, Conference Report, International Conference on Freshwater, Bonn, 3–7 December 2001.

their action on governance, increased funding, and a strengthened role of the international community, capacity building and technology transfer and the consideration of gender issues.[38] A set of Recommendations for Action promote priority actions in the field of governance, mobilization of financial resources and capacity building and knowledge sharing.[39] Among other things, the recommendations call for cooperation, in particular between the private sector, governments and civil society; equitable and sustainable water management with a strong role of the state and participation by all stakeholders; greater commitment to implement commonly agreed principles on water resource management; allocation of water according to human needs; ecosystem integrity; consideration of food requirements while implementing a pro-poor approach, and environmental protection.[40] While supporting water service providers to 'aim for financial sustainability through receiving sufficient income from their customers', and mobilize private investment, the recommendations explicitly state that this should not be 'a barrier to poor people's access to water'.[41] They call for governments to play a more active role in water governance and drive local, national and international water reform, while suggesting the encouragement of more private investment through public–private partnerships. Stating that many governments move away from service provision to facilitation, they encourage the private sector to also play a more active role in water governance, financing, provision and capacity building, however, whilst recognizing co-responsibility for the common good.[42]

Based on a growing acknowledgment of the need to coordinate international environmental, human rights, humanitarian and water law, as well as an awareness of the gaps and limitations of existing instruments, the ILA decided to revise the 1966 Helsinki Rules with the aim of providing a comprehensive, cogent and complete summary of international customary law relevant for water management.[43] In 2004, it adopted the Berlin Rules on Water Management,[44] which initiated a shift from purely transboundary to non-transboundary water law codification. Next to the Bonn instruments, the Berlin Rules probably are the most comprehensive contemporary agreement on water management as they take a holistic approach to water resources management and are of much broader scope and content than the international rules so far in addressing both national

38 *Ibid.*

39 Bonn Recommendations for Action, Conference Report, International Conference on Freshwater, Bonn, 3–7 December 2001.

40 *Ibid.* paras 1, 3, 4, 6 and 8.

41 *Ibid.* para. 15.

42 *Ibid.* paras 22 and 26.

43 Joseph W. Dellapenna, The International Law Associations Berlin Rules on Water Resources (Presentation to IUCN, 2004), available at http://intranet.iucn.org/webfiles/ftp/public/ForumEvents/E1121/Final%20Document/Berlin%20Rules-IUCN.ppt#2.

44 International Law Association, The Berlin Rules on Water Resources, *in* Report of the Seventy-First Conference (London: ILA, 2004).

and international waters.[45] Due to their integration of such principles as participatory, conjunctive and integrated water management, sustainability, minimization of environmental harm – including through precaution and the polluter-pays principle – as well as cooperation, equitable utilization, no harm and participation, the Berlin Rules have been seen to represent a new phase of codification of customary international environmental law.[46] As pointed out by Salman, the Berlin Rules 'have gone beyond what the ILA considered as established principles of international environmental law, and incorporated emerging principles as well'.[47] And, in contrast to the Helsinki Rules or the UN Watercourses Convention, which enshrine the right for equitable and reasonable utilization and the obligation to not cause transboundary harm, the Berlin Rules oblige states to manage waters equitably and reasonably – a new bold approach to the principle of equitable use that has, on the other hand, been criticized by a number of scholars as downgrading the principle and leading to growing confusion concerning the relation between the principles of equitable use and no harm.[48] Although legally non-binding, agreements adopted by the ILA as 'an international law organization of high reputation in the formulation of soft laws' have generally been very influential in water management.[49]

B. The different dimensions of freshwater at the international level

Water is indispensable to human survival and most forms of economic production. In fact, it is essential for nearly every sector of human activity, such as agriculture, industrial production, power generation and transportation and, within that, touches upon a number of different areas of human life, including environment, economy, trade, human rights, conflict and governance. This section analyzes these different dimensions of freshwater and the relevant underlying principles in the light of current water law reforms.

1. The environmental dimension

The environmental dimension of freshwater encompasses the growing international concern regarding the consequences of unsustainable use and pollution of increasingly scarce global freshwater resources. The already dire

45 Alexandre Kiss, Environmental Law Principles 21 (Geneva: UNITAR, 2005).

46 Dellapenna, note 43 above.

47 Salman A.M. Salman, The Helsinki Rules, the Watercourses Convention and the Berlin Rules: Perspectives on International Water Law 15 (Washington D.C.: The World Bank, 2007), available at http://pdfserve.informaworld.com/224138_731202505_783499146.pdf.

48 *Ibid.* 12, 15 and 17. For a dissenting opinion, International Water Law Project, ILA Berlin Conference 2004 – Water Resources Committee Report Dissenting Opinion (2004), available at www.internationalwaterlaw.org/intldocs/ila_berlin_rules_dissent.html.

49 Frederick Lorenz and Edward Erickson, *The Euphrates Triangle: Security Implications of the Southeastern Anatolia Project* (Washington D.C.: National Defense University Press, 1999).

environmental situation of the world's water resources is likely to be further increased by the effects of climate change, thus requiring urgent attention by governments and other decision-makers. The importance of environmental considerations for water management has been reflected in the UNDP's 2006 report, which states that, 'intensive development and the unsustainable depletion of water resources create winner and losers. The environment is a loser every time – while the balance sheet between human users is mixed'.[50]

Water-related environmental concerns have been raised since the 1970, yet almost exclusively on national level. An environmental shift in international water management in the 1990s was reflected in the adoption of several international environmental agreements focusing on, or including, water management, such as Agenda 21, the 1992 Rio Declaration, the UN Watercourses Convention and the Helsinki Convention. In line with a number of soft law instruments adopted since 2000,[51] another important step was taken in 2002 when global leaders committed themselves to sustainable development in the Johannesburg Declaration, and acknowledged the growing depletion of the world's water resources.[52] As stated by the 2006 UNDP Report:

> Sustainable resource use is the proposition for production systems to be managed in a way that the world can live off today's resources, without eroding the asset base to be inherited by future generations. Governments today are widely violating that principle by running down national water assets. Water management therefore has to realign water use with demand at levels that maintain the integrity of the environment.[53]

The following environmental principles can provide a legal framework for such environmental integrity.

International freshwater principles

Originating mainly in international environmental law, a number of specific freshwater principles have developed.[54] The principles of equitable use, no-harm, consultation and cooperation can be seen as the cornerstones of customary international freshwater law. The importance of these three principles for the peaceful

50 UNDP, note 6 above, 143.

51 See, e.g., Ministerial Declaration of the Hague on Water Security in the 21st Century, Second World Water Forum, The Hague (March 2000); Bonn Ministerial Declaration, Conference Report, International Conference on Freshwater, Bonn, 3–7 December 2001 and Bonn Recommendations for Action, Conference Report, International Conference on Freshwater, Bonn, 3–7 December 2001.

52 Johannesburg Declaration on Sustainable Development, 4 September 2002, UN Doc. A/CONF.199/20.

53 UNDP, note 6, 147.

54 Dante A. Caponera, *Principles of Water Law and Administration* 193 (London: Taylor & Francis, 2nd edn 2007).

and effective management of transboundary watercourses has been acknowledged and reflected in various binding and non-binding transboundary water agreements, such as the UN Watercourses Convention. More recently, these principles have gradually been included as key elements of water management on national level. Examples can be seen in the Bonn Declaration, which identified the equitable use of water resources as one of the main challenges for governments, or the Berlin Rules, which not only stress the need for cooperation, but also oblige countries to take an equitable and reasonable approach to water management as well as to avoid transboundary harm.

In particular the principle of equitable use on national level has been widely acknowledged as an indispensable part of effective water management. For example, the achievement of the equitable allocation of water to all, including the poorest of the population, is an express goal of IWRM. As stated by the Global Water Partnership, IWRM is 'a process that promotes the coordinated development and management of water, land and related resources, in order to maximize the resulting economic and social welfare in an equitable manner without compromising the sustainability of vital ecosystems'.[55] Accordingly, UNESCO's Second World Water Development Report mentions equitable use as the basis of the social dimension of water governance, as 'how water quantity and quality and related services is allocated and distributed has direct impact on people's health and their livelihood opportunities.'[56]

Other principles governing the environmental dimension of freshwater

This section sets out some of the most important environmental principles in the sustainable management of water resources, namely the principles of protection, preservation, prevention, precaution, polluter-pays and common, but differentiated, responsibilities (CBDR).

PROTECTION AND PRESERVATION

The principles of protection and preservation lay at the heart of international environmental law and are consequently enshrined in all environmental instruments.[57] Kiss defines them as 'long-term perspective taking into account rights

55 Global Water Partnership, Integrated Water Resources Management 22 (Global Water Partnership Technical Advisory Committee, TAC Background Paper No. 4, 2000).

56 United Nations, *Water – A Shared Responsibility* 46 (Paris: UNESCO, 2006).

57 See e.g., Convention on the Law of the Non-navigational Uses of International Watercourses, New York, 21 May 1997, UN Doc. A/51/869; Convention on the Protection and Use of Transboundary Watercourses and International Lakes, Helsinki, 17 March 1992, UN Doc. ENWA/R.53; the Ramsar Convention on Wetlands of International Importance especially as Waterfowl Habitat, Convention on Biological Diversity; UN Convention to Combat Desertification, and UNEP's Principles of Conduct in the Field of the Environment for the Guidance of States on the Conservation and Harmonious Utilization of National Resources Shared by Two or More States, 1978.

and interests of future generations for whom natural resources should be safeguarded' respectively 'the general principle to abstain from harmful activities and to take into account affirmative measures to ensure deterioration does not occur, including environmental planning and management concerns'.[58] Thus, given the key function of these principles, sustainable development including in water management to maintain environmental integrity cannot be reached without them.

PRECAUTION AND PREVENTION

The principle of precaution is now widely accepted as a fundamental concept of international and national environmental laws and regulations.[59] One of the best known formulations of the principle can be found in Principle 15 of the Rio Declaration which states that 'in order to protect the environment, the precautionary approach shall be widely applied by States according to their capabilities. Where there are threats of serious or irreversible damage, lack of full scientific certainty shall not be used as a reason for postponing cost-effective measures to prevent environmental degradation'.[60] Precaution has a particular meaning in the area of water, as it is impossible to determine precisely what environmental effects changes in the management of water may have on its quality and quantity, while negative effects could have severe and far-reaching consequences on both the national and international levels. Thus, a precautionary approach could help avoid outbreaks of water-related diseases and protect water resources used as sources of drinking water. An example of the positive use of the precautionary principle can be seen in the case of the Indian Supreme Court, which struck down a notification issued by the government of Andhra Pradesh exempting an oil industry located in the vicinity of two major water reservoirs, stating that 'an order of exemption carelessly pursued, ignoring the precaution principle, could be catastrophic'.[61] However, the principle of precaution is still too often neglected in view of assumed economic advantages particularly in many IFI-led water projects. Ironically, next to possible severe environmental consequences, a purely economic approach also ignores the economic advantages of the application of precautionary (and preventive) action, which in general addresses the harm more efficiently and is more cost-effective than remedial action.[62]

The principle of prevention, among others enshrined in the Stockholm Declaration and Principle 11 of the Rio Declaration, requires action to be taken

58 Kiss, note 45 above 28.

59 For example, it is enshrined in the Rio Declaration on Environment and Development, Rio de Janeiro, 14 June 1992, UN Doc. A/CONF.151/26/Rev.l (Vol. l), Convention on Biological Diversity, Rio de Janeiro, 5 June 1992 and the UK Biodiversity Action Plan, 1994.

60 Rio Declaration on Environment and Development, Rio de Janeiro, 14 June 1992, UN Doc. A/CONF.151/26/Rev.l (Vol. l).

61 *AP Pollution Control Board II v. MV Nayudu* 2000(3) SCALE 354, available at www.ielrc.org/content/e0010.pdf

62 For example, according to the World Bank, $1 invested in preparedness saves $7 in remedial action. World Bank, Natural Disasters: Counting the Cost 3 (Washington D.C.: World Bank, 2004).

to protect the environment at an early stage, thus to prevent damages before they occur. Preventive action is an important part of water management, as it requires implementation of environmental impact assessment regimes, phasing out or closing down of polluting production processes, and the application of water-saving agricultural practices. The importance of prevention in water management is reflected in its inclusion in various international water agreements, but also in national water laws such as the 1996 Chinese Law of Water Pollution Prevention or the UK Pollution Prevention and Control Act 1999.[63] However, as with the principle of precaution, a preventive approach is often neglected by governments and IFIs in water projects.[64] This can again be illustrated by the Narmada dam project, where an assessment of the project found severe omissions and violations of environmental principles, such as the lack of an environmental impact assessment,[65] or in the context of the Tirupur private–public partnership project for water supply and sanitation, which, according to Madhav, 'presents a classic case of reductionist engineering, sidelining sustainable water use planning and conservation efforts'.[66]

Common but differentiated responsibilities

The principle of common but differentiated responsibilities (CBDR) is based on the notion of equity. In brief, it requests states that have caused international environmental problems to assist other states to address them.[67] CBDR gained importance after the implementation of the Kyoto Protocol, and it also received growing attention with a view to achieving sustainable development. For example, article 14 of the Plan of Implementation of the 2002 World Summit on Sustainable Development stresses that 'all countries should promote sustainable consumption and production patterns, with the developed countries taking the lead and with all countries benefiting from the process, taking into account the Rio principles, including the principle of common but differentiated responsibilities as set out in principle 7 of the Rio Declaration'.[68] Concerning sustainable water management, the CBDR applies to the provision of water from rich (Northern) countries to

63 China, Law on Prevention and Control of Water Pollution, adopted by the Fifth Session of the Standing Committee of the Sixth National People's Congress on 11 May 1984 (as amended) and Pollution Prevention and Control Act (c.24) United Kingdom.

64 See, e.g., Hoering and Schneider, note 36 above, 21.

65 *Ibid.*

66 Roopa Madhav, 'Tirupur Water Supply and Sanitation Project: A Revolution in Water Resource Management?', in Philippe Cullet, Alix Gowlland Gualtieri, Roopa Madhav and Usha Ramanathan (eds.) *Water Law at the Crossroads National and International Perspectives With Special Emphasis on India* 201, 229 (New Delhi: Cambridge University Press, 2009).

67 Plan of Implementation of the World Summit on Sustainable Development, Report of the World Summit on Sustainable Development, Johannesburg, 26 August–4 September 2002, UN Doc. A/CONF.199/20, para. 13.

68 Rio Declaration on Environment and Development, Rio de Janeiro, 14 June 1992, UN Doc. A/CONF.151/26/Rev.l (Vol. I), Principle 7.

poor (Southern) ones, in particular in the light of growing water scarcity and inequality in access to drinking water. However, until today the CBDR is rarely taken into account in water law reforms, which, on the contrary, encourage (Northern) corporations to make money in providing water in developing countries on the basis of full-cost recovery, privatization and tradable water rights.[69]

Polluter pays principle

The polluter-pays principle is reflected in Principle 16 of the Rio Declaration, requesting '[n]ational authorities [to] endeavour to promote the internalization of environmental costs and the use of economic instruments, taking into account the approach that the polluter should, in principle, bear the cost of pollution, with due regard to the public interest and without distorting international trade and investment'.[70] In theory, effective national government policies and regulations implementing the polluter-pays principle could both enhance the water resource base, and create incentives for new technologies and patterns of intervention.[71] However, the implementation of the polluter-pays principle in water law reforms has led to mixed results. In India, for example, though private companies have introduced technologies that reduce water pollution and increase the availability of downstream users,[72] the enforcement of the principle has proved difficult. As set out in Chapter XII, a lack of strict rules and regulations to ensure compliance with the principle jeopardizes possible advantages, due to the consequences of the required internalization of environmental costs, which often meant that polluters would either not pay for the damages caused by them at all, or would increase the price of their products to account for increased production costs.

Policy approaches as part of the environmental dimension: The relevance of integrated water resources management

Integrated water resources management (IWRM) is a 'framework for governments and other stakeholders to manage their water resources as an integrated part of policies related to water in different sectors of society'.[73] Over the past

69 See Chapter XII.

70 Rio Declaration on Environment and Development, Rio de Janeiro, 14 June 1992, UN Doc. A/CONF.151/26/Rev.l (Vol. l), Principle 16.

71 Craig Langston and Grace Ding, *Sustainable Practices in the Built Environment* 70 (Oxford: Butterworth Heinemann, 2nd edn 2001).

72 Chandra Bhushan, Water Use in Industry, A Down to Earth Supplement (New Delhi: Centre for Science and the Environment, December 2005), available at www.cseindia.org/dte-supplement/industry20040215/non-issue.htm.

73 UNEP Collaborating Centre on Water and Environment, Support for Achieving the 2005 IWRM Target – Accelerating the Process (Horsholm: UNEP Collaborating Centre on Water and Environment, 2005), available at http://new.unep.org/themes/Freshwater/Documents/pdf/IWRMtargetBrochure.pdf.

decade, IWRM has received considerable attention at both national and international levels and was enshrined in various international agreements, most notably Chapter 18 of Agenda 21. As stated by the German Federal Ministry for Economic Cooperation and Development, IWRM has become an internationally recognized paradigm for water policy, which states such as Germany feel bound to apply in their water sector development cooperation.[74] IWRM promotes a holistic approach to the management of water as an all-encompassing, shared natural resource, and according to UN-Water, has been accepted as the way forward for efficient, equitable and sustainable management of the world's limited water resources and for coping with conflicting demands.[75] IWRM attempts to link social and economic development to the protection of natural ecosystems and relates water resources management to regulatory measures concerning other environmental mediums. However, though countries have started to implement IWRM on the national level, there is still a lot of room for improvement.[76]

One of the main barriers for the effectiveness of IWRM has been its focus on water as an economic good and full-cost recovery.[77] This approach has resulted in the strong support of the development and implementation of these principles, including in water law reforms, by IFIs[78] and the World Water Forum. While focusing on the economic dimension of water, the World Water Forum in particular has played a key role in the furthering of the integration of IWRM on the national level. From the Marrakech Declaration of the First World Water Forum in 1997, which stressed the need for a better understanding of water policy issues, better cooperation, protection and preservation and the human need to have access to clean water and sanitation, to the fourth World Water Forum in 2006 which confirmed '[t]he critical importance of water, in particular freshwater, for all aspects on sustainable development'[79] and the need to ensure access to water and sanitation for all, the World Water Forum process reflects the gradual global recognition of the notion of IWRM. However, as a notable gap regarding equitable provision of water for all, the World Water Forum has so far not recognized the concept of a human right to water.[80]

74 German Federal Ministry for Economic Cooperation and Development, The Water Sector in German Development Cooperation (Bonn: German Federal Ministry for Economic Cooperation and Development, 2006), available at www.bmz.de/en/service/infothek/fach/materialien/Materialie153.pdf.

75 UN-Water, Status Report on Integrated Water Resources Management and Water Efficiency Plans – CSD16 at 4 (New York: UN-Water, 2008).

76 *Ibid.* 15.

77 The International Water Association, Industry Sector Report for the World Summit on Sustainable Development (London: International Water Association, 2002), available at www.gdrc.org/uem/water/iwrm/1pager-01.html.

78 For a more detailed analysis of the role of IFIs in IWRM, see Chapter 5 in this volume.

79 Fourth World Water Forum, Ministerial Declaration, Local Actions for a Global Challenge (16–22 March 2006).

80 See, e.g., Sonia Giovanina Sani, Taking Back the World Water Forum – Canadian Perspectives (2006), available at www.blueplanetproject.net/Movement/documents/Taking_Back_the_World_Water_Forum.pdf.

2. The economic and trade dimensions

General principles

The economic dimension of freshwater concerns the availability of water and the economic value connected to it. It is based on the concept of water as a tradable commodity, the price of which should reflect its importance for human life, and not as a public good that should be made available at the lowest cost. Water scarcity is seen as a 'policy-induced outcome flowing from the deeply flawed approach to see water as an infinitely available resource to be diverted, drained or polluted in generating wealth, the predictable consequence of inexhaustible demand chasing an under-priced resource'.[81]

A main implication of the economic perspective is the application of water pricing, including full-cost recovery and trading of water rights, to water law reforms. Supporters of the economic approach to water management call for the cutting of water subsidies and the rethinking of water prices. They argue that if water is treated as a free good, there is no incentive to reduce waste and pollution.[82] This, they argue, can only be ensured by raising the value – thus the price – of water. In addition, economic water strategies of IFIs such as the World Bank also promote full-cost recovery, requiring water providers to pass all costs of production and services onto consumers.[83] However, as seen in many World Bank water projects, increased water pricing often clashes with human rights considerations when implemented without adjacent government policies protecting the poorest of the population who cannot afford higher water prices, for example as seen in the internalization of the costs of water infrastructure in India which often led to formerly free water now being too expensive for rural population.[84]

However, an economic approach would not necessarily be negative. An illustration of an economic option based on the environmental dimension of water is the valuing of ecological water services. Based on the principle of prevention, awareness of economic benefits linked to ecosystem trading should be developed through payments for watershed services in line with an increasing value of water as a productive resource. An example of such valuing can be seen in Costa Rica, where farmers are paid for good land management through the use of an environmentally adjusted water tariff to finance watershed conservation upstream.[85] Thus, as set out in the Ministerial Declaration of The Hague on Water Security in the 21st Century,

81 UNDP, note 6 above,122.

82 See, e.g., Chris J. Perry, Michael Rock and David Seckler, Water as an Economic Good: A Solution or a Problem? 1 (Colombo: International Irrigation Management Institute, Research Report 14, 1994).

83 See, e.g., Matthias Finger and Jeremy Allouche, *Water Privatisation: Trans-National Corporations and the Re-Regulation of the Water Industry* 62 (London: Spon Press, 2002).

84 Hoering and Schneider, note 36 above, 30.

85 Barbara van Koppen, A Gender Performance Indicator for Irrigation: Concepts, Tools and Applications (Colombo: International Water Management Institute, IWMI Research Report 59, 2002).

> valuing water requires the management of water in a way that reflects its economic, social, environmental and cultural values for all its uses, and to move towards pricing water services to reflect the cost of their provision. This approach should take account of the need for equity and the basic needs of the poor and the vulnerable.[86]

Trade in water

Water trade is another facet of the economic dimension. Since the 1980s, international trade law has become a new factor in water management, a shift that has further increased by the new notion of the economic value of water. In times of growing water scarcity, this has led to the economic exploitation of water, notably by multinational corporations, and the development of water as a commodity subject to the rules of pricing, markets and international and regional trade agreements, such as the General Agreement on Tariffs and Trade (GATT),[87] the WTO Agreement on Agriculture and the North American Agreement in Free Trade (NAFTA).[88] In addition discussions are being held to include water services in the General Agreement on Trade in Services (GATS), a move which is strongly opposed by human rights supporters as it would further support the treatment of water as a commodity subject to market rules and powers, and not as a public good affordable and available to all.[89] In some cases, water stress can also be alleviated by virtual water imports, that is, water imported implicitly through the import of food.[90] All these instruments aim for the creation of free, global markets, and call for the retreat of the state from the provision to the sole governance and monitoring of water services.

The application of trade law principles such as non-discrimination, most-favoured-nation and national treatment supports the privatization, commercialization and decentralization of the water sector as promoted by IFIs, and results in the opening of national water markets to foreign investors and international corporations trading. Though a global water market must not necessarily lead to negative consequences, the experiences in the past have shown that water law reforms resulted in severe socio-economic and environmental implications where water was traded globally without appropriate protection measures on national level, in particular for the poorest of the population. Next to increased

86 Ministerial Declaration of the Hague on Water Security in the 21st Century, Second World Water Forum, The Hague (March 2000), para. 3.

87 General Agreement on Tariffs and Trade, 15 April 1994, 1867 *UNTS* 3.

88 North American Free Trade Agreement, 17 December 1992, United States, Mexico and Canada, available at www.nafta-sec-alena.org/DefaultSite/index_e.aspx?DetailID=78.

89 See, e.g., Christina Deckwirth, Water almost out of GATS? – A Corporate Europe Observatory Briefing (March 2006), available at www.corporateeurope.org/water/gatswater2006.pdf.

90 See Chapter IV.

unsustainable exploitation of water resources, water can quickly become too expensive for the poor, as illustrated by examples from Bolivia and Argentina.[91]

The influence of lending policies of international financial institutions (IFIs) on water management

IFIs, in particular the World Bank, but also the International Monetary Fund (IMF), and regional development banks, have been the dominant player in recent water law reforms. The engagement of the World Bank through financial support for water projects and the creation of a multitude of strategies and policy papers have strongly influenced water sector reforms in the past two decades. New strategies with direct impact on water sector reforms, such as the 1993 Water Strategy of the World Bank, have set the course for a fundamental shift in the policy of IFIs regarding the entire water sector, calling for a change from the traditional protectionist approach and the splitting up of the water sector into various individual areas, and abandoning the notion that the governments should provide water free of charge.[92]

Based on the Dublin Principles, the World Bank supports three main principles: privatization, decentralization and IWRM.[93] The implementation of these principles is usually ensured through conditionality clauses in loan agreements such as those set out in the World Bank country strategy stipulating that the level of credits depends on whether the government significantly reduces its budget deficit and extends the role of private enterprises in the infrastructure sector. This can have significant financial consequences, and as such can present strong incentives for governments to implement the conditions. For example, in Sri Lanka, the government received 25 per cent more funds if it complied with the conditionality clause to the Bank's satisfaction.[94] However, overall, these strategies have failed. In theory, an economic approach could improve the quality and effectiveness of water services as the involvement of private, more financially potent providers could reduce the financial burden on states. Experience gained in World Bank projects has shown that 'partial improvements can indeed be attained in the short term'.[95] However, the negligence of governments, IFIs and the private sector towards socio-economic and environmental considerations, in particular the principles of equity, participation, non-discrimination, low-cost recovery and inadequate capacity-building must be seen as the cause of the poor performance

91 See, e.g., Jorge Cuba, 'The Water Battle in Bolivia', *UNESCO Courier* (2000), available at www.unesco.org/courier/2000_12/uk/planet2.htm and Anonymous, 'Water Firm Suez in Argentina Row', *BBC News*, 24 January 2005, available at http://news.bbc.co.uk/2/hi/business/4201461.stm.

92 Hoering & Schneider, note 36 above, 7–8.

93 For further elaboration of these principles, see Chapter V.

94 World Bank, Country Assistance Sector Strategy, Strategic Directions for World Bank Engagement (Washington: World Bank, 2003).

95 Hoering and Schneider, note 36 above, 28.

of most economically driven water projects. Next to the famous failure of privatization projects in Bolivia and Argentina,[96] the case of selling water licenses in India and Sri Lanka is another example which, instead of providing a better and fairer distribution of water, has further widened the gap between rich and poor.[97] Other negative outcomes include the raising of water prices, lack of services in rural areas that are less attractive for private business investment, inequalities in the provision of services, and lack of public participation. This often leads to violent protests among the population.

3. The human rights dimension

The human right to water and a rights-based approach to water management has received growing national and international interest levels, particularly in response to failed attempts to commodify and privatize water markets. Though the human right to water has not been recognized by any of the universal human rights instruments, it has been enshrined in a range of other international instruments, including human rights, environmental and humanitarian agreements.[98] In 2002, General Comment No. 15 on the right to water, adopted by the Committee on Economic, Social and Cultural Rights (CESCR) confirmed the existence of a human right to water on the basis of an analysis of the International Covenant on Economic, Social and Cultural Rights (ICESCR).[99] This approach was followed in the 2007 Annual Report of the United Nations High Commissioner for Human Rights which stated that the development of national strategies on water and sanitation should be seen as a part of the human rights obligations of a state in relation to safe access to drinking water and sanitation.[100]

In order to ensure universal access to water, General Comment No. 15 requires that 'any payment for water services has to be based on the principle of equity, ensuring that these services, whether privately or publicly provided, are affordable for all' and that 'equity demands that poorer households should not be disproportionately burdened with water expenses as compared to richer households'.[101] Equity as the underlying principle of the human right to water is also reflected in the 2006 UNDP Report, which calls for national governments 'to address the inequitable distribution of access to water and sanitation linked to wealth, location, gender and other factors'.[102] An example of the practical

96 Privatization in Argentina and Bolivia led to riots as the population could no longer afford water.

97 Hoering and Schneider, note 36 above 20–21, 29, 32.

98 For a detailed analysis of relevant agreements, see Chapter XI.

99 Committee on Economic, Social and Cultural Rights, General Comment No. 15: The Right to Water (Articles 11 and 12 of the International Covenant on Economic, Social and Cultural Rights), UN Doc. E/C.12/2002/11 (2002).

100 Report of the United Nations High Commissioner for Human Rights on the Scope and Content of the relevant Human Rights Obligations Related to Equitable Access to Safe Drinking Water and Sanitation under International Human Rights Instruments, UN Doc. A/HRC/6/3 (2007).

101 General Comment No. 15, note 199 above, para. 27.

102 UNDP, note 6 above, 149.

importance of a human rights approach can be seen in the Indian Hubli-Dharwad project, where the principle of equity was not sufficiently conceptualized and left a large part of the poorest members of the population without sufficient water supplies.[103] Similar results have been noted in World Bank water projects in Sri Lanka, where 'considerable inequality still prevails. Especially with regard to equal opportunities, access to services and the distribution of natural resources, there are big differences throughout the country'.[104]

As will be further set out in Chapter IX of this book, the recognition of a human right to water is of particular importance for water law reforms as it imposes obligations not only on states, but also on other actors such as international organizations, civil society and private corporations.[105] Obligations to fulfil, protect and respect the human right to water mean that states must refrain from interfering directly or indirectly with the enjoyment of the right to water, including any activities that deny or limit equal access to adequate water, as well as prevent third parties from any form of interference. Moreover, where water services are provided by third parties, states must prevent them from compromising equal, affordable and physical access to safe and acceptable water and facilitate, promote and provide the right to water, including through the adoption of necessary legislation.[106] As such, these obligations can place considerable limitations on the privatization and commercialization of water services, as driven by the World Bank and the private sector.

4. The conflict dimension

Though this book does not directly address the conflict dimension of water, a short overview of relevant issues is given here as the avoidance of conflicts over water uses has indeed always been an issue of water management. In fact, the recognition of water management as a means to ensure national security was originally one of the driving forces of water sector reforms.[107] Today, climate change, growing population, increasing water scarcity and pollution require a whole new understanding of the relationship between environmental degradation, the privatization of water services and conflict. Principles and approaches preventing water conflicts include equitable and sustainable use, participation and transparency, as well as the availability and affordability of sufficient water for all, including the poorest of the population. Neglected tensions over water can

103 Priya Sangameswaran, Roopa Madhav and Clifton D'Rozario, '24/7, "Privatisation" and Water Reform: Insights from Hubli-Dharwad', 48/14 *Economic and Political Weekly* 60, 65 (2008).

104 Hoering and Schneider, note 36 above, 16.

105 General Comment No. 15, note 199 above para. 49. Concerning corporate human rights, see, e.g., Steven R. Ratner, 'Corporations and Human Rights: A Theory of Legal Responsibility', 111 *Yale LJ* 443, 449 (2001).

106 *Ibid.* paras. 21–25.

107 See Chapter II.

occur on local, national, international and global levels.[108] All four levels are interlinked so that intervention affecting one level requires an assessment of potential impacts on other levels.[109] Nevertheless, based on the experience of civil riots due to privatization schemes in Cochabamba or Tucuman,[110] specific focus in water law and water sector reforms should be placed on the local and national levels. In the light of the co-existence of a variety of uses and users, good water governance and management which takes into account the complex interrelationships affecting the dynamics of conflict and peace has been seen as an indispensable supposition to decrease the likelihood of water conflicts.[111]

5. The governance dimension

Last, but not least, international water principles play a role in the governance dimension of water management. Water governance refers to the different political, social and administrative mechanisms that must be in place to develop and manage water resources and the delivery of water services at different levels of society. In a nutshell, it is about striking a balance between competing users on both the international and national levels.

Encompassing principles such as transparency, participation, information and judicial remedies, and inclusive governance have been boosted by the global governance reform process that has been undertaken under the auspices of UNEP aiming at the determination of the future of environmental governance within the UN system.[112] In addition, the adoption of the 1998 UNECE Convention on Access to Information, Public Participation in Environmental Decision-Making and Access to Justice in Environmental Matters has had further impact on good water governance, in particular as it requires parties to apply the Convention's participatory standards when they act in the context of multilateral development banks.[113]

As is further elaborated in different chapters of this book, recent water law reforms have often failed to address governance challenges, including participation, accountability, transparency, effective enforcement mechanisms, single-sector approach, comprehensive policies and adequate legislative frameworks. Of

108 Simon Mason *et al.*, 'Linkages Between Sub-national and International Water Conflicts: The Eastern Nile Basin', in Hans-Guenther Brauch *et al.* (eds.) *Facing Global Environmental Change: Environmental, Human, Energy, Food, Health and Water Security Concepts* 325 (Berlin: Springer Verlag, 2007).

109 Sandra L. Postel and Aaron T. Wolf, 'Dehydrating Conflict', 126 *Foreign Policy* 60 (2001).

110 OECD, Issues Brief: Water and Violent Conflict 3 (2005), available at www.oecd.org/dataoecd/26/5/35785565.pdf.

111 *Ibid.*

112 The Global Governance Project (1998), available at www.environmentalgovernance.org/history/project_summary.php.

113 Convention on Access to Information, Public Participation in Decision Making and Access to Justice in Environmental Matters, Aarhus, 25 June 1998, UN Doc. ECE/CEP/43, art. 3(7).

those challenges, the principle of participation plays a particular role in water management and will therefore be examined in more detail in the following paragraphs.

The importance of participatory rights in water management has been recognized in numerous water law instruments.[114] Moreover, as set out in Chapter I, in response to the growing attention to the values of a participatory approach to water law reforms, some governments have started to improve their governance systems.[115] However, as stated by the global governance project, 'while the scope and scale of environmental concerns have changed and actors in the environmental field have proliferated, the crux of the debate remains the same – how does the international system implement an effective environmental policy that supercedes short-term political concerns'?[116] This is particularly true in economically driven reforms. Participation has been incorporated into recent World Bank operational water policies and procedures.[117] The Bank has also acknowledged improved public participation as an indispensable part of water sector reforms after many of its water projects failed due to their lack of socio-economic and human rights considerations.[118] It has stated that 'involvement of users in all aspects of water supply system development, implementation and O&M improves sustainability' and that '[w]omen's involvement proved to be critical for performance'.[119] However, the implementation practice of these policies has been frequently criticized,[120] including by the Bank's own bodies.[121] For example, concerning its water project in Sri Lanka, the Bank noted that 'though participation increased in the project planning phase, it was marginal in the implementation phase'.[122] Thus, in the words of Cullet, participation is often 'seen as an umbrella term that covers participation from policy planning and project design to the management of water infrastructure. In practice, the focus is on

114 For example the Dublin Statement on Water and Sustainable Development, International Conference on Water and the Environment, Dublin, 31 January 1992; Chapter 18 of Agenda 21; Bonn Recommendations for Action, Conference Report, International Conference on Freshwater, Bonn, 3–7 December 2001; and the 2004 Berlin Rules.

115 Concerning the application of participatory rights in national legislation, see Chapter II in this volume.

116 Global Governance Project, note 112 above.

117 See Chapter V.

118 World Bank, The Bank's New Water Resources Strategy, *in* Environment Matters, Annual Review 18 (Washington D.C.: World Bank, 2003).

119 OED/World Bank: Sri Lanka: Community Water Supply and Sanitation Project, in Impact Evaluation Report 2 (Washington DC: OED/World Bank, 1998).

120 For example in India, where World Bank-led water sector reform projects have shown a striking lack of participation. Hoering and Schneider, note 36 above, 22, 24, 26.

121 World Bank/OED, Efficient, Sustainable Services for All? – An OED Review of the World Bank's Assistance to Water Supply and Sanitation (Washington D.C.: World Bank/OED Report No. 26443, 2003).

122 World Bank, Implementation Completion Report, Sri Lanka, Community Water Supply and Sanitation Project (Colombo: World Bank,1999).

participation at the tail end of the process. In fact, the word participation is some sort of a misnomer'.[123]

CONCLUSION

As seen in this chapter, international aspects play an important role in water law and water sector reforms. There are an impressive number of international instruments and principles applicable to water management and protection. These include not only a range of hard law instruments, including most importantly the UN Watercourses Convention and the Helsinki Convention and its protocols, but also numerous soft law instruments, most notably the Dublin Principles, Helsinki and Berlin Rules, and the Hague and Bonn Declarations. Each of these agreements enshrines various principles of international law underlying the different dimensions of freshwater.

Another important recognition is the fact that over time the strict focus of international water law has gradually shifted to include water management concerns arising on national level, in particular regarding the provision of water for all. This shift is reflected by the increasing recognition of a human right to water, in particular in response to the notion of water as an economic good, which has proved largely unsuccessful in recent water law reforms. On the environmental side, principles such as precaution, prevention and polluter-pays, as well as more generally participation, transparency and accountability, have gained international attention concerning their importance for sustainable, fair and equitable water law and water sector reforms.

Nevertheless, until today, effective water management is hindered not only by overuse, inequitable use and pollution, but also by the fragmented, incoherent and inconsistent nature of existing international instruments. Though the different elements for effective water management appear to be known, and though all actors, in the water sector and outside, agree that the overall goal for water management is the provision of water for all, there is still a lack of a comprehensive international legal framework to guide the achievement of this goal. In addition to fragmentation, most of the existing international instruments are non-binding, which increases the inadequate legal protection and ineffective water management on the national level, mainly at the expense of the poorest and most vulnerable sections of the population, most of whom still do not have access to sufficient and affordable water.

The analysis of international water law instruments shows that the underlying problem is the competing focus of the different dimensions of water in the absence of a comprehensive and coherent legal framework to guide decision makers in providing universal access to water. As a consequence of the support of

123 Philippe Cullet, 'Water Law Reforms: Analysis of Recent Development', 48/2 *J. Indian L. Institute* 206, 209 (2006).

economic water sector reforms by financially powerful IFIs and multinational corporations, the economic dimension has received far more attention than other aspects of water management. Ignoring that 'in the end, water is an inherently public good, that must be subject to a regime of public management that subordinates private uses to policies necessary to meet public needs',[124] the economic focus on water management has moreover resulted in the retreat of the government from the provision of water services to a sole regulatory role. As shown, this development has led to the failure of most IFI-led water projects and water law reforms, and left billions of people without access to sufficient water.

In the light of this failure, the need to change the approach to water management has been widely acknowledged. However, the new notion of a holistic approach to water management as, for example, incorporated in IWRM will only provide a basis for fair, equitable and sustainable water services for all if equally reflecting all dimensions of water and if supported by a comprehensive international legal framework on the basis of the instruments and principles analyzed in this chapter, in particular the human right to water.

124 Joseph W. Dellapenna, 'Transboundary Water Sharing and the Need for Public Management', 133/5 *J. for Water Resources Planning and Management* 397, 397 (2007).

IV. Legal implications of trade in 'real' and 'virtual' water resources

Alix Gowlland Gualtieri

INTRODUCTION

Freshwater is a unique natural resource which is irreplaceable and essential for all life on earth, and which can have strong cultural and religious implications. Since water touches upon many sectors of human activity it can also be addressed under several perspectives, including those of the environment, human rights, economy or trade. While the environmental, social and human rights dimensions of water have been integrated to varying degrees in international law and policy,[1] the international trade and financial institutions have exerted an undeniable influence on the reforms that have been engaged in the water sectors of many countries during the past decade, especially in promoting an economic approach to water-related issues. This has meant that, coupled with growing water scarcity and the increased value put on this resource, water is fast becoming a pure commodity subject to the rules of pricing, markets and international trade.

This chapter focuses on the trade dimension of water. The confluence between international trade rules and water raises two different sets of issues. First, the question of privatization of water services has arisen under the General Agreement on Trade in Services (GATS).[2] While this has real implications for the management of water services in member countries, no country has to date made commitments in terms of water services.[3] Second, as interest is increasing in the potential profits

1 For instance, there has been a growing emphasis on the recognition of an international right to water, most recently illustrated by the inclusion of water on the agenda of the UN Human Rights Council (HRC), see Chapter XI in this volume. See Chapter XII for the environmental dimension of water.

2 General Agreement on Trade in Services, 15 April 1994, Marrakesh Agreement Establishing the World Trade Organization, Annex 1B, Legal Instruments – Results of the Uruguay Round vol. 31, 33 *ILM* 81 (1994).

3 This topic is not further pursued in the present chapter. Further, see, e.g., Mireille Cossy, 'Water Services at the WTO', in Edith Brown Weiss, Laurence Boisson de Chazournes and Nathalie Bernasconi-Osterwalder (eds.) *Fresh Water and International Economic Law* 117 (Oxford: Oxford University Press, 2005); Elisabeth Tuerk, Aaron Ostrovsky and Robert Speed, 'GATS and its Impact on Private Sector Participation in Water Services', in Brown Weiss, Boisson de Chazournes and Bernasconi-Osterwalder 143.

stemming from trading water as a good in itself, water transfers in different forms – through dam-building, pipelines, inter-basin transfers or interlinking of rivers – are becoming increasingly prevalent both within and between states.[4] This in turn raises questions under the General Agreement on Tariffs and Trade (GATT),[5] which have, to date, been relatively little explored. This chapter focuses on issues arising from trade in water resources under the GATT. It aims to explore some of the legal implications of subjecting trade in water resources to the international trade rules, the extent to which these rules take account of the social and environmental values of water, and whether they allow countries to protect water resources from over-exploitation. These implications can be viewed from the point of view both of exporting states wishing to protect domestic water resources and of importing countries wishing to resolve problems of water scarcity through trade.

Section A presents an overview of the modes of trade in water at the international level, taking into consideration the broad shift in international law and policy towards seeing water mainly as an economic good. Section B then turns to the applicability of the GATT rules to trade in water. Section C addresses some of the implications of trade in 'virtual' water – water traded as incorporated in products which either contain significant amounts of water or require large volumes of water to produce. The chapter ends with concluding remarks on the relationship between water and the international economic legal framework.

A. Modes of international trade in water

1. Bottled water and other drinks containing water

The most common form in which water can be traded occurs after its transformation or removal from a natural or bulk state. This concerns most bottled water[6] and other drinks containing water such as soft drinks and juices. An increasingly lucrative international market in bottled water has emerged as a consequence of growing demand. Bottled water consumption has been steadily growing for the past thirty years and consumption increases by an average 12 per cent each year, in spite of its excessively high price compared with tap water and even in countries where there is access to safe public drinking water.[7] While almost half of the

4 On trade in water on the domestic level, especially within developing countries, Mateen Thobani, Tradable Property Rights to Water (Washington D.C.: World Bank, 1995); Mark Rosegrant and Hans Binswanger, 'Markets in Tradable Water Rights: Potential for Efficiency Gains in Developing Country Irrigation', 11 *World Development* 22 (1994).

5 General Agreement on Tariffs and Trade, 15 April 1994, 1867 *UNTS* 3 (hereinafter GATT 1994).

6 The term 'bottled water' subsumes natural mineral water, spring water and purified water. Catherine Ferrier, Bottled Water: Understanding a Social Phenomenon 6–8 (World Wildlife Fund (WWF) Discussion Paper, 2001).

7 Data found on the International Year of Freshwater 2003, available at www.wateryear2003.org. Sales have reached up to US$50 billion worldwide in the past decade and are still growing about 10 percent a year. Peter H. Gleick *et al.*, *The World's Water 2002–2003: the Biennial Report on Freshwater Resources* 43–4 (Washington D.C.: Island Press 2002).

bottled water is consumed by Western Europeans, bottle sales are also increasing in developing countries – in particular Asia – despite the fact that bottled water rarely provides adequate volumes of water for domestic use and that its cost is typically very high.

The bottled water industry is not exempt from environmental impacts, because amongst other things manufacturing, recycling or incinerating bottles requires energy and can release polluting particles into the air and water. Transporting bottled water within or between countries also implies energy needs and fuel combustion. Domestically, the extraction of groundwater necessary for the production of bottled water has caused tensions with local communities. In India, 65 per cent of Indian bottlers simply pump water from any bore well or even municipal water supplies, threatening groundwater resources because of over-exploitation and pollution due to purification and processing, in addition to poor bottled water quality.[8] Communities residing in proximity to Coca-Cola bottling plants have been experiencing severe water shortages allegedly as a result of the company's massive extraction of water from the common groundwater resource.[9] The increase in bottled water consumption can thus represent a threat to domestic water resources, particularly in countries already facing significant problems in providing access to water.

Bottled water and other drinks containing water have for a long time been considered commodities, and they remain the main form in which water is traded between states and on a large scale. The trade boom in these goods that has taken place in recent years, coupled with mounting concerns over scarcer water availability, nevertheless make it relevant to examine the way in which the international trade rules apply to trade in these goods.

2. *Bulk water*

Rather than being transformed and traded in small, defined volumes, or as part of other products, water can be traded in bulk form.[10] While water has always been transferred domestically and between states, mainly through canals and pipelines, the view that bulk water is a commodity that can be traded intensively on the inter-state level is a new phenomenon. Countries were reluctant to engage in the export of this natural resource, particularly because water was for a long time considered a public good which could not be commodified. Technical limitations in transporting water over long distances and in accurately determining the

8 Ferrier, note 6 above, 21.

9 See Chapter IX.

10 Edith Brown Weiss, 'Water Transfers and International Trade Law', in Brown Weiss, Boisson de Chazournes and Bernasconi-Osterwalder, note 3 above, 61, 66–7 (Defining 'bulk water removals' as 'removal by human/made diversions, including canals, taker ships, trucks, or pipeline'. This does not include removal for bottling). This chapter does not deal with water in its natural state, (i.e., shared lakes, rivers and aquifers) which are regulated under international water law in particular as concerns watercourses.

amounts transported also constituted an obstacle to trade in bulk water. New bulk storage and transfer technologies have now been developed to make it possible to move large volumes of water across long distances for commercial purposes, using massive pipelines, supertankers or giant sealed water bags.[11] The growing scarcity of the resource in some countries means that the price of water has increased, making this emerging market all the more appealing for both public and private entities. Opening up the international market-place to water has also been seen as an answer to the inequitable distribution of water in the world between water-rich and water-impoverished countries.

Despite the high cost of transferring water over long distances,[12] adverse environmental impacts,[13] as well as growing public pressure against water trade in exporting countries,[14] there is increasing momentum for the establishment of an international water market by multinational and domestic corporations. These factors contribute to rendering the issue a lot less theoretical than it was a few years ago and give rise to the assumption that in the near future one could witness large-scale water transfers between states.[15] This raises concerns that if trade in water is subject to international trade rules, governments might be prevented from exporting domestic water resources on human rights or environmental grounds.

B. Applicability of the GATT/WTO rules

1. Water as a good or product

Trade in goods is regulated under the GATT/WTO and the applicability of the agreement rests on whether water is defined as a good or a product.[16] Whether or

11 Mark Clayton, 'Forget OPEC – The Next Cartel may Export Drinking Water', *Christian Science Monitor*, 30 December 2004, available at www.csmonitor.com/2004/1230/p13s01-sten.html.

12 It is still less expensive to get fresh water by other means, even by the desalinisation of seawater.

13 The extraction of water for bulk export can involve the disruption of ecosystems, damage to natural habitats, harm to biodiversity, and disruption of aquifers and underground water systems. Transporting bulk water worldwide also has an environmental impact; for instance, supertankers can deplete water supplies along ocean coasts as they cannot venture inland.

14 Water exports are already at the forefront of public concern in some countries like water-rich Canada. Bulk Water Removals and International Trade Considerations: Document from the Canadian Department of Foreign Affairs and International Trade (16 November 1999); International Joint Commission (IJC), Protection of the Waters of the Great Lakes: Final Report to the Governments of Canada and the United States (22 February 2000), available at www.ijc.org/php/publications/html/finalreport.html. See also Scott P. Little, 'Canada's Capacity to Control the Flow: Water Export and the North American Free Trade Agreement', 8 *Pace Int'l L. Rev.* 127 (1996).

15 Relatively few long-term international trades in bulk water have actually taken place. Gleick, note 7 above, 44–5.

16 This is indicated in Articles I, III and XI, GATT 1994, note 5 above. See also Robert Girouard, 'Water Export Restrictions: A Case Study of WTO Dispute Settlement Strategies and Outcomes', 15 *Georgetown Int'l. Envtl. L. Rev.* 247 (2003); John H. Jackson, *The Jurisprudence of GATT and WTO* 58 (Cambridge: Cambridge University Press, 2000).

not water is a good is a question that arises in view of its unique and essential functions which set it apart from any other product. It appears clear that water becomes a good when it is transformed from its natural state, bottled and sold commercially.[17] This is also the case when drinks containing water are sold. The trade in bottled water and drinks containing water, in that it concerns a definite amount of freshwater traded as a separate product, is therefore the most easily recognized application of water as a tradable commodity subject to the rules governing international trade. An indication that member states treat the export and import of bottled water like any other good can be given by the existence of notifications on technical or sanitary regulations and standards concerning bottled water in the framework of the Agreement on Technical Barriers to Trade (TBT).[18] For instance, bottled water exported by Canada to the United States has been expressly recognized as a tradable commodity under the North American Agreement on Free Trade (NAFTA).[19] Since bottled water and other drinks containing water are covered by the international trade regimes, much of the debate and concern has been focused on whether bulk water is a good or a product when it is traded between states.

There is no information on the intent of the parties when negotiating the GATT relevant to the applicability of the Agreement to bulk transfers of water and this question has indeed never been discussed in the framework of the WTO. Several elements are highlighted when addressing the applicability of the Agreement to water, including the absence of an explicit exclusion of water,[20] and the fact that water is included under the GATT Tariff List. Products are identified for tariff purposes under the GATT by the Harmonizing Commodity Description and Coding System (HS), which is developed and maintained by the

17 Mireille Cossy, 'Le statut de l'eau en droit international économique – Principaux aspects au regard des règles de l'Organisation mondiale du commerce', in Laurence Boisson de Chazournes and Salman M.A. Salman (eds.) *Water Resources and International Law* 169 (Leiden: Martinus Nijhoff, 2005).

18 See, e.g., Committee on Technical Barriers to Trade, Notification by the United Kingdom, G/TBT/N/GBR/14 (7 June 2006) (Spring Water and Bottled Drinking Water).

19 North American Free Trade Agreement, 17 December 1992, United States, Mexico and Canada, available at www.nafta-sec-alena.org/DefaultSite/index_e.aspx?DetailID=78. On the topic, Cynthia Baumann, 'Water Wars: Canada's Upstream Battle to Ban Bulk Water Export', 10 *Minnesota J. Global Trade* 114, 119 (2001).

20 Esther de Haan, 'Balancing Free Trade in Water and the Protection of Water Resources in GATT', *in* Edward H.P. Brans *et al.* (eds.) *The Scarcity of Water* 245, 248 (London: Kluwer Law International 1997). The NAFTA member states have however declared that '[u]nless water, in any form, has entered into commerce and becomes a good or product, it is not covered by the provisions of any trade agreement, including the NAFTA. And nothing in the NAFTA would obligate any NAFTA Party to either exploit its water for commercial use, or begin exporting water in any form. Water in its natural state, in lakes, rivers, reservoirs, aquifers, water basins and the like is not a good or product, is not traded, and therefore is not and never has been subject to the terms of any trade agreement'. See Statement by the Governments of Canada, Mexico, and the United States (1993) (on file with the author).

World Customs Organization,[21] and which subsumes water in several forms.[22] The inclusion of water in the Tariff Heading has been interpreted by some to imply not only that water has been transformed from its natural state (in particular bottled water and other beverages containing water) but also that water in any form can be considered a good and can thus be covered by the GATT if traded between states. Such a broad interpretation would lead to the applicability of the GATT to any type of transfer of water that was diverted or pumped in order to cross national borders. The interpretation according to which water is a good since it is included in the Tariff Schedule has been contested by some authors, on the basis that the purpose of the Schedule is not to provide a definition of the goods falling under the GATT/WTO but rather to determine that when water is classified as a good, it falls under a particular tariff heading.[23] The criterion of transformation has also been applied to the determination of whether bulk water is a good or product under the GATT. This would imply that as soon as water is transformed or processed, it can be interpreted more directly as a tradable good. Some authors argue that bulk water transfers by tankers or giant plastic bags is not that different from bottling water, except in the quantities, thus making it a good.[24]

The question of whether or not water in its bulk form can be considered a good or product under the GATT is therefore far from being settled, especially since there have to date been no disputes involving water submitted to the WTO Dispute Settlement Body (DSB). Because of its unique nature, there is a strong argument for excluding water from the scope of the GATT. Trade in such a vital resource should not be subjected to the rules of international trade since states need to retain control over domestic water resources for purposes of allocation, use and protection. In this sense, a distinction should be operated under the GATT/WTO between water and other tradable natural resources such as timber or fossil fuels which are not vital and unique. This argument can be supported by the language of the preamble to the agreement, which acknowledges the intent of the parties to conduct international trade 'while allowing for the optimal use of the world's resources in accordance with the objective of sustainable development'. The importance of determining the applicability of the international trade rules to trade in water lies of course in the possibilities for (a) water-exporting countries to limit or ban exportation of their water resources; and (b) water-importing

21 The HS nomenclature is used as a basis for the collection of customs duties and international trade statistics by almost all countries.

22 World Customs Organization, Harmonized Commodity Description and Coding System: Explanatory Notes (3rd edn 2002), Tariff Heading 2201.90.0000 (Comprising '[w]aters, including natural or artificial mineral water and aerated waters containing added sugar or other sweetening matter or flavoured and other non-alcoholic beverages, not including fruit or vegetable juices of Heading 20.09; [. . .] ice and snow') and Explanatory note A ('ordinary water of all kinds (other than sea water))'.

23 Gleick, note 7 above, 48. The HS is non-binding on member states.

24 Cossy, note 17 above, 7–8.

countries to bring a claim under the WTO DSB against restrictive trade measures applied by water-exporting country. It remains pertinent to examine what the consequences of the application of the GATT to trade in water would be if water were indeed to be considered a good or product under the agreement. Indeed, were a dispute to be submitted to the DSB, it would be likely that water would be considered as falling under the scope of the GATT.

2. *GATT obligations*

As it is well known, the main purpose of the GATT is the liberalization of international trade through the elimination of trade restrictions. The cornerstone principles are couched in Articles I, III and XI of the Agreement. Under Article I, the most-favoured nation principle dictates that all WTO member states must be treated on an equal footing and grant each other equal treatment of like products originating or destined for the territories of all other members. This principle applied to trade in water would mean that countries are prevented from discriminating between importers and exporters. Article III sets out the national treatment principle, which implies that once goods have entered into a market, equal treatment must be applied to foreign and domestic goods. With respect to the import of goods, states party cannot discriminate against imports on the basis of their consumption or production practices under these provisions. This provision would be relevant in the case of states wishing to restrict water imports, for instance on environmental grounds.

The conflict between water export-control measures and trade rules would arise most significantly under Article XI, paragraph 1. The provision reads:

> No prohibitions or restrictions other than duties, taxes or other charges, whether made effective through quotas, import or export licences or other measures, shall be instituted or maintained by any contracting party on the importation of any product of the territory of any other contracting party or on the exportation or sale for export of any product destined for the territory of any other contracting party.

Article XI, paragraph 1, prohibits measures other than duties, taxes or other charges, and states are required to gradually reduce tariffs upon imports and exports of goods meaning that quotas or bans on the export of water for conservation purposes could be challenged as a form of protectionism.[25] This obligation is particularly important with regard to trade in bulk water, since it could be interpreted as constraining WTO member governments from establishing policies, programmes or legislation to regulate or prohibit bulk water exports. If indeed applicable, one question is whether these restrictions would apply only

25 See also Article XIII (non-discriminatory administration of quantitative restrictions), GATT 1994, note 5 above.

to specific, bulk water trades once they have begun, or apply to all potential trading arrangements once trade in bulk water has begun between any WTO signatories.[26]

Few disputes concerning the export of goods under Article XI, paragraph 1, have been submitted to the WTO dispute settlement mechanism by member states. In *Canada Herring*, the Panel found that the measures adopted by Canada prohibiting the exportation or sale for export of certain unprocessed herring and salmon were contrary to Article XI, paragraph 1, and not justified under the GATT exceptions found under Article XI, paragraph 2, and Article XX, paragraph g.[27] Under Article XI, any prohibition maintained by a state to prohibit or to limit water exports from its territory – providing water is a good falling under the agreement – is in principle forbidden, unless it can be justified under one of the GATT exceptions. It thus remains to be seen what other provisions would allow countries to take protective measures for social and/or environmental reasons.

3. GATT exceptions

Article XI, paragraph 2(a)

The GATT contains a first exception which is specifically applicable to Article XI, paragraph 1. Paragraph 2(a) of this provision sets forth a temporary exception to the provisions of paragraph 1 with respect to '[e]xport prohibitions or restrictions temporarily applied to prevent or relieve critical shortages of foodstuffs or other products essential to the exporting contracting party'. When instituting export restrictions or prohibitions to prevent or relieve critical shortages of foodstuff, states must 'give due consideration to the effects of such prohibition or restriction on importing states' food security'. This condition does not apply to restrictions or prohibitions on the basis of the second ground. Since water is undoubtedly essential in preventing or relieving shortages of foodstuffs and can be considered a product essential to the exporting state, water exports could be temporarily excluded from the scope of Article XI, paragraph 1.

The exception found in Article XI, paragraph 2(a) would enable WTO member states to temporarily restrict water exports and justify trade-related environmental measures otherwise inconsistent with GATT principles if there is a critical shortage of water for human consumption or agricultural use. It does not, however, apply to situations in which states seek to ban water exports on environmental grounds, which indeed seems to be one of the most likely scenarios for policy

26 Gleick, note 7 above, 49.

27 Panel Report, Canada – Measures Affecting Exports of Unprocessed Herring and Salmon, BISD 35S/98 (22 March 1988). *See also* Panel Report, Japan – Trade in Semi-conductors, L/6309 – 35S/116 (4 May 1988); Panel Report, Argentina – Measures Affecting the Export of Bovine Hides and the Import of Finished Leather, T/DS155/R (19 December 2000).

decisions to restrict the export of domestic water resources. Were Article XI, paragraph 2(a) to apply, exporting states are still required to balance their own food security interests when applying an export restriction on water as foodstuff with those of the importing state. This is a requirement of Article 12, paragraph 1, of the WTO Agreement on Agriculture, which provides that:

> Where any Member institutes any new export prohibition or restriction on foodstuffs in accordance with paragraph 2(a) of Article XI of GATT 1994, the Member shall observe the following provisions:
>
> > (a) the Member instituting the export prohibition or restriction shall give due consideration to the effects of such prohibition or restriction on importing Members' food security.[28]

Thus, a country applying a water export ban must take into account the impacts of such a measure on the food security situation of the importing country. However, how exactly Articles XI, paragraph 2(a), of the GATT and 12 of the Agreement on Agriculture interrelate in this case is still to be clarified.

Article XX

The GATT contains a general exception to the obligations incumbent upon its member states under Article XX. Thus, should restrictive or prohibitive measures applied to water exports be considered to violate provisions such as Article XI, Article XX could be invoked to justify such measures. This clause provides states with the possibility to derogate from GATT obligations for the protection of certain specific interests.[29]

> Subject to the requirement that such measures are not applied in a manner which would constitute a means of arbitrary or unjustifiable discrimination between countries where the same conditions prevail, or a disguised restriction on international trade, nothing in this Agreement shall be construed to prevent the adoption or enforcement by any contracting party of measures: [. . .]
>
> > (b) necessary for the protection of human, animal or plant life or health; [. . .]

28 Agreement on Agriculture, Annex 1A to the Marrakesh Agreement establishing the World Trade Organization and Final Act, 15 April 1994, available at www.wto.org/english/docs_e/legal_e/14-ag.pdf.

29 See, e.g., Report of the Appellate Body, United States: Import Prohibition of Certain Shrimp and Shrimp Products, WT/DS58/AB/R (12 October 1998), paras 129–130, referring to 'concerns of the community of nations about the protection and conservation of the environment' and the aim of sustainable development found in the Preamble to the WTO Agreement.

> (g) relating to the conservation of exhaustible natural resources, provided that such measures are made effective in conjunction with restrictions on domestic production and consumption.

There is no legal precedent to date regarding the application of Article XX exceptions to trade in water.[30] Moreover, the disputes in which Article XX was considered concerned import restrictions rather than restrictions placed on the export of a domestic resource, leaving many unresolved questions with respect to water trade restrictions. The case of export restrictions would raise additional issues relating to state sovereignty over natural resources, since it would involve a country attempting to protect its domestic resources rather than influence the production processes in another country. The following paragraphs examine how the exceptions found in Article XX could potentially be applicable to trade in water, particularly bulk water.

PARAGRAPH (B)

Paragraph (b) of Article XX provides that states can impose trade restrictions if they are necessary for the protection of human, animal or plant life or health. Measures have to be necessary, that is, the state must use among all available measures the one that is the least trade-restrictive measure for the aim of protecting human, animal or plant life. Indeed, Article XX, paragraph (b), aims to prevent the use of trade measures for protectionist purposes. Case-law by the Appellate Body gives an indication of what constitutes a necessary measure. In *Korea Beef*, the Appellate Body provided (referring to paragraph (d) of Article XX):

> In sum, determination of whether a measure, which is not 'indispensable', may nevertheless be 'necessary' within the contemplation of Article XX(d), involves in every case a process of weighing and balancing a series of factors which prominently include the contribution made by the compliance measure to the enforcement of the law or regulation at issue, the importance of the common interests or values protected by that law or regulation, and the accompanying impact of the law or regulation on imports or exports.[31]

30 In a request for consultations, Canada claimed that Korean regulations on the shelf-life and physical treatment (disinfection) of bottled water were inconsistent with Articles III and XI, GATT 1994, note 5 above, as well as Articles 2 and 5 of the SPS and 2 of the TBT. At the DSB meeting on 24 April 1996, the parties to the dispute announced that they reached a settlement. Korea – Measures Concerning Bottled Water – Request for Consultations by Canada, WT/DS20/1 (22 November 1995).

31 Report of the Appellate Body, Korea – Measures Affecting Imports of Fresh, Chilled and Frozen Beef, WT/DS161/AB/R; WT/DS169/AB/R (11 December 2000), para. 164. *See also* Report of the Appellate Body, European Communities – Measures Affecting Asbestos and Asbestos-Containing Products, WT/DS135/AB/R (12 March 2001); Report of the Appellate Body, United States – Measures Affecting the Cross-border Supply of Gambling and Betting Services, WT/DS285/AB/R (7 April 2005).

The Appellate Body therefore sets out three conditions to determine whether a measure is necessary. Its reference to common interests or values is noteworthy, especially since '[t]he more vital or important those common interests or values are, the easier it would be to accept as "necessary" a measure designed as an enforcement instrument'.[32]

The exception found in paragraph (b) has been used by states in several disputes on the basis of the protection of human or animal health. While these cases concerned restrictions on import, nothing in the text of paragraph (b) suggests that export cannot be concerned. Export restrictions on domestic water resources could be justified under this clause if resulting water scarcity represented a threat to biodiversity, ecosystems, natural habitats, or led to shortages of water for consumption or agricultural purposes. Restrictions would also have to be necessary, that is, aimed at protecting common interests or values; water, as indeed a vital resource, would without doubt fit into this. Consequently, if export restrictions were adopted in a broader plan of national water management aimed at preventing domestic water scarcity, export restrictions could fall within this provision.

PARAGRAPH (G)

Paragraph (g) of Article XX allows states to take restrictive trade measures relating to the conservation of exhaustible natural resources. Any measures adopted under this Article cannot be adopted for protectionist reasons.[33] For trade in water to be restricted for conservation purposes by member states under paragraph (g) of Article XX, water must firstly constitute an exhaustible natural resource in the sense of this provision. Paragraph (g) appears to have been originally drafted in regard to such resources as minerals or fossil fuels. There is no indication in the GATT that water could be considered an exhaustible natural resource, and indeed it is likely that water was not even contemplated during the drafting of the Agreement – not more than other renewable resources such as animals, plants and soil. However, WTO case-law has considered as renewable such biological resources as tuna, salmon, herring, dolphins or sea turtles, as well as non-biological ones (air). In *Shrimp-Turtle*, the Appellate Body moreover applied an evolutionary interpretation to the concept of exhaustible natural resource.[34]

Water could be deemed an exhaustible natural resource under paragraph (g). While many sources of water are renewable resources made available by the hydrologic cycle of the atmospheric–oceanic–terrestrial system,[35] water may also be non-renewable in some situations and some water sources can be made non-renewable through human action.[36] Thus all water resources cannot be treated alike for the purpose of determining whether they are exhaustible or not.

32 *Korea Beef*, note 31 above, para. 162.

33 *Canada Herring*, note 27 above, paras. 4.4–4.7.

34 *Shrimp-Turtle*, note 29 above, paras. 130 and 132.

35 In this sense, continued flows of water are not affected by withdrawals and use of water.

36 Gleick, note 7 above, 39, 49–50.

Paragraph (g) could be argued to apply to water stocks that are not being replenished at a significant rate compared with the rate of use. In this case, there would be a strong argument for applying the exception to restrictions on trade in water where the resources are non-renewable or exhaustible through overuse or abuse. Gleick cites the example of groundwater overdraft in which human extraction exceeds natural replenishments. In the case of exports, some argue that water from the Great Lakes should be considered non-renewable since only a very small part of the lakes are renewable and the vast bulk of the stored water was laid down in geologic times.[37] Thus in some circumstances, exports of bulk water from the lakes could lead to irreversible decline in the lake levels. Similarly, cases where water stocks are contaminated by human action also represent the conversion of renewable water resources in a non-renewable one.

For paragraph (g) to be applicable, the restrictive trade measure must relate to conservation of the resource. It must also be made effective in conjunction with restrictions on domestic production and consumption. Provided export-restrictive measures are part of a state's conservation policies aimed at the protection of national water resources, export restrictions on exhaustible flows of water could fall under the non-renewable mineral resources category for exemption under the provision.[38] The exporting state would also have to restrict domestic consumption of water for conservation purposes in order to meet the conditions of paragraph (g).

CHAPEAU

In addition to meeting the conditions set out in Article XX, paragraphs (b) or (g), a measure must also comply with the chapeau to Article XX, which prohibits the use of trade restrictions when these result in unjustifiable discrimination between states or form a disguised restriction to trade. The interpretation of the chapeau involves the balancing of interests of the member invoking an exception under Article XX and the rights of other members under substantive provisions of the GATT 'so that neither of the competing rights will cancel out the other and thereby distort and nullify or impair the balance of rights and obligations constructed by the Members themselves in that Agreement'.[39] A case-by-case analysis would thus be required in the case of trade restrictions on water to determine whether they comply with the chapeau.

In conclusion, until an actual dispute concerning trade restrictions on water is brought before the DSB, the impact of the WTO case-law is uncertain for the protection of water resources under Article XX, paragraphs (b) and (g). The trade and environment case-law tends to point to the fact that a state wishing to ban

37 1999 Report of the IJC, note 14 above.

38 Renewable flows of water could qualify under paragraph (b) as necessary to protect human, animal or plant life or health.

39 *Shrimp-Turtle case*, note 29 above, para. 159.

water exports as part of its domestic water policy would have difficulty justifying it under the GATT exceptions. The state would have to argue that the restrictions can be justified under the GATT exceptions on environmental or human health grounds, provided the restrictions are implemented in a non-discriminatory manner and do not form a disguised restriction on trade.

C. The case of trade in virtual water

1. The concept of virtual water

Water can also be traded when it is incorporated in products which either contain significant amounts of water or that require large volumes of water to produce. The term 'virtual water' was coined to define the volume of water required to produce a given product.[40] Since the agricultural sector is the largest economic sector which uses water resources at the global level,[41] agricultural products form a significant part of the trade in water-intensive goods and it is mainly with respect to these goods that the discussion on virtual water trade has taken place.[42] The implications of virtual water trade are well illustrated by the fact that the production of one kilogram of grain requires approximately 1,000–2,000 kilograms of water.[43] In many countries, significant amounts of water go towards water-intensive commercial crops grown in arid or semi-arid areas. The virtual water content of products strongly varies from place to place, depending upon the climate, technology adopted for farming and corresponding yields.[44] Trade in virtual water on the regional and international levels has been steadily increasing over the past decades, with about 15 per cent of the water used in the world for

40 J.A. Allan, 'Overall Perspectives on Countries and Regions', in Peter Rogers and Peter Lydon (eds.) *Water in the Arab World: Perspectives and Prognoses* 65 (Cambridge, MA: Harvard University Press, 1994). See also J.A. Allan, 'Virtual Water – the Water, Food, and Trade Nexus: Useful Concept or Misleading Metaphor?' 28 *Water Int'l* 4 (2003).

41 Of the three main uses of water – household/municipal, industrial and agricultural – the latter amounts to 75 per cent. Vital Water Graphics, UNEP 2003; United Nations, *Water for People – Water for Life* (Paris: UNESCO, 2003).

42 According to Hoekstra, 67 per cent of the global virtual water trade is related to international trade in crops, 23 per cent to trade in livestock and livestock products and 10 per cent to trade in industrial products. Arjen Y. Hoekstra (ed.) Virtual Water Trade: Proceedings of the International Expert Meeting on Virtual Water Trade 16 (Value of Water Research Report Series No. 12, February 2003). We do not look here at trade in water-intensive industrial goods.

43 As reported by the World Water Council, www.worldwatercouncil.org/index.php?id=866. Rice has the largest share in the total volume of water used for global crop production, followed by wheat. Moreover, livestock products have a higher virtual water content than crop products. *See* further Arjen Y. Hoekstra and Ashok K. Chapagain, 'Water Footprints of Nations: Water Used by People as a Function of Their Consumption Pattern', 21 *Water Resources Management* 35, 38 (2007).

44 For instance, in regions with a high evaporative demand, the water requirement per unit of crop production is relatively large; water-inefficient agricultural practices mean that water productivity in terms of output per drop of water is relatively low. *Ibid.* 46.

export in virtual form.[45] The United States and Canada are the biggest exporters of virtual water in the world, with India also figuring amongst the top exporters in part due to the large volume of exported agricultural products[46] including cotton, a very water-intensive cash crop mainly grown in dry regions.[47]

The implementation of a sustainable virtual water trade is increasingly being viewed as one solution to further water-use efficiency in the light of existing water scarcity, and global and regional imbalances in water availability. In the discussion on the implications of virtual water trade as a policy option for countries, it has been pointed out that virtual water trade 'should be encouraged to promote water savings for arid countries and at a global level through enhancing food security through appropriate and fair trade agreements'.[48] Since the export of water-intensive products contributes to the loss of significant amounts of domestic water resources, it is argued that water-scarce countries can meet their water demand for food by importing water-intensive agricultural products such as cereals from countries with sufficient or abundant water resources (water-rich countries).[49] In this manner, rather than producing water-intensive products, water-scarce countries can optimize the use of available domestic water resources.[50] For instance,

45 See the facts and figures on virtual water found on the site of the International Year of Freshwater 2003, available at www.wateryear2003.org.

46 Kriti Bardhan Gupta, The Hidden Export of Water through Principal Crops Across Different Regions in India: Social, Economic and Environmental Implications (Paper presented at a National Seminar, Chandigarh, National Institute of Technical Teachers' Training and Research (NITTTR), 2005); O.P. Singh *et al.*, 'Virtual Water Trade in Dairy Economy: Irrigation Water Productivity in Gujarat', 39/31 *Economic and Political Weekly* 3492 (2004). *See also* Shama Perveen, 'Water: The Hidden Export', *India Together*, August 2004, available at www.indiatogether.org/2004/aug/env-virtwater.htm; Sowmya Kerbart Sivakumar, 'It's Called "Virtual Water"', *The Hindu*, June 2004, available at www.hinduonnet.com/mag/2004/06/06/stories/2004060600150200.htm; Amit Bhandari, 'Sold! Waters of India', *The Times of New Delhi*, 21 July 2004, available at www.waterfootprint.org/Reports/Waters_of%20India.pdf.

47 The extensive irrigation of cotton has a severe impact on water resources leading to a depletion of surface or groundwater which can affect the river catchments and the wetlands lying downstream. World Wide Fund for Nature (WWF), The Impact of Cotton on Freshwater Resources and Ecosystems: A Preliminary Synthesis (Washington D.C.: WWF, 1999).

48 Paul van Hofwegen, Virtual Water Trade: Conscious Choices (World Water Council, August 2003). The World Bank also finds that increasing water scarcity in the next two decades will mean that since agriculture consumes by far the largest percentage of water, most countries will take water from agriculture, allocate it to other sectors, and rely on increased food imports to meet their domestic needs. World Bank, Investment Note 1.3: Agricultural Trade, Water, and Food Security. See further Tony Allan, 'Watersheds and Problemsheds: Explaining the Absence of Armed Conflict over Water in the Middle East', 2(1) *Middle East Rev Int'l Affairs J.* 49 (1998).

49 Mark W. Rosegrant, Ximing Cai and Sarah A. Cline, Global Water Outlook to 2025: Averting an Impending Crisis 9 (Washington, D.C.: International Food Policy Research Institute, 2002). Further, *see* J.A. Allan, 'Virtual Water Eliminates Water Wars? A Case Study from the Middle East, Virtual Water Trade' and Jeroen Warner, 'Virtual Water – Virtual Benefits? Scarcity, Distribution, Security and Conflict Reconsidered', in Hoekstra, note 42 above.

50 The virtual water concept is closely linked to the 'water footprint' concept, which aims to determine on a national level the total volume of freshwater that is used to produce the goods and services consumed by the inhabitants of a country. Since not all goods consumed in one particular

Jordan has been seen as having successfully externalized its water footprint by importing wheat and rice products from the United States, which has higher water productivity.[51] Virtual water trade can in this manner represent an alternative to large-scale water transfer schemes which aim to redistribute water on the national or regional levels, often with significant adverse environmental and social impacts. The current debate in India over the inter-linking of rivers project, which aims to remedy water scarcity in the more arid regions, has included discussion of the value of virtual water trade across different states through the trade of principal crops.[52]

On the other hand, there are reasons to be cautious about the value of promoting a virtual water trade.[53] For instance, when a country establishes a national policy for virtual water imports to alleviate its water problem, it will not necessarily develop alternatives in terms of crops or employment to ensure that the livelihoods of farmers are not unduly affected. There can also be significant alterations in the organization and ownership of means of agricultural production within the country. Trade in virtual water might also have environmental costs that outweigh the benefits stemming from more efficient global water use; these include not only the environmental costs of shipping of products but also the risks of over-exploitation of water resources even in water-rich countries where the export of water-intensive crops is encouraged. A further aspect remains whether the countries facing the most severe water scarcity will actually be able to afford importing virtual water.

This chapter does not go into the discussion of the pros and cons of the virtual water trade. The concept is well embedded in the Dublin Principles and other international policy documents that have promoted the economic aspects of water. It rests indeed on the further development of global trade liberalization by developed and developing countries to encourage water-scarce countries to increase imports of agricultural water-intense products and for water-rich regions to grow water-intense products for export worldwide.[54] This is viewed as conducive to increased economic efficiency in the agricultural sector. However, it is clear that virtual water trade should also integrate the social and environmental

country are produced in that country, the water footprint consists both in the use of domestic water resources and use of water outside the borders of the country which implies quantifying the flows of virtual water leaving and entering the country. India (13 per cent), China (12 per cent) and the United States (9 per cent) have been found to be the largest consumers of the global water resources. See Hoekstra and Chapagain, note 43 above, 43.

51 *Ibid.*, at 46.

52 On some of the broader, regional implications, see, e.g., Saila Parveen and I.M. Faisal, 'Trading Virtual Water between Bangladesh and India: A Politico-economic Dilemma', 6 *Water Policy* 1 (2004).

53 See, e.g., M. Dinesh Kumar and O.P. Singh, 'Virtual Water in Global Food and Water Policy Making: Is There a Need for Rethinking?' 19 *Water Resources Management* 759 (2005).

54 The World Bank suggests that the lack of market access for agricultural commodities in high-income countries and the agricultural subsidies in developed countries contribute to supporting water-intensive activities. See World Bank, note 48 above.

values of water. Only in this sense can the concept be viewed as a means of using international and regional trade to forward sustainable water management and conservation goals. Since international trade presently involves a significant part of products whose production is water-intensive, it is important to assess whether the international trade rules allow for the integration of the social and environmental values of water when trade in water-intensive commercial crops is concerned. There has been little research on this topic, and the following section aims to provide some initial thoughts on whether the international trade framework takes into account the water content of agricultural products.

2. *Relationship with the international trade rules*

Since the most important virtual water exports and imports relate to agricultural products, it is within the WTO framework that the link between the emerging discourse on virtual water trade as a means to achieve more sustainable water use and the international trade rules appears. Trade in virtual water can be discussed in relation to several aspects of the WTO work programme that is currently being negotiated. This section focuses on some of the issues arising under the Agreement on Agriculture, which came into force on 1 January 1995.[55] The agreement's broad scope includes basic agricultural products (such as wheat, milk or live animals); products derived from them (butter or meat); and all processed agricultural products.[56] It also applies to fibres such as cotton, wool and silk, and raw animal skins destined for leather production, but excludes fish and fish products as well as forestry products.

While the original GATT provisions on agricultural trade allowed certain non-tariff measures such as import quotas and the use of export subsidies normally prohibited for industrial products, the Agreement on Agriculture sets specific commitments by WTO member states to improve market access and reduce trade-distorting agricultural subsidies. Member states commitments are to be implemented within a six-year period beginning in 1995 for developed countries, and a ten-year period for developing countries.[57]

The Agreement on Agriculture prohibits all agriculture-specific, non-tariff measures, and provides for the eventual reduction in tariffs for agricultural products.[58] The tariff package agreed upon during the Uruguay Round requires the replacement of agriculture-specific non-tariff measures with tariffs that allow equivalent levels of protection.[59] The Schedules of Commitments of individual

55 Generally, see J.A. McMahon, *The WTO Agreement on Agriculture: A Commentary* (Oxford: Oxford University Press, 2006).

56 Article 1, Agreement on Agriculture, note 28 above. See also Annex 1.

57 *Ibid.*, Art. 1(f).

58 *Ibid.*, Arts 4.1 and 4.2. Measures that are to be converted into tariff equivalents are listed in footnote 1 to Art. 4.2.

59 Tariffication does not apply to primary agricultural products, which can impact both goods that are for domestic consumption and those for export purposes. *Ibid.* Annex 5.

WTO member states now include tariffs on virtually all agricultural products traded internationally, meaning that virtually all agricultural protection is currently in the form of tariffs.[60] Since the process of converting quotas and other non-tariff barriers into tariffs might reduce market access where the quota had been small and tariffication produces a high tariff, each member state is required to include minimum and current access commitments for all tariffed products in its Schedules.[61]

WTO member states have also undertaken binding obligations with respect to the reduction and/or elimination of trade-distorting agricultural subsidies in order to promote international trade and improve market access.[62] Subsidies exist under the WTO where 'there is a financial contribution by a government or any public body within the territory of a Member' or if 'there is any form of income or price support'. Subsidies for the agricultural framework are provided for both export and domestic policy purposes. Under the agreement, the use of export subsidies is now prohibited unless they are subject to product-specific reduction commitments within the limits specified in the schedule of the WTO Member concerned.[63] Where they are listed, the agreement requires WTO members to cut both the amount of money they spend on export subsidies and the quantities of exports that receive subsidies.[64]

Reforming the current agricultural domestic support system in developed countries has been one central goal of the Doha Round negotiations. The Agreement on Agriculture requires the reduction of all domestic support that is considered trade-distorting. Trade-distorting measures fall into what is referred to as the Amber Box.[65] These measures are subject to reduction commitments as specified in the schedule of each WTO member providing such support. Article 6

60 For developed countries, the average reduction was to be 36 per cent over six years, whereas for developing members the average reduction was to be 24 per cent to be implemented over a ten-year period. Least developed countries were not required to undertake any reduction commitments, although they were required to bind their tariffs.

61 Further, McMahon, note 55 above, 50.

62 Commitments on subsidies are also found in the Agreement on Subsidies and Countervailing Measures, Annex 1A to the Marrakesh Agreement establishing the World Trade Organization and Final Act, 15 April 1994 and in the GATT 1994. *See* in particular GATT 1994, note 5 above art. XVI. Agricultural subsidies are dealt with in other fora than the WTO, including the Organization for Economic Co-operation and Development (OECD) and the Food and Agriculture Organization (FAO).

63 Agreement on Agriculture, note 28 above, Arts. 3.3, 8 and 10. See also Arts. 9.2(b) and 9.4.

64 Taking averages for 1986–1990 as the base level, developed countries agreed to cut the value of export subsidies by 36 per cent over the six years starting in 1995 (24 per cent over 10 years for developing countries). Developed countries also agreed to reduce the quantities of subsidized exports by 21 per cent over the six years (14 per cent over 10 years for developing countries). Least-developed countries do not need to make any cuts.

65 Commitments regarding the reduction of Amber Box measures are expressed in terms of a total aggregate measurement of support (Total AMS), which includes all product-specific support and non-product-specific support as one single figure. See Agreement on Agriculture, note 28 above Arts. 1(a) and 3.

of the agreement exempts a number of domestic support measures from reduction commitments.[66] It allows, and excludes from reduction commitments, domestic support measures with minimal impact on trade, which are referred to as Green Box policies. For measures to be exempt, they must have no, or little trade-distorting effects or effects on production,[67] and must be provided through a publicly funded government programme not involving transfers from consumers by charging consumers higher prices.[68] Among these are expenditures under environmental programmes, as well as infrastructural services including water supply facilities.[69]

How is the question of trade in virtual water to be addressed in the framework of the Agreement on Agriculture? For the purposes of furthering the sustainability of trade in virtual water, the WTO legal regime should take into account the water-intensive nature of agricultural products as well as more generally the needs of water-scarce member countries. Generally speaking, there are limited provisions in the Agreement on Agriculture on the need to address non-trade concerns, including those that are the closest related to water, namely food security[70] and environmental protection. The preamble to the agreement notes in this regard that:

> Commitments under the reform programme should be made in an equitable way among all Members, having regard to non-trade concerns, including food security and the need to protect the environment; having regard to the agreement that special and differential treatment for developing countries is an integral element of the negotiations, and taking into account the possible negative effects of the implementation of the reform programme on least-developed and net food-importing developing countries.

While the agreement appears to grant special and differential treatment in relation to development, with attention given to the particular situation and problems of least-developed countries,[71] this classification does not take into account the differences in water needs of various member states. The negotiations for continuing the reform process, which were incorporated into the broader negotiating

66 *Ibid.* Annex 2, paras. 2–13.

67 *Ibid.* Annex 2, para. 1.

68 *Ibid.* Annex 2, para. 1(a)–(b).

69 *Ibid.* Annex 2, para. 2(g).

70 See, e.g., *Kevin R. Gray*, Right to Food Principles *vis-à-vis* Rules Governing International Trade 10 *British Institute of International and Comparative Law* (2003).

71 Preamble to the Agreement on Agriculture, note 28 above. Special and differential treatment takes the form of increased flexibility for developing countries to implement reduction commitments with least-developed countries not required to undertake any such commitments. Article 20 permits members to take the non-trade concerns of developing countries into account in the negotiations. Special and differential treatment is also acknowledged in the Doha Declaration as necessary to 'enable developing countries to effectively take into account their development needs, including food security and rural development'.

agenda set at the 2001 Ministerial Conference in Doha, Qatar,[72] have taken note of non-trade concerns, such as environmental protection, food security and rural development reflected in the negotiating proposals already submitted.[73] But it is not likely that water scarcity is one of the non-trade concerns that has been directly mentioned. At any rate, trade negotiations on agriculture were suspended in 2006 after Ministers from the G-6 countries (European Union, United States, Japan, Australia, Brazil and India) failed to reach agreement on the depth and extent of tariff and subsidy cuts in agriculture.[74]

In renewing the market access regime, member states would therefore need to take into account not just trade factors such as competitiveness but also the particular requirements of water-intensive agricultural products. In this manner, tariff levels and minimum access would be determined in light of water conservation requirements. This would mean that market access of water-intensive products to exporting countries would be significantly reduced and importation of water-intensive products by water-scarce countries would be increased.[75] Tariff levels and minimum access for water-intensive products may also be determined by taking account of yearly and seasonal fluctuations of water resources availability. If such a division of labour that takes into account water factors reaches optimum conditions, international trade in water-embedded products might have positive impacts on environmental and other policy objectives related to the efficient and sustainable distribution of water resources. It would allow scarce water resources in water-poor countries to be used for basic human and other needs.

With respect to the regulation of domestic support measures, some agricultural subsidies can have harmful effects on water resources.[76] Irrigation subsidies, which are arguably those that encourage the most waste, pollution and contamination of fresh water on the largest scale, can lead to excessive water and sponsor the planting of water-intensive crops.[77] Examples include the planting of water-intensive crops like cotton, alfalfa and rice in desert-like climates such as California's Central Valley or the drying of the more shallow tube wells and the destruction of

72 *Ibid.* Art. 20.

73 See further the WTO gateway on agriculture, available at www.wto.org/english/tratop_e/agric_e/negoti_e.htm.

74 See further, International Centre for Trade and Sustainable Development (ICTSD), Agriculture Negotiations at the WTO: Developments and Setbacks since the Hong Kong Ministerial Conference, Quarterly Intelligence Report No. 14 (Geneva: ICTSD, August 2006), available at www.agtradepolicy.org/output/resource/Ag_Negs_Aug06_Final.pdf.

75 In this sense, Katsuhiko Mori, 'Virtual Water Trade in Global Governance', in Hoekstra, note 42 above, 119, 120.

76 On the adverse environmental effects of agricultural subsidies, see, e.g., Antonio La Vina *et al.*, *Reforming Agricultural Subsidies* (Washington D.C.: World Resources Institute, 2006); Richard Perkins, *Sustainable Development Should be the Objective for a New Agreement on Agriculture* (WWF, 2001). For an in-depth discussion, *see* Bernasconi-Osterwalder, note 79 above, 208–9.

77 *Ibid.* 210–13. See, e.g., André De Moor and Peter Calamai, *Subsidizing Unsustainable Development: Undermining the Earth with Public Funds* (1997).

freshwater aquifers in India due to massive water pumping for irrigation purposes.[78] This raises the question of whether the international rules on subsidies can be effective in reducing and eliminating agricultural water subsidies that might promote unsustainable virtual water trade, as well as in allowing for subsidies that might be beneficial for these purposes.[79] It appears that in the WTO framework, whether subsidies are environmentally beneficial or not is not taken into consideration by the international trade rules, which only address trade-distorting subsidies. Water-related subsidies, and in particular those that relate to the water-intensiveness of products, are not addressed at all. Since the value of water embedded in a product is not properly recognized and reflected in trading prices for agricultural commodities, it has been suggested that domestic support measures for virtual water trade could be placed in the Green Box category.[80] The classification system of different measures as trade-distorting or not is however particularly nebulous,[81] and according to one author 'may impede an efficient reduction of environmentally harmful water-related subsidies for various reasons'.[82] The existing rules regarding subsidies under the Agreement on Agriculture could potentially play an important role in the effective reduction or elimination of agricultural subsidies that promote, for instance, the growing of water-intensive crops in a water-scarce region. However, for this purpose, the WTO framework would have to adequately allow for non-trade considerations to be taken into account, particularly those regarding protection of water resources.

CONCLUSION

This chapter has attempted to draw out some of the concerns that are emerging with respect to a growing international trade in water resources, whether water is traded directly or as part of the production of traded goods. With trade in water being put forward as one answer to the increasing threat of global water scarcity, the discourse on water still appears dominated by the policies of the international and trade institutions, with the social and environmental values of water often put to the side. The perception of water as a good falling under the international trade rules only strengthens the commodification of water resources over the past two decades and that has largely informed recent water sector and water law reforms in countries such as India. In turn, the legal treatment of water under the international trade rules becomes one additional means through which water can

78 These examples and others are mentioned in Nathalie Bernasconi-Osterwalder, 'Water, Agriculture, and Subsidies in the International Trading System', in Brown Weiss, Boisson de Chazournes and Bernasconi-Osterwalder, note 3 above, 212–13.

79 *Ibid.*, at 207.

80 Mori, note 75 above, 120.

81 In particular, there is no institutional organ supervising the accuracy of a notification of a subsidy as a Green Box subsidy.

82 Bernasconi-Osterwalder, note 78 above, 220.

be more easily treated as an economic and tradable good at the national level, including when drinking water is concerned, and reflects the growing influence of the international economic institutions on water policies in the South.

Trade in water resources presently falls under the WTO framework as the primary forum dealing with these questions. From the point of view of states wishing to protect domestic water resources from export, it is clear that once water enters into commerce, it will be very difficult for it not to fall under the GATT rules. As they stand, these rules do not integrate sufficiently non-trade considerations, in particular those related to water. Greater attention should thus be given to the way in which the GATT exceptions, especially Article XX, could be used to allow for states to adopt restrictive measures on trade in water for social and environmental reasons. On the other hand, since the current case-law does not seem to fully provide protection for states wanting to include water export ban in domestic water conservation policy, it will be important for states wishing to curtail the diversion of bulk water resources from export to put in place national conservation strategies.

While from the point of view of importing countries, especially those water-scarce ones, the possibility to import large amounts of freshwater is certainly attractive as one way to relieve pressure on domestic water problems, it cannot be used as a substitute for the adoption and implementation of national policies on sustainable water management. Conflicts might emerge under the WTO framework between exporter states, which have plentiful water resources on their territory but may wish to restrict exports in order to protect national water resources from over-exploitation, and importer countries, which lack sufficient or adequate freshwater resources and wish to import water.

While the likely impacts of implementing trade obligations on water management will become greater as liberalization on trade advances, greater trade liberalization and market access can be viewed as having some positive outcomes when used in the virtual water trade debate, since declining water availability in some regions may be offset by openness of agricultural markets. Virtual water trade is seen as one solution to enable water-scarce countries to save important quantities of freshwater domestically if it is carried out in a sustainable manner. Nevertheless, the water content of agricultural produce and the water situation of member states have not been addressed in the framework of the current rules on international trade of agricultural goods. This might impede national conservation strategies put in place to further sustainable virtual water trade, lead to over-exploitation of water resources and increase in the degradation to water limits even in water-rich countries. Consequently, adverse environmental and social impacts arising from increases in agricultural market access must be addressed and agricultural rules must more generally be negotiated with the aim to promote sustainable development.

The traditional conflicts which have occurred in the trade regimes between economic and environmental goals thus take on a new dimension when water is concerned, begging the question of whether certain considerations specific to the problem of freshwater resources should constrain the market perspective.

Existing gaps could be filled by entrenching the role of human rights and environmental concerns in the interpretation of WTO agreements, incorporating right-to-water language into WTO requirements, and facilitating better integration of social and environmental concerns into negotiations and implementation of WTO agreements. It is crucial to ensure that '[a]greements concerning trade liberalization should not curtail or inhibit a country's capacity to ensure the full realization of the right to water'.[83]

83 Committee on Economic, Social and Cultural Rights, General Comment No. 15: The Right to Water (Arts. 11 and 12 of the International Covenant on Economic, Social and Cultural Rights), UN Doc. E/C.12/2002/11 (2002), para. 35.

V. The role of the World Bank in water law reforms

Andrés Olleta

INTRODUCTION

The activities of international financial institutions (IFIs) in developing countries have been the subject of intense criticism in recent years.[1] Detractors of the International Monetary Fund (IMF) and the development agencies forming the World Bank Group (WBG) have regularly denounced the inadequacy and inefficiency of the policies recommended by IFIs. Indeed, the disappointing performance of the IMF and WBG in a number of recent economic and social crises around the world has led to questions about the work of these institutions and their relevance on the international scene. In particular, in the wake of the 2001 economic, financial and social collapse in Argentina, the policies of IFIs were scrutinized in such a way that their staff was prompted to assume their share of self-criticism.[2]

Criticism is generally centred on both the content of the policies that IFIs support and the way in which they reach borrowing member states. On the one hand, the policies are usually described as a model, a single set of measures that is invariably forwarded to different borrowing countries regardless of the specificities of each state. It has been contended, as a result, that the measures suggested are inadequate to redress the specific problems of certain members. On the other hand, IFIs are simultaneously perceived as taking advantage of their high bargaining power throughout loan negotiations to impose modifications to core domestic policies of states, thus interfering in the internal affairs of borrowing member countries.[3]

This chapter analyzes the new water policies that the World Bank and regional multilateral development banks (MDBs) have developed over the last couple of

1 See, e.g., Joseph E. Stiglitz, *Globalization and its Discontents* (New York: W.W. Norton, 2003).

2 Christina Daseking *et al.*, Lessons from the Crisis in Argentina (Washington D.C.: IMF Occasional Paper No. 236, 2005) and International Monetary Fund/Independent Evaluation Office, The IMF and Argentina, 1991–2001 (Washington D.C.: IMF, 2004).

3 Jean-Marc Sorel, 'Sur quelques aspects juridiques de la conditionalité du FMI et leur conséquences', 7 *European J. Intl L.* 1, 19–20 (1996).

decades and the ways in which their implementation in developing countries has been encouraged. The policy suggestions of the World Bank, in particular, have had an impact on the water sector of indebted countries and therefore on the lives of millions of persons around the world.

The chapter focuses on water sector reforms that have been implemented at the national level with World Bank approval. It is divided in two main sections. Section A briefly explains the mission and functioning of the WBG and the MDBs, describing their lending mechanisms and shedding some light on how they succeed in introducing their preferred policies in the domestic legal and regulatory frameworks of borrowing countries. Section B takes a look at three main policies that the World Bank favours in water management and water services provision, and analyzes the domestic implementation of these policies in certain jurisdictions. The conclusion attempts to provide a comprehensive picture of how the World Bank has moulded the current water sector reforms that are noticeable worldwide.

A. The role of the World Bank and the regional development banks in water sector reforms

1. Overall mandate

The World Bank and the MDBs are a set of diverse international institutions dedicated to providing financial resources to member countries for specific purposes. They differ in mission, and geographical scope, as well as in the goals that their lending services pursue; however, they all operate through the allocation of resources to members in need, under certain conditions. The main mission of the WBG – and, by extension, of the regional MDBs as well – is to reduce poverty and improve living standards through financial and technical assistance to developing countries. The World Bank's stated mission is to provide members with low-interest loans, interest-free credit, and grants in an effort to help them meet the Millennium Development Goals, one of which is 'to halve the proportion of people who are unable to reach or to afford safe drinking water'.[4] Later, in the Johannesburg Declaration on Sustainable Development, it was additionally agreed to reduce by half the proportion of people without access to basic sanitation.[5] Because of the clear link between access to water and human development,[6] the World Bank has reinforced its participation in water and sanitation projects since the 1980s, and as it will be shown below, it has quickly become 'one of the most, if not the most important actor in the global water sector, be

4 United Nations General Assembly Resolution 55/2, United Nations Millennium Declaration, UN Doc A/RES/55/2 (2000), para. 19.

5 Johannesburg Declaration on Sustainable Development, 4 September 2002 UN Doc. A/CONF.199/20, para. 8 at 11 and para. 25 at 27.

6 United Nations Development Programme, *Human Development Report 2006 – Beyond Scarcity: Power, Poverty and the Global Water Crisis* (New York: UNDP, 2006).

it in terms of financial aid or in terms of general policy-making in the developing countries'.[7]

The MDBs and the World Bank share the same goals and have a similar internal organization and functioning structure, though they have their own independent legal and operational status. Still, with one and the same mandate and a considerable number of members in common, the MDBs and the World Bank maintain a high level of cooperation and are seen as working towards the same ends in water management and water services provision. The three policies that are described in Section B are an integral part of the water management strategies of the World Bank and of all MDBs, including the Asian Development Bank (ADB) of which India is a member.[8]

2. *Common traits in the organization, functioning and lending mechanisms of multilateral development banks*

The internal organization, operation and lending mechanisms of the World Bank and the ADB follow a pattern that is common to all International Financial Institutions. Even if these institutions are international organizations of broad membership, their multilateral character is severely restricted by their organizational structure and the decision-making mechanisms that they have adopted. The IMF, for example, holds a basic share capital formed by the contributions of its members. The contribution of each country entails membership to the organization and grants to the state a number of voting rights that is the same for all members. Yet, extra votes are granted to those that subscribe additional financial contributions to the organization, thus implying that decision-making power is in the end proportional to shareholding, which in turn is a reflection of the economic power of a state in the international scene. Consequently, these banks are literally in the hands of the most developed and powerful states, which have provided a bigger share of the capital of the institution.

The actual weight of borrowers (developing countries) in the governance of these institutions varies among them, but, comparatively speaking, the World Bank is the least balanced organization in this respect. This is odd, since the nature and operation of the biggest multilateral bank should not follow a management model that reflects the investment power of members. The need to change the composition of the World Bank's Executive Board and review the voting power of each country has been debated among governments and civil society for years, in an attempt to increase the influence of borrowing countries.[9] In any case it is unfortunate that the most influential bank in the water sector is

7 Matthias Finger and Jeremy Allouche, *Water Privatisation: Trans-National Corporations and the Re-Regulation of the Water Industry* 62 (London and New York: Spon Press, 2002).

8 For an overview of the ADB's water policies, Asian Development Bank, Water for All – The Water Policy of the Asian Development Bank (Manila: ADB, 2001), available at www.adb.org/Documents/Policies/Water/water-policy.pdf.

9 South Centre, Reform of World Bank Governance Structures (Geneva: South Centre, Analytical Note, 2007).

the one where the countries that are in greater need of financial resources – for improving water resources management and water services provision – have a lesser role in its governance. It would be more helpful if the World Bank followed the example of the Inter-American Development Bank, as its Charter at least foresees that the majority of the shares that form the capital of the institution are held by the borrowing countries as a group.[10]

The traits of the lending mechanisms in these institutions are uniform. Simply put, the process invariably begins with the borrower requesting the necessary funding for the realization of a given project or the implementation of a policy reform plan. The candidate submits a formal proposal for its feasibility to be assessed – known as Letter of Development Policy in the World Bank. Depending on the purpose of the loan, the economic, financial, social and/or environmental viability of the proposal is especially evaluated by those organs within the institution that are in charge of approving the credit. The terms of the repayment will also vary according to these factors. What is a constant in the financial assistance provided by these institutions is the practice of attaching certain conditions for green-lighting the loan, and then monitoring the implementation of the measures that have been recommended.

3. Conditionality

In exchange for the disbursements, borrowers agree to fulfil a set of conditions that the banks bring to the loan negotiation rounds, and also accept that their staff monitors the implementation of the project or policy reform plan that is being funded. Through this practice, MDBs succeed in introducing policy reforms to the water legal and regulatory frameworks of borrowing countries. They are thereby able to incorporate their preferred policies into domestic agendas by attaching them to the loans as conditions that must be fulfilled for the loan to achieve its goal. This practice is known as conditionality, and its main function – according to the World Bank – is to ensure that the commitments undertaken by the country in its formal request for financial resources are respected. The rationale would thus be 'the Bank's due diligence obligation to ensure that its resources are used effectively and responsibly by the borrowing country'.[11]

While it is now clear that conditionality refers to the policies that IFIs expect a member to follow in order to be able to use its general resources,[12] this is not a formal definition of conditionality that can be found in the World Bank's legal framework or operational policies. Indeed, conditionality had not been expressly foreseen in the Articles of Agreement of the International Bank for Reconstruction

10 Agreement Establishing the Inter-American Development Bank, Washington D.C., 8 April 1959, Article VIII, Section 4(b).

11 World Bank/Operations Policy and Country Services, Review of World Bank Conditionality: Modalities of Conditionality, SecM2005-0390/1, 2005, Executive Summary, para. 2.

12 Joseph Gold, Conditionality 1 (Washington D.C.: IMF Pamphlet Series No. 31, 1979).

and Development (IBRD) or of the International Development Association (IDA). However, conditionality seems to have evolved as a practice developed by the IMF's Executive Board, which apparently followed a peculiar interpretation of Article I(v) and of Article V, Section 3(a) of the IMF's Articles of Agreement, which state that the allocation of resources by the institution only proceeds under adequate safeguards.[13] In the case of the World Bank, conditionality emerged simultaneously with the World Bank's decision to financially support policy reform in addition to infrastructure development. Indeed, in the beginning World Bank loans were exclusively destined to finance infrastructure projects that contributed to reach higher standards of development, such as the construction of roads or dams for power generation (project or investment lending). The use of conditionality in this kind of loan is currently discouraged by the World Bank.[14] Conditionality in the World Bank surfaced together with the creation of a second category of loans (policy lending), formed by those disbursements that are destined to 'help a borrower achieve sustainable reductions in poverty through a programme of policy and institutional actions that promote growth and enhance the well-being and increase the incomes of poor people'.[15] It was affirmed that the legal basis for policy lending could be found in the World Bank's Articles of Agreement, which foresee lending under special circumstances. Both the Articles of Agreement of the IBRD and the IDA provide that 'loans made or guaranteed by the World Bank shall, except in special circumstances, be for the purpose of specific projects of reconstruction or development'.[16] Originally, therefore, only project lending was explicitly regulated in the Articles of Agreement.

At the same time, other provisions in the World Bank's Articles were interpreted as giving way to the practice of attaching conditions to loans. As the World Bank's Legal Vice-Presidency explains:

> the Bank's policy-based loans must be in accordance with the "purposes" identified in the Articles. Thus, where certain policy and institutional actions and measures are considered necessary for an operation to achieve the World Bank's development purposes, these "conditions" may be validly justified under the Articles (IBRD and IDA Articles, Article I).[17]

13 Joseph Gold, 'Use of Fund's Resources', in Margaret Garrisen de Vries and J. Keith Horsefield, *The International Monetary Fund 1945–1965, Volume II: Analysis* 522–3 (Washington D.C.: International Monetary Fund, 1969).

14 World Bank/Operations Policy and County Services, Disciplined Use of Conditionality in Lending Operations (Washington D.C.: World Bank, Staff Guidance Note, 2004); and World Bank/Operations Policy and Country Services, Policy Conditions in World Bank Investment Lending: A Stocktaking (Washington D.C.: World Bank, Board Report, 2006).

15 World Bank, Development Policy Lending, OP 8.60, August 2004, para. 2.

16 IBRD Articles of Agreement, Article III, Section (4)(vii) and IDA Articles of Agreement, Article V, Section (1)(b).

17 World Bank/Legal Department, Review of World Bank Conditionality: Legal Aspects of Conditionality in Policy-Based Lending, SecM2005-0390/2, 2005, Executive Summary, para. 4.

Additionally, it has been underlined that 'the IBRD Articles recognize that the institution may provide financing for productive purposes on "suitable conditions", while under its Articles, IDA may provide financing on appropriate terms (IBRD Articles, Article I (ii) and IDA Articles, Article V, Section 2(b))'.[18]

Thus, in the early 1980s, the World Bank started to incorporate conditionality into its policy lending programmes, mirroring the use of conditionality by the IMF in its own structural adjustment programmes. Conditions at first addressed only short-term macroeconomic imbalances and economic distortions. However, by the early 1990s, especially after the World Bank initiated the practice, conditionality was spanning reforms that emphasized improvements in public sector governance. More specifically, loans began supporting governmental efforts to strengthen public financial management, fiduciary arrangements, and public expenditures. To achieve these ends, the IFIs in unison set out a range of distinct policies that included trade liberalization, deregulation, fiscal austerity and privatization – a set of measures under the general umbrella of public sector reform.

Currently, in the World Bank context, there is an express mention of conditionality in the form of paragraph 13 of Operational Policy (OP) 8.60, which identifies three essential requirements for the World Bank to make disbursements in a policy-based loan to a country:

(a) maintenance of an adequate macroeconomic policy framework;
(b) implementation of an overall programme in a manner satisfactory to the World Bank; and
(c) compliance with critical policy and institutional actions.[19]

In the World Bank context, therefore, conditionality can be defined as 'the set of conditions that must be satisfied for the Bank to make disbursements in a development policy operation'.[20] Doubts over the legality and legitimacy of conditionality waned as the acquiescence of members reinforced conditionality as a practice until actual changes were made to the statutory framework of the World Bank in order to incorporate it formally.

B. Multilateral development banks: Policies on water

This section deals with the three main water sector reforms that have reached developing countries through conditionality and technical assistance provided by the World Bank, as well as the ADB in the case of India. It also seeks to discern why these institutions hold these policies as the sole way to meet the MDGs regarding universal access to drinking water and sanitation.

The water sector reforms sponsored by the World Bank and other MDBs are

18 *Ibid.*
19 *Ibid.* Introduction, para. 9.
20 World Bank/Operations Policy and Country Services, note 11 above, para. 9.

the corollary of a global campaign for the formulation of new policies that recognize the economic value of water. The movement was particularly noticeable at a series of international meetings that focused on reforming water resources management to address increasing scarcity of freshwater. A common ground was found on the principles that would guide specific local action in water governance. The participation of the World Bank in these meetings made possible the pairing of the new water management principles with its own policies for domestic public sector reform, in particular the reduction of the role of the state.

The World Bank and the regional MDBs believe that by stressing the importance of water for the productive capacity of the economy, it is possible to achieve a better management of the resource and thus address the difficult situation of many countries facing water scarcity around the world. In this sense, they hold that if market forces decide where water goes according to its economic value, better management of the resource is possible. To them, 'the absence of an economic perspective in the past explains existing unsustainable uses of water'.[21] Therefore it was necessary that water transitioned 'from being a common good and a public service into a commodity that [was] increasingly being managed according to economic principles'.[22] The World Bank claims that the competing uses of water can be reconciled through a holistic approach to water management in which the market allocates shares, and that the recognition of water as an economic good allows for a more sustainable exploitation of the resource, prevents its wastage and pollution, and leads to a financially sustainable provision of water services through the implementation of full cost recovery mechanisms.

The 1992 International Conference on Water and the Environment (Dublin Conference) was the first venue where the economic value of water was recognized and developed into specific policy recommendations. The meeting was convened 'under the aegis' of the United Nations Conference on Environment and Development (UNCED) by the World Meteorological Organization (WMO) on behalf of the more than twenty bodies and agencies of the UN system which operated water-related programmes.[23] The Dublin Statement issued at the end of the Conference included messages that were expected to form the basis for the discussion of water issues at UNCED.[24] In effect, the summary report from the Dublin Conference was directed to the UNCED 'as a contribution to the furtherance of the integrity of the relationship between mankind and the environment'.[25]

21 Philippe Cullet, 'Water Law – Evolving Regulatory Framework', *in* Philippe Cullet, Alix Gowlland Gualtieri, Roopa Madhav and Usha Ramanathan (eds.) *Water Law at the Crossroads – National and International Perspectives with Special Emphasis on India* 26 (New Delhi: Cambridge University Press, 2009).

22 Finger and Allouche, note 7 above, xiii.

23 Draft Outline for the International Conference on Water and the Environment, August 1990 (on file with the author).

24 Dublin Statement on Water and Sustainable Development, International Conference on Water and the Environment, Dublin, 31 January 1992.

25 *Ibid.* Objectives, item 4.

Therefore, the principles included in the Dublin Statement were further elaborated in Rio by the WMO and the United Nations Educational, Scientific and Cultural Organization (UNESCO), resulting in Chapter 18 of Agenda 21.

The Dublin Statement was conceived to replace the 1977 Mar del Plata Action Plan and thus provide 'the main thrusts for policy and action, both internationally and nationally [. . .] for the remainder of this decade and for the start of the next century'.[26] It became a milestone in modern water management policy, since it contained four principles – among them, the recognition of the economic value of water – that would become the basis for the global water sector reforms that would ensue, since the event had been held to answer the call 'for fundamental new approaches to the assessment, development, and management of freshwater resources'.[27]

The Dublin Principles buttress the three main reforms proposed by the World Bank on the water sector: Integrated Water Resources Management (IWRM), decentralization of water management, and privatization of water services. A year after the Dublin Conference and UNCED, the World Bank in turn released its own Policy Paper for water resources management, which elaborated on the Dublin Principles and described the three policies just mentioned.[28] The subsections below explain what is meant by IWRM, decentralization of water resources management and privatization of water services according to the World Bank, and what these concepts imply for water law reforms undertaken at the national level.

1. Integrated water resources management

Integrated Water Resources Management (IWRM) is a new approach to water management that aims at adapting it to water scarcity. It is meant to improve efficiency in management by finding a balance in the allocation of available water among its different uses – domestic, industrial, agricultural, recreational, navigation – according to the most productive use that can be made of the resource.

The recognition of water as an economic good implied in IWRM can be considered as the main contribution of the World Bank to the latter concept, and it is key to understanding water sector reforms and the controversies that they have sparked in countries like India. It might be argued that the recognition of water as an economic good only implies that it is a resource subject to alternative uses that cannot always all be satisfied. This, in the case of water, is an uncontested fact. However, using this argument to treat water as a private good and consequently make allocation and pricing of water dependent on market forces, like the World Bank does, is a very different issue. That is the true element

26 Draft Outline, note 23 above, Background, para. 5.

27 Dublin Statement, note 24 above, Introduction.

28 World Bank, Water Resources Management: A World Bank Policy Paper (Washington D.C.: World Bank, 1993).

of discord that has opposed MDBs and their detractors in policy making for the water sector.

Critics of the MDBs believe that if water becomes a private good, the state would lose control over it and it would go only to those willing and able to pay for it, thus jeopardizing the access to water of the poor. On the contrary, they insist that water be treated as a public good, and believe that its management demands the presence of extra-market forces – most notably, state intervention – in order to see water management serve social objectives effectively and efficiently as well. Aware that access to water services is not only a basic need, but a human right, the concern of this group is that if water allocation was left in the hands of the market, the state would be incapable of fulfilling its human rights-related duties and of guaranteeing access to water. Consequently, water should in priority be considered a social good, and the availability of a minimum amount of water for the satisfaction of elemental needs should be assured by the government to all, regardless of their capacity to pay for it.[29] The World Bank, in turn, acknowledges that market failures exist in water management activities, and that government action or other forms of collective actions are necessary to face inefficiencies that may emerge. Yet it has called for distinguishing areas where market forces could, in their view, be beneficial, such as creating markets for water rights or separating the provision of network facilities from the generation and maintenance of water services. In their view, while the construction of water supply, sanitation and irrigation networks has a significant public good nature, the provision of water services (delivery of drinking water, sanitation and irrigation water) can be subjected to some competitive market forces, such as allowing private operators to bid for a concession contract.[30] Principle 4 of the Dublin Statement – the recognition of water as an economic good – was a sort of compromise between the views of both groups.[31]

The new, comprehensive approach to water management proclaimed in the Dublin Statement quickly became known as IWRM, and the World Bank had an important role in spreading the concept. The institution was quick to state that it would 'support national and international initiatives of member governments and international initiatives in implementing integrated water resources management. [. . .] Its economic and sector work, lending, technical assistance, and participation in international initiatives will aim to promote policy and regulatory reforms; institutional adaptation and capacity building'.[32] In the World Bank's policy papers, IWRM is introduced as the way to achieve optimal management of water. IWRM is presented as the means to reach a series of both social and economic goals, while also ensuring sustainable exploitation of the resource and

29 On the adoption of policies granting a free amount of water to all as part of the realization of the human right to water, especially in South Africa and Belgium, see Chapter XI.

30 World Bank Policy Paper, note 28 above Appendix A at 81–85, in particular 85.

31 Christopher John Perry, Michael Rock and David Seckler, *Water as an Economic Good: A Solution, or a Problem?* 1 (Colombo*: International Irrigation Management Institute Research* Report 14, 1997).

32 World Bank Policy Paper, note 28 above, 65.

participatory decision-making. To the World Bank, it is a remedy to sectoral approaches to water management 'that [have] neglected interdependencies among government agencies and jurisdictions' and to inefficient governance by the public sector, which has 'neglected financial accountability, user participation, and pricing while not delivering services effectively to users and to the poor in particular'.[33] Previous water management strategies, according to the World Bank, had unnecessarily worsened water shortage and aggravated the competition among the possible uses of water. According to the World Bank, 'observing the waste and inefficiencies that have resulted from the frequent failure to use prices and other instruments to manage demand and guide allocation, the new approach stresses the importance of using decentralized implementation processes and market forces to provide water services'.[34]

At the core of this new approach of the World Bank to managing water resources is 'the adoption of a comprehensive policy framework and the treatment of water as an economic good, combined with decentralized management and delivery structures, greater reliance on pricing, and fuller participation by stakeholders'.[35] IWRM is therefore linked to decentralization and can be taken as part of the World Bank's campaign for transferring water management responsibilities from state agencies to new actors, since the World Bank identifies centralized public water management and service provision with inefficiency.[36] In an effort to remedy this, the World Bank connects IWRM to decentralization and privatization as a way of pushing the state to the background, substituting sectoral water management in the hands of centralized institutions for integrated management and services provision in the hands of communities, user associations and private companies at the local level.

The World Bank pushes IWRM because of its strong economic component. First, IWRM enables the allocation of water by the market. The World Bank believes that this is the best way to make water available where it is needed, prevent wastage and ensure a more environmentally friendly management of water through a more rational exploitation of it. The Bank acknowledges that some form of remedial government actions would often be required, but the role envisaged for the state is only subsidiary.[37] Secondly, IWRM stresses cost recovery in water services provision, therefore allowing for the pricing of water as any other private good and for the utilization of full cost recovery as a means to make affordable the maintenance and expansion of improved water services networks.[38] As part of the implementation of full cost recovery, states have abandoned subsidized schemes of pricing in favour of charging users according to their

33 *Ibid.*, at 27.

34 *Ibid.*, at 23.

35 *Ibid.*, Executive Summary 10.

36 *Ibid.*, Appendix C at 100.

37 *Ibid.*, at 26.

38 World Bank, *World Development Report 1994: Investing in Infrastructure* 23 (New York: Oxford University Press, 1994).

actual consumption. This has in fact been deemed 'the main change in the policy adopted by the main World Bank's water specialists'.[39] The Bank claims that full cost recovery is the only way to ensure the financial viability and independence of the entities in charge of water services provision – public or private.

Ever since the Dublin Conference, IWRM has been included in important international declarations and documents. Agenda 21, the programme for action adopted at the UNCED, specified that IWRM is 'based on the perception of water as an integral part of the ecosystem, a natural resource and a social and economic good, whose quantity and quality determine the nature of its utilization'.[40] This phrase reflects once again a compromise between those for and against water management ruled by the market, at the same time making it clear that the availability of water in each jurisdiction is the main determining factor for decision making in water allocation issues. The UNDP soon followed and embraced this holistic new approach to water management, making it a pivotal part of its Safe Water 2000 decade.[41] It further stressed the conviction that competitive market pricing and allocation would improve efficiency in water management, reducing wastage, preventing environmentally harmful uses of water and thus maximizing the benefits that can be derived from it. Notwithstanding the relative normative value of such international statements, it is the World Bank input into IWRM that would have a more concrete impact on domestic water law reforms of borrowing countries.

The reforms that can be expected at the national level as a result of the introduction of IWRM in the water sector policy of indebted countries are varied. On the one hand, amendments to the domestic water legislation and institutional structure for water management are required in order to accommodate the implementation of the new policy. In this line, the first step has been the gradual abolishment of private ownership of water and the concentration of control over water in the hands of the state, which is then in charge of distributing water rights to applicants. Indeed, in order to have integrated management it is vital to prevent the fragmentation of management that the recognition of rights of ownership over water would cause. Instead, by becoming public trustee of the resource or declaring it within the public domain of the state, the state increases its control over water and is able to put in practice IWRM and other policies forwarded by IFIs. However, even if the state is still in charge of developing and implementing water governance, the growing participation of other actors in water management has determined, as it will be shown below, that the state has transitioned in recent years from manager to regulator.

39 Finger and Allouche, note 7 above, 76.

40 Agenda 21, Report of the UNCED, Rio de Janeiro, June 3–14 1992, UN Doc. A/CONF.151/26/Rev.1 (Vol. 1, Annex II), Chapter 18.

41 United Nations Development Programme, Safe Water 2000 3 (New York: United Nations Development Programme, 1990). See also United Nations, A Strategy for the Implementation of the Mar del Plata Plan for the 1990s (New York: United Nations Department of Technical Cooperation, 1991).

The distribution of water rights should be done following priorities clearly set in national legislations that are oriented to preserve the supply needed to satisfy vital domestic needs of the population. The conciliation of the different uses of water sought after by the authority that issues the rights should not be achieved at the price of impacting negatively on the population's access to water. This is particularly pertinent for those jurisdictions that have taken water allocation by the market to the next level by actually introducing markets for water rights. The trade of water rights is an initiative that IFIs have supported in the belief that it is an efficient method for redistributing the excess of water available to a right holder among those in the same basin that demand more water to cover their needs. Such is the case in Chile, Australia or the state of Colorado in the United States of America, among others.[42] How free these markets are depends on many concurrent factors: from the availability of water in each basin to the institutional maturity and capacities of the country that chooses this option, since in many jurisdictions the oversight and intervention of the state might still be necessary to prevent abuses from users in a dominant position.

Secondly, IWRM implies redesigning the institutional structure for water management. By coupling it with public sector reform ideas and decentralization, IFIs substitute sectoral approaches to water management deemed inefficient with integrated management in the hands of state organs at a lower level, while also welcoming other actors. Despite the complexity of a holistic approach to water management, communities and villages are trusted with new functions in the hope that efficiency in water management will increase if transferred to them, to user associations, and to the local private sector. The creation of independent regulatory authorities rounds up the institutional changes adopted at the national level to implement IWRM. These organs, in theory insulated from political interference,[43] are in charge of overseeing water management and perform a wide range of functions: from the regulation of the use of the resource and the distribution of available entitlements to different categories of users, to fixing water charges and tariffs.

Thirdly, IWRM is also about environmental and financial sustainability of water management. In the context of IWRM, both types of sustainability are seen as part of a single strategy, given that charging for consumption and having tariffs that reflect full cost recovery would be the first steps to conserving water and preventing wastage. Environmental concerns are noticeable, for example, in the obligation to produce environmental impact assessments as part of the procedure for applying for a water right, and also in how national water strategies now foresee the creation of water reserves and protected areas in order to ensure that sufficient water is preserved for the sake of biodiversity in watersheds. The concern for financial sustainability is noticeable in how the pricing of water

42 On the topic of water markets, see, e.g., Manuel Mariño and Karin E. Kemper (eds.) Institutional Frameworks in Successful Water Markets: Brazil, Spain and Colorado, USA (Washington D.C.: World Bank Technical Paper No. 427, 1999).

43 For the case of the Maharashtra Authority, Cullet, see note 21 above.

services takes central stage, and in how users are asked to contribute financially to develop infrastructure, particularly for the delivery of water for irrigation in rural areas. IFIs in this sense have succeeded in convincing states of the importance of full cost recovery in water management and water services provision. Though they acknowledge the necessity of assistance to the poor in the form of subsidies or similar strategies, they underline that an important reversal in public policy is needed, making users responsible for the operation and maintenance costs of service provision, which had up to then been heavily subsidized by the public sector. The World Bank in this sense insists that there is willingness to pay as long as there is accountability and transparency on the use of the funds, and users perceive the improvement in the quality of the service.

2. Decentralization

As part of the implementation of IWRM, and as a previous reform to the privatization of water services in urban areas, the World Bank fosters decentralization. It understands that states might be sceptical about delegating water management to an unregulated market; yet, it deplores the fact that they make water management heavily dependent on centralized command and control bureaucracies, and that they rely excessively on government agencies to develop, operate and maintain water systems.[44] To the World Bank, this scheme has proved to be inefficient in many developing countries and decentralization, understood as the delegation of managerial tasks to the lower level possible capable of handling them, is thus proposed as the solution.

The World Bank used to define decentralization as the transfer of political, fiscal and administrative powers to sub-national governments. Although decentralization is indeed a measure oriented to improve, through the reorganization of the public sector, the delivery and quality of tasks and services performed and provided by the state, that definition of decentralization is currently inaccurate in the context of water sector reforms: nowadays, decentralization implies the divestiture of the state of many of its functions in water management and services provision and, in many cases, their transfer to local authorities and user associations, with the participation of private companies. In the words of the World Bank:

> the principle is that nothing should be done at a higher level of government that can be done satisfactorily at a lower level. Thus, where local or private capabilities exist and where an appropriate regulatory system can be established, the World Bank will support central government efforts to decentralize responsibilities to local governments and to transfer service delivery functions to the private sector, to financially autonomous public corporations, and to community organizations such as water user associations.[45]

44 World Bank Policy Paper, note 28 above, 30.
45 *Ibid.*, at 15.

The Bank has thus updated its definition of decentralization, and it is currently understood as 'the transfer of authority and responsibility for public functions from the central government to intermediate and local governments or quasi-independent government organizations and/or the private sector'.[46]

The Bank regards decentralization as critical to its mission of poverty alleviation and achievement of the MDGs, since it considers that to accomplish them, it is vital that the state improves the performance of its public sector. In the view of the MDBs, an optimal public sector is one that delivers quality public services consistent with the preferences of the population and that fosters private market-led growth while managing fiscal resources prudently. Consequently, the World Bank has supported the implementation of programmes in the water sector that include not only political decentralization – the redistribution of functions among state organs in order to improve the provision of public services – but also financial decentralization – the collection of water charges at the district level and their use to improve the local infrastructure and the service delivery. After all, 'decentralized operations can only be successful if finances are also decentralized'.[47] The Bank argues that water charges matching actual operation and maintenance costs are essential for attaining financial sustainability, and that users will be more motivated to pay if they see their money being invested in increasing service quality. Decentralization as a process is not limited to the reorganization of water governance, ascribing new and more important functions to lower levels of the administration. On the contrary, it is an all-encompassing policy that divests the state of part of its role in water management and water services provision, incorporates new actors to the water sector, and brings new chances for addressing the World Bank's main concern for full cost recovery in services provision.

Decentralization is linked to both privatization and user participation. On the one hand, the decentralization of water services provision enables the private sector to distinguish profitable from non-profitable areas, and therefore to decide where to invest in case utilities are privatized. Given that the provision of water services to rural areas is not as profitable for multinational companies, the World Bank fosters the privatization of water services in urban areas and prefers decentralized public provision in rural areas, with the participation at the village level of small local companies in various capacities, if any. On the other hand, decentralization facilitates the participation of users in management and service provision at the local level in rural areas. The participation of users in the water sector has brought along the main legal reforms associated with decentralization. The World Bank has in this sense referred to the democratization of water management, a new approach to water management in rural areas which combines decentralization and user participation by making local user associations responsible for the management of their water resources, the distribution of irrigation water and, sometimes, even the provision of water services. This strategy is

46 World Bank Decentralization home page, available at http://go.worldbank.org/WM37RM8600.
47 World Bank Policy Paper, note 28 above Appendix E at 129.

palpable in the World Bank's participatory irrigation management (PIM) initiatives,[48] which are part of the World Bank's broad community-driven development (CDD) approach to development. PIM entails the involvement of irrigation users in all aspects and all levels of irrigation management, including the initial planning and design of new irrigation projects or improvements, as well as the construction, supervision, financing, operation, maintenance, monitoring, and evaluation of the system.

The idea of involving local executing agents and intended beneficiaries in a project's design, implementation and operation through decentralization is not new.[49] Several donors, including the UNDP, have been involved in community-managed rural water supply projects prior to the release of the 1993 World Bank Policy Paper on Water Management. However, it only became a central policy in the late 1980s, as MDBs promoted the divestiture of the state of many of its functions in an effort to prevent excessive public expenditure and increase efficiency in the delivery of public services.

CDD has increasingly been forwarded to developing countries in the last decade to improve rural water management and the access of rural communities to water and sanitation services, which are among the most urgent needs in the developing world. CDD is described by the World Bank as:

> an approach that gives control over planning decisions and investment resources to community groups and local governments. CDD programmes operate on the principles of local empowerment, participatory governance, demand-responsiveness, administrative autonomy, greater downward accountability, and enhanced local capacity.[50]

Over the last decade the World Bank has regarded CDD as an important strategy for effective poverty reduction and sustainable development, and has focused on lending to CDD programmes to reach local communities directly.[51]

The World Bank stresses that CDD is built on the involvement of civil society and user associations in decision making. However, the chance to participate, as necessary as it is, is of relative importance if unbiased information is not first properly shared with users, so that they are able to make the best decision possible. As certain analysts have pointed out, the basic function of user participation, at least in the World Bank rationale, 'seems to be to make economic and fiscal decentralization acceptable, in particular by (a) seeking the users' consensus on

48 Electronic Handbook on Participatory Irrigation Management at www.worldbank.org/wbi/pimelg/index.htm.

49 See, e.g., World Bank, OMS 1.12: Project Generation and Design – Local Involvement, 1978.

50 World Bank CDD Webpage at http://go.worldbank.org/24K8IHVVS0.

51 For insight into local development as key to decentralisation and improvement of public services, Louis Helling, Rodrigo Serrano and David Warren, Linking Community Empowerment, Decentralized Governance and Public Service Provision through a Local Development Network (Washington D.C: World Bank/CDD SP Discussion Paper No. 0535, 2005).

the overall project and (b) getting them to pay the increased fees at the local level'.[52] In other words, decentralization through CDD would in principle empower communities to choose the project to work on, to contribute to its design, and to build capacity so that users manage it themselves, but these positive points are usually accompanied by lack of information and, above all, by higher prices for services, or at least by the obligation of paying for infrastructure development. CDD projects are allegedly analyzed from an economic and social point of view before they are cleared, yet the mobilization of local revenues to finance local development that the World Bank regards as such an important foundation for sustainable empowerment and governance as well as for service provision can actually become a heavy burden for some sectors of the population, which find themselves forced to financially contribute to the building, maintenance and operation of infrastructure in order to have access to water services.[53]

The Bank indicates that replacing a supply-driven approach for a demand-responsive one, where rural communities identify their needs and the level of services they could manage and for which they are willing to pay, will increase sustainability. However, it is not certain that this should always be the path to follow. Additionally, it has not been conclusively proven that the participatory element of CDD applied to the water sector is determinant in obtaining better results. In fact, the domination of projects by local elites has been an unfortunate constant in CDD projects, to the point that some of them are actually captured by these groups and the intervention of state organs becomes necessary if the right to water is to be secured for all. The evidence so far indicates not only that CDD should not become a model to be hastily proposed in any situation (given the importance of context and of tailor-made solutions for the success of any project) but also that external agents strongly influence the outcome of this kind of venture. Recent World Bank literature indicates that it has come to share this view and now understands that CDD should not become a single blueprint, but rather a set of general principles to be adapted to the specific local context. It is affirmed that water:

> is a complex issue that involves multiple stakeholders from the community level to the international level, so community driven approaches are not the only model to use in all circumstances. The roles and responsibilities of each stakeholder will vary depending on the nature of the water service being delivered and on the level of water availability and scarcity in the area. Similarly, community driven development is an approach to development, not a model. The entry point, sequencing, and rules of the game established by the approach vary according to the nature of the society in which work is taking place.[54]

52 Finger and Allouche, note 7 above, 86.

53 Helling *et al.*, note 51 above, iii.

54 Jacomina P. de Regt, Water in Rural Communities (Note prepared for the International Workshop on African Water Laws: Plural Legislative Frameworks for Rural Water Management in Africa, Johannesburg, South Africa, 26–28 January 2005), available at www.nri.org/projects/waterlaw/AWLworkshop/REGT.doc.

Monitoring and evaluation of the progression of the project, in particular, thus needs to be trusted to state agencies or NGOs and accountability must be a priority to ensure CDD projects function efficiently.[55]

The characteristics of the water law reforms brought along by decentralization and CDD consist firstly in the incorporation of user associations and/or the enhancement of their role in water management, and secondly in the opening of the water sector to local private participation. Through PIM, user groups are incorporated to the institutional structure of water management by becoming participatory, autonomous and financially self-supporting entities, capable of managing water in rural areas. Initially in charge of distributing irrigation water, their responsibilities have been gradually broadened. As the FAO has pointed out, the tasks transferred from state agencies to them now include:

> the operation and maintenance of irrigation systems, [. . .] the management of polders and land drainage schemes, the maintenance of dykes and flood-defence structures, the removal and treatment of waste-water, the supply of water for domestic consumption and other uses, and the management of groundwater resources, usually in cases where an aquifer is under stress due to over-pumping.[56]

User associations are thus supposed not only to manage water for irrigation but also to operate water services, set tariffs, collect fees and manage funds, and ensure continuity in water provision while also addressing environmental concerns. The World Bank is particularly interested in using them as tools for the promotion of full cost recovery, since 'by involving strong water user associations in project management and fee collection at the local level, [governments] can use the capacity of community members to exert social pressures on their neighbors to pay'.[57] When it is time to set charges, the World Bank expects that the associations will employ a financial policy based on the costs of services – in other words, that users will pay more for higher levels of service – and that communities will pay part of the capital cost in proportion to the cost of the facilities, plus all operation and maintenance costs.[58]

Among the countries that have undergone vast decentralization processes under the umbrella of the World Bank, and even implemented CDD in irrigation and Rural Water Supply and Sanitation (RWSS) to the point that the World Bank considers them best-practice examples, are Albania, Ghana and India. The

55 Ghazala Mansuri and Vijayendra Rao, Community-Based (and Driven) Development: A Critical Review (Washington D.C.: World Bank Policy Research Working Paper No. 3209, 2004).

56 Stephen Hodgson, Legislation on Water Users' Associations 1 (Rome: Food and Agriculture Organization, FAO Legislative Study 79, 2003).

57 World Bank Policy Paper, note 28 above, 57.

58 Parameswaran Iyer, Mariana Felicio and David S. Warren, Rural Water Supply and Sanitation Toolkit for Multisector Projects 16 (Washington D.C.: World Bank Rural Water Supply and Sanitation (RWSS) and Social Funds Groups Toolkit, 2004), available at http://go.worldbank.org/7UZUKK9DY0.

implementation of CDD in the water sector of these three countries has brought along institutional arrangements that encompass all levels of government. NGOs and the private sector have been incorporated; the former are in charge of providing technical assistance to communities and of capacity building, while the latter – a group of small local entrepreneurs – supply equipment for the infrastructure. The scheme is completed by the oversight performed by decentralized district level agencies.

To the World Bank, the Swajal Project (Uttar Pradesh, 1996) and the Jal Swaraj project (Maharashtra, 2002) in India are two of the most accomplished examples of implementation of CDD, to the level of actually referring to the guidelines used in them as the Swajal Model.[59] The specifics of CDD in India include the creation of an autonomous agency that coordinates and monitors activities – the Project Management Unit (PMU) – and the involvement of NGOs providing advice on engineering issues and support to the villages, and serving as a link between the communities and the PMU. The PMU and similar institutions introduced in other jurisdictions watch over the local committees, and just like the regulatory entities in charge of issuing water rights mentioned above, have an independent nature and function in ways that make them resemble public corporations instead of state agencies. However, despite the safeguards foreseen in World Bank documentation, the actual project has been marred in practice by significant drawbacks threatening the access to water of the poor.[60]

3. Privatization of water services

The privatization of urban water services has been a controversial aspect of water sector reform, implemented as the result of the support of MDBs given to the participation of multinational companies in the water sector, particularly in the delivery of water services. Even if the participation of local entrepreneurs is also part of CDD projects in rural areas, in truth the role of private companies in the water sector is usually only associated with the operation of water services in urban areas. The following paragraphs focus on these processes of privatization in the urban context.

The privatization of public services in general is a measure that the MDBs have recommended to developing countries since the late 1980s. The MDBs drew attention to the inefficiency of certain states in providing public services and to the waste of resources generated in that effort. Privatization was – and still is – encouraged as a broad measure that simultaneously aims to regularize fiscal accounts and reduce the role of the state in sectors where the MDBs deem its presence dispensable. Simply put, the operation of public services by the private

59 De Regt, note 54 above. See Chapter VIII in this volume.

60 For a critical assessment of CDD in India based on ground research, Preeti Sampat, 'Swa'-jal-dhara or 'Pay'-jal-dhara–Sector Reform and the Right to Drinking Water in Rajasthan and Maharashtra', 3/2 *Law, Environment and Development Journal* 101 (2007), available at www.lead-journal.org/content/07101.pdf.

sector would not only allow for improved service delivery, but would also enable the state to focus its resources on other areas where they are needed. Privatization is thus one of the central reforms sponsored by MDBs to reduce state deficit and stimulate economic growth, and the World Bank has been confident that it would contribute to redressing adverse economic situations or raising development indicators by changing the role of the state in public services, from provider to regulator and monitor.

In support of the incorporation of the private sector, the World Bank argued first that the review of World Bank-funded water projects conducted by its own Operations Evaluation Department had shown that the management of its past projects in public hands had been less than stellar.[61] Additionally, the recognition of water as an economic good and the need for a new approach to water management soon made the participation of the private sector a key component of IWRM. Indeed, the private sector would contribute capacity, technology and investment to the water sector by assuming operation and maintenance of water services from states that were unable to provide quality services due to lack of resources. Specifically, private sector involvement was deemed essential to increase investment in infrastructure and expand network coverage – two of the main goals of the earliest concession contracts.[62] The World Bank claimed that states lacking resources were incapable of modernizing the network to meet current sanitation and environmental standards.[63] In addition to lessening the financial and administrative burden on the state for providing the service, the private sector could allegedly also bring more flexibility, transparency and accountability. To this end, the water sector must be open to the participation of private companies. MDBs have also exhorted governments to eliminate protectionist and regulatory instruments and practices that deter foreign investment.

The furtherance of privatization of public services in general, and of water services in particular, by the MDBs is relatively recent. In fact, the World Bank stood behind the public delivery of water services for decades. From 1960 to 1990, World Bank loans to developing countries were mainly destined for the building, expansion or maintenance of public water utilities, and especially for large projects such as dam construction. The Bank was convinced that investment in public utilities and other infrastructure projects would trigger the development take-off and that water utilities were natural monopolies 'that precluded market competition and therefore required public ownership or government regulation'.[64] This stance was not due to the belief that there were no possible

61 World Bank/Operations Evaluation Department, Water Supply and Sanitation Projects: The Bank's Experience: 1967–1989 para. 36 (Washington D.C. World Bank Report No. 10789, 1992).

62 World Development Report 1994, note 38 above and World Bank, World Development Report 1997: The State in a Changing World (Washington D.C.: World Bank, 1997).

63 World Bank Policy Paper, note 28 above, 27.

64 Public Citizen/Critical Mass Energy Program, Profit Streams: The World Bank & Greedy Global Water Companies 2, available at www.citizen.org/documents/ProfitStreams-World%20Bank.pdf.

alternatives to public management of water, but mainly to the fact that the World Bank was involved in project rather than policy lending, granting loans for the construction and development of infrastructure that would be operated and managed by the state. In addition, private companies themselves did not regard the provision of water services as a profitable venture. However, once the scarcity of water resources came to the forefront of public attention, corporations realized how lucrative the water sector could become and decided to bid for the operation of water services in the main cities of those states that included privatization as part of their public sector reform programmes sponsored by the MDBs.

The role of the MDBs in convincing multinational corporations to invest in the water sector is significant. Through structural adjustment/policy lending programmes, and under the supervision of the World Bank's technical advisory departments – the Private Sector Advisory Services and its Foreign Investment Advisory Service – indebted countries have been given advice on how to improve their investment climates by implementing policies such as privatization, and have thus engaged in substantial reforms in order to turn urban water services into an appealing opportunity for private investment. Since 1993, the World Bank has acknowledged as one of its priorities that 'its economic and sector work, lending, technical assistance, and participation in international initiatives [aim] to promote policy and regulatory reforms' for the implementation of privatization'.[65] The leverage of the World Bank's policy advice is, in this sense, not to be underestimated: together with loan conditionality, policy advice forms a powerful tandem for pushing the implementation of water sector reforms.[66] Additionally, convinced that privatization would entail improving water use, one of the affiliate agencies of the WBG, the IFC, has taken a more practical role by financing the participation of the private sector in water management and water services operation. In a sense, within the WBG, it could be concluded that the World Bank is in charge of developing the policy and then forwarding it to member countries through loan conditionality and technical assistance, while the IFC assumes a more pragmatic role by financially supporting the private sector as needed.

The World Bank encourages the private sector to intervene through service contracts, management contracts, lease contracts or concessions.[67] However, after a number of well-publicized early failures, such as those of Buenos Aires, Cochabamba, Manila and Jakarta, the World Bank has suggested discarding full asset sales or concessions and instead privatize in the form of management or service contracts. This entails that the state keeps its ownership and responsibilities

65 World Bank Policy Paper, note 28 above, 65.

66 Sara Grusky, The IMF, the World Bank and the Global Water Companies: A Shared Agenda 3 (Washington D.C.: International Water Working Group Note, 2001), available at www.citizen.org/documents/sharedagenda.pdf.

67 World Bank Policy Paper, note 28 above, 110.

over infrastructure. The problem with this kind of contract is that the state usually has to expand and maintain the service network at the cost of growing more and more indebted with MDBs, while foreign private companies easily make a profit from the operation of the service and take it abroad. This is an example of how the advice that MDBs offer often seems to be exclusively oriented towards private sector participation in water services, rather than securing a better provision of the service. Similarly, improving service quality seems to actually be a concern for the World Bank only because it would motivate users 'to share in the costs of investments and maintenance or to pay for services'.[68] The same can be said about the importance given to solid legal and regulatory frameworks for the success of a privatization process in a developing country. The World Bank already stated in 1993 that 'a proper regulatory framework must exist for the private sector to be willing to take the risk of investing, even for service contracts', and warned that 'this is perhaps the greatest challenge facing developing countries' but assured them, indicating that it was 'trying to help with research and technical assistance to develop these frameworks'.[69] However, in the light of privatization processes of water services such as the city of Buenos Aires, it is difficult to believe that the focus of the World Bank's assistance to the design of regulatory frameworks was the welfare of users: the failure in Buenos Aires was precisely caused by a combination of a deficient framework that enabled the multinational to modify the contract at its will to the detriment of users, and the institutional immaturity of a country that rushed into implementing privatization and was therefore not capable of properly monitoring the performance of the private company.[70]

Once again, the introduction of privatization to the water sector has mainly entailed institutional changes for states. The domestic legal and regulatory framework must foresee above all an adequate flow of information involving the three main actors concerned: the company operating the service, the government and the users. To this end, the creation of regulatory entities is essential to ensure transparency and accountability, given that they are charged with timely collection of information from the company and sharing it with user associations defending their interests as consumers. The regulatory entity oversees that the company meets its obligations as set out in the privatization contract, fines its underperformance and acts as an intermediary between the concessionaire and user associations, and other governmental organs of control such as parliamentary commissions. Due to its role, it is crucial that this authority counts with highly competent and qualified personnel in water issues, and that its political and financial independence is guaranteed. Additionally, the privatization contract becomes a new component of the regulatory framework. It must be exhaustively

68 *Ibid.*, at 100.

69 *Ibid.*, at 111.

70 On this case, Andrés Olleta 'The World Bank's Influence on Water Privatisation in Argentina – The Experience of the City of Buenos Aires' *in* Cullet *et al.* (eds.) note 21 above, 230.

studied before submitting it to the signature of the private company, in order to make sure that the rights and obligations of each contracting party are adequately stipulated, as a first step towards ensuring that the provisions will be respected. Above all, the contract must be unambiguous in sensitive issues such as tariff setting, deadlines for building infrastructure, and which scenarios call for a modification of the contract.

CONCLUSION

The World Bank and the regional MDBs have played a substantial role in shaping current water sector reforms and the water law changes that states have accordingly made as part of their implementation. In particular, given their high indebtedness, it is relatively easy for the World Bank to push for the implementation of certain policies in developing countries through their use of conditionality and technical advice. By pairing these policies to the loan negotiations, in many cases MDBs succeed in directing the agenda for domestic public sector reform, and this is especially valid in the case of water management and water services provision since the late 1980s.

Ever since the publication of the World Bank's 1993 Water Management Policy Paper, it was clear that the World Bank and many other international organizations working on water-related issues believed that the water crisis was actually a matter of mismanagement of the resource, and that IWRM, decentralization of water management and privatization of water services were the answer to it. To the World Bank, these policies were indispensable to achieve a better management and a smarter use of water resources. However, these policies were undoubtedly too focused on economic factors – specifically on cost recovery – to ensure financial sustainability. The World Bank warned that in individual water-lending operations, 'relevant pricing issues, cost recovery, and financial autonomy and sustainability [would] receive particular attention'; in other words, that these points would be of particular merit in policy lending.[71] Possibly due to this, once put into practice, these policies have encountered some difficulties in implementation and a number of social setbacks have motivated critical reviews of their adequacy, either by civil society or even by the Operations Evaluation Department, which has concluded that the World Bank has failed in assisting in the implementation of effective regulation for the sector (including pricing policies that benefit the poor) and should not, for example, foster privatization of water services as the sole solution.[72]

Nothing negative can be said about attempting to face in a holistic way

71 World Bank Policy Paper, note 28 above, 18.

72 World Bank/Operations Evaluation Department, Efficient, Sustainable Service for All? – An OED Review of the World Bank's Assistance to Water Supply and Sanitation (Washington D.C.: Operations Evaluation Department Report No. 26443, 2003).

the challenges brought by the scarcity of freshwater, as IWRM purports to do. The different uses made of water certainly need to be made compatible while giving priority to domestic use, ensuring an environmentally sustainable exploitation of water resources and preventing wastage and other symptoms of inefficient management. Yet, when the new approach to water management centres on financial sustainability and cost recovery, there is a risk of making decisions that sacrifice the important social goals that proper management of water resources is first and foremost called to achieve. In these times of scarcity, the state certainly needs to eliminate private ownership of water and instead monopolize the control of water, assuming the distribution of water rights among applicants. But the eventual creation of a market for water rights must be a decision carefully taken, since it can only work in selected cases due to the specific characteristics of certain basins. Misjudgements in this sense, like introducing trade of rights where it is not advisable, or making land ownership the basis for obtaining an entitlement, could lead to water being delivered to those that can afford to participate in the trade and thus deprive the poor of access to water.

Similarly, decentralization is a measure full of good intentions such as involving users in decision making, improving transparency and accountability in water management and service provision or increasing the quality of water services. However, the democratization of water management and service provision in the name of efficiency cannot be forwarded without properly informing users of the consequences of the decisions they are called to make, mainly those related to the financial burdens that decentralization entails such as contributing to infrastructure building and to operation and maintenance costs. Once again, stress is put on improving services in order to make users not only pay, but pay charges and tariffs that assure full cost recovery and thus ensure financial sustainability. In some jurisdictions where social aspects have consequently been neglected, decentralization through CDD projects oriented to PIM and water services delivery in rural areas has not met expectations.

Finally, regarding privatization, it is encouraging to see the World Bank requiring that developing countries have an adequate regulatory framework in place before proceeding with the implementation of said policy. However, a sound legal framework is not tailor-made to benefit multinationals and make investment in the water sector an appealing opportunity for them. On the contrary, a sound framework is one that ensures that the interests of all stakeholders are adequately addressed and correctly balanced. Such a framework is especially needed in the case of water services. The most essential of public services should not be left unattended at the mercy of potential excesses of multinational companies seeking easy profits in a context of monopolistic exploitation. Should the state open the door to private investment in the area of public services, it must be ready to face the challenge of assuming its responsibility of overseeing the performance of companies for the sake of the general interest. This is particularly evident in the case of services such as potable water and sanitation, which are about the satisfaction of two of the most vital needs of the population.[73] Still, after disappointing performances around the world, it might be useful to altogether review the necessity and convenience of incorporating the private sector to water services provision in urban areas. While other public services might be more suited to privatization for the reasons argued by the World Bank, the importance of access to water services for human welfare might pose an essential incompatibility between the delivery by private companies and the accomplishment of elemental social objectives such as the realization of the human rights to life and to health of the population. In consequence, it might be necessary in this case that the state does not step back to the extent desired by MDBs, and that it actually intervenes in a capacity other than mere regulator in order to ensure that the service is effectively delivered at affordable prices and with adequate quality. In achieving this goal, MDBs could instead support the public sector and suggest modifications there where they are necessary to correct defects in the public provision of water services.

In summary, MDBs have conducted water sector policy-making worldwide and are in an unmatchable position to forward the agenda for water sector reform in developing countries, through conditionality and technical advice. The challenges for the institutions – the World Bank in particular – working in the water sector are varied. First, the World Bank should assess the appropriateness of policies and the success of individual projects not only from the financial point of view but also increase the analysis of the social impacts of the measures recommended and the reform programmes supported. Ideally, this should be done from the very early stages and not later into the implementation phases. Operations Evaluation Department reports are useful and revelatory but, unfortunately, sometimes they are published somewhat too late for millions of people. The World Bank is a financial institution interested in seeing the measures it recommends have an impact on economic prosperity. It should, however, first act as a development agency that ought to better evaluate the impacts of the application of a liberal policy to the water sector, in particular given that the realization of human rights is at stake.

Secondly, the World Bank should not create models out of positive experiences in certain jurisdictions, to be indiscriminately applied to different countries. Its own publications warn against creating a common recipe that does not take into account the particularities of the situation in each borrowing country. Yet, apparently even against the will of its own staff, it is impossible to ignore the fact that there is always a common denominator in water sector reforms, a set of measures that does not vary as described in this chapter. It is clear that the success of actual water management depends on schemes that are tailor-made to the specific situation and context of countries. It is only possible if rational compromises specific to the country are found between those who emphasize the economic value of water and regard it as a private good and those who stress its social value and

73 In a number of jurisdictions around the world, states have prohibited the privatization of water services and thus ensured that provision remains public. See Chapter II.

regard it as a public good. Actual water policy must thus be 'formulated in terms of multi-objective decision making, recognizing that the relevance and importance of [the positions regarding water as a private or social good] will vary substantially over different conditions of time and place'.[74] The challenge in developing countries has been to achieve the implementation of these new ideas without sacrificing essential social aspects in the process.

74 Perry, Rock and Seckler, note 31 above, 3.

Part 2

Water Law Reforms in India

VI. Context for water sector and water law reforms in India

Roopa Madhav

INTRODUCTION

Reforms to the water sector in India are crucial as water resources and systems in the country are under stress. It has been nearly two decades since reforms have been initiated. Legal and institutional reforms are an integral part of the reform process and have introduced significant changes to the control of and access to water resources. These changes have had an impact on governance models, resource allocation and conservation efforts. Taken together, however, they do not provide a radically different approach to water resource management. Reforms in the past have been primarily sectoral, intermittent and hierarchical, with little or no public participation in decision making. The present set of reforms also adheres to this pattern, seeking to privilege donor expectations and demands over a crucial need for a holistic approach.

Water is crucial for life, and management of this vital resource is critical to ensure food security, livelihood security and ecological security. The increasing conflict over water resources has necessitated a global appraisal and review of water resources and its management. Changing the definition of a vital resource such as water, from a social and public good to an economic good,[1] has an enormous significance for the evolution of law and policy, particularly in developing countries. Issues of justice and equity are emerging as matters of deep concern as they are increasingly sidelined within popular discourse on water policy.

Water resource management is primarily about ensuring water security – both present and future. India is faced with several challenges in the water sector. In the last two decades, policy planners and government institutions have been actively debating reform and restructuring of the water sector. Regional differences of abundance and drought, compounded by institutional challenges, have impacted water management strategies within the country. Presently, water sector reforms in the country are aimed primarily at improving management of systems through

1 Dublin Statement on Water and Environment and Sustainable Development, International Conference on Water and the Environment, Dublin, 31 January 1992.

better governance and efficient cost recovery. Only marginally do they touch upon concerns of equity and sustainability.

This chapter provides a broad overview of the water sector and water law in India, from their pre-reform state to the key reforms being carried out in the country. Section A provides a broad overview of the water resources of the country and the historical developments in the water sector, contextualizing present-day reforms. Section B introduces the specific water law reforms being carried out in three primary water sectors – irrigation, drinking water, and groundwater. In doing so, it introduces some of the principles and concepts informing the reform process. Section C provides a critical analysis of the changing nature of control and access to water resources in the country. Finally, an examination of the future directions for legal reforms vital for the water sector, keeping in mind their innumerable complexities, is attempted in the conclusion.

A. Water sector reforms in context

An examination of India's water law and policy framework requires an historical voyage, particularly into the colonial era, as many of the principles underpinning the water law framework of the country have their origins in common law. The post-independence planning era saw an expansion of state ownership and control over water resources.

1. Characteristics of water resources and water resources management

Geographic elements

Water resources in India are a study in contrasts. The average annual precipitation in volumetric terms is 4,000 billion cubic metres (BCM). The average annual flow out of this is 1,869 BCM, the rest being lost in infiltration and evaporation. The annual replenishable groundwater resources in the country are estimated at 432 BCM. The total availability for use from surface and groundwater sources is thus 1,122 BCM (690 + 432 BCM).[2] While India receives a surface water flow of 1,869 BCM, it can hardly store 180 BCM. Notwithstanding the utilizable water flow of 690 BCM, the storage and irrigation potential are grossly under-used.

The country gets 90 per cent of its rainfall during the monsoon season which lasts from June to September. Further, rainfall in the country is skewed geographically with the average ranging from 100 mm in the West to 1,170 mm in the East. Added to this, approximately 71 per cent of the water resources are concentrated in 36 per cent of the country, namely in the Ganges-Brahmaputra-Meghna

2 Government of India, Irrigation, Flood Control and Command Area, Planning Commission Report, Tenth Plan (2002–2007), chap. 8.1.

Basin and the rivers that flow in the Western Ghats.[3] Thus there are huge inter-basin and inter-region disparities. Every summer, on an annual average, 91 districts are hit by drought and 40 million hectares in 83 districts are submerged in flood waters because of the skewed distribution of rainfall. The hydrological cycle is vastly unpredictable.

Almost 80 per cent of drinking water needs are met from groundwater. Out of the 432 BCM of replenishable groundwater in the country, only 396 is utilizable – 70 BCM for domestic purposes and 326 BCM for irrigation.[4] Availability of water from other sources including the desalination of sea water is negligible. The per capita water availability has been continuously declining from 5,176 m^3 in 1951 to 1,820 m^3 in 2001 and 1,703 m^3 by 2005. This drastic fall over the years in annual per capita availability can be attributed to over-extraction of groundwater, water pollution, climate and other natural causes. It is estimated that the per capita availability will fall to approximately 1,401m^3 by the year 2025.

Post-colonial infrastructure

Two distinct phases mark the development of the water sector in independent India. Firstly, the focused and rapid expansion of irrigation infrastructure that transformed farming practices, impacting rural household income, cropping patterns, landholding patterns and food security in the country. Secondly, the introduction of technology to extract groundwater in the early 1960s that had a far-reaching impact on the rural drinking water supply and ensured on-demand water for irrigation. The adverse fall-out of the latter has been the over-extraction of groundwater, resulting in a severe water crisis in several districts.

The irrigation potential at the time of independence was only 21 million hectares (mha) (1950 figures) but by 2002, this had increased to 93.95 mha (37.05 mha through major and medium irrigation projects and 56.9 mha through minor, including groundwater, schemes) with an annual average growth of 1.4 mha.[5] More than 50 per cent of all public expenditure on agriculture has been on irrigation alone. During the fifty years since independence, the government has spent about Rs 231,400 crore (as 1996–1997 prices) on major, medium and minor irrigation works.[6] Despite such huge investments and growth in capacity, only 40 per cent of agriculture in the country has an assured supply of water from irrigation systems, the other 60 per cent continues to depend on erratic monsoons.

India inherited a large irrigation infrastructure at independence. Irrigation

3 Yoginder K. Alagh, Ganesh Pangare and Biksham Gujja, *Interlinking of Rivers in India* 28 (New Delhi: Academic Foundation, 2006).

4 See Central Water Commission, Data on Water Resources in India (2005) available at www.cwc.nic.in/main/downloads/Water_Data_Complete_Book_2005.pdf.

5 See generally, Government of India, Ministry of Water Resources, Report of the Working Group on Water Resources for the XI Five Year plan (2007–2012).

6 Government of India, Ministry of Water Resources, Report of the Sub-committee on More Crop and Income per drop of Water (2006).

development under British rule began with the renovation, improvement, and extension of existing works. At the time of independence in 1947, the net irrigated area of the Indian sub-continent was approximately 22.8 mha. At the time of partition, major irrigation works on the Sutlej and Indus river basins fell to Pakistan in the West, and irrigation works in the Ganga and Brahmaputra delta regions in the East to Bangladesh. Independent India thus inherited some important productive irrigation works in central India and south India.[7] The administration of irrigation systems was initially a centralized activity, but was divested to the provincial governments by the Government of India Act, 1919. This scheme continued in independent India. Apart from irrigation, further decentralization of the water resources to the lowest tier of governance – panchayats in the rural areas and municipal bodies in urban areas – were achieved through the 73rd and 74th Constitutional Amendments.

Independent India focused on developing and promoting large irrigation works through its planned economy model. Until the end of the 1980s, the Indian economy was characterized by central planning for development. The five-year plans sought to ensure sustained and progressive growth in industry, agriculture and infrastructure. With the development of agriculture as the focus, the planners placed their emphasis on building irrigation projects, and through the five-year plans, enabled the construction and implementation of large irrigation projects such as Bhakra Nangal, Hirakud, Damodar Valley and Nagarajunasagar. This concerted effort resulted in an increase in agricultural production from 50 million tonnes in the 1950s, to about 203 million tonnes in the year 1999–2000.[8] Independent India also shifted focus from irrigation projects to projects that were multi-purpose hydro-power generating units.

Severe droughts in the country during the years 1965–1967 resulted in a shift in focus from major and medium irrigation power projects to groundwater resources. This meant increased budgetary outlays for drilling and blasting equipment, as well as the installation of pump sets and rural electrification to energize irrigation pump sets and tubewells. India's food security has been propelled by the green revolution and the tubewell revolution, although the latter led to long-term damage as unbridled groundwater extraction adversely impacted the hydrological cycle. Although the total replenishable groundwater resource is about 432 BCM, the present level of development in some states is very high, leading to a situation of over-exploitation in certain pockets of the country. Out of 5,711 blocks/mandals/taluks/watersheds in the country, 310 are over-exploited and 160 are critical where the level of exploitation is 85–100 per cent more than the replenishable quantum. This unbridled growth is largely due to the lack of legislation on groundwater extraction, and the provision of free or subsidized power for the agricultural sector.

7 See generally Carolin Arul, *Water Laws and Rights in India*, 20–22 (New Delhi: Indian Society for Promoting Christian Knowledge, 2005).

8 CWC data, note 4 above.

The primary responsibility for providing drinking water facilities in the country rests with the state government. The central government, however, has intervened over the years with technical support, policy directives and crucially, financial assistance. The primary scheme of the Union Government – the Accelerated Rural Water Supply Programme (ARWSP) – has provided funds to state governments for implementing rural water supply schemes. As a result of this and other schemes, a huge network of water supply pipes, hand pumps, mini water supply tanks and standalone pipes crisscross the country, serving both urban and rural populations. With the 73rd and 74th amendments, drinking water and sanitation are included in the list of subjects to be devolved to panchayats, thus delegating powers to the lowest rung of governance.

Not all parts of the country however have piped drinking water supply.[9] In rural areas, Government classifies habitations as Not Covered (NC),[10] Partially Covered (PC) and Fully Covered (FC). A norm of 40 litres per capita per day (lpcd) has been fixed for an FC habitation while an NC habitation has availability in the range of 10 to 40 lpcd. The difficulty in rural water supply is ensuring a minimum supply of safe drinking water as the rural context is riddled with problems of water quality, physical or geographical distances and difficulties in sustaining supply. Serious water quality problems plague the country as surveys indicate the existence of pollutants such as excess fluoride, arsenic, iron and nitrate.[11] Fluoride and arsenic pose a serious health hazard and efforts are being made to address these concerns. Additionally, ARWSP attempts to promote water conservation measures such as rainwater harvesting and water recharge. The urban water supply sector has done better with 91 to 93 per cent of India's urban population having access to water from a protected source.

Administration

The federal structure of the Indian polity mandates that water resources be controlled and managed between the Centre and states. The Centre has the power to legislate in matters relating to the regulation and development of inter-state rivers and river valleys. The states have the power to legislate in respect of water supplies, irrigation and canals, drainage and embankments, water shortage, and water power. While the canal irrigation sector is developed and managed by public agencies, groundwater irrigation is developed and managed by millions of independent farmers and a growing market of informal water traders.

A key feature of the water institutional framework is the administrative fragmentation along sectoral lines for water governance. The Union Ministry of Water Resources (MoWR) is the national organization responsible for the overall

9 On drinking water, see Chapter VIII.

10 A 'not covered' habitation has no public or private drinking water source point within 1.6 km of the habitation if it is located in the plains, and 100-metre elevation if located in the hilly areas.

11 See further Chapter XIII.

planning and management of water resources in the country. Technical support to MoWR is provided by the Central Water Commission (CWC), the Central Groundwater Board and the National Water Development Agency. Research and training support is provided by organizations such as the Water and Land Management Institutes (WALMI), agricultural universities, and other research institutions. The Planning Commission at the national level provides project clearance and approves financial allocation to various water projects in different states.

Two important central organizations have a crucial role in irrigation and groundwater management, respectively. The Command Area Development Authority deals with irrigation projects, where the states have responsibility not only for creating a reservoir, but also the canal networks. Another important central organization dealing with groundwater is the Central Ground Water Board. The Supreme Court in *MC Mehta v. Union of India* directed the setting up of the Central Groundwater Board under the Ministry of Environment and Forest of the central government, under section 3(3) of the Environment (Protection) Act, 1986, and mandated with powers to regulate and control groundwater management and development, including water harvesting.[12]

Multifarious agencies are in charge of drinking water in the country. As regards the institutional structure for rural drinking water supply, the Department of Drinking Water Supply in the Ministry of Rural Development is responsible for planning, policy formulation, direction, financing, monitoring, and reviewing implementation and progress at the central level. At the state level, both rural and urban drinking water involve various government bodies such as the public health engineering departments, panchayati raj departments and water boards. Local governments such as municipalities and panchayats also play an important role in drinking water supply. Donor agencies, such as the World Bank, are attempting to enhance their role in urban and rural water projects as part of their strategy for decentralization. Users and stakeholder groups are also encouraged to get involved in cost recovery and management at the outlet and system levels.

Central and state pollution control boards, operating under the Ministry of Environment and Forests, have the responsibility for water quality aspects. The National Rural Drinking Water Quality Monitoring and Surveillance Programme, initiated in February 2006, envisages the institutionalization of community participation for monitoring and surveillance of drinking water sources at the grassroots level by Gram Panchayats/Village Water & Sanitation Committees. Other central agencies influencing water and environmental issues include the Ministry of Agriculture (MoA), the Ministry of Rural Development and the Ministry of Environment and Forestry (MoEF). The MoA, which has been involved in watershed development since the 1960s, concentrates mainly on agriculture-related issues, such as erosion-prone agricultural lands, optimizing production in rain-fed areas and reclaiming degraded lands.

Important organizational arrangements to achieve inter-state and centre-state coordination include the National Water Resources Council (NWRC) set up in

12 *MC Mehta v. Union of India* (1997) 11 SCC 312.

1983 and the National Water Board (NWB) set up in 1990. The NWRC is an important organ in relation to water resources. Though it does not have a statutory status, it has an important and influential role in evolving water policy. The NWB, which is considered to be the executive arm of the NWRC, is chaired by the secretary of the MoWR and includes the chief secretaries of all the states/ union territories and secretaries of the concerned union ministries.

Decentralization of governance was formally introduced through the 73rd and 74th Constitutional Amendments in 1992. The amendments inserted a new set of functions under the Eleventh Schedule of the Constitution for the panchayats with regard to water[13] and the Twelfth Schedule for Municipalities.[14] Decentralization has thus introduced new actors into the water sector domain, the implications of which are yet to be fully examined. Central-level policies with respect to both sectoral decentralization and privatization do not address the possible conflicts with this third tier of governance.

Multifarious agencies and innumerable actors govern the water sector in the country. Presently, policy formulation lacks the careful consideration that would take these complexities on board and envision the future needs of the sector.

2. *Water policy – An overview*

Until 1987, India did not have a comprehensive policy statement on water. The fragmented approach to water resources which was applied up to 1987, has to be pieced together to enable an analysis of broad policy trends and approaches. The policy attempts to classify surface and groundwater, though no such formal classification is stated in official documents of the earlier period. Classification of water according to source or use is more clearly defined in recent literature.

Surface water

As demonstrated in the Famine Commission Report of 1880 and 1898, the policy of colonial India towards irrigation aimed to provide protective irrigation to alleviate hardship in drought-prone regions. The First Irrigation Commission Report (1901–1903) records the shift in policy from protective to productive, with increased investment in irrigation schemes that were financially viable, yielding revenue for the British Government. Returns on investment and financial viability were reiterated as being the focus of the policy thrust by the Royal Commission on Agriculture, 1928. As the policy prescriptions were designed to improve the

13 Minor irrigation, Water management and Watershed Development, Constitution, Art. 243 G read with Entry 3; Drinking Water, Entry 11; and Health and Sanitation, Entry 23.

14 Water Supply for Domestic, Industrial and Commercial Purposes, Constitution, Art. 243 W read with Entry 5; Public Health, Sanitation Conservancy and Solid Waste Management, Entry 6; and Safeguarding the Interests of Weaker Sections of Society, including the Handicapped and Mentally Retarded, Entry 9.

financial calculus of the system, little emphasis was paid to ensure efficiency in water use.[15]

Until the end of 1980s, the Indian economy was characterized largely by central planning for development. The five-year plans sought to ensure sustained and progressive growth in industry, agriculture and infrastructure. With growing emphasis on agriculture, the planners concentrated on building irrigation projects and enabled the construction and implementation of large irrigation projects through five-year plans. In mapping policy changes, historical records and the plan documents provide an insight into the events that changed the approach to water resources and their management. Severe droughts during 1965–1967 resulted in a shift of focus from major and medium irrigation power projects to groundwater resources. Additionally, green revolution provided the impetus for policies promoting groundwater use through cheap credit, technological innovation, and subsidized electricity. Groundwater gradually took primacy in ensuring water on demand for both irrigation and drinking water.

The administration of irrigation systems was initially a centralized activity, but was divested to the provincial governments by the Government of India Act, 1919. This schema was continued in independent India. Apart from irrigation, further decentralization of the water resources to the panchayat and municipal levels was put in place under the 73rd and 74th Constitutional Amendments. Centralized planning has ensured that the irrigation potential of the country leapt from 21 mha in 1950 to 94 mha by the end of the ninth five-year plan and foodgrain production has increased from 51 mt in 1951 to nearly 213.25 mt in 2003–2004.[16]

The centralizing tendency is evident even in the reform process. The tenth plan links reforms in the water sector to central funding. The accelerated irrigation benefit programme has relaxed lending norms for states agreeing to reforms (4:1 Centre:State lending ratio against 2:1 for non-reforming states). Participatory irrigation management is being further promoted through the Command Area Development Programme, which again is a centralized intervention in the irrigation sector.

Though drinking water has been on the agenda of planners for a long time now, the progress has been slow in ensuring both quantitative and qualitative supplies. Although the Bhore Committee (1946) and the Environmental Committee (1949) recommended that water supply and sanitation should be the focus areas, no immediate measures were adopted by the central government. In 1954 the central government provided assistance to the states to establish special investigation divisions to carry out identification of problem villages.[17] The

15 A. Narayanamoorthy and R.S. Deshpande, *Where Water Seeps! Towards a New Phase in India's Irrigation Reforms* (New Delhi: Academic Foundation, 2005).

16 See CWC data, note 4 above, 14.

17 A problem village is defined as one where no source of safe water is available within a distance of 1.6 km; where water is available at a depth of more than 15 metres; where the water source has excess salinity, iron, fluorides and other toxic materials; or where water is exposed to the risk of cholera or guinea worm.

Accelerated Rural Water Supply Programme (ARWSP) was introduced in 1972–1973 by the central government to accelerate the pace of coverage of problem villages. It was briefly withdrawn in 1974 and reintroduced in 1977. The programme provided assistance to the states and the Union territories with 100 per cent grants to implement the schemes in such villages.

Another milestone in the policy of access to clean drinking water was the United Nations Water Conference at Mar del Plata (Argentina) in 1977, which called for a ten-year campaign by member countries and international agencies to provide access to safe water and sanitation for all and to accelerate political will and investment in the water sector.[18] The Mar del Plata Action Plan made recommendations that covered all the essential components of water management, and discussed the assessment of water use and efficiency; natural hazards, environment, health and pollution control; policy, planning and management; public information, education, training and research; and regional and international cooperation. The years 1981 to 1990 were designated as the International Drinking Water Supply and Sanitation Decade. India, as a signatory, pledged its full support to the Mar del Plata action plan in this regard.

Groundwater

Groundwater and minor irrigation were not focus areas until after the 1970s. The Model Bill on Groundwater, drafted in 1970, sought to introduce a licensing system by the Groundwater Authority, which would notify areas for regulation of groundwater. To improve upon and correct certain defects in the Act, the Model Bill was drafted in 1992. This was amended further in 1996. Several states have enacted laws based on the model bills of 1992 and 1996. In 2005, the Ministry of Water Resources released a further draft bill to control and regulate the use of groundwater.

Groundwater policy has been the most challenging area of the current reforms. Administration of groundwater resources is vested in the state governments, and until recently, very few states sought to set up regulatory authorities that would regulate the use of and protect the aquifers that replenish the groundwater table. In the recent past, there have been a series of laws seeking to establish regulatory authorities, laying down standards and norms for use of groundwater and ensuring conservation of the resources. The state governments of Kerala, Andhra Pradesh, Maharashtra, Tamil Nadu and Karnataka have enacted legislation to regulate and administer groundwater resources.[19] However, these regulations do not address the issues adequately and are designed in a manner that makes implementation difficult. Serious thought needs to be given to de-linking water rights from land rights. Also, discussion on issues of access, equity and recharge of groundwater needs to be prioritized in policy circles.

18 See further Chapter III.
19 See Chapter IX.

At the turn of the century, these planning and policy reforms were revisited in depth, with conservation and management of water resources having gained primacy on the policy agenda. Depleting groundwater had reached a point of crisis, with irrigation also facing challenges of sustainability and viability. Simultaneously, there was a policy shift towards privatization and liberalization of the water sector. The tenth plan document (2002–2007) states that water needs to be managed as an economic asset rather than a free commodity.[20] These dramatic shifts in policy are further reflected and articulated in the National Water Policy, which is examined below.

National Water Policy

As stated earlier, there was no comprehensive policy statement on the water sector until 1987. In 1986, a National Water Resources Council (NWRC) chaired by the Prime Minister was created, which drafted and adopted the first policy document on water. The National Water Policy, 1987 was replaced by the National Water Policy, 2002.[21] The two policies are quite similar in that they both focus on prioritizing drinking water; developing groundwater and irrigation infrastructure; encouraging the participation of stakeholders in water management; monitoring water quality; developing a flood control and management system; maintaining and upgrading the water infrastructure; developing science and technology relevant for water management; and a data bank on water resources.

In 1987, a clear policy choice towards encouraging public–private partnership in the water sector led to a vigorous debate over the ownership of water resources and the need to balance competing claims that emerged in the recent past. The 2002 policy emphasizes the need for improved institutional frameworks with a focus on improving the performance of institutions in place. The national policy has been supplemented by the adoption of a number of state water policies.[22] The national and state policies are based on a set of principles that are broadly similar.

One significant change that is being promoted by these water policies is the introduction of water rights. Water-related rights are not new per se and there is already ample jurisprudence related to control over water. The policies, however, emphasize two specific points. On the one hand, they restate the proposition that the state is the sole owner of the water resources. On the other hand, they propose creation of water rights in favour of users.[23] These rights are said to be the necessary premise for participation in management of water resources, for the

20 See generally, Government of India, Tenth Five-Year Plan 633 (2002–2007).

21 National Water Policy, 1987, available at www.ielrc.org/content/e8701.pdf and National Water Policy, 2002, available at www.ielrc.org/content/e0210.pdf.

22 See, e.g., Rajasthan State Water Policy, 1999, available at www.ielrc.org/content/e9903.pdf; Uttar Pradesh Water Policy, 1999, available at www.ielrc.org/content/e9904.pdf; and Maharashtra State Water Policy, 2003, available at www.ielrc.org/content/e0306.pdf.

23 See Philippe Cullet, 'Water Law Reforms: Analysis of Recent Developments', 48/2 *J. Indian L. Institute* 206 (2006).

setting up of water users associations (WUAs), and for introduction of trading in entitlements. Trading is specifically proposed in certain policies.[24]

Significantly, the policies seek to introduce wide-ranging legal and institutional reforms. These include the introduction of various amendments to existing laws as well as the introduction of new laws. Three main aspects can be singled out. First, a legal framework for the formation of WUAs is set up to decentralize water governance. Second, laws have been adopted that provide for the establishment of a water resources authority whose primary characteristic is to be largely independent from existing irrigation and other water resource departments. The third point is that these policies introduce groundwater regulation.[25]

The essential thrust of the sector reforms is being outlined by the World Bank and other international financial institutions (IFIs). This is evidenced by the comprehensive report on water sector reforms that was produced jointly by the World Bank and the Government of India in 1998.[26] While a more detailed study of the IFIs and their impact on policy is found in another chapter,[27] a brief outline of the policy prescriptions promoted in India can be attempted here. The World Bank's key prescription is to limit the role of the government in planning activities, and de-link it from everyday management responsibilities. Operation and maintenance is to be separated from investments. Water planning, investment and management are to be opened up to greater participation by user groups and private actors. Moreover, the setting up of autonomous regulatory authorities which operate free from political interference is seen as crucial to ensuring a healthy environment for further reforms. Many of these policy prescriptions have been the subject of heated debates among civil society groups and other key government bodies that influence policy at the state and centre levels.

3. *Water law in India*

Although 'water' is a state subject in India's constitutional scheme of separation of powers, independent India has seen an increasing trend towards the centralization of powers in planning, management and financing of water resources. This trend has its origins in colonial India's approach to natural resources. Legislation relating to water is for the most part, an extension of colonial policies, and the importation of common law principles continue to govern both surface and groundwater. Article 372 of the Constitution provides for the continuation of certain laws and legislation enacted prior to the Constitution. It may therefore be useful to trace these colonial laws and legal principles, and their continued application to water resources in the country, while mapping some of the milestones in the growth of water law in independent India.

24 See, e.g., Maharashtra State Water Policy, 2003, sec. 4(2).

25 See, e.g., Karnataka State Water Policy, 2002, para. 7.

26 World Bank, India: Water Resources Management Sector Review, Initiating and Sustaining Water Sector Reforms (Report No 18356-IN, 1998).

27 See Chapter V.

Constitutional framework

Following the adoption of the Constitution, many new laws and rules were enacted. Despite the transformative potential of the Constitution and the complex and intricate legal system, the language and the discourse of law and rights have not necessarily seeped into the governance structures within the country. Traditional structures and power equations, and not only with regard to water, continue to exist alongside formal legal structures which seek to guarantee a liberal egalitarian framework.

Two important legal principles – that of equality and of a basic right to water – are found enshrined and protected under the Constitution. It firstly recognizes the primary principle of equality – and in particular equal access to water – in Article 15, paragraph 2. It explicitly states that no citizen shall 'on grounds only of religion, race, caste, sex, place of birth or any of them' be subject to any disability, liability, restriction or condition with regard to 'the use of wells, tanks, bathing ghats'. Secondly, Article 21, which enshrines the right to life, has also been interpreted by the Supreme Court to include all facets of life and by extension the right to water. In the case of *Narmada Bachao Andolan v. Union of India*, it held that 'water is the basic need for the survival of human beings and is part of the right to life and human rights as enshrined in Article 21 of the Constitution of India and can be served only by providing source of water where there is none'.[28] However, the lack of an explicit provision on the right to water – either in the Constitution or in statutes – will soon need to be addressed.

The directive principle of state policy, which the Constitution in Article 37 declares to be non-justiciable, recognizes the principle of equal access to the material resources of the community. Article 39(b) mandates that 'the State shall, in particular, direct its policy towards securing that the ownership and control of the material resources of the community are so distributed as best to subserve the common good'. Article 51A(g) casts a fundamental duty on every citizen of India 'to protect and improve the natural environment including forests, lakes, rivers, wild life and to have compassion for living creatures'. Thus, states and citizens have an obligation of care towards natural resources within their domain.

In the federal framework, certain powers relating to water resources are vested with the central government. These include development of inter-state rivers and the power to resolve water-related disputes. Entry number fifty-six in the Union List states 'regulation and development of inter-State rivers and river valleys to the extent to which such regulation and development under the control of the Union is declared by Parliament by law to be expedient in the public interest'. In exercise of this power, the Parliament enacted the River Boards Act, 1956, which has been largely inoperative. In accordance with Article 262 of the Constitution, which envisages the creation of an exclusive tribunal for the purpose of resolving inter-state disputes, the Parliament enacted the Inter-State Water Disputes Act, 1956.

28 *Narmada Bachao Andolan v. Union of India* (2000) 10 SCC 664, 767.

Article 252 of the Constitution also empowers the Union Government to pass laws on subjects that are not within its powers, provided the legislatures of two or more states resolve that Parliament should pass such laws. One such enactment is the Water (Prevention and Control of Pollution) Act, 1974 a subject of which the powers vests with the state government.[29] Another significant central legislation relating to water pollution is the Environment (Protection) Act, 1986 under which notifications are issued by the Ministry of Environment and Forests in exercise of the powers under Section 3 of the said Act. Under this Act, the government also set up by a notification the Central Ground Water Authority to study and regulate the indiscriminate use of groundwater in the country.

Another important issue relating to the constitutional framework is dispute resolution. Under Article 262 of the Constitution, the Parliament enacted the Inter-State Water Disputes Act, 1956 which aims to resolve water-related, inter-state disputes. Soon after the adoption of the Constitution, river water disputes began to surface.[30] The more intractable cases have been the Krishna–Godavari waters dispute between Maharashtra, Karnataka, Andhra Pradesh and Orissa; the Cauvery water dispute among Tamil Nadu, Karnataka and Kerala; the Narmada water dispute among Gujarat, Madhya Pradesh, Maharashtra and Rajasthan, all of which bring to fore the inadequacies of the existing legal framework in effectively resolving inter-state water disputes.

Additionally, the gap between supply and demand, and competing use and allocation, aggravate conflicts. For instance, the diversion of irrigation waters for industrial use or drinking water purposes are on the rise. None of the state governments has any law specifying the basis for deciding water allocation between different segments of basins falling within their territories. 'States are left largely free to change the allocation of water between and within particular systems and between uses (e.g., between irrigation and urban water supply) at their discretion without observing any consistent application of clearly defined principles and procedures laid down by law.'[31]

Statutory framework

The British in India introduced several enactments to regulate water resources in the country. These laws include embankment laws to protect and maintain embankments;[32] regulations for ferry, shipping and canal navigation;[33] and

29 Constitution of India, Seventh Schedule, List II (State List), Entry 17.

30 Alan Richards and Nirvikar Singh, Inter State Water Disputes in India: Institutions and Policies (UCSC Department of Economics Working Paper No. 503, 2001), available at http://ssrn.com/abstract=289997 or http://129.3.20.41/eps/dev/papers/0412/0412010.pdf.

31 A. Vaidyanathan, *Water Resource Management: Institutions and Irrigation Development in India* 15 (New Delhi: Oxford University Press, 1999).

32 Bengal Embankment Act, 1855.

33 See, e.g., Northern India Ferries Act, 1878; Bengal Ferries Act, 1885; Inland Vessels Act, 1917; and Merchant Shipping Act, 1958.

regulations for fishing.[34] Irrigation laws introduced during the colonial period prepared the ground for state ownership of water resources. The Northern India Canal and Drainage Act, 1873 which sought to regulate irrigation, navigation and drainage in Northern India recognized in its preamble the right of the Government to 'use and control for public purposes the water of all rivers and streams flowing in natural channels, and of all lakes'. In 1931, Section 26 of the Madhya Pradesh Irrigation Act asserted state control over water as follows: 'All rights in the water of any river, natural stream or natural drainage channel, natural lake or other natural collection of water shall vest in the Government.' The subsequent irrigation laws in independent India went on to consolidate state powers over water resources.

IRRIGATION

The principle of eminent domain is presupposed under Article 31(A) of the Indian Constitution. It is also tacit in the Land Acquisition Act, 1894, the Forest Act, 1927 and other cognate natural resources laws, including the Irrigation Acts. The various Irrigation Acts empower the canal officers to enter any land, remove any obstruction and close any channel in exercise of their powers.[35] In *Secretary of State v. Nageswara Iyer*, the Madras High Court laid down the doctrine that the right of the Government to control the supply and distribution of irrigation waters is not merely a proprietary right but a sovereign one.[36]

In fact, the Mysore Irrigation Act specifically provides that private irrigation works are to be constructed only with the sanction of the state government and subject to conditions which it may impose. The Andhra Pradesh (Andhra Area) Irrigation Works (Repairs Improvement and Construction) Act, 1943 provides for the repair and improvement of private irrigation works, the construction of new irrigation works on private lands and the supply of water from government to private irrigation works. Thus the government was said to have the powers to step into private holdings for repair or improvement of private irrigation works.

The laws relating to surface water and irrigation carried on within the framework set by the British, and the normative order of state ownership of the water resources, remained unquestioned. For instance, the northern states of Uttar Pradesh, Punjab and Haryana draw water from perennial rivers for irrigation and are governed by the Northern India Canal and Drainage Act 1873, which is a comprehensive legislation dealing with irrigation, navigation and drainage. The Rajasthan Irrigation and Drainage Act, 1954 and the Jammu and Kashmir Canal and Drainage Act, 1963 are similar to the Northern India Canal and Drainage

34 Indian Fisheries Act, 1897.

35 These include the Bombay Irrigation Act, 1879; Rajasthan Irrigation and Drainage Act, 1954; Jammu and Kashmir Canal and Drainage Act, 1963; Andhra Pradesh (Telengana) Irrigation Act, 1877 and Mysore Irrigation Act, 1965.

36 *Secretary of State v. Nageswara Iyer* AIR 1936 Mad. 923.

Act. Broadly, these enactments empower the state government to construct works for the use of surface waters in rivers, streams and lakes for irrigation; regulate the supplies of water from the canals; construct watercourses and field channels for carrying water from the outlet of canals to the fields; levy and collect water-rates for the water supplied; and settle disputes regarding the construction and maintenance of water courses and distribution of water among co-sharers.

Similarly, in the eastern regions – West Bengal and Bihar – the Bengal Irrigation Act 1876 provides for the construction, maintenance and regulation of canals, for the supply of water and for the levy of water rates. The Act provides for the levy of water rates on water supplied for irrigation. There are also provisions for cancelling the supply of water to cultivators who are in arrears of water dues. The Act contains elaborate provisions for the construction and maintenance of village channels (watercourses and field channels), but only at the instance of the beneficiaries. Disputes regarding the mutual rights and liabilities of persons in respect of the use, construction or maintenance of such villages are covered by specific provisions.[37]

The Irrigation Commission that was set up to study the irrigation system in the country, in its report of 1972, recommended the consolidation and standardization of the various irrigation laws in different regions of the country. Following up on the recommendation, the Indian Law Institute drafted a model bill. However, states were slow in redesigning or enacting legislation/amendments along the lines suggested by the Model Bill.

GROUNDWATER

There is no comprehensive central legislation to deal with groundwater. The unbridled growth of the groundwater sector has resulted mainly from lack of legislation on the control of groundwater extraction, as well as the provision of free or subsidized power for the agriculture sector. The central government has been urging the states to enact legislation on groundwater, and is even contemplating the adoption of a central legislation on the subject. A Central Groundwater Authority was constituted under the Environmental Protection Act to serve as a watchdog body. The authority does not have policing powers and is dependent on the state governments for implementing its decisions.

The Groundwater (Control and Regulation) Bill was mooted by the government of India in 1970. The draft was then circulated to all the states, with the advice to enact it into an Act, with amendments if necessary. The bill was amended in 1992, 1996 and 2005. However, very few of the states have enacted laws. Gujarat was the first state to introduce a Groundwater Act in 1976. This dealt with the regulation and licensing of tubewell construction, and the control of groundwater use. However, this legislation lacked teeth as it was difficult to enforce.

37 See generally, Alice Jacob and S.N. Singh, *Law Relating to Irrigation* (Bombay: N.M. Tripathi, 1972).

WATER POLLUTION AND WATER QUALITY

During the early years after independence, several states enacted legislation relating to water pollution. These enactments included the Orissa River Pollution Act, 1953, West Bengal Notification No. 7 Regulation – Control of Water Pollution Act, 1957, and the Maharashtra Water Pollution Prevention Act, 1969.

Central legislation has been enacted to ensure that the quality of water is protected.[38] The Water (Prevention and Control of Pollution) Act, 1974 provides for an elaborate administrative scheme through a licensing system to prevent, and protect against, water pollution. The Environmental (Protection) Act, 1986, also provides for water quality through its notifications on permissible quality standards, environmental impact assessments and public hearings. The Coastal Regulation Zone Notification (which may be replaced by the Draft Coastal Management Zone Regulation, 2008) prohibits certain activities such as the discharge of untreated wastes and effluents in coastal areas. In 2000, bottled water was inserted into the Prevention of Food Adulteration Act, 1954. In *MC Mehta v. Kamal Nath*, the Supreme Court held that water is a public trust and that the state is seen as 'the trustee of all natural resources which are by nature meant for public use and enjoyment', thus enabling better protection of water resources.[39]

Common law principles

Presently, the formal legal system assigns property rights to surface water resources to the state, and the rights to groundwater vesting in the owners of land above the subterranean water. Private ownership of water resources is acknowledged as regards artificially constructed tanks and lakes that exist on privately owned land.

RIPARIAN RIGHTS

Riparian owners have a natural right – *ius naturale* – incident to the ownership of the land abutting upon the stream. The establishment and recognition of riparian rights have occurred mainly through the common law method of recognizing customary law and natural rights. In *Hanuman Prasad v. Mendwa*, the Allahabad High Court held that a riparian owner has a right to use the water of a stream which flows past his land equally with other riparian owners, to have the water come to him undiminished in flow, quantity or quality, and to go beyond his land without obstruction.[40] However, this right is subject to the customary rights of the lower riparian to the water flow and no upper riparian can obstruct a public river with a dam.[41]

The doctrine of riparian rights has been rejected wherever serious conflicts

38 See further Chapter XIII.
39 *MC Mehta v. Kamal Nath* (1997) 1 SCC 388.
40 *Hanuman Prasad v. Mendwa* AIR 1935 All 876.
41 *Maung Bya v. Maung Kevi Nyo* AIR 1925 PC 236.

over distribution of water have occurred before river tribunals. The Narmada Water Disputes Tribunal invoked the notion of agreement, close to that of a treaty between two states, as the legal basis for distribution in 1974, so that Rajasthan, which is a non-riparian state, could receive some water. The Narmada Water Disputes Tribunal did accept the doctrine of equitable apportionment as applicable and it acknowledged that the diversion of water of an inter-state river outside the river basin was legal and the need for diversion of water to another basin may be a relevant factor on the question of equitable apportionment, in the circumstances of a particular case. The Eradi, or the Ravi and Beas Water Tribunal similarly rejected the riparian rights doctrine in 1987, when the question of the distribution of water between Punjab and Haryana was addressed.[42]

PRIVATE OWNERSHIP OVER GROUNDWATER

Common law considers groundwater as something attached to the land. Hence, the right to use groundwater was considered as part and parcel of the landowner's right to property. India also inherited this common law tradition. This free-wheeling capture rule combined with the advancement of modern sophisticated technology to extract groundwater from deep aquifers has resulted in over-extraction, and deterioration in quality and quantity of groundwater.[43]

Earlier laws such as the Indian Easements Act, 1882 also followed the common law principle of absolute proprietorship regarding groundwater. The Act recognized the right of every owner of the land to collect and dispose within his own limits of all water under the land which does not pass in a defined channel.[44] It also provides that the rights over groundwater cannot be acquired by prescription.[45] A similar approach is found in earlier case law also.[46] This legal regime of colonial legacy has continued until recently. The state governments have sought through the adoption of legislation to regulate the use of the groundwater without necessarily altering the rights regime over subterranean water. The Single Bench of the Kerala High Court sought in the Plachimada case to review these principles by introducing the public trust doctrine in order to regulate groundwater extraction.[47]

Traditional or customary laws

Much of the formal Indian legal system is of colonial origin, but a parallel, indigenous legal realm (both formal and informal) has survived alongside it. While

42 P.M. Bakshi, Background Paper on Article 262 and Inter State Disputes Relating to Water (Prepared for the National Commission for the Review of the Working of the Constitution, 2002, undated). Available at http://lawmin.nic.in/ncrwc/finalreport/v2b3–6.htm.

43 See further Chapter IX.

44 Indian Easements Act, 1882, sec. 7(b), illustration (g).

45 *Ibid.* sec. 17(d).

46 *Kesava Bhatta v. Krishna Bhatta*, AIR 1946 Mad. 334, 335.

47 *Perumatty Grama Panchayat v. State of Kerala* 2004 (1) *Kerala Law Times* 31.

the historical evolution of water laws from the colonial period is well documented, the legal regime (or regimes) that were overtaken by the colonial structures has, however, not been the subject of comprehensive research.

Colonial rule transformed natural resources management in the country. In 1793, the British introduced permanent settlement, which altered the relationship between government and land resources and introduced the concept of private and individual ownership of resources. All land that was not individually owned was appropriated by the state. A similar approach was adopted as regards forest and water resources. The first major development in water resources regulation came through the Bengal Regulation VI of 1819, which empowered the government to appropriate private rights of ferry through the establishment of public ferry. This marked a radical shift, as it introduced the idea of eminent domain and sought to change earlier practices, which treated water resources as common property for all.[48]

The right to life enshrined in Article 21 of the Constitution has been held to recognize the rights of the community over water.[49] Some statutes also hint at the recognition of community rights. The Indian Easements Act, 1882 recognizes certain customary rights over the use of water. Section 25 of the Limitation Act, 1963 also recognizes the absolute and indefeasible right of use and access to water if it has been enjoyed without interruption for a period of twenty years. The Tribal Self-Rule Act, 1996 and the Panchayats (Extension to the Scheduled Areas) Act, 1996 recognize the right of access of tribal and other communities to natural resources like water.

Section 2 of the Indian Easements Act, 1882 recognizes the absolute right of the state over resources but this absolute right is somewhat curtailed by the recognition of customary rights of all the people. 'The subsequent irrigation laws, however, progressively affirm the state's rights and relegate the customary rights to arenas of lesser and lesser significance.'[50] In India, even before the adoption of the Limitation Act and the Easement Act, various types of servitudes and easements were applied, both in the Hindu and the Muhamadan Law.

The Easements Act, 1882 legitimizes customary rights of the people, and provides two rules for recognition: (a) by long use or prescription and (b) by local customs.[51] Section 2(b) of the Act recognizes customary rights in or over immoveable property which any person, the public, or even the government may possess, irrespective of other immoveable property. Thus, a right may exist by custom in which some people are entitled to take water from another's land. Such a right may be enjoyed by a fluctuating body of persons or even undefined classes. This

48 Iqbal Ahmed Siddiqui, 'History of Water Laws in India', in Chhatrapati Singh (ed.) *Water Law in India* 289, 297 (New Delhi: Indian Law Institute, 1992).

49 *FK Hussain v. Union of India* AIR 1990 (Ker) 321 and *Pazha Karuppiah v. State of Tamil Nadu* AIR 1997 Mad 144.

50 Chhatrapati Singh, *Water Rights and Principles of Water Resources Management* 40 (Bombay: N.M. Tripathi, 1991).

51 Indian Easements Act, 1882, secs. 15 and 18.

includes the rights for water issuing from a well, spring, spout or flowing water in government or public places. According to the Easements Act, a person has, however, no natural or customary right over groundwater, whether collected in a well, or passing through springs or flowing in undefined course.[52] Any diminution of such water by neighbours, hence, gives no ground for action under the Easements Act.

B. Water law reforms

1. General framework for water law reforms

Water sector reforms need to be viewed in the context of the various challenges facing the sector. Investment in surface irrigation systems by the state and the self-provisioning by citizens through groundwater sources to meet all their water needs represent today the primary characteristics of the water sector in India.

Asset maintenance and replacement have consistently been put on the back burner due to poor financial allocation. As a result, the crumbling water infrastructure – both for drinking water and irrigation – requires major investment. Another grave concern, especially in the light of growing scarcity of freshwater, is the scourge of water pollution. Widespread pollution of freshwater sources such as lakes, tanks and rivers are on the rise as aquifers are degraded through human use and saline intrusion. Deteriorating water quality not only impacts the human population, but also ecological and biological diversity.

Inter-sectoral water allocation or the growing conflict between alternative water uses is increasingly prevalent in the country. The need to balance the use of water with its deliberate non-use in order to maintain an environmental balance needs to be prioritized by policy makers. The idea of the interlinking of rivers has been more vigorously debated since 1984, with the setting up of the National Water Development Authority which has been given the task of identifying appropriate inter-basin transfers and examining their feasibility.

Ongoing water sector reforms are informed and guided by certain discourses that have an impact on the law and policy framework in India as well as in many other developing countries. Firstly, the now widely popular argument that water needs to be treated as an economic good, premised on the belief that the inefficient use of water is a direct consequence of the lack of an economic value being attributed to it, has wide policy ramifications.[53] It means that all water services must be based on the principle of full cost recovery. The natural progression of this thinking has been the reduction of subsidies and the increase in water prices for water users (both irrigation and drinking water). In fact, the Swajaldhara Guidelines recommend that water users take up partial responsibility for the

52 *Ibid.*, sec. 2.

53 Dublin Statement on Water and Sustainable Development, International Conference on Water and the Environment, Dublin, 31 January 1992.

capital cost of new drinking water infrastructure, and full responsibility for operation and maintenance.[54]

Secondly, water sector reforms promote decentralization and greater participation of water users. The National Water Policy, 2002 provides a framework for decentralizing decision making to the lowest level and allowing beneficiaries and other stakeholders to be involved from the project planning stage.[55] For instance, the reform initiative launched by the government in the rural drinking water supply sector through the 1999 Sector Reform Project, and since scaled up to cover the entire country through Swajaldhara in December 2002, is based on active community participation in the planning, implementation, operation, maintenance and management of water supply and sanitation schemes.[56] Full ownership of drinking water assets created under Swajaldhara rests with appropriate levels of panchayats. In the case of irrigation, transfer of operation and maintenance is being attempted and users are required to raise the finances to operate and run the irrigation systems.

Thirdly, water sector reforms include several proposals that affect the role that government plays in the water sector. The role of the state under reform process is required to change from that of a service provider to a regulator. This process involves both divesting the government of several functions, while simultaneously vesting it with other powers and functions. Thus on the one hand, state is required to transfer operation, maintenance and management to WUAs to locally manage irrigation schemes. On the other hand, it has entailed the introduction of water regulatory bodies that have vast powers over water resources in each state. With the withdrawal of the state from the role of service provider, new actors are encouraged to step into that role. Incentives are being provided in rural areas for user associations to take on the role, and private sector participation is being promoted in the richer urban areas.

2. *Sector-specific water reforms*

Groundwater

The quantum leap over the years in groundwater use to supplement irrigation, industrial use and drinking water, has been facilitated by technological advancement, liberal funding by government agencies coupled with subsidized or free power supply. It is estimated that the annual groundwater withdrawal for agricultural purposes rose from approximately 20 BCM in 1960 to 230 BCM by the year 2000.[57] The unhindered access to, and withdrawal of, groundwater

54 Ministry of Rural Development, Guidelines on Swajaldhara, 2003, sec. 3(1) available at www.ielrc.org/content/e0212.pdf.

55 See, e.g., National Water Policy, 2002, sec. 6(8).

56 Ministry of Rural Development, Guidelines on Swajaldhara, 2003.

57 Tushaar Shah, 'India's Master Plan for Groundwater Recharge: An Assessment and Some Suggestions for Revision', 43/51 *Economic & Political Weekly* 41 (2008).

has led to the burgeoning of water markets for irrigation and drinking water purposes in water-scarce states such as Gujarat and Tamil Nadu. The water markets are largely informal and unregulated, contributing further to the imbalance in water access and availability. Over-extraction and inadequate recharge has adversely impacted the availability and quality of groundwater, thus necessitating a review of the poor management of groundwater resources. While solutions to enable effective recharge of depleted aquifers are being promoted, the regulation of use of groundwater through administrative and legal reforms is still in its infancy.

The shift towards a better legal and regulatory framework began as early as 1970, when the central government formulated a Model Bill for groundwater in an effort to urge the state governments to adopt the same.[58] The key features of the model bill proposed by the centre include the establishment of a groundwater authority in whom powers are vested for the notification of areas that need registration, control and supervision of groundwater use by the authority. At the central level, other measures were taken under the Environment Protection Act such as the setting up of the Central Groundwater Authority to regulate and control development and management of groundwater resources in the country.[59] Despite several attempts by the central government to persuade states to adopt the model legislation (and the revised versions), progress in this direction was slow until the late 1990s. However, efforts at regulation through restrictions on electricity supply and credit availability has had a limited success. State governments, which have largely been reluctant to legislate have, over the past decade, acknowledged the need for regulation of groundwater use and a number of states have adopted Groundwater Acts.[60]

One of the first states to enact a legislation to regulate groundwater use was Tamil Nadu. The Chennai Metropolitan Area Groundwater (Regulation) Act, 1987 provides for the registration of existing wells, regulation of any wells being sunk and the use of groundwater for non-domestic purposes through the issue of permits by the Chennai Metropolitan Water Supply and Sewerage Board. The Act was amended in 2002 to give the Board more powers to effectively curb misuse by allowing for the seizure of vehicles or other property, and by increasing the penalty payable for any violation.[61] Also, the Andhra Pradesh Water, Land and Trees Act, 2002 seeks to, *inter alia*, promote water conservation and regulate

58 Model Bill to Regulate and Control the Development and Management of Ground Water, 2005, available at www.ielrc.org/content/e0506.pdf.

59 Environment (Protection) Act (1986), MoEF Notification dated January 14 1997.

60 See, e.g., Kerala Ground Water (Control and Regulation) Act, 2002, available at www.ielrc.org/content/e0208.pdf; Andhra Pradesh Act to Promote Water Conservation, and Tree Cover and Regulate the Exploitation and Use of Ground and Surface Water for Protection and Conservation of Water Sources, Land and Environment and Matters, Connected Therewith or Incidental Thereto, 2002, available at www.ielrc.org/content/e0202.pdf; and Goa Ground Water Regulation Act, 2002, available at www.ielrc.org/content/e0201.pdf.

61 See, e.g., Chennai Metropolitan Area Groundwater (Regulation) Amendment Act, 2002, available at www.ielrc.org/content/e0219.pdf

the exploitation and use of groundwater and surface water for protection and conservation of water sources, land, and environment. The legislation states that all groundwater resources in the state shall be regulated by the Authority set up under the Act and requires that the owners of all the wells, including those which are not fitted with power-driven pumps, and water bodies in the state register their wells/water bodies with the Authority.[62]

Besides legislative developments, courts have also addressed the issue of control over groundwater. The most significant recent case in this area is the decision of the High Court of Kerala ruling that groundwater is a private water resource and accepting the proposition that landowners have proprietary rights over groundwater.[63] It also refrained from accepting that panchayats have rights over this essentially private resource. An appeal is pending before the Supreme Court.[64]

Drinking water

The provision of drinking water is primarily the responsibility of the states. Yet, the central government has played a very important role in fashioning the policies that states apply and has provided significant funding to ensure access to water in rural areas. Urban drinking water has witnessed greater corporatization of public utilities and increased participation of the private sector.[65]

The most important body at the national level is the Rajiv Gandhi National Drinking Water Mission (RGNDWM), which functions within the Department of Drinking Water Supply established under the Ministry of Rural Development. It has been the key institution with regard to the development of policies and the administration of the rural drinking water sector. Among the schemes it implements, the Accelerated Rural Water Supply Programme (ARWSP), which is funded by the Government of India and state governments, plays a central role. This was first introduced in 1972, and, with the exception of one period of interruption during the 1970s, it has been a central component of the government's attempts to ensure full coverage of all habitations throughout the country. It continues to provide the basis for the central government's interventions in rural drinking water.

The ARWSP Guidelines provide the core framework used by the RGNDWM to ensure the provision of drinking water to all habitations in the country.[66] The Guidelines provide several key policy elements. Firstly, they define the different levels of coverage. Non-covered habitations are defined as having access to less than 10 lpcd. Partially covered habitations are those having access to 10 to 40 lpcd. Covered habitations are defined as having access to 40 lpcd. The figure of 40 lpcd, used to determine the minimum level of coverage necessary to define a habitation

62 Andhra Pradesh Water, Land and Trees Act, 2002, Chapter. 3.

63 *Hindustan Coca-Cola Beverages v. Perumatty Grama Panchayat* 2005 (2) *Kerala Law Times* 554.

64 See further Chapter IX.

65 See further Chapter VIII. See also Chapter VII.

66 Government of India, Accelerated Rural Water Supply Programme Guidelines.

as covered, has been determined through an amalgamation of figures for different basic minimum uses of water. These include three litres for drinking, five litres for cooking, fifteen litres for bathing, seven litres for washing utensils and the house, and ten litres for ablutions.

Quantity is not the only criterion to determine whether a habitation is covered. The source of water also needs to be within 1.6 km or 100-metre elevation in mountain areas. The water should also not be affected by quality problems even though no specific standards for determining quality are included. Another criterion is that a given public source of water such as a hand-pump should not be used to serve more than 250 people. The ARWSP Guidelines also acknowledge the direct link between drinking water for human beings and water for cattle. Consequently, in a certain number of states especially affected by drought, the guidelines mandate that an additional 30 lpcd should be provided for cattle.[67]

The minimum level of 40 lpcd is acknowledged by the RGNDWM as a minimum level of coverage which should be increased over time. Thus, in states where all habitations have been covered at the level of 40 lpcd, the Government of India has approved that the next level of service should be 55 lpcd within 500 metres of the house or 50-metre elevation in mountain areas.[68]

In the context of the overall water sector reforms, the government adopted in 2002 a new scheme known as Swajaldhara, which proposes to foster new types of intervention to ensure better drinking water availability in villages. Swajaldhara imposes on communities a contribution of at least 10 per cent of the capital costs for a service level of 40 litres per person per day and imposes that they take 100 per cent responsibility for operation and maintenance. It also imposes that the contribution of the community to capital costs should be at least 50 per cent in cash. Further, under Swajaldhara, only individuals or households that make the first 10 per cent contribution will benefit from the schemes being implemented. Others are not part of the scheme.

The Swajaldhara scheme has now been mainstreamed and its principles are applicable to all ARWSP projects. This scheme is a demand-focused strategy, in which people will get the service they are willing to pay for.[69] The basic economic rationale of Swajaldhara is that people should be made to pay for part of the capital costs of drinking water projects, and for the whole cost of operation and maintenance. A basic underlying assumption of Swajaldhara is that the cost sharing will enable participation in implementation and ownership of the assets by communities and a progressive withdrawal of the state from providing drinking water.[70]

67 *Ibid.*, sec. 2.2.2.

68 Rajiv Gandhi National Drinking Water Mission Department of Drinking Water Supply, Submission to the National Advisory Committee (2005).

69 Ministry of Rural Development, Guidelines on Swajaldhara, 2003.

70 Preeti Sampat, 'Swa'-jal-dhara or 'Pay'-jal-dhara – Sector Reform and the Right to Drinking Water in Rajasthan and Maharashtra', 3/2 *Law, Environment and Development Journal* 101 (2007), available at www.lead-journal.org/content/07101.pdf.

Irrigation

In the early years after independence, government policy on water resources focused largely on large and medium irrigation projects with a view to stabilizing and increasing food production, generating hydro-electric power and ensuring flood control of major rivers prone to flooding. The irrigation enactments, primarily enacted by the colonial powers, vest control and ownership over surface water in the government. This state-centric approach is now partially sought to be changed by envisaging a greater role for the farmers in irrigation management, through WUAs.

The setting up of WUAs has been taken up with increasing intensity over the past decade and a number of states have introduced WUA legislation. These range from Andhra Pradesh and Madhya Pradesh to Orissa and Rajasthan.[71] Broadly stated, the enactments provide for the setting up of and functioning of WUAs with a three- or two-tier structure allowing for the federation of the users at highest level. The Acts provide for maintenance and repairs of the irrigation works by the users, for the collection of user charges and the regulation and monitoring of water distribution among WUA members. Presently, the WUAs are under the overall supervision of the irrigation departments, but some of the enactments provide for the irrigation officers to be responsible to the WUAs.

The duplication and *ad hoc* introduction of the enactments pose several problems. These Acts provide for a uniform model of WUAs without taking into account existing arrangements at the local level. Further, these associations are not integrated into existing decentralized frameworks of panchayat raj institutions. The relationship of local panchayats' control over water resources and the powers of other management bodies, particularly WUAs, may need legal clarification. More importantly, since many of these statutes follow a top-down approach to formation of associations, the efficacy and sustainability of the new institutions are being tested.

An essential part of the reform process has been an overhaul of the manner in which the water sector is being managed, amongst other things through the setting up of independent water regulatory authorities in some of the states.[72] The Andhra Pradesh Water Resources Development Corporation Act was enacted during the early days of the reform process in 1997. Much water has flown under the policy bridge since then and a lot more thought has been put in by the

71 Andhra Pradesh Farmers Management of Irrigation Systems Act, 1997, available at www.ielrc.org/content/e9701.pdf: Madhya Pradesh sinchai prabandhan me krishakon ki bhagidari adhiniyam, 1999, available at www.ielrc.org/content/e9902.pdf; Orissa Pani Panchayat Act, 2002, available at www.ielrc.org/content/e0211.pdf; and Rajasthan Farmers' Participation in Management of Irrigation Systems Act, 2000, available at www.ielrc.org/content/e0004.pdf.

72 Philippe Cullet, 'Water Law – Evolving Regulatory Framework', *in* Philippe Cullet, Alix Gowlland Gualtieri, Roopa Madhav & Usha Ramanathan (eds.) *Water Law at the Crossroads – National and International Perspectives with Special Emphasis on India* 26 (New Delhi: Cambridge University Press, 2009).

policy-makers as to the course of the reform agenda. This change of approach can be gauged from the Maharashtra legislation which came about more recently in 2005.[73]

Apart from Maharashtra, the state of Arunachal Pradesh has enacted the Arunachal Pradesh Water Resources Management Authority Act, 2006. Several other states are contemplating similar enactments and others are being pushed by IFIs through clauses in their loan agreements. For instance, Madhya Pradesh is required to establish a State Water Resources Agency and to take certain steps to put in place a State Water Tariff Regulatory Commission under loan conditions laid down by the World Bank.[74] The concept of independent regulatory authorities is nascent in the country and their evolution is evidenced in each new authority being set up.[75]

C. Control over and access to water

The legal framework concerning control over and access to water in India is complex and intricate. While surface water has had a complex history of state and community ownership, groundwater has been deemed to belong to the owner of the land. Control and access issues are determined not only from a legal perspective, but social and economic considerations also have an impact on how water is managed and used.

1. Analysis of evolving trends

Water reforms have an impact on control over and access to water resources. Firstly, the lack of an explicit, basic right to water in the Constitution has an important bearing on future disputes over the resource, although as mentioned above, water has been recognized as a fundamental human right under Article 21. In addition to interpreting the right to life as including the right to water, the Supreme Court has, in the context of water pollution, mandated the cleaning up of water sources including rivers, the coastline and tanks and wells.[76] It has also applied the precautionary principle to prevent potential pollution of drinking water sources consequent upon the setting up of industries in their vicinity.[77] As stress on water resources increase, a more proactive legislative framework applicable to potential challenges to rights to water resources may need to be put in place.

Secondly, it is necessary to recognize water resources as common property in certain clearly defined ways, so as to protect both human and environmental needs. For instance, the continued apathy of successive governments over

73 Maharashtra Management of Irrigation Systems by Farmers Act, 2005, available at www.ielrc.org/content/e0505.pdf.

74 World Bank, Madhya Pradesh Water Sector Restructuring Project 3 (Project Appraisal Document, Report No. 28560-IN, 2004).

75 See further Chapters VII and X.

76 *Vellore Citizen Welfare Forum v. Union of India* AIR 1996 SC 2715.

77 *Indian Council for Enviro-Legal Action v. Union of India* AIR 1996 SC 1446.

groundwater regulation brings to fore the need to engage more vigorously in reforms in this sector. The Supreme Court has also acknowledged that water is a community source which is to be held by the state in public trust in recognition of its duty to respect the principle of inter-generational equity. In *MC Mehta v. Kamal Nath*, it declared that 'our legal system – based on English common law – includes the public trust doctrine as part of its jurisprudence. The state as a trustee is under a legal duty to protect the natural resources. These resources meant for public use cannot be converted into private ownership'.[78]

Thirdly, several areas of conflict also influence control over, and access to, water resources. This is the case with regard to inter-state water disputes over sharing of waters, which have adversely impacted legitimate rights of citizens over water resources. For instance, the dispute between the states of Tamil Nadu and Karnataka over the sharing of the Cauvery River has been long-drawn-out and politically charged. Likewise the disputes between Punjab, Haryana and Himachal Pradesh over the sharing of the waters of Ravi and Beas rivers continue to fester, despite the final award of the tribunal.

Conflicts between local village-level governments and the state government also impact upon the control over and access to water resources. In the case of Tarun Bharat Sangh, a people's initiative at reviving traditional water harvesting systems was turned down by the irrigation department of the state of Rajasthan, which issued a notice to them claiming that the *johads* built in the village of Lava Ka Baas were technically unsafe and illegal, and hence needed to be demolished.[79] The irrigation department's role in sidelining the efforts of the community initiative is indicative of the serious challenge to decentralized governance posed by vested interests within the bureaucracy.

Again, as stated above, groundwater continues to be an area of concern for the future. The unregulated appropriation of groundwater, viewed largely as a private resource, is now beginning to pose a serious threat as communities begin to question this common law legal interpretation. This is best illustrated by the *Plachimada* case, which entails a dispute between a bottling unit of Coca-Cola and the village panchayat of Perumatty in the Palakkad District over groundwater extraction.[80] The case, which is still pending before the Supreme Court, illustrates the concerns existing regarding the control of local governance bodies over local water resources.

Conflicts within communities make up the other big challenge to control over and access to water resources. Segregation and even denial of access to certain water sources is a prevalent practice in many parts of the country. The Scheduled

78 *MC Mehta v. Kamal Nath* (1997) 1 SCC 388.

79 S. Muralidhar, 'The Right to Water: An Overview of the Indian Legal Regime', in Eibe Riedel and Peter Rothen (eds.) *The Human Right to Water* 65 (Berlin: Berliner Wissenschafts-Verlag, 2006).

80 For a more detailed discussion of the case, see Sujith Koonan, 'Groundwater – Legal Aspects of the Plachimada Dispute', in Philippe Cullet, Alix Gowlland Gualtieri, Roopa Madhav and Usha Ramanathan (eds.) *Water Law at the Crossroads – National and International Perspectives with Special Emphasis on India* 159 (New Delhi: Cambridge University Press, 2009).

Castes and Scheduled Tribes (Prevention of Atrocities) Act, 1989 makes it a punishable offence if a person, not being a member of the Scheduled Caste or Scheduled Tribe, corrupts or fouls the water of any spring, reservoir or any other source ordinarily used by the members of the Scheduled Castes or Scheduled Tribes so as to render it less fit for the purpose for which it is ordinarily used.[81]

Moreover, untouchability, although abolished, is still prevalent in several parts of the country. Despite constitutional guarantees, access to basic services such as health, shelter, food and water are determined by social status in India. Dalits, particularly, are subject to inequalities in access to water resources – both for the purposes of agriculture and domestic use. Discrimination takes many forms, for instance, Dalits are served water in separate glasses in public spaces such as hotels and temples; have to stand in their own separate queues at public stand posts; are not allowed to use certain ponds and tanks in a village; or are made subject to separate schedules in water collection. These discriminatory practices have a higher impact on Dalit women, because, as in most households they are responsible for fetching the water for domestic use.

Studies indicate that the poor economic and social status of the Dalit community impacts their right to water. This has a bearing on water sector reforms too. Firstly, it is found that the right to groundwater is closely linked to ownership of land. In the absence of adequate irrigation and drinking water, the only reliable source to tap is groundwater. The small size of land holdings and lack of capital to install bore wells further enhances these inequities.

Secondly, the water sector reforms which aim to provide for drinking water and sanitation require beneficiaries to contribute to the schemes. The economic situation of the Dalits prevents them from effectively participating in these schemes, excluding them from the benefits of these basic services. Thirdly, the new framework for irrigation encourages a participatory approach to water resources management. Essentially, the operation and management of irrigation canals are being handed over to the WUAs in several states. Social inequities further accentuate the differences in participation and decision-making powers of Dalit farmers. The few states that currently have Participatory Irrigation Management laws do not provide for a more inclusive structure, nor do they seek to bridge the gap in gender and caste inequalities within WUAs. These new institutional arrangements thus continue to perpetuate historical wrongs and inequities.

2. *Towards sustainability – An agenda for the future*

The context in which reform is being undertaken is enormously challenging. A clearly identifiable policy thrust, backed largely by external institutions such as the World Bank, has informed the initial phase of reforms. These include, for instance, a demand-led approach, cost sharing for investments, the setting up of

81 Scheduled Castes and Scheduled Tribes (Prevention of Atrocities) Act, 1989, sec. 3(xiii)

WUAs, the establishment of water rights, the reduction of the role of the government in the water sector, and an increase in water charges.

It is hoped that the second phase of reforms and policy changes will reflect in a more realistic manner concerns related to the right to water, while reorienting the debates towards a more sustainable and equitable approach to water sector reforms. Some key identifiable issues are noted here as a possible agenda for the future:

(a) A constitutional amendment is needed to incorporate a fundamental right to water as a specific right.
(b) While decentralization needs to be a continued focus, the challenges of inter-sectoral allocation of water and issues around inter-state water sharing needs a clear mandate which can only be adopted with concerted intervention at the federal level.
(c) The role of the existing panchayat raj institutions needs to be effectively woven into the reform efforts so as to strengthen decentralization and participation being attempted through users associations in irrigation and rural drinking water schemes.
(d) Efforts to regulate groundwater use do not go far enough in seeking to provide equitable and socially just sharing of existing groundwater. Further, new laws do not include sufficient safeguards to ensure that users of groundwater who are poor are not harmed by the new controls and charges. This needs to be factored into the future agenda of reforms for the water sector.
(e) The present phase of reforms introduced through legislation by the water regulatory authorities, enables a gradual rolling back of the state and greater participation of the private sector in the water sector. Reflecting the dominant thinking within policy circles, it regards water as an economic good. The next phase of reforms must broaden and strengthen the understanding of water use and management to include its social, health and environmental aspects.
(f) Targeting of the most vulnerable populations for reform measures and retaining state responsibility is crucial to an effective reform agenda. The paradoxical policy initiatives being adopted by the states need to be reviewed and aligned so as to take forward the broader objectives of a right to water. To enable this, an identification of elements that constitute basic minimum (core) rights of access to water, including safe drinking water, water for sanitation and health, must be undertaken.

CONCLUSION

The reforms initiated in the country need to be viewed as an opportunity to weave in a more holistic approach to the water sector. While addressing issues of efficiency, cost recovery, and better financial, institutional and legal frameworks, it is equally important that the earlier focus on a more equitable and sustainable framework is not lost sight of. In this, emerging debates around a basic right to water is a direct response to the rather limited economic perspective to present day water sector reforms.

Rights and the rights debates are morphing into a crucial component underpinning the future of water as a resource, globally and nationally. Environmental concerns have led to other important aspects of water rights being addressed in the country. Apart from legislation on water – namely The Water (Prevention and Control of Pollution Act), 1974 – judicial concern over water pollution and access to clean water has enlarged the constitutional, tort and criminal jurisprudence on water pollution. It is only since the mid-1990s, with the focused attention on reforms to the water sector, that the scope of the debate on rights expanded to include drinking water, irrigation, groundwater, and water for industrial use and other commercial purposes. Envisioning the next phase of water sector reforms to primarily take forth issues of sustainability, conservation and those related to the human right to water is crucial to the future of water resources in the country.

VII. Institutional reforms for water

Priya Sangameswaran and
Roopa Madhav

INTRODUCTION

Institutional arrangements for water management play a vital role in managing and sustaining water resources. Hence an important part of the water sector reforms being undertaken in India in the last decade is a focused and concerted effort at institutional improvement and institutional restructuring. Efficiency and performance of existing institutions are being evaluated and new institutions introduced through changes in the policy and legal framework. Informed largely by environmental economics and new institutional economics, the changes are promoted and funded by international financial institutions (IFIs). The World Bank, a prime mover of water sector reforms in India, has made funding to governments conditional on the implementation of large-scale institutional reforms. The Bank, however, acknowledges the challenges posed by resistance from what it calls 'political economy factors' and hence has initiated only strategic institutional reforms in the initial phase.[1] We discuss in this chapter both the tentative and the radical new steps in restructuring the water sector institutions in India.

The context within which institutional reforms are being carried out needs to be placed in perspective. A large part of the water sector of the country (roughly 60 per cent) is in the informal sector and is characterized largely by self-provisioning, informal networks and open access or common property water resources.[2] The relatively small degree of formalization of the water sector is borne out by the statistics put out by the surveys of the National Sample Survey Organization.[3]

1 See generally World Bank, Country Strategy for India (Washington D.C.: World Bank, Report No. 29374-IN, 2004).

2 See generally T. Shah, The New Institutional Economics of India's Water Policy (Paper presented at the International Workshop on 'African Water Laws: Plural Legislative Frameworks for Rural Water Management in Africa', Johannesburg, January 26–28 2005), available at www.nri.org/projects/waterlaw/AWLworkshop/SHAH-T.pdf.

3 Initiated in the year 1950, the National Sample Survey, renamed in 1970 as the National Sample Survey Organization (NSSO), is part of the Ministry of Statistics and Programme Implementation of the Government of India, and undertakes continuous (in the form of successive rounds) large-scale nationwide socio-economic surveys.

Only 8.8 per cent of rural households surveyed were linked to a public or community water supply system and only 17.3 per cent of villages surveyed had access to a public irrigation system.[4] In effect, the institutional reforms being discussed in this chapter impact a relatively small proportion of citizens, but have far-reaching implications for institution building in the future.

Three sets of institutional changes – regulation, privatization and decentralization – will be discussed, with a focus on formal institutions in Section A.[5] Three broad sets of issues are taken up for analysis in Section B – the division of labour resulting from the reforms and its implications for accountability, the extent to which the reforms increase rights and powers of users/local bodies, and the compatibility of reforms with a right to water. Since implementation of these reforms has in most cases begun recently, the emphasis will be on the conceptualization of reforms, although wherever possible insights will also be drawn from their working on the ground.

A. Current institutional reforms of water sector

Under the broad heads of regulation, privatization and decentralization, we try and capture the key aspects and concerns of the institutional reforms being undertaken in the water sector in the last decade and a half. While decentralization has been a long-standing demand within the irrigation sector (at least from certain constituencies), regulation and privatization are responses to donor-driven conditions and concerns. Even in the case of decentralization, there are differences between the demands of local groups and water users associations that pre-existed the reforms phase and the form in which it is conceptualized in the current set of enactments being promoted by the reform process. It is also important to note that the legislative domain is now a major arena for bringing in institutional reforms, and both international donors as well as the Government of India encourage state governments to put in place a legislative framework that is conducive to reform in water. For instance, the guidelines for World Bank lending for 2005–2008 point out that the Bank would consider full-scale investment lending in the urban water supply and sanitation sector only if states have an adequate legislative and regulatory framework.[6] Thus access to credit is linked directly to legal and institutional reforms being carried out. Drawing out all implications of this aspect is beyond the scope of this chapter. But one consequence that we would like to mention is that far-reaching changes are codified and taken up on a large scale with one stroke, often without adequate thought about the kind of processes that the changes would put in place.

4 National Sample Survey, Drinking Water, Sanitation, and Hygiene in India (New Delhi: Government of India, Report No. 449, 1999).

5 Adapting institutions to the new mantra of an integrated water resources management framework is also an important aspect of the reform process; however, these will not be discussed in this chapter.

6 World Bank, note 1 above, Annex 5.

1. Regulation

An important new feature of the water sector reforms is the introduction of Water Regulatory Authorities. Presently conceived of as a quasi-judicial body, it seeks to promote a 'fiction of apolitical decision making'.[7] These regulatory authorities or neo-governments are institutions created for better governance that would protect consumer interests, ensure development and regulate markets by acting as an independent umpire, without being weighed down by political or social constraints. The underlying premise of these neo-governments is a perceived need for an independent regulator who can ably balance differing interests,[8] while promoting competition and efficiency.

Regulation may be broadly understood 'as an effort by the state to address social risk, market failure or equity concerns through rule based direction of social and individual action'.[9] In other words, regulation seeks to minimize or manage risks arising out of market volatility, while simultaneously seeking to achieve social objectives of equity and social justice in the provision of basic services. Thus, as stated earlier, the primary aim of regulation is to further efficiency and competition, while also protecting public interest. Until two decades ago, state regulation dominated all spheres of economic life in India. Post liberalization in the early 1990s, regulatory institutions have been set up in utilities such as telecom, power, and securities. Largely ad hoc, the setting up or the conceptualization of these authorities has no uniformity, nor does a cohesive regulatory policy exist in the country.

Three theoretical approaches inform the idea of regulation the world over. The public interest view of regulation sees the regulator as necessary to oversee and protect the interests of the community and enable community-based decision making. The private interest view of regulation regards regulation as handing over of regulatory functions from the state to private actors. The third approach is the institutionalist approach which claims that the public–private divide is overstated and the regulator can play the role of a facilitator between different stakeholders.[10] However, in most developing countries, the regulatory model is being adopted largely at the instance of international financial institutions, even in the absence of a larger normative framework within the country that would assist in the functioning of these regulatory authorities.

The Maharashtra Water Resources Regulatory Authority Act (MWRRA) is the first legislation to constitute a regulatory authority in the water sector. The Andhra Pradesh Government in 1997 enacted the Andhra Pradesh Water

7 Navroz Dubash, 'Independent Regulatory Agencies: A Theoretical Review with Reference to Electricity and Water in India', 43(40) *Economic & Political Weekly* 43 (2008).

8 M.S. Sahoo and C.K.G. Nair, 'Regulatory Independence: Misunderstood and Muddled', *The Economic Times*, 4 July 2008, available at http://economictimes.indiatimes.com/Opinion/Policy/Regulatory-Independence-Misunderstood–muddled/rssarticleshow/3190556.cms?curpg=1.

9 See Planning Commission of India, Approach to Regulation: Issues and Options 2 (New Delhi: Planning Commission of India, 2006), available at http://planningcommission.nic.in/reports/genrep/infra_reglawl.pdf.

10 Dubash, note 7 above, 48.

Resources Development Corporation Act, but the Corporation, though not strictly a regulatory authority, has focused largely on raising funds for completion of irrigation projects. The legislation was a precursor to the MWRRA enactment. The MWRRA Act was enacted in 2005 and the Authority was established in the same year. The MWRRA is unique in many ways. Apart from the powers to regulate and apportion entitlements between different categories of use, it is empowered to fix the criteria for water charges based on the principle of full cost recovery of management, administration, operation and maintenance of irrigation projects. It has the further mandate of monitoring the State Water Plan. The emphasis, like in the case of Andhra Pradesh, is heavily on irrigation finance and cost recovery.

Another important feature of the authority is the power to lay down criteria for the issuance of water entitlements and trading in water entitlements and quotas. The Act defines entitlements as 'any authorization by any River Basin Agency to use the water for the purposes of this Act'.[11] It empowers the regulatory authority to 'fix the criteria for trading water entitlements or quotas on the annual or seasonal basis by a water entitlement holder'.[12] Acknowledging that the creation of a water entitlements system is at the heart of the MWRRA Act, Saleth, one of the prime architects of the World Bank conception of water rights and entitlements, notes that 'while establishment of individual and transferable water entitlements is the long-term strategy, the Bill [now an Act] adopts a politically and administratively pragmatic intermediate strategy of establishing water entitlements for entities such as water user organizations, urban and rural water supply agencies, and industries'.[13] Premised as it is on the logic of ensuring allocation to the highest value user, it is in blatant contradistinction to privileging the needs of socio-economic justice. Small farmers with access to irrigation waters may, in the new scheme of things, find it profitable to sell the water for purposes other than irrigation, where the price of the traded commodity is higher.

The Authority consists of a panel of experts from water resources management, economics, and river basin representatives and is headed by a retired Chief Secretary. The trend towards expert or external regulation and control of water requires governments to give up their discretionary power to decide on several issues, including tariffs. The apolitical nature of the regulatory authorities also carry with them an inherent understanding of their role as furthering the cause of economic efficiency and the creation of greater space for private actors in the water sector.

The MWRRA can only determine water tariffs for various categories of use and does not have the powers to determine tariff rates at the consumer level such as municipal corporations. Also its powers do not extend to the development

11 Maharashtra Water Resources Regulatory Authority Act, 2005, s. 2(1)(i), available at www.lead-journal.org/content/05080.pdf.

12 *Ibid.* Section 11.

13 Maria Saleth, 'Water Rights and Entitlements', in John Briscoe and R.P.S. Malik (eds.) *Handbook of Water Resources in India* 282, 300 (New Delhi: Oxford University Press, 2007).

and management of groundwater resources. Presently, the Water Supply and Sanitation Department and Groundwater Supply Development Authority, in consultation with the MWRRA, which will also function as the State Ground Water Authority, are preparing the draft bill on Maharashtra Groundwater Development and Management.

Apart from Maharashtra, the state of Arunachal Pradesh has enacted the Arunachal Pradesh Water Resources Management Authority Act, 2006. Several other states are contemplating similar enactments. While the Maharashtra and Arunachal enactments are focused on irrigation, Gujarat is contemplating an enactment to regulate drinking water supply. The Gujarat Water Regulatory Commission Bill, 2006, conceives of a regulatory authority whose primary function is to regulate the water supply and sewerage services for the entire state, set the water tariffs, lay down and monitor quality standards of service and adjudicate disputes between service providers and consumers. Similar such enactments are being pushed by IFIs through conditions in their loan agreements. Madhya Pradesh is supposed to establish a State Water Resources Agency and to take certain steps to put in place a State Water Tariff Regulatory Commission as per World Bank loan conditions. As per the conditions of a loan to the State of Uttar Pradesh, the World Bank has sought the establishment of a State Water Tariff Regulatory Commission. Another state that might be putting in place a similar organization is Tamil Nadu. As per the conditions of a loan from the World Bank, the state is supposed to establish the State Water Resource Management Agency.[14] Interestingly, different regulatory functions are being assigned to the newly emerging authorities, in what appears clearly to be a process of experimentation with various kinds of models in different states.

Yet the basic question of how regulation would work for a complex resource such as water still remains. Unlike the telecommunications and the electricity sectors, where regulatory authorities have been introduced in India, the nature of water as a resource (as opposed to a commodity) vital for human life and ecology clearly should transform the conception of a regulator. However, as the discussion in Section 3 shows, it does not seem as if the regulatory reforms have taken into account the unique nature of the resource.

2. *Privatization*

The kinds of privatization that have been encouraged in India in rural and urban drinking water as well as irrigation have been discussed in earlier chapters. In this sub-section, we discuss the particular forms that have already been taken up. But before turning to this discussion, greater clarity about the meaning of privatization is essential.

14 World Bank, Tamil Nadu Irrigated Agriculture Modernization and Water Bodies Restoration and Management Project, 2007, Project ID: P090768, available at www.worldbank.org.in/external/projects/main?pagePK=64283627&piPK=73230&theSitePK=295584&menuPK=295617&Projectid=P090768.

Firstly, there are different views about what precisely constitutes privatization. One opinion is that privatization occurs only when the ownership of the resource and the right to distribute (or not to distribute) in a particular manner is transferred to a private organization. Another view is that transfer of ownership of the resource and the right to distribute it, constitutes just one form of privatization; involvement of private sector parties in specific functions of water provision could also broadly be referred to as privatization. For instance, private involvement in the water sector can be organized as a service contract, a lease contract, a management contract, a Build-Operate-Transfer (BOT) contract, a Build-Operate-Own contract, a concession contract, or a divestiture. Each of these would have different characteristics with respect to asset ownership, capital investment and operations and management, duration of contracts, and who bears the commercial risk.[15] More recently, the model of public–private partnerships has been introduced and new methods of privatizing emerge as investors experiment and innovate. It is in this broad sense of referring to different forms of private sector involvement that the term privatization is used in this chapter. It must be noted that in none of these forms of privatization is the ownership of the water source itself privatized.

Secondly, involvement of private players is not a new phenomenon, be it in drinking water or irrigation. But the current forms of privatization are different on a number of counts. Earlier private operators were mainly individuals – individual farmers in the case of tubewell-based water markets, and contractors in the case of tanker water supply, with the latter often operating under contracts from civic or government authorities.[16] In the new forms of privatization in urban drinking water and irrigation, players are mostly corporations (including MNCs), although in rural drinking water, private players are typically individual actors and small local companies. Also, there is now an increasing push to have consolidated contracts (as against the single-function contracts of the past). These differences need to be kept in mind while considering the policy and legal implications of the current reforms.

India has seen a number of different kinds of privatization attempts in the last decade. These cover surface water and groundwater, as well as different uses of water (drinking, irrigation and industrial). In the realm of surface water, for instance, an Indian company, Radius Water Ltd., entered into a BOOT (Build-Operate-Own-Transfer) agreement with the Madhya Pradesh government in 1998, whereby it would lease in a part of the river Sheonath for a period of twenty-two years and develop water sources (through construction of up to three barrages) in order to supply water to existing and expected industries in the Borai Industrial Growth Centre in Durg District. Although civil society protests eventually led to cancellation of the contract, there have been other attempts at

15 Lyla Mehta with Oriol Mirosa Canal, Financing Water for All: Behind the Border Policy Convergence in Water Management (Brighton: Institute of Development Studies, IDS Working Paper No. 233, 2004), available at www.ids.ac.uk/ids/bookshop/wp/wp233.pdf.

16 Lalitha Sridhar, Water: The Privatization Debate, *India Together*, 2003, available at www.indiatogether.org/2003/nov/env-wtrdebate.htm.

privatization of surface water for industrial use such as the Tiruppur water supply in Tamil Nadu. Private players have also been involved in industrial extraction of groundwater as in the bottled water and soft drinks industry; one example of this is the case of Coca-Cola in Plachimada (Kerala) and Mehdiganj (Uttar Pradesh).

Privatization in irrigation has taken the form of irrigation development corporations, which primarily raise funds through the open market for funding irrigation projects. In Karnataka, for instance, the Krishna Bhagya Jala Nigam was formed under the Companies Act in 1994 with the specific mandate of mobilizing public funds for the development of the Upper Krishna project. The Government of Maharashtra also established five Irrigation Development Corporations in the 1990s in order to accelerate the completion of irrigation projects. More recently, the Maharashtra Krishna Valley Development Corporation invited expressions of interest for private sector participation in the Nira-Deoghar project in the Pune district of Maharashtra.

In the arena of urban drinking water, local bodies are beginning to directly enter into partnerships with private companies to undertake some or all of the functions involved in the provision of drinking water. Examples include the pilot projects under the Karnataka Urban Water Sector Improvement Project in Hubli-Dharwad, Gulbarga and Bijapur where a French company Compagnie Générale des Eaux is involved in the distribution of water, and Chandrapur in eastern Maharashtra where the entire water supply system has been handed over to a private company consisting of local contractors. However, concerns about equity (particularly as a result of the increase in prices that privatization is likely to result in) and also the negative experiences of privatization in other parts of the world have led to protests by civil society groups in many parts of the country. As a result, some initial attempts at privatization in urban water have been aborted or at least postponed, for instance, in Delhi.

It must be noted that the kinds of contracts used vary depending on the functions being privatized. Most of the contracts in industrial water, hydro-electricity as well as sanitation-related fields such as sewerage and solid waste management are of the BOT or BOOT type. Urban water contracts have typically been management or service contracts. Further, the guidelines for sector reforms and successful public–private partnerships in the urban water and sanitation sector, developed by the Ministry of Urban Development and Poverty Alleviation, emphasize that benefits from private sector participation grow as increasing responsibility and risk is transferred to the private sector. That is to say, service and management contracts generate fewer benefits than deeper forms of private sector participation such as leases, concessions and divestitures. However, the former may be preferred in the initial stages – namely until market conditions are better developed – because they are simpler to execute.[17] This in turn implies that over time, one may see a change in the pattern of privatization contracts that are used.

17 Ministry of Urban Development and Poverty Alleviation, Guidelines for Sector Reform and Successful Public–Private Partnerships, 2004 [hereafter 2004 Guidelines].

3. Decentralization

Decentralization has been a core theme in water governance reforms for more than a decade. But it has taken two different forms in the Indian context. One form of decentralization has been the devolution of greater powers to panchayati raj institutions (PRIs), particularly in the aftermath of the 73rd and 74th constitutional amendments in 1993. The second form of decentralization involves the creation of a new set of institutions at the local level (called user groups) and the devolution of certain functions (such as the management of natural resources) to these institutions since the 1980s and particularly in the 1990s. There are two major spheres in which this form of decentralization has taken place: forestry (via Joint Forest Management programmes) and water (via Watershed Management and Participatory Irrigation Management programmes). The two forms of decentralization have often followed distinct trajectories, resulting in two different kinds of institutions emerging – the constitutionally mandated and legally accountable PRIs, on the one hand, and user groups which may or may not have statutory status, on the other hand. Although NGOs may be involved in both kinds of attempts (especially in the arenas of rural drinking water and watershed development), they typically have greater autonomy, powers and/or functions in the case of the second form of decentralization; in the first form, their functioning is dependent on PRIs (at least on paper).

Both PRIs and user groups have their own advantages and disadvantages. One of the major reasons for earlier attempts to privilege user groups over PRIs was that the former were believed to be more apolitical and therefore less subject to control by local elites. Also, certain functions required a decentralized unit that was smaller than a panchayat (as in the case of drinking water) and/or one that was more suitable from a hydrological point of view (as in the case of watershed development). At the same time, many of these groups (like the ones for watershed development) were informal and not registered under any of the formal laws of the state concerned, especially when they were first introduced, and therefore the extent to which they were formally accountable was questionable. This argument became particularly strong in the wake of the 73rd and 74th amendments. Whatever the structural inadequacies of PRIs, they exist, have a strong tradition in many parts of independent India, and have recently acquired constitutional status; hence there is a need to look for spaces within panchayat institutions that can be tapped for water resources management.[18] The result is that there has been an attempt (which has had mixed success) to link these two forms of decentralization since 2000, often by making user groups sub-committees of the gram panchayat and/or accountable to the gram sabha.

We now discuss the specific form that decentralization has taken in different sub-sectors of water, namely rural drinking, urban drinking, and irrigation. Watershed management is not a major area of focus in this chapter; however, we

18 Videh Upadhyay, 'Water Management and Village Groups: Role of Law', 37(49) *Economic & Political Weekly* 4907 (2002).

will briefly discuss its experience with decentralization because it is a reflection of a changing emphasis on the two forms of decentralization.

In terms of chronology, irrigation and watershed management saw the first attempts at decentralization, followed by rural and urban drinking water. Decentralization in the irrigation sector has taken the form of Participatory Irrigation Management (PIM). Although this idea has been supported by the Government of India since the mid-1980s, it is only recently that states have started taking legislative measures to facilitate it, often as part of the restructuring of the water sector.[19] For instance, the Maharashtra government first attempted to put in place water users associations in 1988; this was partly in reaction to the pressures exerted by various civil society groups and partly in response to the growing trend of decentralization.[20] But the formal legislation in this regard, the Maharashtra Management of Irrigation Systems by Farmers Act was enacted only in 2005 and was one of the preconditions for obtaining the World Bank loan for the Maharashtra Water Sector Improvement Project. Similarly, in Orissa, PIM was mentioned in the state water policy of 1994, but was finally undertaken in the context of the World Bank-funded Orissa Water Resources Consolidation Project.[21]

In the case of watershed development, the trend towards decentralization began with the 1994 guidelines of the Government of India, where community organizations such as watershed associations, watershed committees, self-help groups and user groups at the village level were emphasized. No major role was envisaged for the gram panchayats and other PRIs.[22] The 2001 guidelines envisaged some role for PRIs, but the concepts of watershed association and watershed committee were retained for implementing the projects.[23] In the 2003 Hariyali guidelines, however, PRIs were given the onus of implementation of watershed projects.[24]

In the case of rural drinking water, the process of decentralization was first introduced by the centre in 1999 by way of a sector reforms programme in which there was a shift from a supply-driven approach to a demand-driven approach, discussed in Chapter II of this volume. It drew on the lineage of the World Bank-assisted Rural Water Supply and Sanitation project (popularly known as Swajal) in Uttar Pradesh and other water and sanitation projects of the mid- to late-1990s.[25] Initially, demand-driven projects were taken up in sixty-seven pilot districts covering twenty-six states, and were scaled up in 2002 in the form of Swajaldhara.

19 See, e.g., the National Water Policy, 1987, available at www.ielrc.org/content/e8701.pdf.

20 Note that there was already provision for supply of canal water to associations and volumetric pricing in the Maharashtra Irrigation Act, 1976 (Maharashtra Act No. XXXVIII of 1976).

21 Sushanta Kumar Mahapatra, 'Functioning of Water Users Association or Pani Panchayat in Orissa: Principle, Procedure, Performance, and Prospects', 3/2 *Law, Environment and Development Journal* 126 (2007).

22 Ministry of Rural Development, Guidelines for Watershed Development, 1994.

23 Ministry of Rural Development, Guidelines for Watershed Development, 2001.

24 Ministry of Rural Development, Guidelines for Hariyali, 2003, Section 11.

25 M. Venkaiah Naidu, Indian Reform Initiatives in Water Sector (Keynote Address at Water Forum 2002, Washington DC, 6 May 2002), available at http://ddws.nic.in/popups/MWForum.pdf.

Swajaldhara and variants thereof (such as Jalswarajya in Maharashtra) are being undertaken in many parts of the country today. In a bid to remove the dichotomy between community approaches and local government approaches – namely the two forms of decentralization referred to earlier – the Government of India and states are setting up Village Water Supply and Sanitation Committees (VWSCs) as a committee of PRIs.[26] Demand-driven schemes also envisaged a restricted role for NGOs (limited to social mobilization and initial hand-holding of the VWSCs). This was in contrast to NGOs getting direct access to finances and functioning as the principal implementing agency, a trend which had become particularly important since the 1980s. However, neither of the two major institutional changes envisaged in demand-driven schemes – the relation between VWSCs and PRIs and the role of NGOs – has worked as conceptualized.

Similar processes are at work in the field of urban water and sanitation. In the aftermath of the 74th amendment, the range of functions undertaken by urban local bodies (ULBs) has increased and/or the options available to them (in case of functions they cannot or do not want to undertake) have also increased. For instance, ULBs have a choice about where they get their water from: they can develop and maintain a source themselves, get this done by a parastatal body or by a private party, or buy water direct from government agencies. How far ULBs have adopted these changes depends on the extent to which the 74th amendment has been implemented in any given state. The process is very much underway in states such as Maharashtra and Karnataka. Further, there is a major difference between small and large ULBs in the extent to which they have undertaken these reforms.

B. Issues in institutional reforms

In discussing the law and policy debates involved in the three reforms discussed above – regulation, privatization and decentralization – three sets of issues are highlighted: concerns regarding the division of labour and accountability, the effective devolution of powers and rights to users and local bodies, and the extent to which the institutional reforms are compatible with another important legal principle, namely the right to water. The discussion below demonstrates that problems with the reforms are not just those of implementation, but of conceptualization. In some cases, the criteria of the policies themselves or those of other broad constitutional principles and legal doctrines are not met; in other cases, certain normative criteria that would be expected to be met in a democratic country are disregarded.

1. Division of labour and accountability

One major legal and policy issue that emerges from the reforms relates to the large number and variety of players who have emerged in the wake of decentralization

26 *Ibid.* at 8.

and privatization and the implications thereof for questions of division of labour, accountability, and redress mechanisms. These players could be state and parastatal agencies, large corporate bodies, and private individuals or groups of individuals. In urban drinking water, domestic and foreign companies (e.g., Pratibha Industries) have emerged in the arena of water to undertake functions ranging from source development and augmentation to billing and metering. In addition, there are also private players in the provision of finance for water (e.g., Infrastructure Leasing & Financial Services Ltd.), consultants (e.g., Kirloskar Consultants), and water auditors (e.g., Dinesh Rathi and Associates).

In the realm of rural drinking water, the new demand-oriented approach has meant, among other things, that a whole host of technical service providers (as the entities providing the technical services involved in water provision are called) have emerged replacing the earlier system where a single parastatal agency or state board would provide the technical services. Apart from the emergence of new players, the role of older, existing actors (both state and non-state) has also changed. In such a context, structuring new relations between different actors in order to ensure both a clear division of labour as well as mechanisms for accountability and redress becomes critical. However, the experience to date on this count leaves a lot to be desired. Two broad kinds of problems can be distinguished – those that relate to the ways in which relations between actors are attempted to be structured, and those that relate to differential power positions of actors and local dynamics.

Consider the manner in which relations between actors are structured. Often, there is a lack of clarity about where ultimate responsibility vests, especially when there are various levels of contracting and subcontracting, as is increasingly the case, for instance, in urban water supply. Thus in the Karnataka Urban Water Sector Improvement Project, Compagnie Générale des Eaux is the executing and supervising agency for the demonstration works in Hubli-Dharwad, Belgaum and Gulbarga; in turn, it has sub-contracted specific parts to third-party sub-contractors. Given the multiple domestic and international private parties involved, along with the municipal corporations of the respective cities, and the Karnataka Urban Infrastructure Development and Finance Corporation (the nodal agency for the project), pinning responsibility for any specific aspect of the project becomes difficult. Another problem is that the terms of the contract between parties may not facilitate accountability. For instance, if a party to whom a particular water function has been sub-contracted receives a payment that is fixed and non-performance-based, it is difficult to seek redress in case of non-performance.

Sometimes contracts are used as a means to ensure clear division of labour and accountability. This is not an easy task. Although from a strictly legal perspective, it is the letter of the contract that is considered to be binding, the true picture about the division of labour can be understood only if one goes beyond the formal division of labour set out in contracts. For instance, tariff setting may remain with a public body in theory, but if other decisions that affect the tariff-setting process are in the hands of a private entity, then there is very little real autonomy over the tariff-setting process. Similarly, in the Jalswarajya projects for rural drinking water in Maharashtra, the contract between the gram panchayats and the support organizations (SOs) – as the NGOs involved in the projects are called – is for eighteen months, a time that all actors involved in the project uniformly agree is insufficient. This has usually meant that the SOs continue to work after the eighteen-month period without any payment. This is justified by officials either by laying the blame for non-completion of the project at the doors of the SOs or by appeals to SOs to go beyond the text of the contract and do 'social work'.[27]

The second set of issues that complicates questions of division of labour and accountability is the fact that the different actors involved usually have very varying power positions. In turn, this affects the legal and other resources at the disposal of contracting parties, and influences the negotiation process as well as implementation and redress mechanisms, a fact that is not adequately recognized in policy discussions. For instance, management structures, investment patterns, and sharing of risks are influenced by the terms on which governments contract with private parties.[28] What is not emphasized is that many governments – particularly in developing countries – may not even have the legal resources to undertake such negotiations on a level playing field. Similarly, Saleth, in discussing irrigation privatization, emphasizes that the varied private sector parties may have unequal financial, technical and organizational resources and so there is a need to ensure coordination between these groups.[29] But Saleth neither adequately emphasizes the unequal powers of the groups, nor the fact that the government's ability to manage conflicting objectives such as incentives for private investors and protection of farmers' interests is questionable. In fact, in many cases, one finds that even contractual obligations between the government and local institutions are not fulfilled. One instance of this is when water users associations (WUAs) are forced to take over canal systems although the rehabilitation work to be undertaken by the Irrigation Department prior to handing over has not been completed.[30] Hence Mehta and Canal argue that there is no reason to believe that governments who have not succeeded as direct service providers would be able to carry out the regulatory functions necessary for privatization (or for that matter, decentralization) to work out in an equitable fashion.[31] The argument is doubly strengthened when one talks about other entities (such as small local bodies) and marginalized groups who would be even more disadvantaged than state governments in terms of bargaining power and their ability to influence decision making.

The picture is further complicated when people find their own ways of resolving

27 Field visits to Jalswarajya villages in Thane district of Maharashtra in May 2008.

28 United Nations Development Programme, *Human Development Report 2006 – Beyond Scarcity: Power, Poverty and the Global Water Crisis* 133 (New York: UNDP, 2006).

29 Maria R. Saleth, 'Irrigation Privatization in India: Options, Issues and Experience', 34(26) *Economic & Political Weekly* A-86 (1999).

30 Niranjan Pant, 'Some Issues in Participatory Irrigation Management', 43(1) *Economic & Political Weekly* 30 (2008).

31 Mehta and Canal, note 15 above.

the problems raised by the formal division of labour. For instance, a major lacuna of decentralization policies is that they do not adequately deal with the tensions between different decentralized bodies at the local level. Thus in rural drinking water projects such as Swajaldhara, VWSCs co-exist with the constitutionally mandated PRIs. In theory, VWSCs are supposed to be sub-committees of the PRIs, although there are often different rules in operation for different projects. But in cases where the VWSCs have done good work and the schemes have achieved some degree of success, villagers want a clear demarcation between the two. VWSCs are perceived as more just institutions (in part because they are subject to accountability mechanisms such as social audits), while the ability of PRIs to bring in change is viewed with scepticism. In cases where VWSCs have been subject to the same kind of capture by elites that PRIs have traditionally been accused of, decentralization has meant that there is no recourse to supra-local bodies to deal with intra-village inequities and villagers are supposed to deal with 'local politics' themselves.[32]

Accountability in the arena of regulation is attempted within the MWRRA Act. Certain principles have been spelt out as governing the Maharashtra Authority's activities. The Act specifically indicates that the Authority must act within the framework of the state water policy and additional principles found in the Act.[33] For instance, the volumetric amount of water made available to holders of water entitlements is to be fixed according to specific criteria. These include the equitable distribution of water between all landholders, with an emphasis on equity between head-end and tail-end farmers. However, both the Act and the state water policy provide an 'insufficient basis for laying down a proper normative framework' to assist the Authority in decision making.[34] Additionally, the present structure of the regulatory authority provides for appointment of bureaucrats, whose interests invariably are tied in with the state, thus compromising accountability to a wide set of stakeholders.

2. *Effective devolution of powers and rights*

One of the purported aims of the water reforms, and particularly of the process of decentralization, is to grant greater powers and rights to water users and/or to local institutions. This has happened to a limited extent, but two areas of concern need to be stressed. Firstly, a very limited form of decentralization has been conceived. In particular, democratization of decision making has not been included, although lip service is paid to participation by all stakeholders. Secondly, the process of regulation, which has often accompanied the process

32 Field visits to Jalswarajya villages in Thane district of Maharashtra in May 2008.

33 Maharashtra Water Resources Regulatory Act, 2005, note 11 above, Section 12.

34 K.J. Joy, S. Paranjape and S. Kulkarni, Issue note (prepared for the National Consultation on Regulation and the Poor, jointly organised by Prayas (Pune) and the Centre for Study of Law and Governance, Jawaharlal Nehru University, New Delhi, 12–13 July 2007).

of decentralization, pushes in the opposite direction of centralization and is inadequate to protect the rights of users. We consider each of these points in turn.

Firstly, the manner in which decentralization has been conceptualized at the level of both the central and state governments does not result in adequate devolution of powers and rights and empowerment of user groups and local bodies. Decentralization of only a limited number of functions has taken place and WUAs or drinking water committees have little say about many aspects that are relevant for local-level decision making (e.g., allocation of water across different uses in surface water sources).[35] These limitations stem from the objectives of the central policies that propose changes in the water sector. To illustrate, while the National Water Policy, 2002 emphasizes a participatory approach to water resources management, the aim of involving water users associations and local bodies is said to be 'to eventually transfer the management of such facilities to the user groups/local bodies'; there is no mention of ownership of the water facilities by local groups.[36] Similarly, the National Water Policy, 2002 states that the participation of beneficiaries and other stakeholders should be encouraged right from the project planning stage, but the nature of this participation, how and by whom beneficiaries and stakeholders are to be defined, and so on, is unclear. Another major lacuna is that the working of the bureaucracy – and especially the larger social and political context within which bureaucrats work – has not been re-evaluated and reconsidered in the light of the other changes, and this is likely to hamper the realization of the gains of decentralization.

But perhaps what is of greater concern is the lack of democratization of the decision-making process. For instance, even with a participatory, decentralized approach such as PIM, the legal and management regime are never owned by those for whom the law is intended.[37] One manifestation of this is that none of the state PIM laws – except, to some extent, Maharashtra – adopted intensive preparatory procedures before enactment of legislation. A similar argument can be made in the case of rural drinking water, where the initial pilot schemes of the central government were very quickly scaled up even though there had been considerable scepticism expressed about the working of the pilot schemes. A third example is the Maharashtra Water Resources Regulatory Authority's decision to initiate the process of tariff regulation before putting in place a Conduct of Business Regulations.[38]

35 National Institute of Rural Development, *India Panchayati Raj Report 2001: Four Decades of Centralised Governance in Rural India, Volume 1* (Hyderabad: National Institute of Rural Development, 2002).

36 National Water Policy, 2002, Section 12, available at www.ielrc.org/content/e0210.pdf.

37 Videh Upadhyay, 'Canal Irrigation, Water User Associations and Law in India – Emerging Trends in Rights-based Perspective', in Philippe Cullet, Alix Gowlland Gualtieri, Roopa Madhav & Usha Ramanathan (eds.) *Water Law at the Crossroads – National and International Perspectives with Special Emphasis on India* 110 (New Delhi: Cambridge University Press, 2009).

38 Representation by Prayas (a civil society organization based in western India) to Maharashtra Water Resources Regulatory Authority, dated 29 December 2007, accessed at http://prayaspune.org/reli/PUBLICATIONS_files/C_3_6.pdf.

The policy of privatization also provides an example of such a lack of democratization. The decision about privatization is ostensibly decentralized as the state policy (and the supporting legal and regulatory framework) is supposed to provide ULBs with maximum flexibility to select the forms of private sector participation best suited to meet local objectives.[39] At the same time, the decision to move towards greater privatization is itself a centralized one. In fact, even the apparent choice of ULBs is often illusory because although functions are decentralized, decentralization of powers, finances, and functionaries have lagged behind in many states; hence ULBs are often left with no option but to consider some form of private sector involvement.[40] This is compounded by the fact that even the state support that used to be offered earlier – such as state guarantees on loans – is either being reduced or subject to new conditionalities. In general, a major concern has been that important policy decisions – such as decisions about the need for privatization and decentralization and the forms that these should take – have often been policy decisions taken by the executive arm of the government – increasingly, the central government – rather than decisions discussed and endorsed in the legislature even though they involve far-reaching changes in the manner in which water is supplied.

The second major issue of concern in terms of its implication for rights and powers of users and local bodies is the centralizing tendency evident in the reforms. The most important manifestation of this is the setting up of a water regulatory authority which has considerable control over water resources. The resulting tensions are clearly evident in the case of participatory irrigation management. Although the role of the government has been curtailed in irrigation in a number of respects through the setting up of an independent authority, this does not necessarily translate into less regulatory intervention as far as water users are concerned because the Regulatory Authority becomes the new body exercising control over water resources.[41] For example, the Act provides as a general principle that any water from any source can only be used after obtaining an entitlement from the respective river basin agency.[42] This is qualified by a few exceptions such as wells (including bore and tubewells) used for domestic purposes or the grandfathering of existing uses of water for agriculture, at least in an initial phase. The overall impact is therefore as much to reduce the government's role as to transfer and possibly strengthen control over water resources. In general, the powers given to the Authority are extensive and include, among other things, distribution of water entitlements for different categories of use, determination of priorities in distribution of water at different levels (basin, sub-basin, project), and establishment of water tariffs. In fact, the Regulatory Authority not only has the

39 2004 Guidelines, note 17 above, 7.

40 National Institute of Rural Development, note 35 above.

41 Philippe Cullet, 'Water Law Reforms: Analysis of Recent Developments', 48/2 *J. Indian Law Institute* 206, 213 (2006).

42 Maharashtra Water Resources Regulatory Act, 2005, note 11 above, Section 14.

power to make regulations for matters that come under the MWRRA but also for 'all other matters for which provision is [. . .] necessary for the exercise of its powers and the discharge of its functions under this Act'.[43]

As Hardeep Singh notes, the question is not whether regulation is an appropriate model but 'whether a regulatory institution can really take on board all the concerns, especially the concerns related to environment and the poor, and still sustain its independent character. If regulatory authority restricts its role to creating water markets by reconciling the interests of investors/public utilities and the select consumers' satisfaction, then the poor as well as environment concerns may lose in many ways'.[44] The MWRRA, which is set to become the role model for replication by other states, will have to set an example of democratic governance for it to be acceptable to all sections of society.

The need for some kind of centralized mechanism to ensure both coordination and conflict resolution cannot be denied.[45] But what is of concern is the form that this centralization has taken, particularly as manifested in the Regulatory Authority. In particular, there is a rhetoric of depoliticization, that is to say, the Regulatory Authority is supposed to be de-linked from the government, and in that sense, is supposed to be free of politics. However, what this has often meant is that intervention of any kind in its working, even by civil society organizations, is viewed with suspicion, including attempts to get information about the different processes that the Authority is involved in. Hence there are concerns that limited transparency and accountability seem inbuilt in the present regulatory model.

What is needed then is an explicit interface between the regulatory body and democratic/political institutions. There is some hope in this respect given that the regulatory model of governance is still evolving – both nationally and internationally. Some hope is also offered by the ongoing attempts to push the regulatory body towards greater transparency and scrutiny through the active intervention of civil society organizations and citizens. This is currently being tested; the MWRRA is hearing a petition by civil society organizations challenging the Maharashtra government's move to call for tenders from private actors in the Nira Deogarh case in Maharashtra. But one also needs to recognize the fact that the Constitution does not recognize these newer models of governance. This in turn raises a number of questions: In the present constitutional framework, where do these new institutions fit? Is there a need for an amendment to the constitution to explicitly recognize and provide for an appropriate autonomy–accountability framework?

The tension between centralizing and decentralizing tendencies in irrigation is further complicated by the view taken by a number of analysts that WUAs should

43 *Ibid.* Section 31.

44 Hardeep Singh, Regulation in the Water Sector and the Poor (Paper presented at the National Consultation on Regulation and the Poor, jointly organised by Prayas (Pune) and the Centre for Study of Law and Governance, Jawaharlal Nehru University, New Delhi, 12–13 July 2007).

45 R. Maria Saleth and Ariel Dinar, *The Institutional Economics of Water: A Cross-Country Analysis of Institutions and Performance* (Cheltenham: Edward Elgar, 2004).

also be regarded as private entities, so that the trend towards decentralization is seen as being congruous with the trend towards privatization.[46] While concerns that WUAs may have to function on a commercial basis and become self-sufficient entities are not misplaced, to regard them as private entities would be to disregard the history of struggles in many locales to ensure access to water, and may well end up putting farmers' associations on an equal footing with corporate bodies. The latter is evident for instance in the statement that 'there is a critical role for the private sector whether this is represented by community organizations, small water vendors, or by operators of water companies', where there is no distinction made between these different kinds of private players.[47] A similar argument would also be applicable in the drinking water arena.

If an excessively totalitarian form of regulation is the problem in the case of irrigation, in other realms such as urban drinking water, the problem is a complete lack of regulation over the actions of private parties with whom urban local bodies may enter into contracts (e.g., with respect to disconnections and tariff setting). Thus forms of private sector participation/public–private partnership have already been adopted either on a pilot basis or by way of long-term contracts, even though there has not been sufficient groundwork done in terms of what forms of regulation would be appropriate. The caveat to this is that the mere establishment of a regulatory framework does not ensure accountability, and bodies that are responsible for regulation may be subject to the same kind of pressures (e.g., from international bodies) as governments; further, regulation that merely focuses on efficiency and growth may not necessarily be committed to ensuring access to basic services or protecting access of the poor to services.[48]

In the case of rural drinking water, the National Council of Applied Economic Research recommends that the task of regulation of water services, which would become more critical as privatization proceeds and more local providers are involved in village-level water and sanitation services, may be handled by panchayati raj institutions given their proximity to users and providers.[49] However, this is potentially likely to result in tension between decentralization in regulation and decentralization in service provision.

3. Institutional reforms and right to water

Ongoing water law reforms are narrowly conceived in terms of the interests and concerns that are taken into account, and broader principles such as the human right to water, water as a public trust, and environmental and health aspects (precautionary principle) are not taken into account.[50] Here we focus specifically

46 See, e.g., Saleth, note 29 above, A-87.
47 Naidu, note 25 above, 15.
48 Mehta, note 15 above, 16.
49 National Council of Applied Economic Research, India Rural Infrastructure Report (New Delhi: Sage Publications, 2007).
50 Cullet, note 41 above.

on how selected aspects of privatization, decentralization, and regulation are compatible (or not) with a right to water.

The reform which has been most controversial in terms of its implication for a right to water has been privatization. International discussions of right to water are relatively flexible about the system of water delivery and do not take an *a priori* stand for or against privatization.[51] For instance, General Comment 15 provides for a major role for the state; it also emphasizes that non-state actors should take necessary steps to realize the right to water (or at least not to thwart it).[52] It gives states flexibility in choosing their economic system, but it does emphasize the need for democratic private sector participation, as well as for regulation by the state. However, the drawback of the General Comment and the Guissé Report is that there is not enough discussion of the controversial aspects of ownership, delivery, and pricing of water, all of which are aspects that are critical for private sector participation in any form and for the fulfilment of a right to water. The tension between meeting costs and financing a right to water, on the one hand, and affordability to all groups, on the other hand, is not considered. There is a similar lacuna in the Indian context. There is judicial support for the right to water under Article 21, the right to life, which is a fundamental right in the Constitution. Although this right has been restated in a number of judicial judgments, many aspects of the current reforms are incompatible with the right to water, and often with other dimensions of existing law.

Consider first the question of privatization and right to water. In the context of drinking water, judicial decisions – such as the 1980 *Municipal Council Ratlam v. Vardhichand* – hold that the state cannot claim insufficient funds as a reason to not carry out its duties.[53] Similarly, water and sanitation are clearly within the mandate of the local bodies as per the 73rd and 74th amendments; there are also other functions such as public health which are with the local bodies and are dependent on the provision of water. Hence it becomes critical for ULBs to continue to bear responsibility for issues related to water. But even if one accepts the basic point about the ULB/state's responsibility to ensure provision of adequate water, what exactly such responsibility entails is not clear. For instance, it is not clear whether critical tasks involved in providing water to people (such as tariff making) can be delegated to any private body or only that such private bodies should be subject to regulation by the state. Further, when

51 See, e.g., Committee on Economic, Social and Cultural Rights, General Comment No. 15: The Right to Water (Articles 11 and 12 of the International Covenant on Economic, Social and Cultural Rights), UN Doc. E/C.12/2002/11 (2002) and Realization of the Right to Drinking Water and Sanitation, Report of the Special Rapporteur, El Hadji Guissé, United Nations Commission on Human Rights, Sub-Commission on the Promotion and Protection of Human Rights, Fifty-seventh session, Economic, Social And Cultural Rights, UN Doc. E/CN.4/Sub.2/2005/25 (2005).

52 General Comment No. 15, note 51 above, Part VI.

53 Sanjay Upadhyay and Videh Upadhyay, *Handbook on Environmental Law Volume III: Environmental Protection, Land and Energy Laws* (LexisNexis: New Delhi, 2002).

the paucity of financial resources with the state is a major justification used for involvement of the private sector, it is not clear to what extent the assorted concessions, guarantees, rates of return and escrow mechanisms used to ensure private sector involvement are acceptable. Similarly, the traditionally asserted state ownership over surface waters has formed the basis for BOT contracts in surface water, such as in the Sheonath case. But if irrigation law is read in conjunction with recent judicial judgments that declare surface water to be a public trust, the type of rights and privileges that can be claimed over surface water would be limited.[54] Disconnection is another big threat that is appended to the process of privatization. In South Africa, for instance, disconnections to non-paying consumers (especially in urban areas) are not uncommon, and have been linked by some to outbreaks of cholera and other gastrointestinal infections, a clear instance of infringement of basic rights.[55] However, in 2008, there was a landmark judgment by the High Court of South Africa when it ruled that the City of Johannesburg's practice of forced installation of prepaid water meters in Phiri, Soweto was unconstitutional and unlawful, and directed the City authorities to provide residents with the option of an ordinary credit metered water supply.[56]

A second aspect of water reforms that is controversial from the point of view of a right to water is the principle of cost recovery and the mechanisms being put in place to ensure this in the course of processes of regulation and decentralization. Thus the MWRRA Act lays down that the water charges shall reflect the full recovery of the cost of irrigation management, administration, operation and maintenance of water resources projects. This in turn has led to fears that the new institution of regulator being introduced will seek to regulate water markets by reconciling interests of private competitors or public utilities and consumers and not take on board the concerns of the poor and the marginalized, or of the environment. That such fears are not unfounded is evident, for instance, in the fact that the landless are excluded from the provision of water entitlements.[57]

Similarly, in the wake of decentralization and the shift to demand-oriented projects, ULBs have had to diversify their sources of finance.[58] But such finances often come with conditions such as newer accounting norms which could ensure greater transparency. These new norms could also serve to impose greater vigilance in transfer of funds across different sub-sectors/uses of water, thereby reducing or eliminating an important means of providing for services like public

54 Cullet, note 41 above.

55 Ann-Christin Sjolander Holland, *The Water Business: Corporations Versus People* (Books for Change: Bangalore, 2005).

56 *Lindiwe Mazibuko & ors v. the City of Johannesburg & ors*, Case no. 06/13885, High Court of South Africa (Witwatersrand Local Division), Judgment of 30 April 2008.

57 Maharashtra Management of Irrigation Systems by Farmers Act, 2005, Chapter 1(2).

58 National Institute of Urban Affairs, Urban Governance Decentralization in India – A Review (New Delhi: National Institute of Urban Affairs, Research Studies No. 95, 2004).

standposts where it would not be possible to meet costs.[59] This example also serves to highlight the fact that the emphasis on commercialization and profit-making and the changes it brings in its wake are not restricted to private players. In fact, given that privatization is encountering a number of obstacles – such as opposition by local groups, unwillingness of water companies to enter into functions perceived as loss-making – there is increasing emphasis on changing the working of state and parastatal bodies by introducing commercial principles. This is important to take into account in any discussion of the relation between reforms and right to water.

In the rural drinking water context, the cost recovery mechanisms in place have led to the question of concerns about the fairness and practicality of expecting all users to contribute 10 per cent of capital costs. But there is also another important area of concern. The cost contribution measure has often given rise to new power centres in rural areas – for instance, individual contractors or NGOs who pay the popular contribution of 10 per cent capital costs on behalf of villagers in return for other favours.[60] Although this is an institutional change with potential negative equity implications, there are no adequate guidelines or mechanisms in place to deal with this. On the contrary, such changes could serve to facilitate the withdrawal of the state from provision of water and therefore not even be viewed as negative.

The third point that is important from the point of view of right to water is that the existing and potential conflicts between basic uses and economic uses of water, as also between the right to water and other rights, have not been considered adequately. In fact, this is a lacuna that is also evident in General Comment 15 and the National Water Policy, 2002 and stems at least partly from not taking the multi-faceted nature of water into account. For instance, in the case of canal water, the fact that there may be uses of water that involve primary activities other than agriculture as well as the use of water for drinking/domestic purposes is often not taken into account in policies of participatory irrigation management. In such a scenario, the trading of water entitlements (described earlier) might lead to high prices for water and make it available only for particular users.[61] Similarly, whether one has access to public drinking water in cities is often linked to the right to housing and whether or not one is legally recognized as a citizen; any reform in water in such a context would necessarily have to deal with laws regarding housing and other basic amenities as well as citizenship norms.

59 For a discussion of similar processes in the specific case of Maharashtra, see Priya Sangameswaran, 'Noeliberalism and Water Reforms in Western India: Commercialization, Self-Sufficiency and Regulatory Bodies', 40 in *Geoforum* 195 (2009).

60 Preeti Sampat, "Swajaldhara or 'Pay'-jal-dhara: Right to Drinking Water in Rajasthan', 42(52) *Economic and Political Weekly* 102 (2007)".

61 Gaurav Dwivedi, Rehmat, and Shripad Dharmadhikary, *Water: Private, Limited* (Badwani: Manthan Adhyayan Kendra, 2006).

CONCLUSION

This chapter brings out some of the concerns raised by the recent institutional reforms for water. The manner in which relations between different actors are sought to be structured, as well as their differential power positions and local dynamics, indicates considerable ambiguity about questions of division of labour and accountability. Reforms such as decentralization have not succeeded in increasing the rights and powers of users and local bodies because, among other things, they do not include transfer of ownership rights or attempt to democratize the decision-making process. There is also a tension between the process of decentralization and the process of regulation with its thrust towards centralizing power and authority. Finally, many aspects of the institutional reforms are in conflict with different dimensions of the right to water, which has judicial support under the Constitution. Given the range and depth of these concerns, there is clearly a need to take stock before moving any further in terms of changing the water sector. In fact, the material and non-material costs and consequences of even the reforms already in place would need to be constantly monitored and reviewed. All this is, of course, easier said than done. But we would like to conclude this chapter by emphasizing a number of areas in which further work is needed in order to critically think through the question of institutional reforms in water.

Firstly, the formalization/legalization process in general and legislative changes in particular have often not taken into account the already existing situation with respect to different actors involved, methods of water provision or sharing, and rights regimes. While this chapter did discuss some aspects of this – such as the tensions between different kinds of decentralized bodies, and conflicts with a right to water – work is needed, for instance, on the informal sector and the role it plays/can play. In fact, such work is often available, as in the case of informal groundwater markets, but this would have to be extended to the current reforms.[62]

Secondly, there is need for research that considers the social, political, historical, and cultural aspects of water at various levels in order to better understand how specific reforms would actually work out as well as how different actors would modify its working. There are some examples of this, but this is in general a big lacuna.[63]

Thirdly, in much of the literature on water reforms, there is no clear distinction made between means and goals. For instance, Saleth and Dinar point out that 'the major goals of institutional initiatives in the water sector are transparent [. . .] to treat water as an economic good, strengthen allocation capabilities, increase reliance on market forces, revive the payment culture, ensure financial self-sufficiency, promote decentralized decision structures, and encourage the use of modern

62 See, e.g., Navroz K. Dubash, *Tubewell Capitalism: Groundwater Development and Agrarian Change in Gujarat* (New Delhi: Oxford University Press, 2002).

63 See, e.g., David Mosse, *The Rule of Water: Statecraft, Ecology and Collective Action in South India* (New Delhi: Oxford University Press, 2003).

technology and information inputs'.[64] Irrespective of whether one believes or not that water should be treated as an economic good or that there should be increased reliance on markets, these are really means to achieve particular goals. That is to say, if one starts with goals, such as providing everyone with a certain quantity and quality of water or ensuring that there are adequate funds to cover investment in the water sector, then one can have a debate about the different means to reach these goals, which need not necessarily involve market forces or treating water as an economic good. This in turn would help to question the need and rationale for many of the institutional reforms.

Finally, binding principles – both legal and normative – to anchor the reforms or to even arrive at a set of common goals are lacking. A more informed, vibrant and inclusive process of public debate could potentially help in this, as also broadly in ensuring democratic governance and institutional wellbeing.

64 Saleth and Dinar, note 45 above, 302.

VIII. Drinking water reforms

Philippe Cullet

INTRODUCTION

Drinking water is one of the most crucial components of any water law and policy framework. This is due to the intrinsic link with the realization of the human right to water. It also deserves separate treatment in the context of water law reforms because it has not been given significant attention in the new laws examined in other chapters of this section of the book. This is not because other specific laws have been adopted for drinking water in recent years, but because drinking water has in fact not been a real priority in ongoing law reforms. While all documents linked to water sector reforms reiterate the position that drinking water is the first priority of the water sector, the laws that have been and are being adopted are not specifically concerned with drinking water. Yet, this does not mean that no reforms have been undertaken concerning drinking water. On the contrary, since at least the middle of the 1990s, sweeping reforms have been introduced for drinking water supply in rural areas. The reform of the policy context for the provision of drinking water in rural areas has, however, never been discussed in the context of any legislation. There has thus been no parliamentary oversee of the process. In the case of urban areas, reform of water supply services has been proposed at the level of individual cities. This may be part of a stand-alone project focusing on the privatization of water services or be part of a broader urban reform project.[1]

A. Policy framework for drinking water

Access to safe water sources has dramatically improved over the past few decades. It has been estimated that coverage in rural areas increased from 18 per cent in

1 Two of the most important schemes are the Jawaharlal Nehru National Urban Renewal Mission (JNNURM) for big cities and the Urban Infrastructure Development Scheme for Small and Medium Towns (UIDSSMT).

1974 to 43 per cent in 1996 and 94 per cent by 2004.[2] The latest figures given by the UNDP put the percentage of people with sustainable access to an improved water source at 86 per cent.[3] Yet, the extent of the progress remains unclear. Thus a background document for the eleventh plan puts the figure at 95 per cent coverage, but acknowledges that nearly 20 per cent of residences have slipped back to being only partially covered under the norms in force.[4] This is acknowledged as one of the biggest problems in the rural drinking water sector and can be due to a variety of reasons from a decrease in groundwater level to new quality concerns or new habitations resulting in reduced per capita availability.[5] Additionally, another 15 per cent of habitations suffer from problems with water quality, among which 4 per cent suffer from serious salinity, arsenic and fluoride contamination.[6] These figures may underestimate the problem as a survey undertaken in 2003 found that 16.9 per cent of habitations were not covered and an additional 25.8 per cent were only partially covered.[7] Even under the lower figures, since water quality is one of the criteria used to assess coverage, this means that at least 35 per cent of habitations are not receiving 40 litres per capita per day (lpcd) or are suffering from water quality problems that cannot be categorized as being fully covered under existing norms. Additionally, existing norms clearly indicate the definition of coverage only implies a basic minimum, which is a first step towards expanded targets. In other words, partially covered habitations are those that are not covered according to the norms in force, but have some water supply.

These figures need to be put in perspective. Indeed a covered habitation is one that has a drinking water source within 1.6 km or 100-metre elevation in hilly areas. These may be either private or common sources of water.[8] Additionally, the definition of a habitation is not that of a house. The official definition is of a locality within a village where a cluster of around twenty families (or 100 people) reside.[9] This indicates that even when 100 per cent coverage of habitations is achieved, more isolated houses may still remain uncovered. This has

2 Maggie Black with Rupert Talbot, *Water – A Matter of Life and Health* (New Delhi: Oxford University Press, 2005), A.J. James, 'From Sector Reform to Swajaldhara – Scaling up in India', 23/2 *Waterlines* 11 (2004) and World Bank, Staff Appraisal Report – Uttar Pradesh Rural Water Supply and Environmental Sanitation Project (Report No. 15516-IN, 1996).

3 United Nations Development Programme, *Human Development Report 2006 – Beyond Scarcity: Power, Poverty and the Global Water Crisis* 307 (New York: UNDP, 2006).

4 Planning Commission, Towards Faster and More Inclusive Growth: An Approach to the Eleventh Five-Year Plan 2007–2012 at 69 (New Delhi: Government of India, 2006).

5 Agenda Note for Conference of State Ministers In-charge of Rural Water Supply and Sanitation, Sustainable Drinking Water Supply Sanitation for All 2012 (New Delhi: Rajiv Gandhi National Drinking Water Mission, 2007).

6 Planning Commission, Towards Faster and More Inclusive Growth: An Approach to the Eleventh Five-Year Plan 2007–2012 at 69 (New Delhi: Government of India, 2006).

7 Department of Drinking Water Supply, National Habitation Survey 2003 (New Delhi: Government of India, 2003).

8 Accelerated Rural Water Supply Programme Guidelines, s. 2(3).

9 Norms for Providing Potable Drinking Water in Rural Areas, available at http://megphed.gov.in/knowledge/standards/guiderural.pdf.

been recognized as a problem and under the eleventh plan, an effort is being made in the case of SC/ST habitations to cover them under the Accelerated Rural Water Supply Programme (ARWSP) even if there are fewer than 100 people. Additionally, it has been proposed to reduce distance criteria to 0.5 kilometres and a 30-metre elevation.[10]

An additional issue is that of the quality of the water that can be accessed. Indeed, the definition of drinking water implies that water must be salubrious from the chemical, physical and microbiological points of view. This is still problematic. Approximately 217,000 habitations suffer from water quality problems: 118,088 habitations suffer from excess iron, 31,306 from excess fluoride, 23,495 from excess salinity, 13,958 from excess nitrate and 5,029 from excess arsenic.[11] In Gujarat, for instance, water in as many as 38 per cent of habitations did not comply with WHO guidelines with respect to fluoride content and 23.6 per cent were above the maximum permissible limit.[12]

Access to water has rapidly changed over time. In rural areas, groundwater provides around 85 per cent of domestic water supply.[13] In urban areas, groundwater is the main source of water for about a quarter of households but most (70 per cent) depend on municipal water supply.[14] Yet the majority (about 61 per cent) do not have access to water within their dwelling and have to transport it from the main source. Of households that have access to tap water, 54 per cent usually depend on a tap which is not inside the house.[15]

1. *Law and policy context*

Drinking water supply is of primary importance in the context of water regulation.[16] Yet, neither the Union nor the states have been particularly proactive in developing legal frameworks that elaborate on the right of individuals and groups to have access to water, the quantities involved and the quality. In fact, the provision of domestic water is, from a regulatory point of view, a patchwork of policy documents at different levels and rules and regulations often adopted in the context of specific legislation defining the rights and duties of specific municipal corporations.

Different issues arise in the regulation of drinking water. Water quality is key to ensuring the delivery of water that is not harmful to health. Existing standards are

10 Rural Water Supply and Sanitation, Eleventh Five-Year Plan – Approach Paper 13 (2006).

11 Planning Commission – Government of India, *Eleventh Five-Year Plan 2007–12 – Volume II – Social Sector* 164 (New Delhi: Oxford University Press, 2008).

12 Indira Hirway, 'Ensuring Drinking Water to All: A Study in Gujarat', in K.V. Raju (ed.) *Elixir of Life – The Socio-Ecological Governance of Drinking Water* 74, 78 (Bangalore: Books for Change, 2007).

13 Rajiv Gandhi National Drinking Water Mission Department of Drinking Water Supply, Submission to the National Advisory Committee (2005).

14 P. Bajpai and L. Bhandari, 'Ensuring Access to Water in Urban Households', 36/39 *Economic & Political Weekly* 3774 (2001).

15 *Ibid.* 3775.

16 *Delhi Water Supply and Sewage Disposal Undertaking v. State of Haryana* (1996) 2 SCC 572 (Supreme Court of India, 1996).

in principle applicable to all water supplies. These include, for instance, criteria for bacteriological examination whose result must yield no E. coli.[17] These standards are, however, not absolute since they are only compulsory in the case of piped supply of water.[18] Since there is a broad dichotomy between rural and urban areas with regard to piped supply, the implication is that standards applicable in rural areas differ more often than not from the ones applicable in cities. Additionally, existing standards are not generally binding. Rather, they constitute best practices and desirable aims for all agencies concerned with drinking water supply. Thus, while water quality standards are central to the supply of potable water, the regulatory framework remains sketchy. Standards exist and practitioners all seem to have a reference point set by the Bureau of Indian Standards, the Central Public Health and Environmental Engineering Organization (CPHEEO) or the WHO. It would be difficult to hold actors accountable for the violation of a specific standard. This important gap has been recognized by policy-makers who have made a first attempt at introducing a binding framework for water quality. The existing proposal is to make state water supply authorities responsible for ensuring that water intended for human consumption does not constitute a danger to public health and complies with Indian Standard 10500 highlighted above.[19]

Whereas quality standards are generally similar for all areas, drinking water policies differ between urban and rural areas. As a result, the framework for water supply in urban and rural water is examined separately.

2. *The situation in urban areas*

Cities have generally been considered separately from rural areas. This is partly due to the fact that cities were until relatively recently the exception to the norm, which was life in rural areas. Cities have also always been centres of power from the local to the national level and the introduction of a different set of measures for the provision of drinking water in urban areas is thus relatively unremarkable in the policy context in which they developed. Yet, the separate – and more favourable – treatment of cities raises a number of questions. This has become even more important because of rapid urbanization and the increasing pressure that cities put on the environment and natural resources in general, and water in particular.

Water consumption in cities is generally much higher than in rural areas. Thus, a study of some large and medium-sized cities found that people consumed an average of 91.6 litres per day, with a range among the cities studied going from 77 in Kanpur, Uttar Pradesh to 116 in Kolkata.[20] While this is very high in

17 Indian Standard – Drinking Water – Specification (First Revision) (New Delhi: Bureau of Indian Standards, IS 10500: 1991) s. 3(2)(1).

18 *Ibid.*, s. 3(2)(2).

19 Department of Drinking Water Supply, Draft Guidelines for Preparation of Legislation for Framing Drinking Water Regulations (2007) ss. 45–6. See further Chapter XIII in this volume.

20 Abdul Shaban and R.N. Sharma, 'Water Consumption Patterns in Domestic Households in Major Cities', 42/23 *Economic & Political Weekly* 2190, 2192 (2007).

comparison with the rest of India, it is lower than the recommended 100 to 300 litres proposed by the WHO for optimal access.[21] With a 100-litre benchmark, the same study found that 65 per cent of households were water deficient.[22] Water deficiency in cities clearly correlates with socio-economic status and the level of education. The water consumption activities of households show that water is used for a variety of basic purposes. These include bathing (28 per cent of daily use), followed by toilet use (20 per cent), clothes washing (18.6 per cent), cleaning utensils (16.3 per cent) and drinking and cooking (less than 10 per cent).[23]

As against these figures collected from surveys, the policy framework of the Government of India has recognized 70 lpcd as the absolute minimum level that needs to be provided even in times of drought and in the poorest colonies.[24] This is divided into 10 litres for cooking and drinking, 30 litres for bathing and sanitation and 30 litres for washing utensils and clothes. Besides the minimum level of 70 lpcd, a gradation has been provided among different kinds of towns. Small towns with piped water supply but without a sewerage system should receive 70 lpcd. Cities with a sewerage system should receive 135 lpcd, while big metropolitan cities should receive 150 lpcd.[25] The CPHEEO Manual makes it clear that these figures only apply to households with individual piped connections. Where water is provided through public standposts (e.g., in villages) the norm is 40 lpcd.[26] Further, different norms are proposed by different bodies. Thus, the Bureau of Indian Standards suggests that 200 lpcd should be provided in cities with full flushing systems while the National Commission for Integrated Water Resource Development Plan suggested that by 2050 cities should receive 220 lpcd and rural areas 150 lpcd.[27] Additionally, different states interpret these norms differently. In Karnataka, for instance, the water policy provides that towns should get 70 lpcd, municipal council areas should get 100 lpcd and the biggest city corporations 135 lpcd.[28]

The delivery of water by the government in urban areas is premised on two different principles. On the one hand, urban dwellers benefiting from individual connections have been charged a fee for the service that is provided.[29] In most

21 Guy Howard and Jamie Bartram, Domestic Water Quantity, Service Level and Health 22 (Geneva: World Health Organization, 2003).

22 Shaban and Sharma, note 20 above, 2193.

23 *Ibid.*

24 Government of India, Report of the National Commission on Urbanization – Volume II at 294 (1988).

25 National Commission for Integrated Water Resource Development Plan, Report 63 (New Delhi: Ministry of Water Resources, 1999).

26 Central Public Health and Environmental Engineering Organization, Manual on Water Supply and Treatment 11 (New Delhi: Ministry of Urban Development, 1999).

27 Shaban and Sharma, note 20 above, 2191 and National Commission for Integrated Water Resource Development Plan, Report 64 (New Delhi: Ministry of Water Resources, 1999).

28 Karnataka State Water Policy, 2002, s. 4.

29 Lyla Mehta, 'Problems of Publicness and Access Rights: Perspectives from the Water Domain', in Inge Kaul *et al.* (eds.) *Providing Global Public Goods – Managing Globalization* 556, 562 (Oxford: Oxford University Press, 2003).

cases, the fee charged does not cover the overall cost of providing the water, but this is changing fast in the context of ongoing reforms. On the other hand, free community taps and/or hand-pumps have been provided in a variety of places. In fact, some acts specifically provide that the relevant authority has the power to provide public standposts free of charge.[30]

Water supply in cities is in principle the responsibility of urban local bodies. These are regulated in a variety of ways. Certain major cities such as Kolkata have their own water supply and sanitation act. In other cases such as Uttar Pradesh, water supply in cities is regulated in part by the Water Supply and Sewerage Act (UP Water Act) and in part by the specific regulations applying to the type of cities, usually categorized according to population size. There is thus no uniformity in the treatment of water supply in cities throughout the country.

The central obligation that is imposed on municipalities is to provide drinking water. For example, in Uttar Pradesh municipalities have a duty to provide 'sufficient supply of pure and wholesome water'.[31] In addition, they must also maintain public wells.[32] The main responsibility for water delivery in cities is in the hands of city-specific jal sansthans whose main task is to provide 'wholesome water' to city dwellers.[33] While the obligation to provide water is central to the functions of municipalities with regard to drinking water, it is not necessarily absolute. Thus, in New Delhi, while the Council has a similar obligation to provide a sufficient supply of pure and wholesome water, this is qualified by the fact that it must only do what is practicable and at a reasonable cost.[34]

Water supply providers are funded at least in part through cost recovery from water users. In Uttar Pradesh, jal sansthans are funded through two different routes. Firstly, they charge a water tax on each property (e.g., this is 8 per cent in Aligarh).[35] The only exceptions to this rule are for properties which are not within a specified distance, often 200 metres, from the nearest standpost or other waterworks maintained by the jal sansthan.[36] Secondly, a water rate for actual consumption of water is also imposed. This is, for instance, Rs 100 per month in Aligarh. The power to charge water users is associated with a number of duties. These include the obligation to maintain a system of water supply through pipes and to provide water at a prescribed pressure during prescribed hours.[37]

An important feature of a number of regulatory frameworks is the possibility for the authority in charge to disconnect a private water supply. In Uttar Pradesh,

30 Calcutta Metropolitan Water Supply and Sanitation Authority Act, 1966, s. 45(2) and New Delhi Municipal Council Act, 1994, s. 154.

31 Uttar Pradesh Municipalities Act, 1916, s. 7(j).

32 *Ibid.*, s. 7(jj).

33 Uttar Pradesh Water Supply and Sewerage Act, 1975, s. 24.

34 New Delhi Municipal Council Act, 1994, ss. 11 and 147.

35 Personal communication with Mr Dwivedi, Mukhya Nagar Adhikari, Municipal Corporation, Aligarh.

36 Uttar Pradesh Water Supply and Sewerage Act, 1975, s. 55.

37 Uttar Pradesh Municipalities Act 1916, s. 228.

jal sansthans have the power to cut water off where users fail to pay their bills. While there is no uniformity in the answer that officials give concerning the carrying out of disconnections, the act provides that the jal sansthan has the power to cut off water to anyone who fails to pay a bill within 15 days of receiving it and it appears to be enforced at least in some places.[38] This possibility takes additional significance in a context where households with an individual piped connection are prohibited from taking water away from the premises unless water supply is charged by meter, thereby preventing solidarity among neighbours.[39] The power to disconnect private water supply is a regular feature in other acts.[40]

In addition to state-specific or city-specific regulation, certain cities are governed by more than one legal regime concerning water supply. A situation which is covered by a central act is that of cantonment areas. A remnant of an older colonial legislation, the new Cantonments Act 2006 still provides for the separate governance of cantonment areas by a cantonment board. Under the Act, one of the duties of the board is to provide or arrange for the provision of water supply.[41] This includes a duty to ensure 'as far as possible' that the supply is continuous throughout the year and that the water is 'fit for human consumption'.[42] Surprisingly, the 2006 Act which reproduces nearly word for word certain sections of the colonial Act has dispensed with an important adjective which imposed on the board a duty to provide water that is 'pure and fit for human consumption'.[43] While no definition of purity was given under the old act, this can be assumed to mean potable since in the first sub-section of section 186, the new act has replaced pure with potable. This omission would tend to indicate that the duties of the board are being narrowed down over time. They are also limited to the extent that the board is not liable for any failure of supply arising from accidents or so-called unavoidable circumstances such as drought.[44]

The existence of different legal regimes governing different parts of the same city, which like in Delhi includes three distinct areas, is problematic in principle and in practice. Firstly, drinking water being a fundamental right, there is no justification for treating different urban residents of the same city differently. Secondly, in an immense city like Delhi, sub-dividing the responsibility for water supply by area could in principle be useful. The current scheme is, however, not based on a rational decision to improve water supply delivery for the poorest or the areas of the city suffering from much lower access to water. Thirdly, the existence of a separate framework for the New Delhi area, which benefits from

38 Uttar Pradesh Water Supply and Sewerage Act 1975, s. 72.
39 Uttar Pradesh Municipal Corporations Adhiniyam 1959, s. 270(2).
40 See, e.g., New Delhi Municipal Council Act 1994, s. 169 and Mumbai Municipal Corporation Act, s. 279.
41 Cantonments Act, 2006, s. 186.
42 *Ibid*, s. 186(2).
43 Cantonments Act, 1924, s. 217(2).
44 Cantonments Act, 2006, s. 194.

much higher per capita availability of water than any other area of the city, can only reinforce existing inequalities in access to water. These inequalities in consumption are, for instance, highlighted by the fact that the New Delhi Municipal Council area is supplied with 462 lpcd while the neighbouring wealthy area of South Delhi has a supply of 148 lpcd.[45]

3. The situation in rural areas

The provision of drinking water in rural areas has been a priority for governments for several decades. This is due both to the fact that at independence, the drinking water situation was comparatively worse in villages than in towns and that an overwhelming percentage of the population lived in rural areas. In terms of coverage, there have been dramatic improvements over time even though many issues remain to be solved, from uncovered habitations to habitations that slip back to being uncovered or only partially covered, and the fact that partially covered habitations are for all practical purposes not covered as the minimum level for identification as partially covered is only 10 lpcd.

The provision of drinking water is primarily the responsibility of states. Yet, the central government has played a very important role in fashioning the policies that states apply and has provided significant funding to ensure access to water in rural areas. The most important body at the national level is the Rajiv Gandhi National Drinking Water Mission (RGNDWM) that functions within the Department of Drinking Water Supply established in 1999 under the Ministry of Rural Development. It has been the key institution with regard to the development of policies and the administration of the rural drinking water sector. Among the schemes it implements, the ARWSP – which is funded both by the Government of India and state governments – plays a central role. The ARWSP was first introduced in 1972. Apart from an interruption during the 1970s, it has been a central component of the government's attempts to ensure full coverage of all habitations throughout the country. It continues to provide the basis for the central government's interventions in rural drinking water.

The ARWSP Guidelines provide the core framework used by the RGNDWM in ensuring the provision of drinking water to all habitations in the country.[46] The Guidelines provide several key policy elements. Firstly, they define the different levels of coverage. Non-covered habitations are defined as having access to less than 10 lpcd. Partially covered habitations are those having access to 10 to 40 lpcd. Covered habitations are defined as having access to 40 lpcd. The figure of 40 lpcd is used to determine the minimum level of coverage necessary to define a habitation as covered. It has been determined through an amalgamation of figures for different basic minimum uses of water. These include three litres for drinking, five

45 Government of Delhi, Economic Survey of Delhi 2001–2002, 115.

46 Government of India, Accelerated Rural Water Supply Programme Guidelines (1999–2000) (hereafter the ARWSP Guidelines).

litres for cooking, fifteen litres for bathing, seven litres for washing utensils and the house and ten litres for ablutions.

Quantity itself is not the only criterion to determine whether a habitation is covered. The source of water also needs to be within 1.6 km or 100-metre elevation in mountain areas. The water should also not be affected by quality problems even though no specific standards for determining quality are included and must thus be indirectly inferred from existing standards. Another criterion is that a given public source of water such as a hand-pump should not be used to serve more than 250 people.

The ARWSP Guidelines also acknowledge the direct link between drinking water for human beings and water for cattle. Consequently, in a certain number of states especially affected by drought, the guidelines mandate that an additional 30 lpcd should be provided for cattle.[47]

The minimum level of 40 lpcd is acknowledged by the RGNDWM as a minimum level of coverage which should be increased over time. Thus, in states where all habitations have been covered at the level of 40 lpcd, the Government of India has approved that the next level of service should be 55 lpcd within 500 metres of the house or 50-metre elevation in mountain areas.[48] Further, some states have long-term objectives which go far beyond these minimums. Thus, Gujarat's Vision 2010 envisages the supply of 80 lpcd in rural areas, 120 lpcd in urban towns without underground drainage and 180 lpcd for cities with drainage.[49]

B. Drinking water reforms in rural areas

The conceptual framework for drinking water policy has comprehensively changed over the past decade. A number of initiatives have been taken in different contexts from the Union level to international funding agencies' projects and state-level measures. While each can be analyzed separately – and this section highlights two particular initiatives, the Swajal project and the Swajaldhara guidelines – the pattern which emerges is overwhelmingly consistent. In other words, while there are different problems in different parts of the country, while different actors have been involved in policy changes, the response given by policy makers at all levels is substantially the same. This implies that at the level of formal policy making, there is in large part a consensus of the basic problems affecting drinking water in rural areas and basic solutions that need to be brought.

Changes in rural drinking water policies have been brought in a number of ways. These include changes in the Government of India's existing policies, adoption of new policies at the Union and state level as well as development projects

47 ARWSP Guidelines, note 46 above s. 2(2)(2).

48 Rajiv Gandhi National Drinking Water Mission Department of Drinking Water Supply, Submission to the National Advisory Committee (2005).

49 K.V. Raju, Keshab Das and S. Manasi, 'Emerging Trends in Rural Water Supply: A Comparative Analysis of Karnataka and Gujarat', in K.V. Raju (ed.) *Elixir of Life – The Socio-Ecological Governance of Drinking Water* 1 (Bangalore: Books for Change, 2007).

such as World Bank projects. This section first highlights some general changes in rural water policy and then specifically examines the case of the Swajal project and the Swajaldhara guidelines, two landmarks in the development of a new rural water policy framework that is still unfolding today.

At the Union level, one of the first important noticeable signs of the new conceptual framework is found in the 1999–2000 version of the Accelerated Rural Water Supply Programme Guidelines. They specifically highlight that one of the reasons for the existence of villages still uncovered includes the non-involvement of people in operation and maintenance.[50] The revision of the guidelines was specifically undertaken with a view to achieving full coverage of all rural habitations during the ninth plan (1997–2002). Three of the guiding principles are worth highlighting. These are the call for an increase in people's participation, the need to treat water as a socio-economic good and the use of 20 per cent of available funds for states promoting reforms along these lines. The revised guidelines make it clear that it is necessary to move away from the perception of water as a social right and rather manage water as a socio-economic good to ensure its effective use.[51]

The guidelines put significant emphasis on the need for people participation as a way to move away from supply-led to demand-led schemes. A number of key conditions for the introduction of demand-led projects are identified. These include ownership of assets and involvement in the setting up of the infrastructure. More significantly, the guidelines recognize that demand-led schemes require an imposition on people to pay for operation and maintenance and the knowledge that the government will not maintain the assets.[52] The message has recently been reinforced with the eleventh five-year plan specifically calling for state support to panchayats for operation and maintenance as a 'hand-holding support for the first few years before the local bodies become self-sustainable'.[53]

The sector reforms put in place require all state and district authorities to impose at least 10 per cent capital cost payment by villagers.[54] Additionally, the ARSWP Guidelines clearly lay down that the contribution of people must increase with the level of service provided. Thus, where villages want to increase their supply from 40 lpcd to 55 lpcd, they now have to pay at least 20 per cent of the capital cost on top of all operation and maintenance expenses.[55] The form of the contribution has been an ongoing debate. While in certain cases, a choice of cash, labour or materials is provided, some documents suggest a full cash contribution.[56]

50 ARWSP Guidelines, note 46 above, s. 1(3).

51 *Ibid.*, s. 3(1).

52 *Ibid.*

53 Planning Commission – Government of India, *Eleventh Five-Year Plan 2007–12 – Volume II – Social Sector* 166 (New Delhi: Oxford University Press, 2008).

54 Planning Commission, Report of Working Group on Tenth Plan for Drinking Water Supply & Sanitation 2002–2007 at 4.

55 ARWSP Guidelines, note 46 above s. 2(3)(1).

56 See, e.g., Rajiv Gandhi National Drinking Water Mission Department of Drinking Water Supply, Submission to the National Advisory Committee (2005) s. 2(2). *Note* that BPL, SC and ST families are excluded.

The ARSWP guidelines have significant influence on the policies followed in states, partly because the states need to put their own money in projects funded under the ARSWP. Yet, some states have gone further than others in adopting reforms in this field. Thus, Maharashtra has been particularly eager in this context. The Government of Maharashtra was thus the first state to adopt demand-driven and participatory approaches for all its rural water supply and sanitation service delivery already in 2000.[57] Besides, it also took the decision to phase out all government subsidies to local bodies for operation and maintenance of water supply.[58] The Government of Maharashtra has also gone much further than the ARWSP Guidelines in requiring not only 10 per cent capital cost contribution for a level of service of 40 lpcd (5 per cent for tribal settlements) but also 100 per cent contribution from the villagers for an increase to 55 lpcd, including for tribal settlements.[59] Maharashtra has also imposed that even the rehabilitation of existing drinking water supply schemes attracts a 10 per cent community contribution up to 40 lpcd and 100 per cent above that level. Other states have taken similar initiatives. Uttar Pradesh has, for instance, implemented the handing over of operation and maintenance of all drinking water schemes to gram panchayats. Rajasthan also has long-term plans to transfer ownership and responsibilities for the management of public water and sanitation assets to panchayati raj institutions as well as imposing 100 per cent responsibility for operation and maintenance by users by 2012.[60]

Rural drinking water policy reforms have taken place at different levels. Initiatives taken at the union or state level constitute two important elements of the overall reform process. Their effort must nevertheless be understood in the broader context of a string of water-related development projects funded, in particular, by the World Bank. Indeed, not only have World Bank projects been instrumental in pushing forward the new policy agenda but also the World Bank has been closely associated with the policy changes taken at the state and union levels. It advocated, for instance, already a decade ago that '[s]ubsidized water and highly centralized water management in the rural sector have resulted in poor water service at high cost' and that this undermined efforts to promote a more efficient and sustainable use of water.[61] It further advised the government that cost recovery was the only option to ensure that universal access to drinking water would not remain an 'unattainable dream'.[62] It also specifically called for the

57 World Bank, Project Appraisal Document – Maharashtra Rural Water Supply and Sanitation 'Jalswarajya' Project 5 (Report No. 26247-IN, 2003).

58 *Ibid.*, at 9.

59 *Ibid.*, at 9.

60 Government of Rajasthan, Sector Policy for Rural Drinking Water and Sanitation (Draft, August 2005).

61 World Bank, India – Water Resources Management Sector Review – Initiating and Sustaining Water Sector Reforms 25 (Report No. 18356-IN, 1998).

62 World Bank, India – Water Resources Management Sector Review – Rural Water Supply and Sanitation Report viii (Report No. 18323, 1998).

immediate imposition of operation and maintenance to users and the progressive implementation of capital cost recovery with the introduction of a 10 per cent contribution during what it saw as a transition period towards full cost recovery.[63]

1. The Swajal project

Since the mid-1990s, a number of initiatives have been taken to foster better water supply and access to water in rural areas by following a new policy framework. It has by now been implemented in the context of several different schemes. While the instruments to deliver the new policy framework have evolved, the basic principles introduced in the mid-1990s continue to guide the overall policy framework for drinking water today.

One of the first formal steps towards introducing a new policy framework for drinking water in rural areas was the implementation of the World Bank Uttar Pradesh Rural Water Supply and Environmental Project (Swajal project). This was a pilot project carried out in two regions of the then undivided Uttar Pradesh, Uttarakhand and Bundelkhand, both facing severe – but different – water supply problems. With funding of $63 million, it covered about 1,200 villages in nineteen districts between 1996 and 2002.[64]

The Swajal project was based around a string of important policy propositions. It sought to introduce a demand-driven approach to replace the supply-driven approach deemed to result in 'inefficient service delivery and poor quality of construction'.[65] The Swajal project thus sought to introduce participation by 'users' allowing them to determine their own contributions to the scheme and to manage operation and maintenance.[66] Under Swajal, participation by villagers should not be understood in the sense of democratic governance at the local level. Indeed, participation was conceived both in terms of ensuring people's control over schemes at the local level and in terms of introducing new obligations and responsibilities that villagers need to shoulder. This is linked to the fact that participation cannot be seen in isolation from another basic principle – cost recovery – introduced under Swajal.[67] Thus, decentralization comes in the form of people's control over some aspects of locally implemented schemes together with the imposition on villagers of the need to shoulder 10 per cent of the capital costs of new projects and the full costs of the operation and the maintenance of those schemes.

The ultimate rationale of the principle of cost recovery is that all projects should be fully self-sufficient. At the outset, project proponents determined that the full

63 *Ibid.*, at viii.

64 Note that the project size was reduced from an original $71 million after the 1999 mid-term review.

65 World Bank, Staff Appraisal Report – Uttar Pradesh Rural Water Supply and Environmental Sanitation Project 6 (Report No 15516-IN, 1996).

66 *Ibid.*, at 7.

67 *Ibid.*, at 7–8.

cost recovery should only be imposed with regard to operation and maintenance. This was linked to the perception that that there would be sufficient 'demand' for this service while poverty might preclude demand for new expensive schemes in favour of maintaining or repairing existing infrastructure. In the meantime, the transfer of the responsibility of operation and maintenance to villagers has been mainstreamed and gram panchayats are now responsible for the operation and maintenance of the whole drinking water infrastructure.

With regard to capital costs, a different strategy was adopted. In view of the political and practical impossibility to impose the full cost recovery on villagers, it was decided to proceed incrementally. Thus, under the Swajal project a community contribution of 10 per cent of the capital costs was made mandatory. This contribution was divided into two components. The first was a 2 per cent cash contribution (1 per cent in the hill districts of Uttarakhand). The rest of the contribution could be in the form of labour, cash or a combination of the two. Anyone seeking an individual connection had to pay a cash contribution of Rs 1,000, half of which had to be paid upfront.[68] This 10 per cent contribution was an arbitrary figure chosen as a way to introduce the principle of cost recovery rather than as a goal in itself. Thus, by 2002 a report prepared for the Planning Commission suggested that in all government projects 50 per cent of capital costs should be recovered.[69]

Another important facet of the Swajal project is that it sought to create new institutional capacity at the village level through the setting up of Village Water Supply and Sanitation Committees (VWSCs). These were first set up as sub-committees of the gram panchayat to which the panchayat had to delegate its powers to allow the committee to fulfil its responsibilities under the project.[70] The membership of the VWSC was in principle from seven to twelve members, but the number was not fixed in the relevant government order. As a result, in the first phase of the Swajal project, while the VWSCs were legally set up in the framework of the panchayati raj institutions, in practice they were independent from the gram panchayat. This was linked to the fact the World Bank did not think that panchayati raj institutions had the 'capacity and inclination to facilitate a demand-responsive approach'.[71] Rather, it was proposed that the project beneficiaries should be the ones directly involved in the schemes.[72] Efforts were made to ensure the representation of diverse constituencies by imposing, for instance, 20 per cent representation of SC/ST members and 30 per cent of women.

68 World Bank, Staff Appraisal Report – Uttar Pradesh Rural Water Supply and Environmental Sanitation Project 142 (Report No. 15516-IN, 1996).

69 G.N. Kathpalia and R. Kapoor, Water Policy and Action Plan for India 2020: An Alternative 24 (Delhi: Alternative Futures, 2002).

70 Government of Uttar Pradesh, Order on Village Water and Sanitation Committee, No: 4430/33-1-95-373/95, 15 December 1995.

71 World Bank, India: The Swajal Project, Uttar Pradesh 2 (on file with the author, 2003).

72 World Bank, Staff Appraisal Report – Uttar Pradesh Rural Water Supply and Environmental Sanitation Project 15 (Report No. 15516-IN, 1996).

However, on the whole, communities were to determine how to organize themselves.[73] The process of excluding the panchayati raj institutions and letting villagers organize themselves is problematic. Thus, allowing communities to determine how they want to organize themselves is more likely than not a way to reproduce existing patterns of power which tend to go along with caste and wealth. In a context marked by vast and even extreme inequalities, letting communities fend for themselves is likely to lead to further marginalizing of the weaker and poorer communities. Further, sidelining the panchayati raj institutions is also problematic because project beneficiaries are the people who have been able to pay the 10 per cent contribution at the outset. In practice, this largely restricts the number of project beneficiaries to the better-off members of the village. Nevertheless, it is the setting up of VWSCs that was the visible face of participation in the first phase of the Swajal project.

The relationship between the VWSCs and the panchayati raj institutions calls for further comments. Indeed, while there was at first an attempt to link the VWSCs to the panchayati raj institutions, in 1998 the Panchayat Act was amended and the VWSCs were then derecognized. This was partly linked to the perception that only committees made of elected panchayat members should fall under the panchayat and partly to the fact that the project proponents thought the user committees would function better if they were completely independent from elected bodies. Severing all links between the VWSCs and panchayati raj institutions quickly proved odd. Indeed, the constitutional mandate of the 73rd Amendment provides that panchayats have control over water supply at the local level.[74] In furtherance of this mandate, the Panchayat Raj Act was amended in 1994 and the functions of the gram panchayat now include the '[c]onstruction, repair and maintenance of public wells, tanks and ponds for supply of water for drinking, washing, bathing purposes and regulation of sources of water supply for drinking purposes'.[75] A water management committee has been provided to foster the realization of these functions. Since this did not leave any legitimacy for the committees constituted under the Swajal project, an additional layer of institutional complexity was added by allowing 'special invitees' to become *de facto* part of the relevant committee even though they are not formally elected. There can be up to seven invitees which brings the total number of people sitting in the committee to fourteen.[76] These invitees do not have the right to vote.[77] In other words, the non-elected user committees indirectly become part of the

73 Government of Uttar Pradesh – UP Rural Water Supply and Environmental Sanitation Project (The Swajal Project), Description of Activities (DOA) of Support Organizations for the Batch 2 Planning Phase (on file with the author, undated) para. 6(3).

74 Constitution, Art. 243G and Eleventh Schedule.

75 Uttar Pradesh Panchayat Raj Act 1947, s. 15(xi).

76 Uttar Pradesh Panchayat Raj (Constitution of Committees of Gram Panchayats for Assistance in Performance of their Functions) Rules 2002, UP Gazette, Extra., Part 4, Section (Kha) (13 September 2002) s. 5.

77 *Ibid.*, s. 5(2).

constitutionally sanctioned water committees of the panchayat. This is justified by the fact that 'users' are more directly involved in the relevant issues than the elected members of the panchayat committee. This is the same line of argument which is used to restrict WUAs to landowners, but does not have any legitimacy in the context of drinking water supply, which is without doubt the concern of each and every individual living in the panchayat. A similar position has been adopted in the ARWSP Guidelines that request the setting up of VWSCs for each water project in reform mode made of members of the panchayat as well as co-opted members and other stakeholders.[78] The Guidelines also specify that this is a committee of the gram panchayat, regardless of its membership.

Swajal also initiated the progressive move towards reducing the government's contribution in drinking water schemes from being a provider to being a facilitator. Under Swajal, two initiatives were taken. Firstly, the Swajal project decided not to use existing government departments to implement the project but rather to use a Project Management Unit (PMU), an autonomous body established specifically to coordinate and monitor its implementation.[79] Secondly, the main interface between the PMU and the VWSC was so-called support organizations. The non-governmental organizations chosen to function as support organizations took on the tasks of facilitating specific schemes and working in individual villages. In practice, the support organizations played a determining role in the implementation of the Swajal project.

Official assessments of demand-led reforms were generally positive.[80] From the Union Government's perspective, this can be assessed through two related initiatives. In 1999 the Union Government had decided to broaden the Swajal experiment throughout the country. It started the so-called Sector Reform Project which sought to implement in sixty-seven districts of the whole country the key principles of the Swajal Project.[81] This was then extended to the whole country in the guise of the Swajaldhara Guidelines immediately after the completion of the Swajal project. Subsequently, Sector Reform Project schemes were clubbed together with Swajaldhara projects.[82] From the point of view of the World Bank, the Implementation Completion Report stated, for instance, that the establishment of the VWSC has allowed 'the village communities to fully participate in the process and gain a sense of ownership of the infrastructure schemes constructed

78 ARWSP Guidelines, note 46 above, s. 4(2)(3)(iii).

79 Memorandum of Association and Rules – Uttar Pradesh Rural Water Supply and Environmental Sanitation Project Management Unit (PMU), in World Bank, Staff Appraisal Report – Uttar Pradesh Rural Water Supply and Environmental Sanitation Project 101 (Report No. 15516-IN, 1996).

80 See, e.g., Ministry of Rural Development, All India Evaluation Study – Sector Reforms Projects in Rural Drinking Water Supply (2005) which provides recommendations on how to improve demand-driven rural water supply but does not call for a paradigm shift.

81 World Bank, Implementation Completion Report – Uttar Pradesh and Uttaranchal Rural Water Supply and Environmental Sanitation (Swajal) Project 4 (Report No: 27288, 2003).

82 Rajiv Gandhi National Drinking Water Mission Department of Drinking Water Supply, Submission to the National Advisory Committee 21 (2005).

under the Project'.[83] This success has led the Bank to implement several other projects based on the Swajal project philosophy in the past few years.[84] Yet, even according to official assessments, a number of problems surfaced in the first phases of the reforms. Thus, in Uttar Pradesh, while the Swajal Project was deemed successful, jal nidhi (the name given to the Sector Reform Project in Uttar Pradesh) proved unsuccessful. This has been explained as being due to the fact that in the plain districts where jal nidhi was implemented, demand for the schemes could not be created.

2. The Swajaldhara Guidelines

The Union Government's positive assessment of the Sector Reform Project and the Swajal project led to the formulation of the Swajaldhara Guidelines which extended the key principles of the Swajal project to the whole country during the Tenth plan. Twenty per cent of funds allocated to the ARWSP were directed to reform projects under the Swajaldhara Guidelines during this period.[85]

The Ministry of Rural Development spearheaded the introduction of Swajaldhara through the adoption of the Guidelines on Swajaldhara. The conceptual background is directly derived from the Swajal project. It first sets out to demonstrate that while water has been perceived as a social right, it should, in fact, be seen as a socio-economic good. Additionally, the delivery of the social right has been through the government which has not sufficiently taken into account the preferences of users and has been ineffective in ensuring the carrying out of operation and maintenance activities. This thus calls for a demand-led approach seeing water as an economic good.[86] The second paragraph of the background is even more revealing. It specifically links the transformation of a supply-driven system to a demand-driven system taking into account the preferences of users, 'where users get the service they want and are willing to pay for'.[87] This is taken one step further by indicating that it is the imposition of full cost recovery of operations and maintenance and replacement costs on the communities which are expected to generate a sense of ownership and ensure the financial viability and sustainability of the schemes.

The Swajaldhara principles are remarkably similar to the ones introduced under the Swajal project. Firstly, Swajaldhara provides for the adoption of a demand-led approach that includes participation of the community from the choice of the drinking water scheme up to its implementation. Secondly, the guidelines seek a form of decentralization and request that drinking water assets should be owned by

83 World Bank, Implementation Completion Report – Uttar Pradesh and Uttaranchal Rural Water Supply and Environmental Sanitation (Swajal) Project 4 (Report No. 27288, 2003).

84 See, e.g., World Bank, Project Appraisal Document, Maharashtra Rural Water Supply and Sanitation 'Jalswarajya' Project (Report No. 26247-IN, 2003).

85 Ministry of Rural Development, Guidelines on Swajaldhara, 2003, s. 15(1).

86 Ministry of Rural Development, Guidelines on Swajaldhara, 2003, s. 1(1).

87 *Ibid.*, s. 1(2).

the relevant panchayat and that the communities should have the power to plan, implement and operate all drinking water schemes. Thirdly, the participation and decentralization elements are brought together in the context of the financial principles which are a compromised version of full-cost recovery. Thus, while users have to bear the entire responsibility for the operation and maintenance of drinking water schemes, their contribution to capital costs is limited. In practice, this was first set at 10 per cent for a service level of 40 lpcd but,[88] in a number of situations, this percentage has already been exceeded. Under Swajaldhara, at least half of the 10 per cent contribution must be in cash, a significant increase over the 20 per cent under the Swajal project.[89] Exceptions have, for instance, been provided for scheduled tribes' areas where the cash contribution was first reduced to one-quarter of the community contribution.[90] Subsequently, in 2006 an amendment to the guidelines provided that the contribution in the case of villages where scheduled tribes or scheduled castes constituted more than half of all habitations could be in any form without any stipulation of a contribution in cash.[91] Fourthly, from an institutional perspective, one of the consequences of a demand-led perspective is the rethinking of the role of the government. The guidelines here specifically provide that the aim is to shift the government's role from 'direct service delivery' to only supporting a limited number of activities such as planning, policy formulation, monitoring and evaluation.

An important aspect of the Swajaldhara scheme is that it was undertaken at the union level without specific Parliamentary mandate. The Union decided to proceed in two ways. Firstly, it decided to provide full funding for the scheme, a departure from the usual ARWSP norm where the centre and the states each share half of the costs. Secondly, it proposed that the states interested in taking up Swajaldhara funding should sign up to a Memorandum of Understanding with the Centre. The intent of the model Memorandum circulated to states was to ensure that the reform principles would be, as far as possible, mainstreamed.[92] Apart from this general commitment to reforms of the drinking water sector, the states were, for instance, also called upon to hand over all existing drinking water schemes to gram panchayats for operation and maintenance.[93]

The process of decentralization and participation takes different forms under ongoing reforms. On the one hand, some of the proposed measures go towards ensuring that operation and maintenance of schemes is more successful. Thus, panchayats are for instance allowed to contract any required person where the

88 Ministry of Rural Development, Guidelines on Swajaldhara, 2003, s. 5(1).

89 *Ibid.*, s. 5(3).

90 Lok Sabha, Unstarred Question No. 2451, Swajaldhara Yojana, Answered by Minister of State in the Ministry of Rural Development (Shri A. Narendra) on 18 March 2005.

91 Government of India Department of Drinking Water Supply, Office Memorandum – Amendment to Swajaldhara Guidelines, Doc. No. W-11021/2/2003-TM.IV (SW), 15 May 2006.

92 Draft Memorandum of Understanding between the State Government of ___ and the Department of Drinking Water Supply, Ministry of Rural Development, Government of India (2003) s. 5.

93 *Ibid.*, s. 8.

government does not make its people available.[94] In principle, this should provide ways to ensure that any bottlenecks in the government do not affect actual operation and maintenance. Villagers complain, however, that only government officials have access to original spare parts and that private traders always provide sub-standard quality. On the other hand, the same set of reforms propose measures which are likely to lead to even more widespread inequalities in access to water within panchayats and particularly within different districts and states. Thus, the proposal seeking to allow panchayats to fix and collect water tariffs is fraught with difficulties.[95] If tariffs are fixed at the level of each and every panchayat, the most likely consequence is that villages that suffer the most from water scarcity – for instance, because the water table is very low or water quality is low – will have to bear all the costs themselves and will thus pay much more than villages that happen to be endowed with more or better water. Similarly, where costs are fixed within each panchayat without government overseeing, the likelihood is that Dalits and other marginalized communities will be further marginalized because their say in the decision-making process will not only be low, as it has been traditionally, but more so because the reforms propose a decision-making process based on the notion of 'users' which excludes most poor people. As identified in the context of the Swajal project, where the communities are left to collect tariffs themselves, public taps are rapidly switched off. Whether the real reason is the one given by the users that money is not being paid or whether it is a decision of a more political nature, the result is the same for people who are denied access to water. Given the nature of water, and drinking water in particular, any policy which does not attempt to redistribute the costs of getting access to water across social classes and geographical areas is bound to fail from an equity perspective.

C. Swajaldhara and its aftermath

The implementation of the Swajaldhara Guidelines in its first few years of operation led the Government to rethink the scheme for the eleventh plan. This led to a series of documents proposing different ways to take the reforms forward.

In the first place, it was suggested that the Swajaldhara scheme would be discontinued at the end of 2007. This decision did not imply that the principles underlying the reforms were being abandoned. In fact, it had been proposed since at least 2005 to extend the reform principles of Swajaldhara to ARWSP from 2007 onwards.[96] In other words, the idea was to progressively mainstream the reform principles beyond the Swajaldhara projects to all rural drinking water projects. At the same time the break with Swajaldhara signalled a desire to rethink parts of the reforms in view of some perceived failures and resistance to the reforms. Indeed, official assessments of Swajaldhara found that it had been plagued by 'constraints'

94 *Ibid.*, s. 20.

95 *Ibid.*, s. 8(vii).

96 Rajiv Gandhi National Drinking Water Mission – Department of Drinking Water Supply, Submission to the National Advisory Committee 26 (2005).

because government officials were slow to adopt reforms and because panchayats lack finances and skills to take up the responsibility immediately.[97] The government also conceded that difficulties in collecting the community contribution were a factor impeding the success of Swajaldhara.[98]

The combination of these two different assessments led to suggesting a revamped reform effort that sought to address some of the shortcomings identified by the Government in the implementation of demand-led reforms. Firstly, while the principles of Swajaldhara were to be upheld, the financing of projects was to change to the pattern implemented under ARWSP where states and the Union each contribute 50 per cent of the funding.[99] Further, the states were to be given the discretion to determine how and whether to foster community contribution. The incentives given to the states were that the community contribution would be deducted from the share of the state. Another relaxation concerned the community contribution. In villages where more than half the population is SC/ST, the stipulation of a cash contribution was to be abandoned.[100] An indirect way to relax the element of community contribution is to, for instance, allow MPs and MLAs to use development funds at their disposal for this purpose.[101]

The gist of the new scheme was to tone down the reform rhetoric and to demand only 'an element of token community contribution and involvement of user groups/panchayats in the selection and implementation of the schemes and for subsequent [operation and maintenance]'.[102] The government also indicated that it was ready to exempt communities from making a contribution in 'exceptional cases of hardship'.[103] Further, communities could choose the mode of contribution. This removed the previous insistence on a cash contribution.[104]

One explanation for these suggested changes was the fact that Swajaldhara had failed at two different levels. Firstly, the demand-led mode has not functioned as hoped by its promoters. This is in part due to the fact that they have not been able to create enough demand. Indeed, where people have adequate access to water, they are loath to adopt the proposed reforms. Thus, it is in the areas where there is least availability of water that people are most responsive to the new scheme. This explains, for instance, why the pilot areas chosen for the Swajal project were areas facing tremendous drinking water problems. Secondly, there has been resistance from within the government to the new schemes. These have been perceived as eroding the power of existing departments. Thus, Swajaldhara was, for instance,

97 Rural Water Supply and Sanitation, Eleventh Five-Year Plan – Approach Paper 3 (2006).

98 Lok Sabha, Unstarred Question No 1550 – Slow Pace of Swajaldhara Yojana, Answer of Minister of State in the Ministry of Rural Development (Shri Chandra Sekhar Sahu), 9 March 2007.

99 Department of Drinking Water Supply, DO No. W-11012/15/2007/DWS-III, 30 August 2007.

100 Office Memorandum – Amendment to Swajaldhara Guidelines, No W-11021/2/2003-TM.IV (SW), 15 May 2006.

101 See, e.g., Uttar Pradesh State Rural Drinking Water Policy, s. 3(2).

102 Rural Water Supply and Sanitation, Eleventh Five-Year Plan – Approach Paper (2006) 7.

103 *Ibid.*

104 *Ibid.*

disliked because it seemed to make the technical expertise of government departments irrelevant since their services are not necessarily required on a compulsory basis in the choice and design of schemes.[105] One of the new proposed features in reaction to this was that a Junior Engineer would be specifically charged with providing three to four gram panchayats with technical assistance.[106]

This rethinking of the reform principles has eventually not been carried forward. This is illustrated by the fact that the commitment to demand-led reforms during the eleventh plan is more or less intact.[107] Indeed, the government has not shown any inclination to abandon the reforms at this stage. The idea is in fact to ensure that further reform measures are taken during the eleventh plan. The concept of a minimum community contribution of 10 per cent is, for instance, reasserted as something that needs to be part of all drinking water supply schemes.[108] Similarly, the operation and maintenance of all new single village schemes is to be borne by the community and further states' governments are to progressively transfer all existing schemes to gram panchayats.[109] This has now been translated in instructions from the Union Government that seek to carry Swajaldhara forward as originally defined. In other words, the government is back to suggesting that 20 per cent of projects should be implemented under Swajaldhara principles and these projects are financed entirely by the Union Government rather than shared between the state and the Union.[110] This does indicate that the time is not yet ripe for mainstreaming Swajaldhara to all ARWSP projects, but also shows that the commitment to a community contribution is maintained. Indeed, the report of a meeting of State Secretaries on rural drinking water states that from 2008 onwards, 'the Swajaldhara scheme is going to be expedited drastically'.[111]

This happens to coincide with a similar commitment to the principles of the Swajal project by the World Bank. Thus, the Uttaranchal Rural Water Supply and Sanitation Project is, for instance, premised on the 'success of demand-driven community participatory approach'.[112] The Bank anticipates that the current project can contribute to the replication of the model in other states and the mainstreaming of the approach by the Union Government that the Bank could then support.[113] This project seeks to go beyond Swajal in strengthening the involvement

105 Remarks of RM Tripathi, Executive Engineer, Jal Nigam 29 August 2007.

106 Rural Water Supply and Sanitation, Eleventh Five-Year Plan – Approach Paper 7 (2006).

107 Planning Commission – Government of India, *Eleventh Five Year Plan 2007–12 – Volume II – Social Sector* 163 (New Delhi: Oxford University Press, 2008).

108 *Ibid.*, at 48.

109 *Ibid.*

110 Rajiv Gandhi National Drinking Water Mission, Allocation of funds under Accelerated Rural Water Supply Programme (ARWSP) during 2008–2009, No. G–11011/5/2008–DWS.I (2008).

111 Minutes of the State Secretaries' Conference on Rural Drinking Water and Sanitation held on 13 and 14 May 2008, Surajkund, Haryana, 4.

112 World Bank, India – Uttaranchal Rural Water Supply and Sanitation Project, Project Appraisal Document 2 (Report No. 35464-IN, 2006).

113 *Ibid.*, at 3.

of the panchayati raj institutions whose limited involvement in the earlier scheme was seen as detrimental to the sustainability of the schemes.[114] In the context of this World Bank project, the cost-sharing principle has not been abandoned and communities will bear 10 per cent of the cost of all new investments. The only relaxation is that only 2 per cent of the cost must be in the form of cash while the other 8 per cent can be in the form of labour or cash.[115]

CONCLUSION

Drinking water supply policy has witnessed the introduction of dramatic changes over the past fifteen years. This includes, in particular, a complete rethink of the basic principles underlying public sector activities in this sector. These changes are linked to the broader paradigm shift brought about in the context of water sector reforms, in particular the introduction of the notion that water is an economic good. In the context of drinking water, the politically sensitive nature of the problem and the immediate social consequences of any changes have ensured that the basic reform principles proposed as part of water sector reforms have not (yet) been fully implemented. Thus, while the policy goal is to impose full capital costs on drinking water users in villages, this is currently politically and economically impractical. This explains that rural communities are made to pay in a first phase – and for a minimum level of supply – 10 per cent of the capital costs even though policy documents clearly indicate that the eventual objective is much higher.

The new policy framework, which provides that supply-led drinking water management should be changed for a demand-led paradigm, has significant social consequences. In particular, the poor who cannot afford the 10 per cent upfront capital contribution are left out of the new schemes altogether and thus only have access to existing infrastructure. This is without negative consequences for them as long as this existing infrastructure is repaired. Policy reforms include, however, the decentralization of operation and maintenance to the panchayat level. Once this is fully implemented, if operation and maintenance funds are preferentially allocated to the new schemes implemented, for instance, under Swajaldhara, the poor will then directly suffer because the infrastructure they use to access water may not be functional any more. If this happens, the reforms will raise human rights issues with regard to the fulfilment of the human right to water.

One of the concerns raised by ongoing reforms is the process that leads to their adoption. Indeed, drinking water policy reforms are the only significant reforms in the water sector that have not been introduced through the adoption of legislation. While there has been a spate of water users association legislation and groundwater legislation, drinking water is mostly conspicuous by its invisibility in legislative assemblies or Parliament. Indeed, the single biggest reform introduced

114 *Ibid.*, at 7.
115 *Ibid.*, at 14.

over the past decade in the drinking water sector, the Swajaldhara Guidelines, was never considered by elected representatives of the people. This is all the more cause for concern when the policy framework that is adopted is directly derived from a pilot World Bank project, the Swajal project.

Policy reforms in the drinking water sector warrant significant consideration because of the far-reaching impacts they are likely to have. Yet, this should not detract the attention from two broader issues that are not specifically linked to ongoing reforms. Firstly, regardless of any policy reforms, the primary duty of the government is the realization of the human right to water. This is completely independent of any policy changes that the executive may want to introduce. Secondly, beyond the issue of the consideration of significant policy reforms by elected representatives, the broader issue that arises is the lack of drinking water legislation. In a context where water law has developed and will continue to develop in a sectoral manner, it is at the very least surprising that drinking water, which is the acknowledged first priority, would be the only major sector of water law that has never been provided with a legislative framework. Ongoing reforms that seek to completely change operative principles make the adoption of a legislation setting down all the basic legal principles guiding the provision of drinking water much more urgent.

IX. Legal regime governing groundwater

Sujith Koonan

INTRODUCTION

Groundwater plays a crucial role as a source of freshwater in India. It accounts for around 58 per cent of the total irrigated area in the country, whereas the share of groundwater in meeting the drinking water needs is much higher at around 80 per cent.[1] One of the reasons for the unprecedented reliance on groundwater is believed to be the poor functioning of the public irrigation system.[2] With large areas of India having substantial aquifers, people who have access to land and financial resources are able to ignore the inconvenience of poorly functioning public systems and became self-reliant using groundwater.[3] As a result, the extent of groundwater extraction is increasing exponentially in India. This is seen in the growth in the number of dug wells and tubewells since independence. In the period 1951–1997, the number of dug wells and tubewells has increased from 3,865 thousand to 17,334 thousand.[4] Dependence on groundwater as a freshwater source is likely to increase in the coming years. A recent Planning Commission report noted that by the year 2025, the demand for industrial and domestic use is projected to rise to 29 billion cubic metres (BCM) from the current demand of 18 BCM.[5]

1 Planning Commission of India, Report of the Expert Group on Ground Water Management and Ownership (Delhi: Government of India, Planning Commission, 2007); M. Dinesh Kumar and Tushaar Shah, Groundwater Pollution and Contamination in India: The Emerging Challenge (IWMI–TATA Water Policy Briefing Paper, 2006) and National Academy of Agricultural Sciences, Emerging Issues in Water Management – The Question of Ownership (New Delhi: National Academy of Agricultural Sciences, Policy Paper No. 32, 2005).

2 A. Narayanamoorthy and R.S. Deshpande, *Where Water Seeps! Towards a New Phase in India's Irrigation Reforms* 37 (New Delhi: Academic Foundation, 2005).

3 Maria R. Saleth, 'Groundwater Markets in India: A Legal and Institutional Perspective', 29(2) *Indian Economic Review* 157 (1994) and S. Janakarajan, Wells and Welfare: An Overview of Groundwater Use and Abuse in Tamil Nadu, South India (Discussion paper prepared for the IWMI-Tata Program Annual Partners' Meet 2002), available at www.iwmi.cgiar.org/iwmi-tata_html/PartnersMeet/pdf/Photocopy%20006%20-%20JanakRajan.pdf.

4 Planning Commission of India, note 1 above, 4.

5 *Ibid.*, at 5.

As a result of the indiscriminate exploitation of groundwater for various purposes, the number of areas considered to be semi-critical, critical and over-exploited from the point of view of groundwater availability has increased. Out of the 5,723 assessment units (Blocks/Taluks/Mandals/Districts) assessed jointly by the state groundwater departments and the Central Ground Water Board with respect to groundwater availability, 4,078 are safe (71 per cent), 550 are semi-critical (10 per cent), 226 are critical (4 per cent) and 839 are over-exploited (15 per cent).[6] These aggregates do not give the real scenario and its implications. The degree of exploitation varies widely across the country. In the states of Gujarat, Haryana, Punjab, Maharashtra, Rajasthan and Tamil Nadu, the number of over-exploited and critical areas are significantly higher.[7] The percentage of over-exploited areas has increased over the past few decades from 4 per cent to 15 per cent.[8] At the same time, it should also be noted that there is huge reserve of groundwater in the deep aquifers that has not been fully utilized. The under-utilization of the deep aquifer groundwater is also likely to result in its eventual deterioration in quality.[9]

One of the major causes for over-exploitation was the existing legal framework regulating groundwater use. Except for the experiment with public tubewells in north Indian states, groundwater largely remained in the domain of private property.[10] Apart from the common law principle which permitted landowners to draw as much groundwater as they wished, there was no legal regime governing groundwater.[11] The electricity policy, the lending policy of nationalized banks, the fertilizer subsidies and price support policies have also exacerbated the over-exploitation of groundwater.[12] The system of providing free power to owners of agricultural pump sets, low-interest loans for deepening existing wells, constructing new wells and for purchasing pumps and other equipment seems to have promoted the indiscriminate exploitation of groundwater. At the same time little attention has been paid to maintaining the already existing traditional irrigation sources such as canals, tanks and spring channels, which were good sources for recharging groundwater.[13]

6 Central Ground Water Board, Dynamic Ground Water Resources in India 19 (Faridabad: CGWB, 2006).

7 *Ibid.*, at 19.

8 Planning Commission of India, note 1 above, 8 and R.C. Purohit and Virendra Kumar, 'Ownership of Ground Water and Its Pricing: Rajasthan Perspective', in Saleem Romani *et al.* (eds.) *Groundwater Governance – Ownership of Groundwater and its Pricing* 354 (New Delhi: Capital Publishing Company, 2007).

9 Planning Commission of India, note 1 above, 8.

10 M.S. Rathore, 'Water Rights and Other Alternative to Groundwater Management in India', *in* Romani *et al.* (eds.), note 8 above, 333.

11 Chhatrapati Singh, *Water Rights and Principles of Water Resource Management* (Bombay: N.M. Tripathi, 1991).

12 A.S. Bhullar and R.S. Sidhu, 'Integrated Land and Water Use: A Case Study of Punjab', 42 (52) *Economic & Political Weekly* 5353 (2006).

13 S. Janakarajan and Marcus Moench, 'Are Wells a Potential Threat to Farmer's Well Being? Case of Deteriorating Groundwater Irrigation in Tamil Nadu', 41(37) *Economic & Political Weekly* 3977 (2006).

The mounting problem of groundwater depletion and pollution induced the central government to circulate a Model Groundwater Bill to guide the state governments while they enact separate groundwater laws.[14] The need for groundwater regulation was also recognized in the National Water Policy, 1987 and National Water Policy, 2002.[15] State governments have started responding to the policy initiatives of the central government in the last few years. As a result, a few state governments have enacted separate groundwater laws. Therefore, the legal regime governing groundwater in India today is a combination of the traditional common law-based legal regime and recent legal reforms. With this background, this chapter analyses the legal regime governing groundwater in India as a whole, with a special emphasis on ongoing reforms. Instead of adopting a comparative analysis of the different legal frameworks in various states, an analysis at a conceptual level is adopted, with special reference to specific state laws wherever necessary.

A. The legal regime governing groundwater: An overview

The legal regime governing groundwater in India is multifarious. It includes traditional common law rules, various principles and doctrines evolved primarily by the judiciary as part of environmental jurisprudence and laws specifically relating to groundwater passed recently in the course of water law reforms. This part of the chapter analyses the common law rule and the two important principles of environmental law – namely the public trust doctrine and the precautionary principle – which are significant in the regulation of groundwater use. Being a major legal change specific to groundwater management, the evolution of specific groundwater laws requires special scrutiny. Therefore, the evolution of groundwater laws is discussed separately in Section B.

1. Common law rules

As per the Constitution, unless and until legislators use their powers to enact contrary laws, pre-constitution laws remain in force.[16] Consequently, common law principles continued to govern the regulation of groundwater use even after independence. This legal regime has undergone some changes in the few states where new groundwater laws have been enacted in the last few years. However, the traditional common law rule still continues to apply to groundwater extraction in many states.

14 Model Bill to Regulate and Control the Development and Management of Ground Water, 2005, available at www.ielrc.org/content/e0506.pdf.

15 National Water Policy, 1987, available at www.ielrc.org/content/e8701.pdf. See also National Water Policy, 2002, available at www.ielrc.org/content/e0210.pdf.

16 Constitution of India, Article 372, in P.M. Bakshi, *The Constitution of India* (Delhi: Universal Law Publishing, 7th edn 2006).

Under common law, groundwater is considered to be part and parcel of the land. Groundwater is viewed as a chattel attached to the land without having a distinctive character from the land, and without there being any separate title of ownership over it.[17] Further, under common law, groundwater is considered to be distinct from the surface water resources such as streams and rivers. Therefore, principles and doctrines developed with regard to surface water resources are not applicable to groundwater under the common law.[18]

Early case law in this regard asserted and established the rule that, 'percolating water below the surface of earth is a common reservoir in which nobody has any property but of which everybody has (as far as he can) the right of appropriating the whole'.[19] Common law does not recognize any natural or prescriptive right in groundwater flowing in undefined channels. It can be said that under common law, groundwater is considered a 'common supply' and, therefore, the 'absolute property' of any occupier by whom it is appropriated.[20] These were the norms that were customarily followed in India with respect to groundwater extraction.[21] Individuals have always considered water fetched from wells located on their private lands to be their private property.[22]

The legal consequence of the application of this common law rule by the courts is that the owner of the land can dig well(s) in his land and extract as much groundwater as he wants or is available. Landowners are entitled to extract groundwater from their land according to their free will and pleasure. Landowners are not legally liable for any damage caused to the water resources of their neighbours as a result of over-extraction, even if they have over-exploited groundwater with malicious intention to cause injury to their neighbours' wells. Common law jurisprudence dismisses such a problem with the curt observation that such a result is not actionable in law.[23]

The colonial Acts in India followed this legal position of groundwater rights being regarded as part of land rights. This common law right of landowners was recognized by early statutes in India. For instance, the Indian Easements Act of 1882 recognizes the right of every landowner to collect and dispose within their own limits all the water under the land which does not pass in a defined channel.[24] The Indian Easements Act further lays down that a prescriptive right cannot be

17 *Roath v. Driscoll*, 20 Conn. 533, 541 (1850) and *Chatfield v. Wilson*, 28 Vt. 49, 53 (1855) as cited in Robert Emmet Clark (ed.) *Water and Water Rights, Vol. I* at 71 (Indiana: The Allen Smith Company Publishers, 1967).

18 *Acton v. Blundell* (1843) 12 M&W 324 as cited in G.C. Mathur (ed.) *Amin and Sastry's Law of Easements* 434 (Lucknow: Eastern Book Company, 1984).

19 *Chasemore v. Richards* (1859) 7 HLC 349324 and *English v. Metropolitan Water Board* (1907) 1 K.B. 588324 as cited in Mathur (ed.) note 18 above.

20 V. Sitarama Rao, *Law Relating to Water Rights* 185 (Hyderabad: Asia Law House, 1996).

21 Rathore, note 10 above, 339.

22 *Ibid.*, at 339.

23 Clark (ed.), note 17 above, 71.

24 Indian Easements Act, 1882, s. 7.

acquired over underground water not passing in a defined channel.[25] Early case law also asserts this legal position.[26]

The historical reason for the evolution of these rules could be the lack of knowledge of hydrology. This could be considered one of the major reasons for not subjecting groundwater use to legal regulation.[27] Further, since the mechanisms for tapping groundwater had not improved much, there was little chance of extracting too much groundwater, and as such, groundwater extraction was unlikely to cause any serious social problem requiring mediation through law. Both these reasons have now become obsolete. The characteristics of groundwater, such as its movement, are now within the reach of human knowledge. The very fact that groundwater is always in a state of movement makes the private property approach, as reflected in the common law rule, untenable.[28] In fact, the present qualitative and quantitative condition of groundwater necessitates legal intervention.

Moreover, the human rights jurisprudence developed by the Indian judiciary does not support a legal regime favouring uncontrolled exploitation of groundwater by individuals. The right to pollution-free water has been declared to be a part of the fundamental right to life under Article 21 of the Constitution of India.[29] Since over-exploitation by one person is likely to affect the availability of groundwater for another person, in terms of both quality and quantity, Article 21-based jurisprudence seems to support the regulation of groundwater use. Despite this, the common law principle still dominates the legal regime governing

25 *Ibid.*, s. 17(d). This provision appears to exclude groundwater 'passing in a defined channel'. However, it is not clear how to determine the fixed path of groundwater in a particular area. Also *Malayam Patel v. Lakka Narayan Reddi*, AIR 1934 Mad. 284 and *Het Singh v. Anar Singh*, AIR 1982 All. 468.

26 *Mst. Manturabai v. Ithal Chiman* AIR 1954 Nag. 103 as cited in K.K. Singh (ed.) *Amin and Shastri's The Law of Easements* 126 (Lucknow: Eastern Book Company, 4th edn 1970). The observance of this common law tradition in the early days is indicated in the judgment of Chandra Shekhara Aiyar, J. in *Kesava Bhatta v. Krishna Bhatta* AIR 1946 Mad. 334 wherein it is categorically stated at para. 335: 'the general rule is that the owner of a land has got a natural right to all the water that percolates or flows in undefined channels within his land and that even if his object in digging a well or a pond be to cause damage to his neighbour by abstracting water from his field or land it does not in the least matter because it is the act and not the motive which must be regarded. No action lies for the obstruction or diversion of percolating water even if the result of such abstraction is to diminish or take away the water from a neighbouring well in an adjoining land'.

27 Lawrence J. MacDonnell, 'Rules Guiding Groundwater Use in the United States', 1 *Indian Juridical Review* 43, 46 (2005) and Sanjiv Phansalkar and Vivek Kher, 'A Decade of the Maharashtra Groundwater Legislation: Analysis of the Implementation Process', 2(1) *Law, Environment and Development Journal* 67 (2006).

28 T.N. Narasimhan, 'Groundwater Management and Ownership', 43(7) *Economic & Political Weekly* 21 (2008).

29 *Indian Council for Enviro-Legal Action and Ors v. Union of India* (1996) 3 SCC 212; *Venkatagiriyappa v. Karnataka Electricity Board* 1999(4) *Karnataka Law Journal* 482; *Attakoya Thangal* v. *Union of India*, 1990 (1) *Kerala Law Times* 580 and *F.K. Hussain v. Union of India*, AIR 1990 Ker. 321.

groundwater use in many states that do not have separate statutory provisions to regulate groundwater use.[30]

2. *Environmental law principles*

The development of environmental law in India, particularly over the last couple of decades, has resulted in the incorporation of some important environmental principles by the Indian judiciary. Two important principles relevant to the regulation of groundwater use are the public trust doctrine and the precautionary principle.[31] These principles are relevant to groundwater regulation in at least two ways. Firstly, these principles can provide an important basis for groundwater regulation in states where separate groundwater laws do not exist. Secondly, these principles could form the foundation of the evolving groundwater regulatory framework.

Public trust doctrine

The public trust doctrine views vital natural resources as being vested in public trust. Being the trustee, the government has the responsibility to protect and preserve it for and on behalf of the beneficiaries, that is, the people. The public trust doctrine does not approve of any kind of private appropriation of vital natural resources.[32] The public trust doctrine attempts to redefine the rights and duties of the government and the people *vis-à-vis* natural resources. Regarding the role of the government, the adoption of the public trust doctrine marks a shift from the position of 'sovereign rights' to that of the state as 'trustee'.[33] This means that

30 Recently a division bench of the Kerala High Court, by reversing a single judge decision, upheld the common law rule on groundwater. *Perumatty Grama Panchayat v. State of Kerala*, 2005 (2) *Kerala Law Times* 554.

31 There are other environmental law principles, such as polluter-pays principle, which can be discussed in the context of groundwater regulation. However, there is no clarity on the applicability of the polluter-pays principle to situations other than pollution caused by hazardous substances and industries. For a detailed discussion on the scope of the polluter-pays principle in India, see *Vellore Citizens' Welfare Forum v. Union of India* (1996) 5 SCC 647 and *Indian Council for Enviro-Legal Action v. Union of India* (1996)3 SCC 212.

32 Singh, note 11 above, 76.

33 An analysis of the colonial and post-colonial laws on irrigation reveals that the government has exercised complete control over surface water resources, mainly by exercising control over the construction of irrigation structures, water supply and cropping patterns. Early statutes such as the Indian Easements Act, 1882 also save the right of the government in the water of natural rivers and streams and other public irrigation works. For details, Alice Jacob and S.N. Singh, *Law Relating to Irrigation* 1–14 (Bombay: N.M. Tripathi, 1972) and Iqbal Ahmed Siddiqui, 'History of Water Laws in India', in Chhatrapati Singh (ed.) *Water Law in India* 289 (Delhi: Indian Law Institute, 1992).

neither the government nor individuals can exercise absolute rights over such natural resources.[34]

The public trust doctrine, in principle, is a part of environmental jurisprudence in India. Though it is not an express part of any environmental statute, it has been incorporated into environmental jurisprudence by the Indian higher judiciary in the *Kamal Nath* case.[35] The Supreme Court of India has categorically stated that the state is the trustee of all natural resources which are by nature meant for public use and enjoyment. The public at large is the beneficiary of the sea-shore, running water, air, forests and ecologically fragile lands.[36] This view has been recognized and reiterated in a number of cases.[37]

There can be little disagreement on the applicability of the public trust doctrine in the case of surface water resources such as rivers and streams. However, the application of the public trust doctrine *vis-à-vis* groundwater is not clear. This issue came before discussion, recently, in the *Plachimada* case.[38] At the first instance, the single judge of the Kerala High Court decided in favour of the applicability of the public trust doctrine to groundwater resources in private property.[39] The single judge relied upon the Supreme Court decision in the *Kamal Nath* case and held that groundwater resources in private property would come under the purview of the public trust doctrine. The single bench decision was reversed by the division bench in appeal.[40] Presently, the appeal is pending before the Supreme Court of India and, therefore, finality on this issue is yet to be arrived at.

The common law-based legal regime could be considered one of the reasons for the uncontrolled over-exploitation of groundwater in the last several decades. Given the adverse human rights and environmental implications of such a regime, a proper regulatory framework is imperative. The public trust doctrine ought to be an underlying principle in such a regulatory framework as it will provide a justification for state interference in private property rights over groundwater as recognized by common law. A public trust doctrine-based regulatory framework would further entail the abolishment of all private property right claims, as well as sovereign right claims of the government, over ground-

34 For a description of the origin and development of the public trust doctrine, see Sujith Koonan, 'Groundwater: Legal Aspects of the Plachimada Dispute', in Philippe Cullet, Alix Gowlland Gualtieri, Roopa Madhav and Usha Ramanathan (eds.) *Water Law at the Crossroads – National and International Perspectives with Special Emphasis on India* 159, 166–169 (New Delhi: Cambridge University Press, 2009).

35 *MC Mehta v. Kamal Nath* (1997) 1 SCC 388.

36 *Ibid.*, at para. 34.

37 *MI Builders v. Radhey Shyam Sahu* (1999) 6 SCC 464; *Intellectual Forum v. State of Andhra Pradesh* (2006) 3 SCC 549; *Karnataka Industrial Area Development Board v. Kenchappan* (2006) 6 SCC 371.

38 Reference to the *Plachimada* case is to the decision of the Kerala High Court on the issue of the power of the local bodies to regulate the over-exploitation of groundwater in their jurisdiction. For an analysis of the *Plachimada* case, see Koonan, note 34 above, 179–189.

39 *Perumatty Grama Panchayat v. State of Kerala* 2004 (1) *Kerala Law Times* 31.

40 *Hindustan Coca-Cola Beverages v. Perumatty Grama Panchayat* 2005 (2) *Kerala Law Times* 554.

water, thus changing the very basis of the legal regime governing groundwater. Even though some states have enacted separate groundwater laws, none of the groundwater laws expressly recognize the public trust doctrine as a cardinal principle.[41]

The need for a public trust doctrine-based regulatory framework needs to be further examined, particularly in the context of over-exploitation of groundwater by big multinational corporations in various parts of the country. It should be noted that the common law rule is often cited in order to establish that such over-exploitation is not legally actionable. Precisely this argument was raised by the Coca-Cola Company in the *Plachimada* case.

Precautionary principle

The precautionary principle is one of the cardinal principles of international environmental law.[42] It has been endorsed expressly in a number of environmental treaties and declarations in the last two decades.[43] By virtue of the Supreme Court decision in the *Vellore Citizens' Welfare Forum* case, the precautionary principle is also a part of environmental jurisprudence in India.[44] The precautionary principle, as defined by the Supreme Court of India, casts a duty upon the state to take measures to 'anticipate, prevent and attack the causes of environmental degradation'.[45] As per the precautionary principle, precautionary measures cannot be delayed on grounds of scientific uncertainty.[46]

A regulatory regime recognizing the precautionary principle is particularly relevant so far as groundwater resources are concerned, because once polluted, it is very difficult or it takes a long time to reinstate these resources.[47] Moreover, it may be difficult to establish the cause–effect relationship in the case of depletion or pollution of groundwater. The precautionary principle provides the means to overcome these regulatory hurdles. If the precautionary principle is applied, it is not required that conclusive scientific evidence should be presented or an exact cause–effect relationship be established in order to limit groundwater

41 There could be another argument based on the Supreme Court decision in the *Kamal Nath* case. As per Art. 141 of the Constitution of India, the law declared by the Supreme Court is considered to be the law of the land and is to be followed. In this respect, the public trust doctrine could be, in principle, considered to be an implied part of all environmental laws in the country and this should be respected by all other courts and by the executive wing as well.

42 Nicolas de Sadeleer, *Environmental Principles: From Political Slogans to Legal Rules* 96 (Oxford: Oxford University Press, 2002).

43 *Ibid.*, at 96.

44 *Vellore Citizen's Welfare Forum v. Union of India* (1996) 5 SCC 647. In the instant case, the Supreme Court said at para. 14 that 'we have no hesitation in holding that the precautionary principle and polluter-pays principle are part of the environmental law of the country'.

45 *Ibid.*, at para. 11.

46 *Ibid.*

47 J. Chilton, 'Groundwater', in Deborah Chapman, *Water Quality Assessments – A Guide to Use of Biota, Sediments and Water in Environmental Monitoring* 412 (London: Taylor & Francis, 2nd edn 1996).

usage. The precautionary principle further calls for a shift in burden of proof. It is the polluter who has to disprove the cause–effect relationship.[48]

The precautionary principle could be considered to be a part of Indian environmental jurisprudence. However, it has not effectively been put into practice. Take, for instance, the *Plachimada* case. This dealt at length with the cause–effect relationship with respect to groundwater depletion and the role of the Coca-Cola factory in the same. The Kerala High Court, while absolving the Coca-Cola factory, relied upon government reports which described poor rainfall as the major reason for groundwater depletion.[49] Had the precautionary principle been applied, the nature of the discussion would have been quite different. Moreover, the precautionary principle could have provided a strong legal basis for the government agencies to take action even in the absence of conclusive scientific evidence against the Coca-Cola factory.

The *Plachimada* example, therefore, points to the need for express recognition of the precautionary principle in the groundwater regulatory framework. It further exposes the need for strict application of the principle by institutions established for the implementation of environmental laws and by the judiciary as well.

3. Model Groundwater Bill: A reform initiative

Groundwater in India has been indiscriminately exploited over the last several decades. This is one of the major reasons for groundwater depletion and pollution. The common law principle seems to have provided a legal basis for its uncontrolled exploitation by those who have access to land and financial resources. Though principles such as the public trust doctrine and precautionary principle have evolved subsequently, these principles have hardly changed the legal regime which functions on the basis of recognition of private property rights. At the same time, the deteriorating condition of groundwater resources in the country has provoked a significant change in the legal framework with a view to regulating its indiscriminate exploitation.

A change in the legal approach was initiated by the Ministry of Water Resources, Government of India by framing and circulating the Model Groundwater Bill in 1970.[50] The Model Bill was revised three times in 1992, 1996 and 2005 respectively. It was intended to promote legal reforms for the regulation of groundwater use.[51] In fact, the schemes and approaches envisaged by the Model Bill have been, by and large, followed by most of the states while framing groundwater laws.

48 *Vellore Citizen's Welfare Forum v. Union of India* (1996) 5 SCC 647 Para. 11.

49 For details, *see* Koonan, note 34 above, 179–189.

50 As per the constitutional scheme, the subject of 'water resources' comes within the purview of state legislature. Therefore, the role of the central government is minimal in this regard. Constitution of India, Schedule VII, List II, Entry 17.

51 For an analysis of the Model Groundwater Bill of 2005, see Philippe Cullet, 'Water Law Reforms – An Analysis of Recent Developments', 48 *Journal of Indian Law Institute* 206 (2006).

As such, the Model Bill has set in motion legal reforms in some states in India, and a new regulatory approach has been initiated which recognizes and justifies the government regulation of groundwater use by individuals and companies.[52] While the common law-based regime protected only the right of landowners to withdraw groundwater from their land, the Model Bill introduced a new legal framework accommodating public interest and conservation objectives. This can be seen as the beginning of a significant shift from the common law-based regime, which supported uncontrolled extraction by private persons, to a regime of government regulation.

B. Legal reforms: An analysis

1. Newly evolving groundwater laws: An introduction

Over the last few years, a number of states have enacted their own groundwater laws.[53] This has led to major reforms in the legal regime governing groundwater in India. Broadly speaking, the laws evolving of late redefine the rights, duties and roles of the government, as well as those of individuals, *vis-à-vis* groundwater resources. These laws have also resulted in institutional reforms and have also

Table 9.1 Status of groundwater legislation in India

States/UTs that have passed enactment on groundwater	*States/UTs that have adopted Bills on groundwater but have not passed the same*	*States/UTs that have initiated the process of enacting laws on groundwater*	*States/UTs that feel it unnecessary to enact any legislation on groundwater*
• Andhra Pradesh	• Maharashtra	• Assam	• Nagaland
• Goa	• Gujarat	• Bihar	• Sikkim
• Tamil Nadu	• Karnataka	• Haryana	• Tripura
• Lakshadweep	• NCT Delhi	• Jammu and Kashmir	• Punjab
• Kerala	• Uttar Pradesh	• Mizoram	• Chandigarh
• Puducherry		• Orissa	• Manipur
• West Bengal		• Rajasthan	• Arunachal Pradesh
• Himachal Pradesh		• Daman and Diu	
		• Jharkhand	
		• Meghalaya	
		• Dadra & Nagar Haveli	
		• Madhya Pradesh	
		• Uttarakhand	
		• Andaman & Nicobar	
		• Chhattisgarh	

Source: The statement of the Minister of Water Resources in the Parliament on 4 December 2006, available at http://164.100.24.208/lsq14/quest.asp?qref=37130.

52 Model Bill to Regulate and Control the Development and Management of Ground Water, 2005.

53 For the list of states having separate groundwater law, see Table 9.1.

attempted to incorporate into groundwater regulation important norms of environmental law such as conservation and sustainable use.

Evolving groundwater laws seek to vest in the concerned state government the power to regulate and control the use of groundwater by private individuals. Evolving groundwater laws, generally, authorize the government to regulate groundwater use wherever it is felt necessary, in the interests of 'conservation, management and development' of the resource. This essentially indicates that these laws provide for government action founded on the 'public interest' rationale.[54] In this way, ongoing legal reforms make a considerable departure from the traditional legal approach wherein rights over groundwater were considered to be 'part and parcel' of land rights.

This shift within the legal regime has significant implications. Most importantly, the right to use groundwater, in principle, can now be severed from land rights. The argument in favour of severing groundwater rights from land rights has been in existence for at least two decades, the rationale being that conflation of land rights with the right to use groundwater results in the denial of the latter to the landless.[55] This situation is of special relevance to a country like India where the number of landless people is not insignificant. Moreover, conflation of the two rights would conceptually justify privatization of groundwater resources.

Another important legal change is the incorporation of objectives of conservation and development of the resource. Most of the evolving groundwater statutes emphasize conservation and development as major objectives. For instance, the Kerala Act recognizes the need for conservation in its preamble.[56] The title of the Andhra Pradesh Act also indicates the emphasis on 'protection and conservation'.[57] The translation of such environmental concerns into legal norms would have been very difficult, if not impossible, within the traditional legal regime.

The evolving legal regime envisages new institutional mechanisms, in the form of groundwater authorities, for proper and effective implementation of the law. Groundwater authorities are vested with the responsibility of enforcing the regulatory tools provided by the relevant statutes to ensure sustainable use. By and large, the institutional mechanisms provided under the various state groundwater laws follow a similar structure and perform similar functions. However, there are some variations across states. For example, West Bengal has put in place a decentralized structure by providing three levels of groundwater authorities – state level, district level and corporation level.[58]

54 For instance, preamble of the Kerala Ground Water (Regulation and Control) Act, 2002, available at www.ielrc.org/content/e0208.pdf.

55 Singh, note 11 above, 39.

56 Kerala Ground Water (Regulation and Control) Act, 2002.

57 Andhra Pradesh Water, Land and Trees Act, 2002, available at www.ielrc.org/content/e0202.pdf.

58 West Bengal Ground Water Resources (Management, Control and Regulation) Act, 2005, available at www.ielrc.org/content/e0502.pdf.

2. *Regulatory approach*

Various state acts have adopted the licensing system as a regulatory tool (i.e., a permit or registration-based system). Almost all the state groundwater laws envisage the classification of areas on the basis of the condition of the groundwater.[59] These may be notified formally by the state government. All users in such notified areas are required to register their wells. Potential users are required to seek prior permission. The control over groundwater use is sought to be effectuated by imposing conditions specified in the permit or certificate of registration.[60] The granting of permits and the conditions specified in the permit are as per the condition of the groundwater in the concerned area as estimated by the groundwater authority. The permit may be denied if it is likely to endanger the existing use of groundwater in that area.[61]

This regulatory framework also provides scope for subsequent alterations in the conditions specified in the permit or certificate of registration.[62] Cancellation of the permit or certificate of registration is also envisaged. Cancellation could be made under various grounds, such as failure of the holder to comply with the specified conditions, or the need for more stringent regulation.[63]

The extent of regulation could range from mere monitoring of groundwater use through the registration process to complete prohibition or closing down of wells. The nature and extent of regulation depends upon the quality and quantity of groundwater in a particular area. For instance, there could be a complete prohibition on new wells in areas designated as 'over-exploited'.[64]

One of the important features of the evolving regulatory framework is the priority given to drinking water, particularly the special protection given to public drinking water sources. This prioritization is sometimes manifested in the form of provisions prescribing the distance required to be maintained by new wells from public drinking water sources[65] or may be manifested in the form of exemptions from prohibitive provisions. For instance, the Goa Act prohibits sinking of new wells in areas designated as over-exploited.[66] However, the sinking of a well for its use as a public drinking water resource is exempted from this prohibitive provision.[67]

59 See, e.g., Kerala Ground Water (Regulation and Control) Act, 2002 and West Bengal Ground Water Resources (Management, Control and Regulation) Act, 2005.

60 Groundwater authorities have the power to specify conditions in the permit or certificate of registration. See, e.g., Himachal Pradesh Ground Water (Regulation and Control of Development and Management) Act, 2005, s. 7(3), available at www.ielrc.org/content/e0507.pdf.

61 See, e.g., Kerala Ground Water (Regulation and Control) Act, 2002, Section 7.

62 *Ibid.*, s. 11.

63 *Ibid.*, s. 12.

64 See, e.g., Goa Ground Water Regulation Act, 2002, s. 9, available at www.ielrc.org/content/e0201.pdf and Andhra Pradesh Water, Land and Trees Act, 2002, s. 11.

65 See, e.g., Andhra Pradesh Water, Land and Trees Act, 2002, s. 10.

66 See, e.g., Goa Ground Water Regulation Act, 2002, s. 9.

67 *Ibid.*

Generally, the regulatory framework envisaged by the various state groundwater laws seems to follow a compartmentalized approach, wherein groundwater is treated as a separate unit governed by a separate regulatory framework. One of the basic flaws in such a regulatory approach is that it does not take into account the natural link between surface water and groundwater sources. This link is so close that any distinction is highly artificial. Moreover, a compartmentalized regulatory framework does not help achieve desired results. A case study observed that the poor maintenance of tanks in Tamil Nadu over the last few decades has adversely affected groundwater recharging, resulting in the lowering of the groundwater table.[68] This indicates the need for an integrated regulatory approach, a long-term perspective, which takes into consideration the link between surface water and groundwater.

3. Institutional framework

Institutional reforms are a major part of water law reforms in India. The setting up of separate groundwater authorities can be considered to be a major reform in the institutional structure. In addition to the emergence of new state groundwater authorities, there are other statutory authorities, such as the Central Ground Water Authority and pollution control boards, having direct and indirect roles in groundwater regulation. This section of the chapter is divided into two parts. The first part describes the nature and function of the newly emerged groundwater authorities; the second analyses their role in the context of already existing institutions.

Groundwater authorities

The development of groundwater laws has made significant changes to the institutional framework of groundwater regulation. All state groundwater laws provide for the constitution of a groundwater authority. These groundwater authorities constitute the major organizational mechanism for implementation and enforcement of the new groundwater regulations.

Groundwater authorities are the agencies primarily responsible for the effective and proper use of regulatory tools provided by the statutes. It is the responsibility of the groundwater authorities to issue permits and certificates of registration to the users. They have the power to specify terms and conditions in the permit or certificate of registration. These authorities can also give directions to the users in furtherance of the objects of the statute. The powers of these authorities also include the power to enter property to inspect wells, to order closure of wells and

68 A. Gurunathan and C.R. Shanmugham, 'Customary Rights and their Relevance in Modern Tank Management: Selected Cases in Tamil Nadu', in Philippe Cullet, Alix Gowlland Gualtieri, Roopa Madhav and Usha Ramanathan (eds.) *Water Law at the Crossroads – National and International Perspectives with Special Emphasis on India* 129 (New Delhi: Cambridge University Press, 2009).

to seize any device used for the extraction of groundwater. However, the regulatory framework provided under the various existing state laws is generally applicable only to the 'notified areas'. The power to notify an area vests with the state government.[69] The power of the groundwater authority in this regard is, therefore, merely advisory in nature.

The institutional framework set up by the various state laws follows a centralized command and control approach. The various groundwater laws, generally, provide for the setting up of groundwater authorities at the state level.[70] This is probably an approach inspired by the Model Groundwater Bill formulated in 1970. This institutional approach marks a significant departure from the institutional reforms being undertaken with respect to other water laws. For instance, one of the important institutional changes introduced in the course of irrigation law reforms in India is the setting up of water users associations, which are vested with the responsibility of management and allocation of surface water.[71] In fact, the institutional framework adopted by the new groundwater laws follows a diametrically opposite approach to that promoted by the other water law reforms taking place over the past decade.

Institutional problems

Two main criticisms are levied against the institutional framework adopted by the new groundwater laws. The first is regarding the basic institutional approach, the second is the issue of institutional multiplicity.

It has already been stated that the institutional framework envisaged in the course of groundwater law reforms follows a centralized command and control approach. Given the ever shifting nature of groundwater, a centralized institutional mechanism would be ineffective. Tremendous organizational capacity is required to monitor the behaviour of individuals and it would be extremely difficult and expensive for a centralized institution to carry out the same and to effectively implement the law. Moreover, this centralized regulatory approach does not seem to respect or recognize the decentralization principle envisaged by the Constitution of India, which promotes the devolution of more and more powers to local bodies. The ongoing groundwater legal reforms by providing a centralized institutional mechanism appear to neglect this constitutional aspiration.[72]

69 See, e.g., Kerala Ground Water (Regulation and Control) Act, 2002, s. 6.

70 Andhra Pradesh Water, Land and Trees Act, 2002 and West Bengal Ground Water Resources (Management, Control and Regulation) Act, 2005, ss. 4 and 5 also provide for institutions at the local level.

71 See, e.g., Andhra Pradesh Farmers' Management of Irrigation Systems Act, 1997, available at www.ielrc.org/content/e9701.pdf.

72 For instance, though the Kerala Government enacted the Kerala Panchayat Raj Act in 1994, the Kerala Groundwater Act, 2002 does not contain any provisions according requisite powers to the local bodies.

Further, the setting up of groundwater authorities has added one more institution to the list of those already involved in groundwater regulation. This raises the question of institutional multiplicity and thus the issue of co-operation and co-ordination between these institutions having overlapping mandates. The mandate of the various state groundwater authorities and the Central Groundwater Authority overlap in many areas. Both these institutions have almost identical regulatory powers but under different statutes.[73] State authorities are constituted under the various state groundwater laws, whereas the Central Groundwater Authority is constituted under a central legislation, the Environment (Protection) Act, 1986. In states where separate groundwater legislation does not exist, the Central Groundwater Authority has a significant or primary role in the regulation of groundwater use. However, in states where groundwater authorities exist, there arises the problem of lack of institutional co-ordination and that of overlapping mandates.

Similar issues arise between the groundwater authorities and pollution control boards so far as groundwater pollution is concerned. Pollution control boards have been set up in almost all states. They are the agencies primarily responsible for addressing pollution issues in general, and particularly water pollution under the Water (Prevention and Control of Pollution) Act, 1974. Being the specialized authorities with respect to groundwater resources, groundwater authorities also have the mandate of addressing pollution of groundwater and its sources. These overlapping mandates might result in one authority shifting the responsibility to the other, with both parties ending up with either no results or delayed results.

C. Gaps and the way forward

Reforms in groundwater laws in India mark a significant departure from what was being followed for a long time. The traditional regime founded on private property rights cannot be justified on human rights, equity and environmental grounds. Instead of treating groundwater as a vital natural resource to be conserved and used discriminately, the traditional legal principles treated it as part of private property rights. The legal reforms introduced considerable changes in this regime by justifying government regulation. Normally, this can be seen as the recognition of groundwater as a natural resource vital to the whole society, in opposition to the traditional view which considered it to be *de facto* private property.

Given the adverse implications of the traditional legal approach towards groundwater, the ongoing legal reforms constitute a much needed change. The enactment of separate groundwater laws should be welcomed and viewed as a move to fill in the gaps that have long been present in the legal regime governing groundwater in India. However, there is much to be improved on. By and

73 For the role of the central authority, Central Groundwater Authority, Policy Guidelines of CGWA, available at http://cgwb.gov.in/GroundWater/CGWA%20guidelines%20for%20NOC%201.pdf. As on 2 December 2006, the CGWA had notified 43 areas for groundwater regulation, mostly in states where groundwater regulation does not exist. The list of notified areas is available at http://cgwb.gov.in/GroundWater/authority_area.htm.

large, the reforms follow the framework introduced by the Model Bill framed in 1970. Therefore, several environmental and other legal norms evolved over the last couple of decades have not been incorporated or considered in the new regulatory framework; examples being the public trust doctrine and decentralization principle. Thus a number of norms need to be incorporated into the new regulatory framework. Some of the important normative gaps in the new regulatory framework are discussed here.

1. Human rights concerns

Life cannot exist without water. Therefore, the right to water must be recognized as a fundamental human right.[74] In fact, the human right to water has been recognized under international human rights law.[75] Various national laws also have recognized the human right to water in so far as drinking water is concerned.[76] It can also be argued that even in the absence of an express recognition of the human right to water, the same is an indispensable part of the existing human rights jurisprudence because the realization of other well-recognized human rights is difficult without access to sufficient potable water.[77]

The human right to water is, in principle, a part of Indian law. The Indian judiciary has expanded the scope of the right to life under Article 21 of the Constitution of India to include right to water as part of the right to life.[78] Given the fact that groundwater is largely being used as a source of drinking water, access to the same would logically fall within the purview of Article 21 of the Constitution of India. Moreover, it has been expressly recognized by case law that human rights jurisprudence is applicable to groundwater also.[79] Since the human right to water is a part of Indian law, the legal regime governing groundwater also must necessarily implement this human right.

Recognition of the human right to water casts a duty upon the government to regulate the over-exploitation and pollution of groundwater. Failure to do so would be tantamount to the violation of the right to life which is enforceable through constitutional remedies.[80] It further justifies restricting the uncontrolled exploitation of groundwater by individuals who rely on the common law rule.

74 For more details on the human right to water, see Chapter XI.

75 Committee on Economic, Social and Cultural Rights, General Comment No. 15: The Right to Water (Arts. 11 and 12 of the International Covenant on Economic, Social and Cultural Rights), UN Doc. E/C.12/2002/11 (2002).

76 Cullet, note 51 above.

77 Peter Gleick, 'The Human Right to Water', 1 *Water Policy* 487 (1999).

78 *Narmada Bachao Andolan v. Union of India* (2000) 10 SCC 664; *A.P. Pollution Control Board II v. M.V. Nayudu* 2000(3) SCALE 354, available at www.ielrc.org/content/e0010.pdf; *F.K. Hussain v. Union of India* AIR 1990 Ker. 321 and *Venkatagiriappa v. Karnataka Electricity Board* 1999 (4) *Karnataka Law Journal* 482.

79 *Puttappa Honnappa Talwar v. The Deputy Commissioner* AIR 1998 Kant. 10.

80 An individual can approach the Supreme Court or the High Court directly under Arts. 32 and 226 respectively in case of violation or infringement of a fundamental right.

However, these normative grounds have not been invoked properly in the case of groundwater resources so far. In a recent case (the *Plachimada* case), the Kerala High Court refused to apply this human rights norm.[81] In this case, the division bench of the Kerala High Court upheld the private property right in view of the absence of any express statutory authorization to regulate individual use. One of the reasons for this situation could be the derivative nature and ambiguity of the human right to water. Therefore, clarity needs to be achieved through the express recognition of the human right to water in groundwater statutes. The human right to water should be seen as an express normative basis for groundwater regulation.

A human rights approach further requires proper and express recognition of a principle of prioritization by the legal framework. That is, the first and foremost priority must be accorded to drinking water. The need to prioritize drinking water has been expressly recognized by the judiciary in a couple of cases.[82] The norm of according priority to drinking water can also be seen in some groundwater laws.[83] This is mainly by way of providing protection to public drinking water resources.[84] Apart from the prioritization of drinking water, the state groundwater laws do not mention subsequent priorities. Though the use of groundwater for food crop cultivation also needs to be viewed from a human rights perspective in the larger interest of food security, such an emphasis has not found manifestation in the ongoing legal reforms. Therefore, an expanded list of priorities needs to be provided in the various groundwater laws.[85] Guidance in this regard has already been provided by the National Water Policy, 2002.[86]

2. The decentralization principle

The Constitution of India, through the 73rd and the 74th amendments, envisages decentralization as a principle of governance. The Constitution provides for the devolution of powers and responsibilities at the local level. The decentralization principle suggests a bottom-up approach to governance in the place of the centralized top-down approach that has been followed for a long time. If the decentralization principle is applied, the management and control of groundwater is required to be within the purview of local bodies.[87] Moreover, the

81 *Hindustan Coca-Cola Beverages v. Perumatty Grama Panchayat* 2005 (2) *Kerala Law Times* 554.

82 *FK Hussain v. Union of India* AIR 1990 Ker. 321 and *Venkatagiriappa v. Karnataka Electricity Board* 1999 (4) *Karnataka Law Journal* 482.

83 See, e.g., Maharashtra Groundwater (Regulation for Drinking Water Purposes) Act, 1993, available at www.ielrc.org/content/e9301.pdf and Karnataka Ground Water (Regulation for Protection of Sources of Drinking Water) Act, 1999, available at www.ielrc.org/content/e9905.pdf.

84 See, e.g., Himachal Pradesh Ground Water (Regulation and Control of Development and Management) Act, 2005, s. 7(3).

85 The need for proper incorporation of the norm of priority to drinking water is to be viewed in the context of recent allegations of groundwater depletion due to over-exploitation for commercial use. Koonan, note 34 above, 159.

86 National Water Policy, 2002, para. 7(8).

87 Constitution of India, Eleventh Schedule, Entry 3 and 11.

people are expected to be given a more prominent role in the decision-making process.

The constitutional scheme encourages devolution of powers and responsibilities at the local level with respect to groundwater management and control. The state legislatures, by virtue of the 73rd and 74th amendments, are morally obliged to incorporate the policy of decentralization into their laws. In the context of groundwater, the state legislatures need to incorporate this principle into the existing or evolving groundwater laws. In fact, most of the state legislatures have enacted laws which give effect to the decentralization principle.[88] This constitutional aspiration and its wide legislative manifestations need to be considered by the judiciary also while interpreting and deciding any legal issues related to groundwater.

However, ongoing legal reforms in groundwater regulation, by adopting a method of centralized command and control, have neglected the constitutional aspiration of decentralization. The evolving groundwater laws generally do not recognize the role of local bodies in groundwater regulation. Moreover, some of the recent legal changes, particularly some laws enacted with the object of promoting development and investment, completely disregard the decentralization principle as envisaged by the Constitution. For instance, the Kerala State Single Window Clearance Boards and Industrial Township Area Development Act, 1999 expressly takes away the regulatory powers of local bodies over designated industrial areas.[89] The issue of the power of local bodies to regulate groundwater use in such industrial areas has been discussed by the Kerala High Court recently in the *Pepsico* case.[90] In the instant case, the power of the panchayat was not upheld in the light of the express statutory provision omitting the jurisdiction of the panchayat in industrial areas.[91]

Moreover, vesting more regulatory powers in local bodies would promote the idea of people's participation. Further, the problem of institutional multiplicity would also be relieved if the primary regulatory powers are vested in the local bodies, with technical bodies such as groundwater authorities and pollution control boards act as facilitating agencies. Given the high costs and operational complications involved in a centralized regulatory approach, a decentralized regulatory approach is likely to be more feasible and efficient.

3. Groundwater use and energy and land utilization policies

Groundwater regulation has a significant link with energy and land utilization policies. The provision of energy for pumping groundwater at subsidized rates,

88 Government of India, Annual Report of the Ministry of Rural Development 2002–2003 (New Delhi: Ministry of Rural Development, 2003).

89 Kerala State Single Window Clearance Boards and Industrial Township Area Development Act, 1999, s. 6.

90 *Pepsico India Holdings v. State of Kerala*, Kerala High Court, 2008(1) *Kerala Law Journal* 218.

91 *Ibid.*

particularly for agricultural purposes, is often considered to be one of the major reasons for indiscriminate exploitation of groundwater.[92] Land use patterns also have significant implications for the quality and quantity of groundwater. Therefore, it is argued that sustainable groundwater use can be achieved only if groundwater regulation is effectively complemented by a proper energy and land utilization policy.

Large-scale groundwater extraction in India is fuelled by availability of cheap electricity and diesel. About one-third of the total electricity supplied is estimated to be used for pumping groundwater for irrigation as well as for the domestic water supply in urban and rural areas.[93] Some studies show that around 75 per cent of irrigated lands are served by groundwater wells.[94] It has been further observed that groundwater-based irrigation uses up a significant portion (around 20 per cent) of the total electricity produced.[95] The number of electric and diesel pump sets, estimated to be 12.5 and 6 million respectively, gives a clearer picture as to the close link between energy policy and groundwater exploitation.[96]

Both electricity and diesel are provided at subsidized rates to farmers by state governments. The rise in the price of energy is mostly absorbed by the government.[97] In fact some states have provided electricity to farmers free of cost.[98] As such, the exploitation of groundwater is supported by a sympathetic energy policy.

Therefore, a proper energy policy could be an effective regulatory tool for groundwater use. Factors such as the gap between wells or the distance of a private well from a public drinking water source ought to be taken into consideration before providing an electricity connection. Charging the consumer the actual energy cost could also be a possible tool for regulation of groundwater

92 M. Dinesh Kumar, 'Impact of Electricity Prices and Volumetric Water Allocation on Energy and Groundwater Demand: Analysis from Western India', in V. Ratna Reddy and S. Mahendra Dev (eds.) *Managing Water Resources: Policies, Institutions, and Technology* 118 (New Delhi: Oxford University Press, 2006). At the same time, it is also to be noted that extensive groundwater-based irrigation has produced positive results in the form of significant increase in food production. Tushaar Shah *et al.*, 'Energy-Irrigation Nexus in South Asia: Improving Groundwater Conservation and Power Sector Viability', in Mark Giordano and Karen G. Villholth (eds.) *The Agricultural Groundwater Revolution: Opportunities and Threats to Development* 211 (Wallingford: CABI International, 2007).

93 Ramesh Bhatia, 'Water and Energy Interactions', in John Briscoe and R.P.S. Malik (eds.) *Handbook of Water Resources in India* 206 (New Delhi: Oxford University Press, 2007).

94 Shah *et al.*, note 92 above, 215.

95 *Ibid.*, at 214.

96 Bhatia, note 93 above, 219.

97 Tingju Zhu, Claudia Ringler and Ximing Cai, Energy Price and Groundwater Extraction for Agriculture: Exploring the Energy Water Food Nexus at the Global and Basin Levels (International conference organised by the International Water Management Institute (IWMI) and Food and Agriculture Organization (FAO), Hyderabad, 29–30 January 2007), available at www.lk.iwmi.org/EWMA/files/papers/Energyprice_GW.pdf.

98 Navroz K. Dubash, 'The Electricity-Groundwater Conundrum: Case for a Political Solution to a Political Problem', 42(52) *Economic & Political Weekly* 45 (2007).

use.[99] At the same time, charging the actual energy cost would most likely affect the poor and marginal farmers. Hence, reforms in this regard should follow a policy whereby poor and marginal farmers and domestic users should be least affected. This means the pricing policy should be formulated in such a manner that the public utility shall charge actual costs only in the case of commercial and industrial users and shall retain the government subsidy for poor and marginal farmers and domestic users.

Land utilization policies also have significant implications for the quality and quantity of groundwater. Various land-based activities, such as the application of fertilizers and utilization of land for dumping wastes, pollute groundwater resources. These quality issues need to be addressed on an urgent basis because unlike in the case of surface water resources, it is difficult to identify a groundwater pollutant at the initial stages, largely due to its invisible nature. Once an aquifer has become polluted, it is extremely difficult to clean up and it is rarely possible to return it to a pristine condition.[100] This issue is critical given the fact that groundwater is increasingly becoming a major source of drinking water. Therefore a preventive and precautionary solution through the regulation of land use on the basis of the availability of groundwater resources is imperative.

Environment impact assessments should also be undertaken to enable precautionary measures to be taken. These regulatory tools should be applied by respecting the norm of priority to drinking water. For instance, in an area mainly used for food crop cultivation, or in an area where the majority of the population rely on groundwater for domestic purposes, water-based industries intending to exploit groundwater should not be permitted as they are likely to adversely affect the existing agricultural and domestic consumption.

CONCLUSION

The legal regime governing groundwater in India is a combination of English common law principles, evolving state-level groundwater enactments, the Indian Easements Act, 1882, and various pollution control laws. Falling within the purview of the legislative competence of state governments, the law governing groundwater varies from state to state. However, most states have adopted a form of centralized command and control as the basis of regulation. In the states which have not adopted separate groundwater laws, the legal regime governing groundwater use mainly comprises the traditional English common law principle as recognized in early enactments, such as the Easements Act.

99 Charging farmers the actual cost of electricity has been suggested as a regulatory tool to restrict indiscriminate exploitation of groundwater. Navroz K. Dubash, *Tubewell Capitalism: Groundwater Development and Agrarian Change in Gujarat* (New Delhi: Oxford University Press, 2002) and Karin Erika Kemper, 'Instruments and Institutions for Groundwater Management', in Giordano and Villholth (eds.) note 92 above, 153.

100 Chilton, note 47 above, 451.

Table 9.2 Main features of state-level groundwater enactments

State	*Application*	*Regulatory tools*	*Priority*	*Pollution*	*Other features*
Andhra Pradesh, 2002	Applicable to all ground-water resources in the state and to all users, including government bodies	Registration mandatory for all wells [s. 8(2)] Statutory authority can impose prohibition on use of wells to prevent deterioration of or damage to groundwater resource [s. 9(1)] Statutory authority can require sealing or closure of wells [s. 15(1)] Regulation of distance and depth of wells [s. 13] Permission from the electricity authority is required for sinking wells in the vicinity of public drinking water sources [s. 10(2)] Statutory authority may declare certain areas to be 'over-exploited areas' where absolute prohibition of groundwater extraction applies for all uses except for drinking purpose [s. 11].	Priority accorded to drinking water usage, particularly in the form of protection of public drinking water resources [s. 10(1)]. Most of the regulatory tools are meant for the protection of drinking water resources	Groundwater contamination in any manner and by anyone is prohibited [s. 19].	Decentralized institutional structure [s. 6(h)] Integrated regulatory approach [s. 6(b)] Express recognition of equity in access [s. 6(f)] Government may take over wells by paying compensation [s. 12(5)] Mandatory requirement of rain water harvesting for buildings [s. 17(2)].
Kerala, 2002	Applicable to notified areas [s. 6(1)].	Permit/certificate of registration system with conditions as prescribed by the Authority [ss. 7(1) &8(1)].	Priority accorded to (public) drinking water usage [s. 10].	Maintenance of quality of groundwater is a criterion required to be considered while granting permit/certificate of registration [ss. 8(5)(d) & 7(7)(d)].	Centralized institutional mechanism [s. 3(1)].
West Bengal, 2005	Applicable to all areas and to all users [s. 2(j)].	Permit/certificate of registration system with conditions as prescribed by the Authority [ss. 7(1) & 8(1)].	Priority accorded to irrigation and domestic usage [ss. 7(1) & 8(1)].	Maintenance of quality of groundwater is an objective stated in the preamble and is to be considered while granting permit/certificate of registration. [ss. 7(3) (a) & 8(2) (a)]	Decentralized institutional mechanism [ss. 4(1) & 5(1)] Emphasis on preparation of district-wide groundwater profile periodically [s. 9(a)].

Karnataka, 1999 (Legislation specific to drinking water)	Applicable to the whole state and to all users except the government [s. 3(1)].	Regulation of sinking wells within 500 metres of public drinking water sources [s. 5(1)] Declaration of certain areas to be 'water-scarce areas' and 'over-exploited watersheds' and restriction on or prohibition of extraction of groundwater from such declared areas [ss. 4 and 7].	Specifically for the protection of public drinking water sources	Does not directly address the quality issue	Definition of 'drinking water purpose' is wide enough to include cooking, washing, bathing and all other domestic purposes and water for live-stock as well [s. 2(2)] Compensation is granted in case of closure of wells [s. 12(1)].
Goa, 2002	Applicable to all areas and to all users	Declaration of certain areas to be scheduled areas, water-scarce and over-exploited areas and restriction on or prohibition of groundwater use in declared areas [ss. 4, 5 and 6].	Extraction of groundwater for drinking purpose is exempted from regulations [s. 9(1) (a)].	Aspects related to preservation of quality of groundwater are not expressly addressed. However, the same is required to be considered while granting permission to transport groundwater from the scheduled areas [s. 6(5)(e)] Provision to close toilets/septic tanks/soak pits if they cause pollution of groundwater [s. 13(1)(m)].	Major emphasis on quantitative control; provision for compensation in case of closure of wells [s. 12] Regulation through electricity connections [s. 13(1)(l)].
Himachal Pradesh, 2005	Application limited to notified areas	Permit/registration system put in place in notified areas [ss. 5, 7(1) and 8(1)].	First priority accorded to drinking water usage while granting permit/ registration [s. 8(2)].	Maintenance of quality of groundwater is to be considered while granting permits [ss. 7(5)(d) and 8(4)(d)].	Royalty to be paid to the State Government by users in notified areas [s. 12(1)].

English common law principles and the property-right approach as envisaged in the Easements Act evolved at a time when the need for groundwater regulation was not taken seriously. This is no longer the case, given the fact that the over-exploitation of groundwater over the past few decades has caused a deterioration in quality and quantity across the country. As indicated by the *Plachimada* case, the uncontrolled exploitation of groundwater for industrial and commercial purposes threatens access to drinking water – a basic human need. In this context, the traditional common law approach is too dated. However, these common law principles still remain in force in some parts of the country where specialized groundwater regulatory mechanisms have not been put in place.

An alternative regulatory approach has been provided by the Supreme Court of India in the form of the public trust doctrine. A single judge of the Kerala High Court, in the *Plachimada* case, applied this doctrine to the case of groundwater resources. The decision of the single judge was subsequently reversed on appeal by the division bench. Since the case is now pending before the Supreme Court of India, the scope of application of the public trust doctrine to groundwater resources is unclear, or is yet to be decided.

The groundwater laws, which have been evolved by the states, do not address contemporary challenges properly. Being largely based on the central government's Model Groundwater Bill of 1970, these state laws follow the traditional centralized command and control regulatory approach. Since the 1970s a significant change has occurred in the legal system by the introduction of the decentralization principle through constitutional amendments in 1992. Despite this change, the newly evolved and evolving state groundwater laws neglect the decentralization principle. Local bodies and the local community have not been assigned any express roles in groundwater regulation in the evolving legal framework.

The human right to water has, in principle, been recognized by the higher judiciary of the country. Despite repeated assertions by the judiciary, a human rights approach has not found explicit expression in the evolving groundwater laws. Though the norm of priority to drinking water has found recognition to some extent in the form of the protection of drinking water sources, proper statutory recognition of this norm is still to be done.

Hence, it is reiterated that the legal framework governing groundwater in India needs to be improved so as to address contemporary challenges. The basic regulatory approach and underlying legal principles need to be reviewed in the light of the need to incorporate a human rights approach, the precautionary principle, decentralization principle and the norm of according foremost priority to drinking water. Further, the regulatory framework should recognize the interface between groundwater and surface water.

X. Law and policy reforms for irrigation

Roopa Madhav

INTRODUCTION

Irrigation is vital in a semi-arid, agriculture-based economy, such as India. Only 43 per cent of the net sown area in the country is irrigated.[1] An increase in area under irrigation provides direct benefits to the users, ensures food security and contributes to the overall development of the economy. The rise in agricultural productivity contributes to increased incomes, indirect employment generation and increased consumption. However, growing demand to irrigate multiple crops and water-intensive crops has increased the pressure for supply, allocation and use of irrigated waters. Additionally, demand for irrigation water for other uses, including industry, human consumption, inland fisheries and water tourism, has added to the irrigation water stress in the country.

Surface irrigation waters are supplemented by a corresponding volume of water from groundwater sources across the country. At the last count, an estimated 70 to 75 per cent of agriculture in India currently depends on groundwater for irrigation.[2] The key issues in surface irrigation water are discussed in this chapter while those relating to groundwater irrigation are discussed in Chapter IX. On the other hand, traditional surface irrigation such as tanks and ponds are in decline as government support and funding have been reduced to a trickle. Some initiative is now being taken to support and revive tanks and other minor irrigation works in several states in the country.[3]

Greater decentralization is the primary thrust of the reforms in the irrigation

1 This amounts to approximately 142 million hectares. Planning Commission – Government of India, *Eleventh Five-Year Plan 2007–12 – Volume 3* (New Delhi: Oxford University Press, 2008).

2 See generally Planning Commission of India, Report of the Expert Group on Ground Water Management and Ownership (New Delhi: Planning Commission of India, 2007). This chapter discusses the key issues in surface irrigation water; for an examination of those relating to groundwater irrigation, see Chapter IX in this volume.

3 Fifteen thousand minor irrigation works are sought to be revived in the State of Andhra Pradesh with the help of World Bank funding. Anonymous, 'AP to have 10 New Circles for Minor Irrigation', *The Hindu Business Line*, 18 June 2008, available at www.thehindubusinessline.com/2008/06/18/stories/2008061851002100.htm.

sector. The reform agenda is being actively carried out by a few states in the country, while some others have initiated tentative changes. Broadly speaking, these changes seek to increase the productivity of water and infrastructure; set up WUAs; reform formal irrigation institutions and their framework; improve wider basin efficiency through integrated water management systems; recognize and manage water rights; and improve efficiency to reduce groundwater exploitation. Decentralization, it is hoped, will enable greater participation in governance and enable localized solutions to problems plaguing the sector.

Section A provides a broad overview, contextualizing the irrigation sector reforms in the country. In particular, it outlines the policy, legal and institutional framework of the sector at the time the reform process was initiated. Section B introduces the ideas at the core of the current irrigation sector reforms, namely greater user participation in operation and maintenance through WUAs. It examines the history of the reforms and its present conceptualization through a slew of new enactments. Section C reviews the new institutional paradigms being introduced, highlighting their impact on key issues of water rights and community ownership of water resources.

A. Law and policy reforms in irrigation – An overview

1. Context for institutional and legal reforms

India has the largest irrigated area in the world. In the early years after independence, the irrigated area in the country was only 22.6 million hectare (Mha) according to figures for 1950–51.[4] Since food production was much below the country's needs, the initial years of the planned economy focused on the expansion of irrigation. Until the end of the Ninth Five-Year Plan (1997–2002),[5] an irrigation potential of 93.95 Mha was created through surface and groundwater irrigation schemes. It is anticipated that 8.82 Mha potential would be created by the end of the Tenth Five-Year Plan (2002–2007), bringing the total irrigation potential to about 100 Mha.[6]

The ultimate irrigation potential of India has been estimated at 139.9 Mha, of which 58.46 Mha is from major and medium-sized irrigation projects, 17.38 Mha from minor surface water schemes, and 64.05 Mha from minor groundwater schemes including privately owned shallow wells, tubewells and bore wells[7]. Against an ultimate irrigation potential of 58.46 Mha from major and

4 See, e.g., C.J. Perry, 'Role of Water Pricing in Irrigation: The Example of India', in Kamta Prasad ed., *Water Resources and Sustainable Development* 362 (New Delhi: Shipra, 2003).

5 India, post-independence adopted the five-year planning process to develop its economy. The Planning Commission has been entrusted with the responsibility of the creation, development and execution of India's five-year plans. The first five-year plan was formulated for the execution between 1951 to 1956 and the eleventh five-year plan is currently underway.

6 See generally Government of India, note 1 above.

7 This estimate is subject to the caveat of changes in technology and related developments.

medium-sized projects, 35 Mha (60 per cent) has been developed so far, though the undertaking of 375 major projects and 1,118 medium-sized projects.[8] The asymmetrical rainfall pattern in India requires large storage capacities to tide over lean rainfall periods.

Presently, nearly half the land in the country is irrigated with groundwater. This rapid expansion aimed primarily at irrigating land not covered by surface irrigation systems is alarming and no longer sustainable. The over-emphasis on mega-irrigation projects and groundwater irrigation has led to widespread neglect of traditional irrigation, such as tanks. For centuries, tanks have been central to agriculture in several regions, especially in the states of Andhra Pradesh, Karnataka, Tamil Nadu and Chhattisgarh.

Independent India was bequeathed a vast irrigation surface system. Parallel to the formal irrigation structures erected by the government, the traditional systems initiated and managed by communities survive in several parts of the country.[9] Thus, existing irrigation systems can be classified into three distinct categories:

(a) community systems, operated and maintained by the water users themselves and/or their representatives;
(b) bureaucratically managed systems, fully administered by an agency of the government; or
(c) jointly managed systems, in which some functions are performed by the irrigation agency while others are the responsibility of one or more water-user communities.[10]

The purpose of the irrigation systems, whether community owned or state run, is to enable an effective system of water management in an equitable manner, thus furthering the interest of all.

The structure of irrigation management over the years has been characterized as statist, bureaucratic and top-down in nature. Formal irrigation systems are planned, designed, financed, constructed, and maintained by the state. The five-year plans initiated in 1951 began to focus on irrigation and in the initial years, investments were mostly targeted at creation of large-scale surface irrigation works, often as part of multi-purpose hydro-power projects.[11] Minor irrigation

8 International Commission on Irrigation and Drainage, Water Policy Issues of India: Study Outcomes and Suggested Policy Interventions (New Delhi: International Commission on Irrigation and Drainage, Country Policy Support Programme, 2005).

9 M.S. Vani, *Role of Panchayat Institutions in Irrigation Management* 17–23 (New Delhi: Indian Law Institute, 1992) and A. Agarwal and S. Narain, *Dying Wisdom: The Rise, Fall and Potential of India's Traditional Water Harvesting Systems* (New Delhi: Centre for Science and Environment, 2001).

10 E. Walter Coward Jr., 'Irrigation Development: Institutional and Organizational Issues', in E. Walter Coward Jr. (ed.) *Irrigation and Agricultural Development in Asia: Perspectives from the Social Sciences* 27 (Ithaca: Cornell University Press, 1980).

11 See, generally A. Vaidyanathan, *India's Water Resources, Contemporary Issues on Irrigation* (New Delhi: Oxford University Press, 2006).

projects received significant attention through the 1960s under the Third and Fourth Plans and the use of groundwater was encouraged to raise productivity. Funds for maintaining and constructing the irrigation systems come entirely from public funds with the provision for recovering them through betterment levies and irrigation charges.[12]

In the last few decades, the focus on investing and expanding large irrigation projects is on the wane, and the total plan expenditure on irrigation has been reduced over the years. Though it increased from Rs 441.8 crore in the First Plan to 95,743.42 crore (outlay) in the Tenth Plan, it decreased from 23 per cent in the First Plan to 6.3 per cent in the Tenth Plan.[13] Even as fulfilling the needs of the large population of the country continues to be a challenge,[14] the irrigation sector in India has been plagued by many ills for several decades now. Though large, complex and numerous, there are several recurrent problems:

(a) irrigation projects are rarely completed on time, resulting in escalating costs of infrastructure;
(b) projects fall short of original expectations of planners and optimum utilization is rarely realized;
(c) irrigation is a poor revenue earner because of low water rates and still lower collections;
(d) lack of funds, combined with the preference for construction of new projects over the maintenance of existing ones, results in grossly inadequate allocation for the latter; and
(e) unresponsive and uninspired bureaucratic functioning rendering many irrigation projects inefficient and unreliable, particularly in delivery and utilization of water at the farm level.

Critics of the present system for water management point to it being more administrative and revenue oriented with little or no direct involvement of individual farmers. They note that the existing system of irrigation management expects each farmer to deal directly with the irrigation department, thus imposing a tremendous pressure on the department to manage all levels of the system, namely maintaining the main system, planning water distribution and preparing the schedule for individual farmers and overall poor utilization of irrigation potential.[15] Notably, the state's capacity to maintain and operate large irrigation systems has been questioned by the National Commission for Integrated Water Resources Development Plan. The Commission acknowledged the growing need

12 Kameshwar Choudhary, Historical Overview of Social Movements in India and Water Management (Draft paper for the eighth Biennial Conference of the IASCP, Indiana University), available at http://dlc.dlib.indiana.edu/archive/00000237/00/choudharyk062100.pdf.

13 Government of India, note 1 above, 46.

14 *Ibid.*

15 See generally Kamta Prasad, *Farmers' Participation in Irrigation Management in India* (New Delhi: Uppal Publishing House, 2001).

for the role of the state to be reduced and the area of management by the people enlarged through the mechanism of participatory irrigation management (PIM) or irrigation management transfer (IMT).[16]

The PIM/IMT discourse has dominated the reform agenda in the last decade.[17] Proponents of the discourse believe that integrated management frameworks generate a number of important benefits and result in greater efficiency in the use of a given amount of water through appropriate and equitable allocation, conjunctive management of ground and surface water, more intensive recycling and re-utilization of water, adoption of more socially and environmentally appropriate policies in water management, thus contributing to reduced water conflicts, and inclusion of a wider range of basin-level stakeholders in decision making.

Thus, steady demand for irrigation sector reforms has been growing within the country. Over the past five decades, the reform debates have spanned a gamut of issues – financial viability and sustenance, efficiency in water use, institutional issues in administration and future of water resource development.[18] In recent years, pressure from external sources has begun to determine the course of the reforms. The reforms discourse has also been steadily influenced by donor agencies such as the World Bank, which have substantial interest in the irrigation sector of most developing countries. In laying the road map for the future of water sector, the World Bank has declared the key reform process as being the 'articulation and operationalizing of a modern institutional model that unbundles the bulk infrastructure from the distribution infrastructure, separates the public and private roles (legal framework and regulation) and private (profit and non-profit) roles for service delivery'.[19]

2. *Legal and institutional framework for irrigation*

To understand present-day irrigation reforms, it is important to have a broad overview of the legal and institutional framework governing the irrigation sector.

Legal framework

According to the Indian Constitution, state governments are responsible for the development of irrigation. The central government, however, plays a key role in

16 Government of India, National Commission for Integrated Water Resource Development Plan 19 (New Delhi: Government of India, Ministry of Water Resources, 1999).

17 PIM usually refers to the level, mode and intensity of user group participation that would increase farmer responsibility in the management process. IMT is a more specialized term that refers to the process of shifting basic irrigation management functions from a public agency or state government to a local or private sector entity. David Groendfelt and Mark Svendson (eds.), Case Studies in Participatory Irrigation Management (Washington: World Bank Institute, 2000).

18 See generally A. Narayanamoorthy and R.S. Deshpande, *Where Water Seeps! Towards a New Phase in India's Irrigation Reforms* (New Delhi: Academic Foundation, 2005).

19 World Bank, Water Resources Sector Strategy: Strategic directions for World Bank Engagement 14–15 (Washington, D.C.: World Bank, 2004).

planning and financing irrigation projects recommended by states. It has powers to regulate and develop inter-state rivers and river valleys when declared by Parliament as involving public interest.[20] Also, the Environment Protection Act, 1986 and notifications issued by the Ministry of Environment and Forests require states to obtain central government sanction for major irrigation projects.

Large irrigation projects require the acquisition and submergence of large tracts of land by the state through the exercise of its eminent domain powers. The principle of eminent domain refers to the powers of acquisition of private property by the state after payment of just compensation. It is presupposed in Article 31(A) of the Constitution and tacitly recognized in the Land Acquisition Act, 1894, the Forest Act, 1927 and other similar natural resources laws, including irrigation acts. According to one author:

> This power of eminent domain is further consolidated in Indian laws by making almost all laws concerning land resources unjusticiable in any court by enlisting them in the Ninth Schedule of the Constitution. Article 318 of the Constitution makes these laws concerning land resources unjusticiable – in terms of whether or not they give effect to the policy advocated by the Directive Principles. Article 31 C debars any claim that they are inconsistent with fundamental rights, such as the right to reside in any part of India, including on river banks.[21]

The nature of the rights exercised by the state in natural waters is all-pervasive and embodied in the different irrigation acts. Introduced during the colonial era, the extensive state powers were mechanically incorporated into post-independence legislation. One of the earliest enactments on irrigation, the Northern India Canal and Drainage Act, 1873 seeks to regulate irrigation, navigation and drainage in Northern India. The preamble states that the provincial government is entitled to the use and control for public purposes of the waters of all rivers and streams flowing in natural channels, and of all lakes and other natural collections of still water. Similarly, the Andhra Pradesh (Andhra Area) Irrigation Works (Repairs Improvement and Construction) Act, 1943 provides for the repair and improvement of private irrigation works, the construction of new irrigation works on private lands and the supply of water from government to private irrigation works.

A series of court rulings have also upheld the sovereign right of the state over surface irrigated water. As far back as 1918, it was held that:

> It is impossible for a person to acquire an easement in water in a Government channel when that water has been supplied by the Government for the

20 Constitution of India, Art. 262.

21 Chhatrapati Singh, *Water Rights and Principles of Water Resource Management* 82 (Bombay: N.M. Tripathi, 1991).

purpose of irrigating land and in fulfilment of the duty cast upon them to supply water to lands which they classify as wet. Therefore no easement can possibly be acquired.[22]

In *Secretary of State v. Nageswara Iyer*, the court laid down the doctrine that the right of the government to control the supply and distribution of irrigation waters is not merely a proprietary right, but is a sovereign right when it stated that '(a) right by prescription can be acquired as against the proprietary right of another but not as against the sovereign right which under the Indian law the state possesses to regulate'.[23]

Consequently, irrigation laws in most states are of colonial origin. Typically, they provide for grant of usufructuary rights over water to landowners in the command area. Landowners who fall within the command area of an irrigation project are registered by the project authorities and are entitled to get water for specified seasons/crops. These entitlements, however, are conferred unilaterally by the state, are not legally enforceable rights and can be withdrawn or changed at the authorities' discretion.[24] Jacob and Singh review the Indian irrigation laws according to regions, providing an effective overview of these enactments.

In the northern states of Uttar Pradesh, Punjab and Haryana, irrigation canals draw water from perennial rivers, and these states are governed by the Northern India Canal and Drainage Act, 1873. This comprehensive legislation which deals with irrigation, navigation and drainage empowers the state government to construct works for the use of surface waters in rivers, streams and lakes for irrigation. The Rajasthan Irrigation and Drainage Act, 1954 and the Jammu and Kashmir Canal and Drainage Act, 1963 are similar to the Northern India Canal and Drainage Act.

Likewise, the system of irrigation management prevalent in the southern states of Tamil Nadu, Andhra Pradesh, Mysore and Kerala is based on the 'dry' and 'wet' classification of lands. The localization of crops is practised in these states under which a soil survey of the command area is carried out, with the area most suitable for different seasonal and perennial crops determined once and for all. Only those crops allowed for an area can be grown in it. The Mysore Irrigation Act, 1965 is a comprehensive legislation which highlights state control of irrigation management. No irrigation work is to be constructed, controlled and maintained by a person without the previous authorization of the state government. The state government may control the construction of wells in any area in the interest of proper irrigation. With regard to the construction of field channels, the Act confers powers on the irrigation officers to decide in the public interest

22 *Ibid.* at 82.

23 *Secretary of State v. Nageswara Iyer*, AIR 1936 Mad. 923, 926.

24 A. Vaidyanathan, 'Role of Law in Water Resources Management', 1 *Indian Juridical Review* 12, 14 (2004).

where the construction of channels should be and what the most suitable alignment is for such channels.[25]

Institutional framework

At the central level, the highest body is the Ministry of Water Resources. The Central Water Commission (CWC), formerly known as the Central Waterways, Irrigation and Navigation Commission, plays a central role in that it functions as the highest expert body on all problems related to the development of water resources. The CWC provides expert advice to the central government, the Planning Commission and to state governments. Although states have legal rights to develop their water resources, it has been found that proposals that are tailored to the orientation of the CWC have better chances of being approved.

Irrigation projects in India are classified into three categories, namely major, medium and minor projects. Projects which have a cultivable command area (CCA) of more than 10,000 ha are termed as major projects, those which have a CCA between 2,000 ha – 10,000 ha are termed as medium projects, and those which have a CCA of less than 2,000 ha are classified as minor projects. It is worth noting, however, that this classification misses out on the larger picture – nearly 75 per cent of Indian agriculture depends on groundwater and rainwater.

A Command Area Development Authority was set up in 1974 on the recommendations of the National Irrigation Commission, 1972 with the objective of bridging the gap between irrigation potential created and utilized through micro-level infrastructure development and efficient farm water management by farmers' associations. In 2004, the programme was restructured and renamed the Command Area Development and Water Management Programme. Under this centrally sponsored scheme, associations were formed at the lowest level of the canal network to promote greater participation of farmers in operation and maintenance of the canals. The inadequate and unreliable supply of water resulted in the failure of the scheme, except perhaps in a few states like Gujarat (pani panchayats) and Maharashtra.

3. New paradigms and policy reforms

Before focusing on the reforms currently being carried out in the irrigation sector, it is useful to note some milestones in irrigation policy and development. Tracing the history of the reform debates in India until the 1990s, Mollinga and Bolding identify the triggers for each phase of reform.[26] They trace the reforms in the

25 For a more detailed discussion, Alice Jacob and S.N. Singh, *Law Relating to Irrigation* (Bombay: N.M. Tripathi, 1972).

26 Peter P. Mollinga and Alex Bolding (eds.) *The Politics of Irrigation Reform: Contested Policy Formulation and Implementation in Asia, Africa and Latin America* 246 (Aldershot: Ashgate, 2004).

mid-1960s to being largely a result of experience gained from the field. This had revealed performance problems, particularly with regard to the canal irrigation systems where not all areas were actually irrigated; yields that did not match projections; below-maintenance standards; widespread waterlogging and salinity; and non-recovery of costs. To address these issues, a series of intervention programmes were designed to improve performance. One such approach aimed at improving individual farm development by making the canal system reach every farm and improving irrigation methods. Thus the focus in this early period was to improve the use of water at farmer level.

The next phase of reforms occurred with the realization that farmer-level problems are caused partly by main system management. Accordingly, the scope of the debate and intervention during the 1980s gradually moved up the canal system to include secondary and primary canals. According to Mollinga and Bolding, the debate and intervention shifted up one more level during the 1990s, where it was felt that the irrigation sector requires reorganization at the agency, policy and legal levels, to enable the solution of performance problems at system and farmer levels.[27] Reform therefore sought to focus on establishing river basin management organizations for integrated water resources management (IWRM). Hence the changing debates have also enlarged in scope from merely technical improvements to increased farmer participation to the present-day interventions for greater self-governance and accountability.

The five-year plan documents provided the blueprint for reforms. During the early years after independence, the focus of the irrigation sector was primarily on the expansion of major irrigation and reviewing its progress. The second phase (1970–1985) marked the review and improvement of the efficiencies of the irrigation infrastructure. During the 1970s, the idea of integrated water management gradually gained ascendancy, as organizations like the FAO emphasized the need for greater participation of beneficiaries. In India, the introduction of the second-generation approach to irrigation began in the mid-1970s. The Command Area Development Authority (CADA) introduced during the period envisaged more autonomy for states in planning, execution and management of irrigation projects with active involvement of beneficiaries at every stage, from launching of the project to its completion.[28] In order to further capacity building and extension activities, water and land management institutions were also set up around the country. This phase also saw a spurt in groundwater extraction to supplement irrigation needs and rapid growth in minor irrigation.

The third phase (1986–1995) saw a more intensive search into identifying problem areas and testing solutions. More specifically, several states experimented with pilot projects that examined the feasibility of WUAs at the outlet level, with

27 *Ibid.*

28 C.J. Joseph, Beneficiary Participation in Irrigation Water Management: The Kerala Experience (Thiruvananthapuram: Centre for Development Studies, Discussion Paper No. 36, 2001), available at www.krpcds.org/w36.pdf.

varying degrees of success. The government examined the issue of water pricing and enabled the fixation of water rates with the help of Central guidelines. Another issue that influenced irrigation policy was the over-exploitation of groundwater that in certain areas was identified as critical. A task force to examine the groundwater situation was also set up. This period is also significant as the first National Water Policy document was released in 1987.[29]

During this phase, a noteworthy development was the questioning of the dominant discourse on big dams and large multi-purpose projects. Large irrigation projects that simultaneously addressed concerns of food security and generated hydro-electricity were unconditionally promoted by technocrats in the nation and donor agencies internationally. While there are many benefits to large irrigation projects, their social, economic and environment costs are rarely acknowledged in popular literature on the subject. A brief period of rethink occurred in the early 1990s when the World Bank withdrew from the Sardar Sarovar Project under severe pressure from the anti-dam movement. This subsequently led to the setting up of the World Commission on Dams, which, while acknowledging the need to address the costs of human displacement and environmental damage, reformulated the World Bank policy in favour of promoting big dams.[30]

The fifth, or present phase, has taken forward the IMT/PIM discourse in a very active manner. This is reflected in various policy statements and financing arrangements. The Tenth Plan attempts to link reforms in the water sector to central funding. The Accelerated Irrigation Benefits Programme has relaxed lending norms for states agreeing to reforms (4:1 Centre: State lending ratio against 2:1 for non-reforming states).[31] PIM is being further promoted through the Command Area Development Programme. Under the Restructured Command Area Development and Water Management Programme, payment of central assistance to states is linked with the formation of WUAs. Farmers will now have to contribute 10 per cent of the cost of construction of field channels, reclamation of waterlogged areas, de-silting and renovation of minor irrigation tanks. The management subsidy rate to encourage formation of WUAs has been increased under the restructured programme to Rs 600 (earlier rates were Rs 275, which were increased to Rs 500) at the rate of 270:270:60 to be shared by centre, state and farmers respectively.[32] The states are also promoting the setting up of WUAs through several concessions and monetary incentives. For instance, the state of

29 See generally Narayanamoorthy and Deshpande, note 18 above.

30 Himanshu Thakkar, Assessment of Irrigation in India, Contributing Paper to the World Commission of Dams (2000), available at www.wii.gov.in/eianew/eia/dams per cent20and per cent20 development/kbase/contrib/opt161.pdf.

31 Government of India, Ministry of Water Resources, Accelerated Irrigation Benefits Programme Guidelines, 2006.

32 Government of India, Status of Participatory Irrigation Management in India – Policy Initiatives Taken and Emerging Issues (New Delhi: Government of India, Ministry of Water Resources, 2000), available at http://wrmin.nic.in/writereaddata/mainlinkFile/File421.pdf.

Maharashtra is aggressively establishing WUAs through a number of concessions and incentives.[33]

To sum up, until the end of the 1980s, the Indian economy was characterized by central planning for development. Planning ensured sustained and progressive growth in industry, agriculture and infrastructure. In improving agriculture, the planners placed special emphasis on irrigation projects and through the five-year plans enabled the construction and implementation of large irrigation projects such as Bhakra Nangal, Hirakud, Damodar Valley and Nagarajunasagar. This concerted effort resulted in an increase in agricultural production, from 50 million tonnes in the 1950s to about 203 million tonnes in the year 1999–2000.[34] In mapping the policy changes, the historical records and the plan documents provide an insight into the events that changed the approach to water resources and their management. Broadly, the recurring themes in these policy statements are financial viability and sustainability of water resources, impact assessment of irrigation projects, administrative and institutional efficiency, water-use efficiency and the future of water resources management.

The shift in paradigms reflects the shifting focus within the sector. The initial years focused on providing protective irrigation and large-scale expansion of technology. The new paradigm needs for irrigation are articulated in the latest policy document, the National Water Policy, 2002, which states that:

> Irrigation planning either in an individual project or in a basin as a whole, should take into account the irrigability of land, cost-effective irrigation options possible from all available sources of water and appropriate irrigation techniques for optimizing water use efficiency. Irrigation intensity should be such as to extend the benefits of irrigation to as large a number of farm families as possible, keeping in view the need to maximize production.[35]

Interestingly, the policy identifies and emphasizes the need to link water-use and land-use policies.[36] The National Water Policy, 2002 stresses the need for a participatory approach to water resources management, 'by involving not only the users and other stakeholders, in an effective and decisive manner, in various aspects of planning, design, development and management of the water resources schemes'.[37] In briefly laying out the road map for a participatory approach, the National Water Policy, 2002 stresses the need for legal and institutional provisions at the appropriate levels, ensuring an appropriate role for women, and enabling the involvement of WUAs and local bodies such as municipalities and gram

33 Niranjan Pant, 'Some Issues in Participatory Irrigation Management', 43(1) *Economic & Political Weekly* 30 (2008).

34 See generally National Water Policy of 1987, available at www.ielrc.org/content/e8701.pdf. See also National Water Policy of 2002, available at www.ielrc.org/content/e0210.pdf.

35 National Water Policy, 2002, s. 9.1.

36 *Ibid.*, s.9.2.

37 *Ibid.*, s. 12.

panchayats in the operation, maintenance and management of water infrastructures/facilities at the appropriate levels.

In taking forward the objectives of IMT/PIM and to strengthen IWRM, it is believed that a robust regulatory framework for the governance of WUAs, and for resource management by them, is a fundamental ingredient. The former will provide the configuration for healthy functioning of WUAs, while the latter will concern itself with the way organizations obtain and manage the water and land resources they need to operate.[38] This institutional innovation of regulatory authorities is now being experimented in India. This chapter deals at length with WUAs; water regulatory authorities are discussed in detail in Chapter VII.

B. Water user legislation in India

The PIM discourse originated as a response to the existence of multiple institutional problems in irrigation, more particularly the irrigation bureaucracy's stagnation and the need to ensure better financial management. PIM involves redefinition of the role of state in the area of irrigation so as to allow for a more genuine and greater participation of water users in the integrated and joint management of irrigation systems. The setting up of WUAs is envisaged in conventional irrigation systems and in larger tank and lift irrigation systems. The broad approach is to work in tandem with government agencies to increase water-use and allocation efficiencies, improve equity, sustainability and conservation of irrigation systems.[39]

Prior to IMT/PIM discourse capturing the water reform agenda and the thrust towards specific enactments that aid in their formation, WUAs, formed either as pilot projects or as civil society initiatives, were registered under the Central Societies Registration Act of 1860 (for instance Bihar and Haryana), various state Societies Registration Acts (Gujarat and Maharashtra), or the Cooperative Societies Registration Acts (Kerala and Tamil Nadu). Some states sought to amend existing irrigation laws to provide for WUAs,[40] while others achieved the same objective through policy statements and executive orders.[41] In 1989, the Government of Gujarat experimented with Gujarat Irrigation and Drainage Ordinance, 1989 to define the roles and functions of WUAs, which subsequently was not approved by the Gujarat Assembly.[42]

38 Stephen Hodgson, Legislation on Water Users' Organizations iii (Rome: Food and Agriculture Organization, Legislative Study No. 79, 2003).

39 See generally Rakesh Hooja, *Management of Water for Agriculture* (New Delhi: Rawat Publications, 2006).

40 The Madhya Pradesh Irrigation Act, 1931; The Maharashtra Irrigation Act, 1976; The Kerala Command Areas Development Act, 1986.

41 The Government of Tamil Nadu in 1994 and the of Maharashtra in 1994.

42 J. Brewer *et al.* (eds.) *Irrigation Management Transfer in India: Policies, Processes and Performance* 69 (New Delhi: Oxford and IBH, 1999).

1. Early experiences in India

In India, the major irrigation acts, now over 100 years old, did not consider farmers' involvement in irrigation as either important or desirable. Irrigation administration was retained in the hands of a technocratic bureaucracy, who prefer to distance themselves from the farmers, even today. The new enactments and statutes in post-independence India followed the same tradition, occasionally recognizing the importance of farmers but never giving them the authority, powers or resources to deal with irrigation administration in ways farmers may deem fit.

Traditional irrigation systems have been successfully owned and managed by communities in India. The better known and documented experiences are the Phad system of Maharashtra, the tank irrigation systems in southern India, the Khuls of Himachal Pradesh and Jammu and Kashmir. However, the experiences with government-owned and established formal irrigation systems have been erratic and inconsistent. In the past, irrigation committees have been set up in some states such as Andhra Pradesh (called pipe committees), Uttar Pradesh (kolaba samitis), Maharashtra (pani panchayats) and in Gujarat. These efforts have displayed a varying degree of success and in large measure they have been marked by a high degree of unsustainability, poor maintenance, lack of autonomy and self-sufficiency.[43] However, the traditional systems have neither provided the impetus nor the design for the modern-day PIM ventures.

Niranjan Pant classifies the growth of the PIM in the country into four distinct phases.[44] He notes that in the first phase, that is, the decade from 1975–1985, focused on creating outlet-based water user organizations leading up to research on PIM as a solution for equitable distribution of water among irrigators, maintenance of lower-end canal systems and resolution of conflicts amongst water users. The second phase from 1985–1990 saw the shift from PIM research to field experiments on a pilot basis, funded largely by the Ford Foundation and the World Bank. It may be noted here that in the first two phases, up to the late 1980s, the irrigation bureaucracy was hostile to the stepping up of WUAs, arguing that it was a futile exercise. This resistance is largely muted, though not entirely missing within the lower ranks of the department.

The third phase starting from the early 1990s saw the emergence of the idea of a handover/turnover of irrigation systems in the case of smaller systems, and a devolution of management of sub-systems, particularly at the level of distributaries/minors in the case of larger systems, to the irrigating farmers. The first enactment was by Andhra Pradesh in 1997 and several states (Andhra Pradesh, Madhya Pradesh, Chhattisgarh, Rajasthan, Karnataka and Orissa) have now enacted legislation that makes PIM a statutory requirement to get access to irrigation water. According to Pant,

43 K.K. Singh (ed.) *Farmers in the Management of Irrigation Systems* (New Delhi: Sterling Publishers, 1991).

44 Pant, note 33 above, 30.

> The fourth phase starting from 1997 marks the emergence of donor-funding as a bait for restructuring India's irrigation sector with PIM/IMT as a core programme. The two available models in India are the Andhra Pradesh model and the Maharashtra Model. The former represents a top-down approach where the act was passed in a relatively short time and a large number of WUAs were established with a bang. However, the autonomy and sustainability of WUAs are being questioned both by academics and activists.[45]

The Maharashtra model has had a more organic growth as the process of setting up of users association, has a long history and very effective engagement of civil society, farmers and the government.

The National Commission for Integrated Water Resources Development Plan notes that pilot projects on PIM have demonstrated many positive gains, such as substantial increase in the area under irrigation, more equitable distribution of water and increase in agricultural incomes. It further states that:

> if organized in groups such as WUAs or farmers' organizations the potential of group management for furthering collective interests could be realized. WUAs are known to raise funds to meet current and future farm needs, including investment on farm machinery, development and marketing and the like.[46]

Beginning in 1973, a coordinated approach to the development of irrigated agriculture was implemented by the newly created command area development (CAD) authorities. These initial efforts were not received with any enthusiasm by the farmers. When CAD authorities made a determined attempt to provide water to each farmer holding in the early 1980s, emphasis shifted to rotational water supply along with on-farm development. The farmers initially responded with enthusiasm. However, the uncertainty in water supply and the frequent non-compliance with the prescribed rotational water supply schedule did not bring about any marked improvement in water utilization. On the whole, CAD continued to be seen as a government programme imposed from the top. When critically evaluating the programme, some project administrators argued that implementation and water utilization could probably be improved if farmers were given the responsibility for irrigation management. In 1987, the Ministry of Water Resources, Government of India issued guidelines for the formation of the WUAs.

Early enactments

The precursor to the present-day water user legislation can be found in a few earlier enactments that introduced the concept in a tentative manner. Under the

45 *Ibid.*, at 30–31.

46 Government of India, note 16 above, 233.

Madhya Pradesh Irrigation Act, 1931, an irrigation panchayat was established for every village.[47] The collector had the discretion to establish an irrigation panchayat for the irrigation unit – a *mahal* or *chak* – in which the permanent holders and occupiers of land entered into an agreement. The functions of these panchayats were to assist the state irrigation officials in detecting and preventing encroachment on canal lands, to prevent damage to irrigation works and to arrange for the construction and repair of watercourses, to settle disputes, and to collect canal revenues and remit these revenues to the Treasury.[48]

In Uttar Pradesh under the Uttar Pradesh kshetra samitis and zilla parishads adhiniyam, 1961, one of the functions of the kshetra samitis was the construction, repair and maintenance of minor irrigation works, as well as the supply of water from them and from other sources subject to financial limits fixed by the state government.[49] The Northern India Canal and Drainage Act, 1873 as applicable to Uttar Pradesh was amended in 1963 to provide for the active cooperation of the gaon sabhas in the construction of watercourse and field channels. The UP Command Area Development Authority was set up under the Uttar Pradesh Command Area Development Act, 1976. District Committees were set up and at the lowest level was the Chak Sabha, a general body of the outlet responsible for on-farm development and maintenance.

In Bihar, under the provisions of the Bihar Irrigation Field Channels Act, 1965, the duty to construct field channels was primarily with the gram panchayats. If the collector on the report of an officer of the irrigation department found it necessary to construct field channels within any gram panchayat, he could direct the panchayat samiti concerned to get field channels constructed within six months. Under the Travancore-Cochin Irrigation Act, 1956 the local panchayats had been entrusted with the work of construction, repair and maintenance of all petty works. These works include all irrigation works, useful for the drainage or protection of an extent of not more than five acres. For financing the works, the panchayat had been empowered by the Act with the prior permission of state government to levy an annual cess on areas benefited by the irrigation work constructed, repaired or maintained at its cost.

Pipe committees were set up under the Andhra Pradesh Irrigation Utilization and Command Area Development Act, 1984. The pipe committee consisted of five to seven members representing all zones. Zones are contiguous sub-units of a Chak or Chaks. Likewise, the Maharashtra Irrigation Act, 1976 which was enacted with a view to unifying and amending the law relating to irrigation in the State of Maharashtra, envisaged the formation of water committees. The Kerala Command Area Development Act, 1986 permits the formation of outlet WUAs called Beneficiary Farmers' Associations and their functions include construction,

47 Madhya Pradesh Irrigation Act, 1931, s. 62(1).

48 *Ibid.*, ss. 62(2) and 69.

49 Uttar Pradesh kshetra samitis and zilla parishads adhiniyam, 1961, s. 32 read with Schedule I(i), (ii) and (iv).

maintenance, repair and upkeep of the irrigation systems under the outlet; regulation of water supply for irrigation by turns or rotation according to the schedule and in the manner approved by CADA; regulation and control of water supply for irrigation by volumetric measurement; and supervision of the irrigation system with a view to prevent waste of water and damage to the system.[50]

2. *Water user enactments post-1990s*

Since the mid-1980s, and driven largely by financial pressures, there has been an upsurge in efforts by governments around the world to transfer the management of irrigation systems from government entities to farmers. This has occurred both in developed and in developing countries, though the form in which it is introduced differs. Whether the transfer includes all or part of management functions, full or only partial responsibility, an entire irrigation system (or tubewell command) or only a part of a system, organizations of irrigation water users have been increasingly assuming a pivotal role in irrigation management.

Several states have now set up WUAs transferring a large part of the responsibilities of operation and maintenance at the primary level of the canal system. While some have done so through enabling legislation, others have amended existing laws or effected the formation of WUAs through government resolutions.[51] The new state enactments include the Andhra Pradesh Farmers' Management of Irrigation Systems Act, 1997, Goa Command Area Development Act, 1997 and Madhya Pradesh sinchai prabandhan me krishkon ki bhagidari adhiniyam, 1999; the Tamil Nadu Farmers' Management of Irrigation System Act, 2000; the Kerala Irrigation and Water Conservation Act, 2003; the Orissa Pani Panchayat Act, 2002 and the Maharashtra Management of Irrigation System by Farmers Act, 2005; and the Chhattisgarh sinchai prabandhan me krishkon ki bhagidari adhiniyam, 2006.

Many of these enactments are replicas of each other and hence some broad generalizations of the content of the enactment can be made. The large or major irrigation projects provide for a three-tier structure for WUAs, also referred to as farmers' organizations by the enactments. The WUA or the pani panchayat are at the primary level, the distributory committee at the secondary level and the project committee at the project level. Some enactments also provide for a state committee at the state level.[52] The criterion for delineating the command area is based on a hydraulic basis; an administratively viable unit is considered to be between 300–600 hectares. In the case of lift irrigation, the entire command area is usually considered to be one unit.

50 See generally Jacob and Singh, note 25 above.

51 See, e.g., Bihar Irrigation, Flood Management and Drainage Rules 2003 under the Bihar Irrigation Act, 1997. The state of Karnataka promulgated an Ordinance on 7 June 2000 amending the Karnataka Irrigation Act, 1957. Government resolutions relating to PIM have also been issued by the State of Maharashtra in 2001 and Gujarat in 1995.

52 See, e.g., Orissa Pani Panchayat Act, 2002, s. 9, available at www.ielrc.org/content/e0211.pdf.

Every WUA consists of all water users who are landowners in the delineated area,[53] and all the members constitute its general body. A managing committee for every WUA is necessary, and the members of the managing committee are directly elected from among its members by the method of secret ballot. All the presidents of the primary level WUA constitute the general body of the distributory committee, which also elects its president and members. At the third or top level is the project committee, the general body of which is constituted by all the presidents of the distributory committee, who elect their own president and secretary. The tenure of WUAs and their elected representatives vary in the different states, with the maximum being five years. The enactments provide for recall of leaders on grounds of non-performance or any other violations, which is an innovative step with the potential to strengthen democracy. While this is the broad institutional format in most states, minor variations depending on local and administrative needs exist.

Typically the functions of the WUAs at the three different levels include:

(a) preparation of an operational plan based on entitlement, area, soil, cropping pattern at the beginning of each season. In other words the task of preparing the warabandi schedule is now a primary task of the WUAs;
(b) preparation of a plan for the maintenance of irrigation system in the area of its operation at the end of each crop season and carry out maintenance works with the funds allotted to it;
(c) regulating the use of water among the various outlets under its area of operation;
(d) assisting in the collection of water rates;
(e) maintaining various inventories of the irrigation system in their respective jurisdictions; and
(f) other tasks include annual audit, monitoring flow of water, resolving disputes, water budgeting and periodical social auditing and assisting in the conduct of election to the various committees.

In the remaining part of this section, an attempt is made to critically examine the key features of the enactments that have been introduced to enable participatory irrigation management.

Water rights of users

Although not always explicitly stated, all surface water is owned by the state and the state can decide the allocation of water. Most irrigation laws in the country

53 The Andhra Pradesh Farmer Management of Irrigation Systems Act, 1997, available at www.ielrc.org/content/e9701.pdf, includes the tenant of the landholder as also being entitled to be a member. See also Chhattisgarh sinchai prabandhan me krishkon ki bhagidari adhiniyam, 2006, available at www.ielrc.org/content/e0605.pdf.

link water rights to land holdings and hence rights over surface water are transferred along with land rights. The water users enactments unquestioningly adopt these two premises. Though the definition of 'water user' is wide enough to include all users, in creating the WUA, it is necessarily linked to land. For instance, the Orissa Pani Panchayat Act, 2002 defines a water user to mean 'any individual or body corporate or a society using water for agriculture and domestic purpose from an irrigation work'.[54] However, in constituting the lowest rung of WUA, that is, the pani panchayat, the Act states that 'every pani panchayat shall consist of all the water users who are landholders in the area of a Pani Panchayat',[55] effectively excluding water users who do not own land from the decision-making process.

The exclusion is of particular significance as the discourse around tradable water rights gains precedence in policy circles. Tradable water rights are rights to use water that can be transferred in whole or in part, separately from the transfer of land. Water rights can be leased for a season, a year, or many years. It is widely believed that there are many benefits linked to the establishing of markets in tradable water rights such as empowerment of water users, provision of investment incentives, improved water efficiency, improved equity in the provision and financing of water services and increased flexibility in resource allocation. It is argued that assigning tradable water rights to individuals within WUAs, or to the WUAs, enhances the control of these groups over water resources, ensuring better access to water. In such a scenario, the interests of the non-land-owning classes and their water use rights needs to be protected, through protective clauses in the water user enactments.

Interface with irrigation departments

A key component of the PIM model is the effective interface between the irrigation departments and newly set-up WUAs. The enactments provide for the linkage between the irrigation department and WUAs through the appointment of officers who are responsible for implementation and execution of all decisions taken by the WUA. At the WUA level, an assistant engineer/assistant executive engineer is the competent authority; the deputy executive engineer is the competent authority at the distributory committee level and the superintending engineer/chief engineer is competent at the project committee level.[56]

Despite the statutory mandate, sharing of information critical to the effective functioning of the new institutions continues to be inadequate. To assist the WUAs in their multiple activities, the irrigation and revenue departments are required to provide them with necessary information. Reports from the ground indicate that the newly elected members of WUAs are not receiving adequate

54 Orissa Pani Panchayat Act, 2002, s. 2(x).
55 *Ibid.*, s. 3(4).
56 See, e.g., Andhra Pradesh Farmer Management of Irrigation Systems Act, 1997.

support and cooperation from irrigation and revenue departments.[57] According to the Act, WUAs are required to maintain a number of registers and records for which they need the help of irrigation and revenue department staff. Presently, several WUAs are not maintaining these records, due largely to the general attitude and lack of interest of local irrigation and revenue staff in extending cooperation.

Greater accountability of the irrigation agencies towards farmers, and the sharing of rights and powers with regard to delivery of water to WUAs in government-managed irrigation systems, are essential for successful irrigation management transfer to work. These features ought to be reflected in law. However, the new crop of enactments lacks clarity in this regard. The enactments only provide for linkage between the irrigation department and WUAs through the appointment of officers who are responsible for the implementation and execution of all decisions taken by the WUA.

The initial set of enactments enacted in early 2000 provides for several functions to be carried out by the WUAs with little or no statutory powers to implement them. Likewise, the irrigation department does not have a corresponding duty to provide the WUAs with necessary support to carry out their functions. The continuing reluctance of the irrigation department in granting real powers to the WUAs is illustrated in many field accounts that record their obstructive behaviour. Several officers of the irrigation department continue to express their strong reservations and doubts about the capability of WUAs to perform the tasks entrusted to them and the viability of the programme in the long run. In their opinion, the WUAs are more interested in receiving funds from the government and maintenance activities than water regulation and equitable distribution among all the members.[58]

Operation and maintenance

At the heart of the irrigation reforms is the transfer of operation and maintenance to WUAs. For government systems, state irrigation laws generally classify the distribution system into canals, watercourses, field channels and drainage works. Separate sections deal with each one of them. Prior to the reform process, farmers were not generally held responsible for construction, but they were collectively responsible for maintenance of watercourses and field channels and sometimes of drainage works. By implication, the irrigation agency was responsible for all or almost all construction and maintenance of canals.

Now a large number of functions have been divested to the WUAs. For instance, the Andhra Pradesh Farmers' Management of Irrigation Systems Act,

57 Davuluri Venkateswarlu, 'Politics of Irrigation Management Reforms in Andhra Pradesh', in Rakesh Hooja, Ganesh Pangare and K.V. Raju (eds.) *Users in Water Management: The Andhra Model and its Replicability in India* 159, 175–176 (New Delhi: Rawat Publications, 2002).

58 *Ibid.*, at 176.

1997 states that the objects of the WUAs shall be to promote and secure distribution of water among its users, ensure maintenance of the irrigation system, ensure efficient and economical utilization of water to optimize agricultural production, as well as to protect the environment and ensure ecological balance by involving the farmers, thus inculcating a sense of ownership of the irrigation system. More specifically, the WUAs (at all three tiers) are broadly required to perform the functions of planning, regulating water use, collecting water charges and raising resources, maintaining records and resolving disputes. Additionally, the WUAs are also required by the act to encourage avenue plantation on canal bunds and tank bunds, to conduct regular water budgeting and to conduct periodical social audits.

Financial sustainability

In addition to improving water use and efficiency, the other major driver for the present reforms is financial viability. Irrigation revenue is not uniform and differs from state to state. It includes water charges, occasional water charges (including charges for unauthorized use and water wastage), charges for benefit by percolation and leakage from canals, betterment tax, and dues in arrears and penalties. The WUA enactments broadly provide for funds for the WUA as being sourced primarily from:

(a) government grants as a share of the water tax collected;
(b) funds granted by the state and central government for the development of the area of operation;
(c) resources raised from any financing agency for undertaking economic development activities in its area of operation;
(d) income from the properties and assets attached to the irrigation system within its area of operation; and
(e) fees collected by the farmers' organization for the services rendered in better management of the irrigation system.

Studies show that for the most part, the utilization of funds indicates a number of improvements in the physical structures of the irrigation systems handed over to the WUAs. Others indicate on the other hand that the inflow of cash has replaced the middlemen/contractors, with the presidents of WUAs doubling up as contractors to carry out minor maintenance works. It is also reported that the WUAs are unable to recover the 15 per cent farmer contribution to undertake minor works. Charges of corruption have also been corroborated by field studies in Andhra Pradesh.[59]

59 See generally Venkateshwaralu, note 57 above.

Conditions for sustainability of WUAs

Studies have been conducted to examine the feasibility and the functioning of the new institution of farmers' participation in irrigation management. In a study of the early years of PIM in Andhra Pradesh, both the positive and negative impacts of the water sector reforms in that state are documented. Highlighting the positive aspects, the study finds that in the last few years PIM has resulted in inculcating a sense of ownership and belonging among farmers in the irrigation systems; reducing irrigation disputes among farmers; reducing irrigation offences by preventing damage and tampering with the irrigation structures by farmers, particularly in the tail-end areas; enabling farmers to take up minimum rehabilitation and operation and maintenance works according to their needs and choices; improving water supplies to undeserved areas, particularly at the tail ends; and improving capacity building and empowerment of these organizations in decision making and execution of works. However, this study also highlighted several inadequacies. These include the fact that some of the WUA presidents were found to assume the role of contractors by undertaking the physical work in their own name; poor quality execution of works; corruption; and appropriation of command areas and structures by WUA presidents. Further, it was also found that the WUAs were performing only limited duties such as executing works but not complying with their statutory responsibilities including maintenance of records and accounts, conduct of financial audits, and organization of general body and managing committee meetings.[60] All of this points to the importance of adapting conceptual frameworks to local conditions, to ensure their sustainability and their eventual success.

Additionally, irrigation systems are largely managed on the basis of command area of the project and may cover more than one local governance institutional jurisdiction. Rural local government was vested with independent powers and resources. Importantly, the 73rd amendment to the Constitution in 1993 granted local panchayats the power to control all natural resources within their jurisdiction, including water. This leads to a key question of the powers of local panchayats over irrigation. At the federating levels, the WUAs would benefit immensely from the involvement of panchayat raj institutions. This can be further enabled by providing for representation of representatives of PRIs in the local WUAs.

3. *Conceptual framework of WUA enactments – An analysis*

Several variables influence the effective functioning of WUAs. Effective leadership, group cohesiveness, capacity building and collective ownership are a vital, but amorphous, element of a successful WUA. The more tangible key ingredient is well-crafted laws, rules and crucially the powers to effectively implement them.

60 K.V. Raju, 'Participatory Irrigation Management in Andhra Pradesh', in Hooja, Pangare and Raju (eds), note 57 above, 84.

Time and again, it has been found that the legislation enacted fails on this count and this is true of the new acts being introduced as part of the water law reforms in the irrigation sector.

The benefits of devolution and effective decentralization are undoubtedly immense. Management transfers resulting in users managing portions of larger canal irrigation systems or entire tank and lift irrigation schemes enable the creation of collective goods. 'Access to improved irrigation enables users to reduce risks of crop losses or increase their incomes substantially.'[61] However, in conceptualizing a top-down approach to institutionalizing WUAs, the legislation pre-empts the voluntary nature of the WUA. Given its mandated formation, water user legislation must be creative to devise innovative methods that recreate a community that is both inspired and has the incentive to engage collectively. The current crop of legislation is overly focused on ensuring the collection of water rates from farmers and passing on the backlog of maintenance works to farmers' organizations, inspiring little confidence in the genuineness of the devolution of powers to the WUAs.

Another troubled area has been the reluctance of the bureaucracy to surrender its powers and functions to farmers. The enactments uniformly fail to visualize an innovative strategy to dismantle the bureaucratic stranglehold to the extent that certain states are keen on retaining the idea of WUAs only on paper, not enabling an effective devolution of powers, finances and functions. For instance, the Maharashtra Irrigation Act, 1976 empowered the Irrigation Department to provide water on a volumetric basis with certain stipulations to water committees. The section officer or his nominee could also be a member of the water committee, covering all villages in the command area. The functions of the water committee, *inter alia*, included assisting the canal officer in receiving complaints, preventing unauthorized use of water and assisting the canal officer in discharge of his duties. The Maharashtra Farmer Management of Irrigation Systems Act, 2005 repeals these provisions relating to water committees. A clear distinction is thus evident in the approach. While the earlier enactment saw farmers' organizations as assisting the irrigation department, the new enactments clearly provide for the bureaucrats to assist the WUAs in carrying out their functions.

In reality, however, very few WUAs are empowered enough to demand that the department and its officials play a subordinate role to the farmers' collective. Likewise the officials find it difficult to take orders from the farmers, over whom they ruled for many decades. The primary challenge for those setting up WUAs is empowering farmers' organizations to control them and run them effectively. It has consequently been stated that there 'is a need to treat the minor level operation as a preparatory school for becoming familiar with water use planning and principles'.[62]

61 Brewer, note 42 above, 89.

62 Society for Promoting Participative Ecosystem Management (SOPPECOM), Participatory Irrigation Management: An Overview of Issues and the Way Ahead (Theme paper presented at Water and Land Management Institute, Aurangabad, 2007), available at http://soppecom.org/walmi_theme_paper.htm.

These enactments have other far-reaching consequences. Firstly, it has an impact on the usufructuary rights of water users. It vests usufructuary rights in a collective, but the enactments do not create any new rights. It is important, however, to examine the nature of the usufructuary rights vested in the WUAs. The first generation of irrigation laws post-independence clearly vested all rights over the water in the command area and the canals in the state. Individual farmers were only granted water-use rights if they paid a fee, but the timeliness and adequacy of the water that reached the farmers was not guaranteed or protected by legislation. The new legal framework in the various states does not alter this status. It seeks only to transfer the proprietary usufructuary right of use to a collective called the WUA. Significantly, it does not explicitly provide that the WUAs' usufructuary rights are justiciable. Disputes that might occur intra- or inter-WUA should be resolved internally or by the unit immediately preceding it in the hierarchy. However, if that were to fail, it is unclear whether an aggrieved individual farmer can use the provisions of the enactment to assert a 'rights' claim, as against the WUA or even the state.

Secondly, there is a clear need to shift the emphasis from pricing and 'participatory irrigation management' to concerns of water as a human right. The content of the rights discourse needs to incorporate issues of equity, sustainability and inclusiveness, alongside socio-political concerns of effective decentralization and substantive empowerment of all water users. A 1999 World Bank technical report discusses water rights as:

> A WUA could negotiate a larger or smaller share of water with other WUAs or with another bulk consumer such as an industry or municipality. The negotiated price would reflect a lease or sale value mutually attractive and beneficial to both renter and leaser of water. This could facilitate water going to its most viable use.[63]

The Bank thus exteriorizes water rights as an antecedent to enabling an economy which uses the resource as a tradable commodity.

Thirdly, the legislation has an impact on the democratic participatory rights of water users. As noted above, with the exception of the state of Chhattisgarh,[64] the enactments which set up the WUAs do not seek to redress the possible 'rights' claims that may be denied to political, socio-economic, gender-specific, religious and caste compositions in access to water and investment decisions at the local level, thus determining the sustainability and effectiveness of the WUAs. The widest possible definition is incorporated in the definition of the water user in the Chhattisgarh Act. Without these rights being formalized through a statute, it is unlikely that the heterogeneous groupings of WUAs will

63 Keith Oblitas and J. Raymond Peter, Transferring Irrigation Management to Farmers in Andhra Pradesh, India 59 (Washington D.C.: World Bank, Technical Report No. 449, 1999).

64 It provides for reservation for scheduled castes, scheduled tribes, backward classes and women.

succeed in ensuring equity in the governance structures, thus making them truly participatory.

Fourthly, the legislation does not address the issue of equity and the sustainability of water distribution. Linked to the issue of equitable and participatory governance structures is the issue of de-linking water rights from land rights, so as to give a voice to the landless in water management structures. This is vital if irrigation waters are being put to uses other than their primary goal of agricultural production. Inherent in the irrigation system design is the concern over the equitable distribution of water, which impacts the rights of water users. The resource distribution between the head-reaches and the tail-enders has been a recurring concern of many committees and commissions that have been set up to study the implementation of the irrigation works.[65] Water availability is progressively reduced as one heads from the head end to the tail end of a canal system. Equity is a matter of concern not only for the engineers and architects, but also for the law reformers grappling with rights issues.

Fifthly, the reforms do not take into account the traditional water rights regimes and their interface with the formal legal system. While much of the debate in popular literature on irrigation focuses on water rights as essential for the larger goal of commodification of water, it is equally important to highlight water rights from a human rights perspective. The new enactments do not specify any linkages between the newly constituted statutory WUAs and the traditional institutions associated with traditional water harvesting systems. The traditional functionaries such as Neergantis are not referred to, and their experience and that of these functionaries are not taken into account.[66] The inevitable tension between traditional rights, common property rights and rights acquired 'due to access to resources'[67] highlights the need for a clearly defined rights framework, not from a 'markets' prism but taking on board all social needs and complexities. The ongoing debates on human rights within the national and international frameworks need an irrigation-specific focus.

C. Irrigation sector reforms – A critical appraisal

Although the focus in this chapter is primarily on irrigation management transfer to farmers, the broader picture of irrigation reforms in the country needs to be explained and analyzed. Legal reforms in the irrigation sector consist primarily of the enactment of a range of state-level legislation that sets up WUAs.

65 Head refers to head of the system, head of the distributory and head of the minor. Similarly, tail end refers to the end of a channel – main canal, distributory and minor.

66 V. Gangadhar, 'Riches of Rayalaseema', in Anil Agarwal, Sunita Narain and Indira Khurana, *Making Water Everybody's Business* 33, 42 (New Delhi: Centre for Science and Environment, 2001).

67 A. Rajagopal and S. Janakarajan, Water Rights and Participatory Irrigation Management in India: The Case of Surface Water Sector in Tamilnadu State (Chennai: MIDS, Working Paper No. 171, 2002), available at www.water-2001.de/datenbank/546076235.41368.14/WATER%20RIGHT%20PAP.doc.

Significantly, the other important set of legislative changes being introduced is the setting up of water regulatory authorities in a few states that have an impact on the irrigation sector.[68] However, this does not capture all the other developments simultaneously being carried out in the irrigation sector.

Water sector planning and development in the country is centralized and the irrigation sector continues to be dominated by the 'mega project' discourse. Despite the policy thrust towards greater participation and decentralization, a more holistic approach to incorporate it into all other current initiatives is not evident. On the contrary, the government continues to promote and encourage large dams and interlinking of rivers, effectively divesting local control over resources and reinforcing the centralizing bias. Also missing from the reform agenda is a serious engagement with the alternative models of rainwater conservation, crucial in irrigating large tracts of arid land. Over half the irrigated areas and two-thirds of production from irrigated areas comes from land irrigated by tanks, ponds, watershed programmes and groundwater.[69]

In keeping with the mega project model of development, the Ministry of Water Resources has prepared a National Perspective for Water Development, which proposes to link rivers in different basins throughout the country with the objective of transferring excess water from water-rich basins to water-deficient ones. Despite opposition,[70] this project received a boost when in 2002 the Supreme Court ordered the government of India to complete the task of linking rivers by the year 2014.[71] The first scheme attempted under this massive project was the interlinking of the Ken and Betwa rivers. The interlinking of rivers project aims to link fourteen rivers in the Himalayan region and sixteen rivers in the peninsular region of the country at a minimum estimated cost of 5,600 trillion rupees.[72]

Another noteworthy development in recent years has been an active support to privatization of irrigation projects. The first major attempt at privatization for industrial use was of a surface water source (Sheonath river in Chhattisgarh). Attempts have also been made to privatize the irrigation system in the Krishna Valley. Privatization in irrigation has taken the form of irrigation development corporations. The Maharashtra Krishna Valley Development Corporation invited expressions of interest for private sector participation in the Nira-Deoghar project.

68 See, e.g., Maharashtra Water Resources Regulatory Act, 2005, available at www.lead-journal.org/content/05080.pdf.

69 See generally Himanshu Thakkar, Restoring Sanity in Water Use, 2004, available at www.india-seminar.com/2004/541/541 per cent20himanshu per cent20thakkar.htm.

70 Several arguments against such interlinking of rivers have been raised. It is argued in particular that a river does not have surplus water; that such a project is not cost-effective and indeed hugely expensive; that it will lead to sub-optimal use of water resources; that it will change the structures of channels thereby leading to increasing the salt gradient, water loss, seepage and saline pollution of soil in the transporting section.

71 *In Re: Networking of Rivers*, Writ Petition (Civil) No. 512/2002, order dated 31-10-2002.

72 Vandana Shiva, Resisting Water Privatisation, Building Water Democracy (Paper presented at the World Water Forum, Mexico, March 2006), available at www.globalternative.org/downloads/shiva-water.pdf.

The matter was contested before the Maharashtra Water Resources Regulatory Authority (MWRRA), with one of the main arguments brought before the regulatory authority being the fact that irrigation tariffs would rise as a result of privatization, since water charges were a key component for the private developer to recover his investment. The MWRRA passed an order directing the withdrawal of the tender notification, citing lack of clarity in water privatization policy.[73] In Karnataka, the Krishna Bhagya Jal Nigam was formed under the Companies Act in 1994 with a specific purpose of mobilizing public funds for the development of the Upper Krishna project.

Water regulatory authorities are now being experimented with in the country. States like Maharashtra, Arunachal Pradesh and Uttar Pradesh have already enacted state laws for the establishment of independent regulatory authorities in the water sector. The MWRRA was the first of its kind in the water sector in India. It was established in 2005 and supported through the institutional restructuring component of the World Bank loan for the Maharashtra Water Sector Improvement Project.[74] The Planning Commission Working Group on Water Resources has in its recent report recommended that all states set up a water regulatory authority along the lines of the one in Maharashtra.[75] In carrying forward this policy prescription, the Government of Karnataka recently passed a government order for the constitution of the Karnataka Water Resources Authority.[76]

1. New institutional paradigm – A critique

The irrigation reforms have largely been influenced by economic theories of new institutionalism, which is based on theoretical insights from neoliberal economics and political theory. In the context of natural resources management policies, it primarily focuses on the creation of financial and other incentives by markets and institutions so as to improve the economic efficiency of resource use.

Cost recovery, accountability, and financial autonomy are key solutions advocated by new institutionalists for what is perceived as poor performance and perverse subsidies of irrigation governance. Irrigation agencies should recover from the users management costs and eventually part of the investment costs.[77] The treatment of water as an economic good, the tradability of water rights and water pricing are often cited as means to save water and use it more efficiently.

73 Shripad Dharmadhikary, Water Privatisation: Start Again, *India Together*, 18 December 2008, available at www.indiatogether.org/2008/dec/env-mwrra.htm.

74 World Bank, Project Appraisal Document for Maharashtra Water Sector Improvement Project (Washington D.C.: World Bank, Report No: 31997-IN, 2005).

75 Government of India, Report of the Working Group on Water Resources for the XI Five Year Plan: 2007–2012 at 46 (New Delhi: Government of India, Ministry of Water Resources, 2006).

76 Karnataka Government Order G.O. No. WRD 85 MBI 2008, 8 September 2008.

77 See generally Jasveen Jairath, *Water User Associations in Andhra Pradesh* (New Delhi: Concept Publishing Company, 2001).

Lack of accountability is often identified as one of the key causes of low performance in irrigation systems. Financial autonomy, together with cost recovery, is the most important mechanism (as identified by new institutionalists) to bring about accountability. State subsidies, accordingly, need to be shunned to take forward this logic.

The importance of well-defined and enforceable water rights is undisputed in the new institutionalist paradigm of irrigation management, and is identified as a necessary condition for the emergence of water markets. Water rights are seen as the legal complement to technical infrastructure and institutions, the three of which are necessary for an effective allocation and distribution of water. Existing water rights systems are seen and analyzed against this 'ideal' model, and generally judged not to fulfil the necessary conditions for well-defined property rights to water. The existence of legal pluralism and their obvious strengths is rarely recognized and developed.

Interestingly, the solutions put forward by the new institutional economics is not entirely new. David Mosse finds parallels between the contemporary reforms and the proposals put forward a century ago in the 1870s by the colonial state divesting itself of the responsibility and expense of repairing 'ruined' tanks by making these over to the ryots or to private proprietors.[78]

Further, Mosse notes that popular rhetoric represents PIM as a radical transformation of the role of the state in irrigation, from doer to facilitator, from centre to periphery. Questioning this representation he asks the vital question:

> Why should the state – characterized by 150 years of accumulating authority, proprietary rights and legitimacy – be interested in decentralizing and devolving real power? Can the assumption of a strong state committed to just and sustainable social development and convinced of the economic, social and environmental benefits of devolving control over water resources and rationally experimenting with different approaches, really be accepted at face value?[79]

Interestingly, colonial technological interventions led to the replacement of community management by the irrigation juggernaut. The imperative of canal building led to the total bureaucratization of the irrigation system and exclusion of the vast majority of the local community from irrigation management. The bureaucratization of canal management was to a large extent comparable to the existing bureaucratization of civil administration. The contrast in approach of the colonial rulers to failing tank irrigation systems and new technological interventions is being replayed and replicated by the neo-liberal state as is evidenced by the decentralization (WUAs) and centralizing (water regulatory authorities)

78 David Mosse, *The Rule of Water: Statecraft, Ecology and Collective Action in South India* 269 (New Delhi: Oxford University Press, 2003).

79 *Ibid.*, 270.

strategy. Deriving from this, one is tempted to ask the question: do states – colonial or modern – resort to the community ownership paradigm only when they stand to gain from such a transfer? In other words, when the space for private and state investment is limited, the arena is enthusiastically thrown open to community ownership.

2. *Legal framework – Strengthening existing regimes*

According to Hooja, the goal of PIM is 'improved and integrated management of WUAs at various levels with the concerned State Governmental agencies'.[80] Implicit in the idea of WUAs is the principle of voluntary democratic collectives with the power to design their own rules and bye-laws. The enactments pre-empt this voluntary nature by making it mandatory for all water users to be members of the association. Though touted as PIM, what the legislation in reality does is put in place mandated groups of people with the commonality of being water users to ensure efficient and sustainable distribution of irrigation water. The enactments do not necessarily focus on 'rights' or even equitable distribution of the water resources. A related question that needs to be raised is whether the decentralization envisaged under the new crop of WUA legislations is truly democratic. Does it take into account the lack of homogeneity amongst water users (gender, class and caste divisions) so as to be truly representative? How does one incorporate some of the lessons from field experiences of WUAs into the legal framework?

Community involvement and participatory management implies arrangements wherein farmers can function as coequals in a participatory mode. For this, a number of changes will have to be made in the conceptualization of the laws and rules. Some essential changes are crucial to give direction to the next generation of laws for WUAs. Firstly, farmers should have the right to water allocation through the agreement, thus ensuring reliability, timeliness and assured supply of water. Without this, an effective and strong PIM framework may indeed prove unworkable. In addition, the overriding principles of accountability and transparency need to be incorporated both in letter and spirit. Farmers are given many responsibilities, but not the authority or the resources to enforce decisions. They ought to have the powers to hold irrigation officials accountable for non-delivery of water at the outlet point.

Secondly, the WUAs should have the right to receive information and thereby, the right to demand information related to water availability. Thirdly, they require an independent resource base and an enabling organizational structure which represents the various interest groups. Fourthly, incorporating key principles of water conservation and conjunctive use of surface and groundwater need to be

80 Rakesh Hooja, 'Below the Third Tier: Water Users Associations and Participatory Irrigation Management in India', 1 *Indian Journal of Federal Studies* (2004), available at www.jamiahamdard.edu/cfs/jour4-1_4.htm.

woven innovatively as guiding principles into the legislation. Finally, and most importantly, the PIM framework should create a sense of ownership in respect of the irrigation system. All the above measures are essential to achieve this objective.

In drawing up an agenda for the future, laws need to be drafted not exclusively by legal experts or economists. Community organization such as the WUAs would benefit immensely from social scientists being involved. As Alice Jacob notes, legislation is only one of the means for initiating social change:

> For laws to be successful as initiators of social change, the target group for whose benefit the laws create institutions for achieving their participation, must be made aware of the existence of such laws and the rules made thereunder. A great onus lies on the irrigation development administrators to enlist the active and willing cooperation of the farmers whose destiny the irrigation development plans to change. For this, the attitude and perception of the irrigation departments and other state agencies must undergo a sea change. Their attitude must be one of giving 'power to the people' rather than one of exercising 'power over the people'.[81]

CONCLUSION

It is important to take a broad view of the legal changes that the reforms seek to achieve. In focusing on the areas that need reforms, more particularly on the water rights framework, a distinction between private and public goods that accrue as a result of surface irrigation works is necessary. 'This is so even when water for irrigation is viewed as an economic good, to be priced in relation to volume consumed or benefit derived thereof.'[82] It is important that the 'public goods' aspect finds prominence in the policy statements of the legal reforms.

The problem of equity continues to haunt the sector. It is imperative that newer institutions are envisioned with a clear strategy for their integration into existing socio-political structures. Presently, the reality of the irrigation reforms is characterized by ill-informed, unmotivated WUAs, an elite capture of power of the newly formed decentralized units, a bureaucracy that is opposed to change and an overall top-down approach to changes that are being effected. It is important that we look beyond the limited donor-driven vision of reforms to incorporate issues of democratic participation, human rights, conservation, environment and sustainability in tandem with the goals of participatory management, tradable water rights and cost-recovery concerns. It must be firmly stated here that the traditional goals of irrigation, namely flood and famine control, food security and

81 Jacob and Singh, note 25 above, 232.

82 S.J. Phansalkar, Private Sector Participation in Financing and Managing Surface Irrigation: Chasing a Mirage? (New Delhi: International Development Enterprises (India), Working Paper Series on Agriculture and the Poor No. 38, 2005).

water management for agricultural purposes are goals that have been largely unmet and continue to be relevant, even today.

Finally, the reform process, including the legal reforms, must reflect greater diversity. Donor-driven reforms must have a more participatory approach with greater engagement of the community and civil society. Legal reforms in the last decade have focused on cost recovery, transfer of irrigation management to farmers and corporatization of public sector agencies. While better governance is a desirable goal, issues of equity, sustainability and social justice need to be an essential component in the next stages of the reform process. Greater participation, transparency and accountability have been a continuing demand of civil society in all post-independence reform processes. That demand is germane to the ongoing irrigation sector reforms too. More importantly, in the zeal to experiment with newer paradigms of development and growth, the wisdom of traditional knowledge systems and insights gained over several decades of functioning of the surface irrigation systems ought not to be lost sight of.

Part 3

Human Rights, Social, Health and Environmental Aspects

XI. International human rights aspects of water law reforms

Alix Gowlland Gualtieri

INTRODUCTION

As seen in previous chapters of this book, water law is made up of a number of different elements comprising economic, environmental, agricultural as well as human rights dimensions. In that '[a]ccess to water for life is a basic human need and a fundamental human right',[1] the human rights dimension of water appears as one of the core concerns from a legal perspective.[2] On the international level, the rising interest in entrenching water as a basic and self-standing human right has seemingly paralleled the trend towards what can be generally referred to as the 'commodification' of the water sector observed in many countries including India.[3] Indeed, the water sector and water law reforms initiated by states in the face of perceived scarcity of water resources[4] have typically aimed to attain efficiency and financial cost recovery while giving more marginal notice to social and human rights dimensions of water. In India, while the new water law and policy frameworks have resulted in sweeping reforms to domestic water provision, both in urban and rural areas, these frameworks have so far failed in ensuring access to safe and accessible water for all parts of the population.[5] It has become necessary to determine more precisely what the human right to water says about the impacts of these reforms on access to safe drinking water since a rights-based approach is a fundamentally effective means to fulfil the water needs of populations as a whole, including the poorest and most marginalized groups. Emblematic of this renewed interest, the United Nations Human Rights Council (HRC) held its first

1 United Nations Development Programme, *Human Development Report 2006 – Beyond Scarcity: Power, Poverty and the Global Water Crisis* v (New York: UNDP, 2006).

2 Accordingly, emphasis is put in this chapter on water for personal and domestic uses (i.e., water for drinking, washing clothes, food preparation and for personal and household hygiene; drinking water is a particularly fundamental aspect of the human right to water).

3 For some of the earlier attempts to do so, see, e.g., Peter Gleick, 'The Human Right to Water', 1 *Water Policy* 487 (1999); Stephen McCaffrey, 'A Human Right to Water: Domestic and International Implications', 5 *Georgetown Intl Envtl L. Rev.* 1 (1992).

4 See, e.g., UNDP Human Development Report, note 1 above 2–3.

5 See Chapter VIII.

session dedicated to the human right to water in September 2007. In her report, the High Commissioner for Human Rights declared:

> It is now time to consider access to safe drinking water and sanitation[6] as a human right, defined as the right to equal and non-discriminatory access to a sufficient amount of safe drinking water for personal and professional uses – drinking, personal sanitation, washing of clothes, food preparation and personal and household hygiene – to sustain life and health.[7]

To date, there has been little discussion of the practical implications of the recognition of a right to water from the perspective of the state and of the compatibility of water law reforms with international human rights obligations. This chapter focuses on the way in which the international human rights framework deals with those most pressing and problematic issues that have emerged in recent water law reforms. It is composed of three main sections. Section A briefly presents the international legal framework for a right to water. Section B examines the contours of the scope of the right as set out in existing instruments. Section C then turns to more specific issues that have arisen in recent water law reform, namely their economic thrust; the participation of private actors; some of the implications of cost recovery policies such as water services disconnections; the devolution of certain state responsibilities to local authorities and other communities of users; and the question of prioritization of domestic uses. It sets these issues against the backdrop of the right to water in an international framework.

A. The international legal framework

1. *Treaties*

The fact that the main human rights treaties – the Universal Declaration of Human Rights,[8] the International Covenant on Civil and Political Rights (ICCPR),[9] and the International Covenant on Economic, Social and Cultural Rights (ICESCR)[10] – do not explicitly mention a right to water initially generated a debate over its legal status. In view of the close relationship that exists between access to water and their fulfilment, the existence of the right to water has,

6 We are concentrating here on the human right to water. However, this right is often linked to the right to adequate sanitation.

7 Report of the United Nations High Commissioner for Human Rights on the Scope and Content of the Relevant Human Rights Obligations Related to Equitable Access to Safe Drinking Water and Sanitation Under International Human Rights Instruments, UN Doc. A/HRC/6/3 (2007), para. 66.

8 Universal Declaration of Human Rights, UN General Assembly Resolution 217 (III), UN GAOR, 3d Sess., Supp. No. 13, UN Doc. A/810 (1948) 71.

9 International Covenant on Civil and Political Rights, UN Doc. A/6316 (1966), 999 *UNTS* 171.

10 International Covenant on Economic, Social and Cultural Rights UN Doc. A/6316 (1966), 993 *UNTS* 3.

however, been extrapolated from a range of other rights found in human rights treaties. The nexus with the right to life, protected under Article 6(1) of the ICCPR, is a particularly fundamental one in view of the acute link between lack of access to adequate water and disease – some 1.8 million children die each year as a result of diarrhoea due to unclean water.[11] In this regard, the UN Human Rights Committee has stressed that, in addition to protecting against the active taking of life, the right to life also requires states to adopt positive measures, in particular to reduce infant mortality, increase life expectancy, and eliminate malnutrition and epidemics.[12] A second fundamental connection exists between water and economic, social and cultural rights protected under the ICESCR. General Comment No. 15 on the right to water, adopted by the UN Committee on Economic, Social and Cultural Rights at its twenty-ninth session in 2002, underlines that water constitutes a fundamental element for securing amongst other rights those to an adequate standard of living and to the highest attainable standard of health.[13] Access to safe drinking water also constitutes an important element for the realization of the right to food as evidenced by the mandate of the Special Rapporteur on the right to food which was extended to also cover water.[14] One can also mention here the right to adequate housing[15] and the right to education,[16] both of which have been deemed to rely on the fulfilment of the right to water.

This recognition of the existence of a right to water under the two Covenants is significant in terms of imposing a single standard of implementation to which all states can be held and to ensure its enforceability. This is particularly noteworthy in view of the recent adoption by the UN General Assembly of the Optional Protocol to the ICESCR, which makes the Committee on Economic, Social and Cultural Rights competent to receive and consider individual communications.[17] However, the justiciability of the right in international and domestic courts is more effective when based on an explicit right to water than one merely implied. An illustration of this can be drawn from the national context: the South African legal system explicitly confers a justiciable, affirmative right of access to adequate

11 UNDP Human Development Report, note 1 above, 3.

12 Committee on Economic, Social and Cultural Rights, General Comment No. 6, The Right to Life (Art. 6) UN Doc. HRI/GEN/1/Rev.8 (1982), para. 5. See also Concluding Observations of the Human Rights Committee: Honduras, UN Doc. CCPR/C/HND/CO/1 (2006), para. 15; Report of the Committee against Torture, GAOR Supplement No. 44 (A/56/44), paras. 181 and 183.

13 Committee on Economic, Social and Cultural Rights, General Comment No. 15: The Right to Water (Articles 11 and 12 of the International Covenant on Economic, Social and Cultural Rights), UN Doc. E/C.12/2002/11 (2002). See Section I(B)(1) below.

14 Commission on Human Rights, Resolution 2001/25, The Right to Food (2001), para. 9.

15 Committee on Economic, Social and Cultural Rights, General Comment No. 4, The Right to Adequate Housing (Art. 11(1) of the Covenant), UN Doc. HRI/GEN/1/Rev.8 (1991), para. 8(b).

16 Committee on Economic, Social and Cultural Rights, General Comment No. 13, The Right to Education (Art. 13), UN Doc. E/C.12/1999/10 (1999), para. 6(a).

17 UN General Assembly Resolution 63/117, Optional Protocol to the International Covenant on Economic, Social and Cultural Rights, UN Doc. A/RES/63/117 (2008).

water which has been upheld by the Constitutional Court.[18] In contrast,[19] the Indian model has derived an implicit, justiciable right to water from the broader 'right to life' embodied in Article 21 of the country's Constitution.[20] In both cases an economic and social right has been made justiciable, but in the former the right to water is directly acknowledged at the constitutional level while in the latter it has been implicitly derived.

Explicit obligations in relation to access to safe drinking water have been increasingly included in more recent human rights treaties, such as those on the rights of the child,[21] the elimination of discrimination against women,[22] and the rights of persons with disabilities.[23] It is noteworthy that several civil society groups have been pressuring states to adopt an international treaty on the right to water.[24] With regard to regional human rights instruments, the right to water is typically found in treaties adopted under the auspices of the Organization of African Unity (OAU).[25] Humanitarian law treaties also entail specific provisions related to access to water. The Geneva Conventions and their additional protocols outline the fundamental importance of access to safe drinking water for health and survival in situations of international and non-international armed conflicts.[26]

Water is a cross-cutting issues, and consequently it falls also in the realm of

18 *South Africa v. Grootboom* 2000 (11) BCLR 1169 (CC). See further Sandra Liebenberg, 'The Value of Human Dignity in Interpreting Socio-Economic Rights', 21 *South African J. Human Rts* 1 (2005); Mia Swart, 'Left out in the Cold? Crafting Constitutional Remedies For the Poorest of the Poor', 21 *South African J. Human Rts* 215 (2005).

19 Comparing both systems, Anonymous, 'What Price for the Priceless? Implementing the Justiciability of the Right to Water', 120 *Harvard L Rev* 1067 (2007).

20 See, e.g., *Francis Coralie Mullin v. Administrator*, Union Territory of Delhi, (1981) 2 SCR 516; *Attakoya Thangal v. Union of India* (1990) 1 KLT. 580, 583; *A.P. Pollution Control Board II v. M.V. Nayudu* 2000(3) SCALE 354, available at www.ielrc.org/content/e0010.pdf; *Vellore Citizens' Welfare Forum v. Union of India* (1996) 5 SCC 647, 660. See further S. Muralidhar, 'The Right to Water: An Overview of the Indian Legal Regime', in Eibe Riedel and Peter Rothen (eds.) *The Human Right to Water* 65 (Berlin: Berliner Wissenschafts-Verlag, 2006).

21 Convention on the Rights of the Child, 20 November 1989, 1577 *UNTS* 3, Arts. 1 and 2(c).

22 Convention on the Elimination of All Forms of Discrimination against Women, 18 December 1979, 1249 *UNTS* 13, Art. 14(2)(h).

23 Convention on the Rights of Persons with Disabilities, UN Doc. A/Res/61/106 (13 December 2006), Art. 28(2)(a).

24 Fundamental Principles for a Framework Convention on the Right to Water (May 2005), available at www.watertreaty.org/convention.php.

25 African Charter on the Rights and Welfare of the Child, OAU Doc. CAB/LEG/24.9/49 (1990), Arts. 1 and 2(c); Protocol to the African Charter on Human and Peoples' Rights on the Rights of Women in Africa, CAB/LEG/66.6 (2000), Art 15(a).

26 Geneva Convention (III) relative to the Treatment of Prisoners of War, 1949, 75 *UNTS* 135, Arts. 20, 26, 29 and 46; Geneva Convention (IV) relative to the Protection of Civilian Persons in Time of War, 1949, 75 *UNTS* 287, Arts. 89 and 127; Protocol Additional to the Geneva Conventions of 12 August 1949, and relating to the Protection of Victims of International Armed Conflicts (Protocol I), 1977, 1125 *UNTS* 3, Arts. 54 and 69; Protocol Additional to the Geneva Conventions of 12 August 1949, and relating to the Protection of Victims of Non-International Armed Conflicts (Protocol II), 1977, 1125 *UNTS* 609, Arts. 14 and 18.

environmental law and policy.[27] Two international environmental treaties refer to the right to water although they do not provide much guidance on the issue. The UN Watercourses Convention mentions that in the event of a conflict between uses of an international watercourse, specific regard should be given to the requirements of vital human needs, defined as water to sustain human life, including drinking water and water required for production of food in order to prevent starvation.[28] This can be interpreted as the mention of an explicit right to water.[29] On the regional level, the Protocol on Water and Health adopted by the UN Economic Commission for Europe (UNECE) provides that states parties have the obligation to take appropriate measures to provide access to drinking water and sanitation and to protect drinking water from pollution.[30]

Thus, human rights treaties entail both explicit and implicit obligations regarding the right to water and sanitation, and obligations are also found in humanitarian and environmental law treaties. As further explored in the next sections, the right to water has been acknowledged by several expert human rights bodies, most notably the Committee on Economic, Social and Cultural Rights, the Sub-Commission for the Promotion and Protection of Human Rights and the HRC. It is also acknowledged by states in several resolutions, declarations and plans of action.

2. *Contributions of United Nations human rights bodies*

General Comment No. 15

Several instruments adopted on the international level in recent years – especially by UN human rights bodies – have firmly acknowledged the existence of the right to water,[31] and access to water is now considered a fundamental right, essential for human life and dignity.[32] Those instruments that set out the right

27 See Chapter XII.

28 United Nations Convention on the Law of the Non-navigational Uses of International Watercourses, 36 *ILM* 715 (1997) and Report of the Sixth Committee convening as the Working Group of the Whole, UN Doc. A/51/869 (1997), Art. 10(2).

29 Malgosia Fitzmaurice, 'General Principles Governing the Cooperation between States in Relation to Non-Navigational Uses of International Watercourses', 14 *Yb. Int'l Envtl. L.* 3, n. 544 (2005).

30 Protocol on Water and Health to the 1992 Convention on the Protection and Use of Watercourses and International Lakes, UN Doc. MP.WAT/2000/1, EUR/ICP/EHCO 020205/8Fin (1999).

31 In particular Malcolm Langford, 'Ambition that Overleaps Itself? A Response to Stephen Tully's "Critique" of the General Comment on the Right to Water', 26 *Netherlands Quarterly of Human Rts* 433 (2006); Stephen McCaffrey, 'The Human Right to Water', in Edith Brown Weiss, Laurence Boisson de Chazournes and Nathalie Bernasconi-Osterwalder (eds.) *Fresh Water and International Economic Law* 93 (Oxford: Oxford University Press, 2005).

32 See, e.g., Pierre-Marie Dupuy, 'Le droit à l'eau, un droit international?', in Guillaume Grisel (ed.), *La mise en oeuvre du droit à l'eau – Actes du XXIXème Congrès ordinaire de l'IDEF* 277 (Genève: Schulthess, 2006). Explicit rights language with regard to water is found also in national jurisdictions where a growing number of countries are incorporating the right to water in constitutions and legislative frameworks. Chapter II in this volume.

to water as a self-standing right are summarily introduced in the next section, with its scope and content taken up in more detail in subsequent parts of the chapter.

General Comment No. 15 represented the first official recognition by a UN human rights expert body of the human right to water and undoubtedly makes a step towards its transformation into a customary right.[33] It also heralded the increasing weight given to the right to water as a self-standing human right in policy statements of states and international organizations, as well as civil society organizations and the private sector. For instance, the Millennium Project Task Force has recommended that the international community explore ways to use General Comment No. 15 when setting up national policies on water.[34]

General Comment No. 15 represents an authoritative interpretation by the Committee of Articles 11, paragraph 1, and 12, paragraph 1, of the ICESCR (rights to an adequate standard of living and to the highest attainable standard of health). It also takes into account references to the right to water and substantive obligations thereunder contained in a variety of international treaties and declarations, including international environmental and humanitarian law treaties. Briefly, the Comment provides that the right to water entitles everyone to 'sufficient, safe, acceptable, physically accessible and affordable water for personal and domestic uses' and 'access to adequate sanitation'.[35] It specifies that state parties have a core obligation to ensure 'the satisfaction of, at the very least, minimum essential levels' in the provision of water services.[36] Some measures can be adopted by states immediately, such as putting in place a concrete plan of action with timelines for achievement and to eliminate discrimination. Other steps are to be realized progressively. Governments are obliged to use available resources effectively in a concrete and targeted manner in programmes to ensure that all have access to water.[37] Where there is deliberate movement backwards on achieving the rights, a government must prove that this occurred after all alternatives were considered and that it had used its available resources to the maximum extent.[38] While the Comment is non-legally binding, the Committee on Economic, Social and Cultural Rights, however, expects states to indicate in reporting on their implementation of the ICESCR the progress made in achieving the full realization of the right to water.

33 See further Eibe Riedel, 'The Human Right to Water and General Comment No. 15', in Eibe Riedel and Peter Rothen, *The Human Right to Water* 19 (Berlin: Berliner Wissenschafts-Verlag, 2006). But see Stephen Tully, 'A Human Right to Access Water? A Critique of General Comment No. 15', 23 *Netherlands Quarterly of Human Rts* 35, 37 (2005).

34 Millennium Project Task Force, note 53 above, 178.

35 General Comment No. 15, note 13 above, para. 2.

36 *Ibid.*, at para. 37.

37 General Comment No. 15, note 13 above, paras. 17–18.

38 *Ibid.*, at para. 19.

Reports of the United Nations Sub-Commission on the Promotion and Protection of Human Rights

In 2004, the UN Sub-Commission on the Promotion and Protection of Human Rights conducted a study of the legal basis and implications of the right to water and sanitation.[39] The Report provides in particular that '[t]he right to drinking water and sanitation constitutes a part of internationally recognized human rights and may be considered as a prerequisite to the realization of other rights'.[40]

Two years later, the Sub-Commission adopted its Guidelines for the Realization of the Right to Drinking Water Supply and Sanitation.[41] They are intended to assist governments, policymakers, international agencies and members of civil society to implement the right to water. They are consistent with General Comment No. 15 and provide concrete guidance on the implementation of the right to water.

Procedures under the United Nations Human Rights Council

One of the first decisions of the HRC was to request the Office of the High Commissioner for Human Rights to carry out a study on human rights obligations related to access to safe drinking water and sanitation.[42] The resulting report was adopted by the Council in September 2007.[43] At its seventh session in March 2008, the HRC reiterated that 'international human rights law instruments [. . .] entail obligations in relation to access to safe drinking water'.[44] The draft resolution proposed by Germany and Spain, identifying water as a global human right, was turned down in favour of a more watered-down text[45] which replaced the rights-based terminology with that of 'human rights and access to

39 UN Sub-Commission on the Promotion and Protection of Human Rights, Final report of the Special Rapporteur on the relationship between the enjoyment of economic, social and cultural rights and the promotion of the realization of the right to drinking water supply and sanitation (14 July 2004), UN Doc. E/CN.4/Sub.2/2004/20. Other special rapporteurs have looked at the relevance to their mandates of the issue of access to safe drinking water, including those on the right to food, the right to health, and the right to adequate housing.

40 *Ibid.*, at 2.

41 UN Sub-Commission on the Promotion and Protection of Human Rights, Res. 2006/10, Promotion of the realization of the right to drinking water and sanitation, 24 August 2006, UN Doc. A/HRC/Sub.1/58/L11, adopting the Draft Guidelines for the realization of the right to drinking water and sanitation, UN Doc. E/CN.4/Sub.2/2005/25 (2005).

42 HRC, Implementation of General Assembly Resolution 60/251, Human Rights and Access to Water, UN Doc. A/HRC/2/L.3/Rev.3 (21 November 2006).

43 UNHCHR Report, see note 7 above.

44 HRC, Human Rights and Access to Safe Drinking Water and Sanitation, UN Doc. A/HRC/7/L.16 (2008).

45 Centre on Housing Rights and Evictions (COHRE), Press Release – UN Human Rights Council Moves Forward on the Right to Water and Sanitation (2008), available at www.cohre.org/store/attachments/RWP%20%20Information_note_on_HRC_2008_Resolution%2031%2003%2008_Final.pdf.

safe drinking water'. In the resolution, the Council decided to appoint an Independent Expert to investigate the issues of 'human rights obligations related to access to safe drinking water and sanitation' over the next three years. The expert's task is to undertake a study on best practices and human rights obligations related to water and sanitation, and submit a final report to the Council at its tenth session.[46] This is a significant development, firstly because it creates within the UN human rights system a separate mechanism dealing exclusively with issues related to the right to water, and secondly because it reinforces states' obligations to ensure access to safe drinking water under international human rights law.

3. Declarations, resolutions and other non-binding instruments

International law sources of human rights obligations include declarations, resolutions, principles and guidelines. While these do not have binding force, they may contain elements that contribute to, or may feed into, the customary law process. Also relevant in this regard are international plans of action and documents adopted under the auspices of UN conferences.

The first document adopted at an international conference to acknowledge the right to water was the Mar del Plata Action Plan of 1977. Its preamble provides that irrespective of the level of development, all people 'have the right to have access to drinking water in quantities and of a quality equal to their basic needs'.[47] Agenda 21, adopted at the UN Conference on Environment and Development (UNCED) in 1992 provides that the commonly agreed premise was that 'all peoples, whatever their stage of development and their social and economic conditions, have the right to have access to drinking water in quantities and of a quality equal to their basic needs'.[48] The right to water is also found in the Programme of Action of the International Conference on Population and Development[49] and the UN–Habitat Plan of Action.[50] The UN General Assembly has also referred to safe drinking water as a human right.[51] Conversely, the right to water in itself is not mentioned in the UN Millennium Declaration,[52]

46 The Independent Expert, Catarina de Albuquerque, commenced work in November 2008.

47 Mar del Plata Action Plan, UN Water Conference, Mar del Plata, Argentina (1977).

48 Agenda 21, in Report of the United Nations Conference on Environment and Development, Rio de Janeiro, UN Doc. A/CONF.151/26/Rev.1 (Vol. 1), Annex II (1992), Chapter 18, para. 47.

49 Report of the International Conference on Population and Development, Annex, Programme of Action, Cairo, Egypt, UN Doc. A/CONF.171/13 (1994), princ. 2.

50 The Habitat Agenda, Commitments and the Global Plan of Action, UN Conference on Human Settlements (Habitat II), UN Doc. A/CONF.165/14 (1996), para. 11.

51 See, e.g., UN General Assembly, The Right to Development, UN Doc. A/RES/54/175 (15 February 2000), para. 12(a).

52 UN General Assembly, United Nations Millennium Declaration, UN Doc. A/RES/55/2 (18 September 2000), para. 19.

although the UN Millennium Project Task Force on Water and Sanitation has noted the importance of the right to water and sanitation for achieving the Millennium Development Goals (MDGs).[53]

The 2006 UNDP Human Development Report underlines the importance of adopting a rights-based approach to the provision of safe drinking water and stresses that access to water for life is a basic human need and a fundamental human right. Obligations related to the right to water are found in other international guidelines and policy documents which highlight the fact that access to water in sufficient quantity and quality for all is essential for life and health.[54] On the regional level, the Committee of Ministers of the Council of Europe has stated that everyone has the right to a sufficient quantity of water for his or her basic needs.[55] In November 2008, at the Third South Asian Conference on Sanitation (SACOSAN III) held in New Delhi, the eight SACOSAN member states[56] adopted the Delhi Declaration which reaffirms that access to safe drinking water constitutes a basic human right.[57] This is in contrast to the eThekwini Declaration adopted at the Second African Conference on Sanitation and Hygiene (AfricaSan+5) in February 2008.[58]

B. Content and scope

1. Content

General Comment No. 15 and the other international instruments described above set out several components of the right to water. Each person must have access to water that is sufficient, safe and accessible, both physically and economically.

53 UN Millennium Project Task Force on Water and Sanitation, 'Health, Dignity and Development: What Will it Take?' xiv (2005), available at www.unmillenniumproject.org/documents/WaterComplete-lowres.pdf.

54 See, e.g., First United Nations Congress on the Prevention of Crime and the Treatment of Offenders, Standard Minimum Rules for the Treatment of Prisoners (1955), Arts. 15 and 20 (2); UN General Assembly Res. 45/113, United Nations Rules for the Protection of Juveniles Deprived of their Liberty (14 December 1990), Arts. 34 and 37; Food and Agriculture Organization (FAO) Council, 127th Session, Voluntary Guidelines to Support the Progressive Realization of the Right to Adequate Food in the Context of National Security (November 2004), paras. 3.6 and 8.1.

55 Recommendation 14 of the Committee of Ministers of the Council of Europe to member states on the European Charter on Water Resources (17 October 2001), para. 5.

56 India, Afghanistan, Bangladesh, Bhutan, Maldives, Nepal, Pakistan and Sri Lanka.

57 The Delhi Declaration, Third South Asian Conference on Sanitation (SACOSAN), 'Sanitation for Dignity and Health', New Delhi, India (16–21 November 2008), para. 1.

58 The eThekwini Declaration, Second African Conference on Sanitation and Hygiene (AfricaSan+5), Durban, South Africa (18–20 February 2008).

Availability

According to General Comment No. 15, the availability of water means that each person must benefit from a sufficient and continuous water supply for personal and domestic uses.[59] These uses ordinarily include drinking, personal sanitation, washing of clothes, food preparation, personal and household hygiene. Some individuals and groups may require additional water due to health, climate, and work conditions. While international opinion has not settled on a precise amount that would satisfy the human right to water,[60] World Health Organization (WHO) guidelines recommend the amount of 100–300 litres per capita per day to ensure optimal access.[61] In practice, countries have determined themselves the minimum reasonable amount of water needed to cover personal and domestic uses. South Africa's water policy, for instance, provides each person with the basic amount of 25 litres of free water per day.[62] This amount has been judged by many experts too low to fulfil basic needs, and in 2008, the High Court of South Africa requested that the city of Johannesburg provided residents of Phiri, Soweto, with 50 litres of free basic water per person per day.[63] The Court noted that the City had the available water and financial resources to provide this amount, including funds provided by the national government for water provision that the City had so far chosen not to use for the benefit of the poor.

Quality

The human right to water also means the water must be of sufficient quality, which is essential to ensure that the water provided does not adversely impact human health. Water must be safe, free from micro-organisms, chemical substances and radiological hazards that constitute a threat to a person's health.[64] Furthermore, water should be of an acceptable colour, odour and taste for each personal or domestic use. The United Nations High Commissioner for Human Rights (UNHCHR) Report offers a reminder that access to adequate sanitation constitutes one of the principal mechanisms for protecting the quality of drinking water.[65]

59 General Comment No. 15, note 13 above, para. 12(a).

60 UNHCHR Report, note 7 above, para. 15.

61 Guy Howard and James Bartram, Domestic Water Quantity, Service Level and Health, World Health Organization 22 (Geneva: World Health Organization, 2003).

62 South Africa, Department of Water Affairs and Forestry (DWAF), Free Basic Water Implementation Strategy Document (2001). *See further* Alix Gowlland Gualtieri, 'South Africa's Law and Policy Water Framework: Implications for the Right to Water', in Philippe Cullet, Alix Gowlland Gualtieri, Roopa Madhav and Usha Ramanathan (eds.) *Water Law at the Crossroads – National and International Perspectives with Special Emphasis on India* 389 (New Delhi: Cambridge University Press, 2009); Michael Kidd, 'Not a Drop to Drink: Disconnection of Water Services for Non-Payment and the Right of Access to Water', 20 *South African J. Human Rts* 119 (2004).

63 *Lindiwe Mazibuko v. City of Johannesburg*, Case No. 06/13865, High Court of South Africa (Witswatersrand Local Division) (2008), para. 183.

64 General Comment No. 15, note 13 above, para. 12(b).

65 UNHCHR Report, note 7 above, para. 16.

Other international instruments include commitments by states regarding specific quality requirements. According to the WHO, water must represent no health risks, and must be free of toxic chemical substances and microbiological pathogens.[66] The pollution or contamination of water resources that will affect human health is prohibited.

In order to ensure the salubrity of water in rural and urban areas alike, it is essential that quality standards are contained in binding instruments. In India, the quality of water remains problematic in many cases. In Gujarat, for instance, water in as many as 38 per cent of habitations was found not to comply with WHO Guidelines with respect to fluoride content –23.6 per cent was found to be above the maximum permissible limit.[67] Despite this, existing standards are not generally binding and often constitute only best practice for agencies concerned with the supply of potable water.[68]

Accessibility

The source of safe, basic water must be accessible to all. Accessibility takes several forms. First, water must be physically accessible, which requires that water be provided in, or in close proximity to, the home, school or workplace, and that accessing that water poses no threat to physical security.[69] According to the WHO, a water source with the capacity to provide sufficient, safe and regular water should normally be within less than 1,000 metres of the household and collection time for around 20 litres a day of water should not exceed thirty minutes.[70]

The UNHCHR Report specifies that human rights obligations do not require states to provide access in every house, but that water and sanitation facilities have to be in close proximity to each house, educational institution and workplace and must be within safe reach for all sectors of the population, taking into account the needs of particular groups, including persons with disabilities, children, the elderly, and women.[71] Water reforms in India have not improved the physical accessibility of drinking water in urban areas, since a majority of households do not have access to water within their dwelling and have to transport it from the main source.[72]

Second, water must be economically accessible, that is, affordable and not reducing a person's capacity to buy other essential goods.[73] The UNHCHR Report

66 WHO 3rd edn Guidelines for Drinking-Water Quality (Geneva: WHO, 2006).

67 See Chapter VIII.

68 See also Chapter XIII..

69 General Comment No. 15, note 13 above, para. 12(c)(i); UNHCHR Report, note 7 above, para. 25.

70 WHO 2003, note 61 above, 22.

71 *Ibid.* See also Sub-Commission Guidelines, note 41 above, para. 1.3 (a)–(c).

72 See further Chapter VIII.

73 General Comment No. 15, note 13 above, para. 12(c)(ii); Sub-Commission Guidelines, note 41 above, para. 1.3(d).

specifies that '[a]ffordability requires that direct and indirect costs related to water and sanitation should not prevent a person from accessing safe drinking water and should not compromise his or her ability to enjoy other rights, such as the right to food, housing, health and education. These costs include both connection and delivery costs'.[74] This condition has become particularly critical in the face of the tendency of water law reforms to implement full cost recovery policies (see section III(1) below).

Third, water must be equally accessible.[75] The UNHCHR Report finds that considerations of equity are often broader than the question of how to distribute water expenses, and that equitable access considered within a human rights framework refers to equal and non-discriminatory access.[76] Discriminatory policies, laws or practices that affect individuals' access to safe drinking water, including discrimination on the basis of housing or land status, are thus prohibited, and states should ensure that no population group is excluded.[77]

Finally, the human right to water requires democratic accessibility, that is, accessibility to information concerning water and water services, whether held by public authorities or third parties and public participation in decision making regarding water provision.[78] When individuals or groups have been denied their right to water, they should have access to effective judicial or other appropriate remedies, including adequate reparation (restitution, compensation, satisfaction or guarantees of non-repetition).[79] In particular, the poor and members of vulnerable groups are frequently excluded from decision making regarding water, which results in their inequitable access to water services. Participatory and procedural safeguards should thus be integrated in the design and implementation of water laws and policies.[80] In the municipality of Porto Alegre, Brazil, citizens can have a say on which new investments should be made first.[81] This model has contributed to dramatic increases in access to water by poor communities.

74 UNHCHR Report, note 7 above, para. 28.

75 General Comment No. 15, note 13 above, paras. 12(c)(iii) and 13–16; Sub-Commission Guidelines, note 41 above, para. 3. See also ICESCR, note 10 above, Art. 2(2).

76 UNHCHR Report, note 7 above, para. 22.

77 See, e.g., United Kingdom, Water Industry Act 1999, Part I, s. 5(3)(a) (Stating that subsidies can be provided by private suppliers to certain vulnerable groups on the basis of age, ill-health or financial circumstances).

78 General Comment No. 15, note 13 above paras. 12 (c)(iv) and 48; Sub-Commission Guidelines, note 41 above, paras. 8.1–8.3. See further Philippe Cullet and Alix Gowlland Gualtieri, 'Local Communities and Water Investments', in Brown Weiss, Boisson de Chazournes and Bernasconi-Osterwalder note 31 above, 303.

79 General Comment No. 15, note 13 above, paras. 55–59; Sub-Commission Guidelines, note 41 above, para. 9.

80 *Ibid.*, para. 48.

81 Centre on Housing Rights and Evictions (COHRE), Case Studies on Efforts to Implement the Right to Water and Sanitation in Urban Areas: Brazil, Kenya, Sri Lanka and South Africa 9–10 (Geneva: COHRE, 2008), available at www.cohre.org/store/attachments/RWP%20%20Case%20studies%20four%20countries%2020%2003%2008.pdf.

2. *Duty bearers*

The Committee on Economic, Social and Cultural Rights has identified three categories of obligations incumbent upon states having ratified the ICESCR, namely those to respect, protect and fulfil. The obligation to respect requires states to refrain from arbitrarily interfering with a person's access to water.[82] It includes, for instance, the obligation to avoid engaging in any practice or activity that denies or limits access to safe drinking water, or that pollutes water.[83]

The obligation to protect requires states to prevent third parties from interfering in any way with the enjoyment of the right to water, unless such interference is permitted by law and includes appropriate procedural protection.[84] Third parties include individuals, groups, corporations and other entities as well as agents acting under their authority. This obligation includes, for instance, the adoption of necessary and effective legislative and other measures to prevent third parties from denying access to safe drinking water.[85] The obligation to fulfil requires that governments adopt the necessary measures directed towards the full realization of the right to water to the maximum available resources.[86] In this sense, it is of a more positive nature than the two categories' obligations mentioned above, which essentially prohibit governments and other parties from infringing upon the right to water. The obligation to fulfil implies that governments must provide a basic quantity of water regardless of the expense involved or the recipients' ability to pays. As part of the obligation to fulfil, the Committee on Economic, Social and Cultural Rights has highlighted those procedural obligations mentioned above.[87]

There is a tendency for national legislative systems to focus on the obligations to respect and protect social and economic rights rather than the positive obligation to fulfil. For instance, recent Indian case law on water has focused on protecting the right of clean drinking water when there is a situation of pollution of water sources by industrial causes.[88] The positive obligations of governments to fulfil the right to water has proven more difficult to uphold in domestic courts. South Africa is one exception to this, with the justiciability of water rights claims embedded in the Constitution itself.[89]

While human rights are principally concerned with obligations of governments, General Comment No. 15 calls upon states to ensure that non-state actors including civil society, non-governmental organizations (NGOs), the private sector and international organizations, are aware of, and consider the importance of, the right to water in pursuing their activities.[90] States have a responsibility to

82 General Comment No. 15, note 13 above, para. 21.
83 UNHCHR Report, note 7 above, para. 36.
84 General Comment No. 15, note 13 above, para. 23. See also Section III (2) below.
85 UNHCHR Report, note 7 above, para. 37.
86 General Comment No. 15, note 13 above, para. 25.
87 See notes 78 ff above and related text.
88 See, e.g., *D.D. Vyas v. Ghaziabad Development Authority* A.I.R. 1993 All. 57.
89 South Africa, Constitution of 1996 (Constitution Act 108, 1996), sec. 27.
90 General Comment No. 15, note 13 above, para. 49.

regulate non-state actors to ensure they do not have any negative impacts on the right to water.[91] States, as members of international organizations, should also ensure that the international organizations apply human rights in all aspects of their work, and the organizations themselves have direct obligations to take into account the right to water. General Comment No. 15 indeed notes that UN agencies and other international organizations concerned with water, sanitation, trade and humanitarian intervention should cooperate effectively with states to implement the right to water and sanitation at the national level.[92] In particular, international financial institutions should take into account the right to water in their lending policies, credit agreements, structural adjustments and other development policies.[93]

C. Implications for water law reforms

International recognition of the right to water does not negate the fact that many fundamental questions pertaining to its scope and content remain, especially in relation to the water law reforms that have been taking place in recent years. As has been detailed elsewhere in this book, these reforms have been characterized by an emphasis on the economic dimension of water, a greater role given to the private sector and generally a retreat of the state in providing water, with the result that often considerations of equity and sustainability get put aside.[94] This results in a fundamental conflict between an economic approach to water and a rights-based one. In view of this state of affairs, it is urgent to examine more closely how the challenges posed by water reforms can be addressed in relation to state obligations regarding the right to water.

1. Integration of economic policies in the provision of water

The international community first proposed treating water as an economic good in an attempt to minimize inefficiencies in the system through pricing techniques.[95] The problem is that if water is perceived solely as an economic good, then access may be determined based purely upon market forces, without regard to equity or need. From the commodification of water logically follows the application of full-cost recovery policies, that is, a system where users must pay for all aspects of water infrastructure and management. Water as an economic good rather than a public good or public trust resource that must be protected by the

91 *Ibid.*, paras. 23–24.

92 *Ibid.*, paras. 36 and 60.

93 Salman M.A. Salman and Siobhán Alice McInerey-Lankford, The Human Right to Water: Legal and Policy Dimensions 70–2 (Washington, D.C: The World Bank, 2004).

94 See further Chapters III, VI, and VIII.

95 Erik B. Bluemel, 'The Implications of Formulating a Human Right to Water', 31 *Ecology L.Q.* 957, 962 (2004). See also Dublin Statement on Water and Sustainable Development, International Conference on Water and the Environment, Dublin, 31 January 1992.

government has informed many of the ongoing water reforms, in particular in India.[96] In the rural drinking water context, for instance, the cost recovery mechanisms in place have generated concerns about the fairness and practicality of expecting all users to contribute 10 per cent of capital costs.[97] One extreme example of the application of cost recovery principles has been the institution of prepaid meters which guarantee payment for water use. In South Africa,[98] the forced installation of water meters in Phiri has been deemed unconstitutional and unlawful in a recent case of the South African High Court.[99]

The concept that water is an economic good thus often leads to inequities in the supply of, and access to, this basic resource. Clearly, then, this concept must be limited by the concept of water as a human right in order to ensure equitable distribution of, and access to, this basic and essential resource. International human rights instruments are, however, not very specific on the treatment of economic approaches to water, and limit themselves to dealing with issues related to the pricing of water under the general umbrella of 'affordability'.[100] Early on, Agenda 21 underlined that beyond the provision of safe drinking water for the satisfaction of basic needs, water users should be charged appropriately,[101] and the Johannesburg Plan of Implementation later made clear that cost recovery objectives should not become a barrier to access to safe drinking water by poor people.[102] As does General Comment No. 15,[103] the UNHCHR Report provides that the affordability requirement is not incompatible with the principle of cost recovery for water services.[104] It merely sets limits to cost recovery and highlights the fact that it should not become a barrier to access to safe drinking water, notably by the poor. It also provides that the obligation to protect the right implies safeguarding people from interference by others, such as for instance in the case of unaffordable price increases by corporations. Beyond that, the current human

96 See, e.g., Government of India, Ministry of Water Resources, *National Water Policy*, New Delhi, April, 2002, sec. 11; World Bank, India – Water Resources Management Sector Review: Report on the Irrigation Sector (Report No. 18416 IN, 1998). See further Chapter VIII.

97 See Chapter VII.

98 See, e.g., Public Citizen, Orange Farm, South Africa: The Forced Implementation of Prepaid Water Meters (June 2004).

99 *Lindiwe Mazibuko v. City of Johannesburg*, Case No. 06/13865, High Court of South Africa (Witswatersrand Local Division) (2008), para. 183.

100 The affordability requirement has been implemented quite broadly in some countries, such as in each of the three regions of Belgium. For an overview, *see* Bart Ouvry, Deputy Permanent Representative of Belgium to the United Nations, 'Access to water as a basic human right, different ways to fulfil this vital commitment, the case of Belgium and its Regions', Speaking notes for the UN Economic Commission for Europe (UNECE) Roundtable on the Human Right to Water and the Protocol on Water and Health (Geneva: UNECE, January 2007), available at http://unece.org/env/water/whmop1/roundtable/Ouvry.pdf.

101 Agenda 21, note 48 above, Chap. 18, para. 8.

102 Plan of Implementation of the World Summit on Sustainable Development, para. 26(b), UN Doc. A/CONF.199/20 (2002).

103 General Comment No. 15, note 13 above, para. 12 (c)(ii).

104 UNHCHR Report, note 7 above, para. 28.

rights framework does not provide any further indication of how cost recovery policies can be reconciled with the right to water requirements, nor does it adequately address the contradictions between cost recovery policies and the principle of affordability of water. Moreover, the focus on the 'affordability' of water carries the implication that water is priced, although in some cases the only way that water may be 'affordable' to the poor is that they receive basic water supplies for free. Thus far, while the human rights framework contemplates the possibility that safe drinking water should be provided for free in certain circumstances, in general it does not set this as a rule.[105] The idea of free water is, however, making its way on the international agenda. The recent Draft Guiding Principles on extreme poverty and human rights adopted by the UN Sub-Commission on the Promotion and Protection of Human Rights, for instance, provide that '[p]ersons living in extreme poverty have a right to drinkable water, and the State is duty-bound to provide this service to them *free of charge*'.[106] On the national level, the concept of free basic water is found in South Africa's water policy.[107]

2. *Private provision of water services*

Another characteristic of water law reforms has been the involvement of the private sector in the provision of water.[108] A related aspect is what has been referred to as the 'corporatization' of public bodies, that is, when water services owned and operated by the local governments are restructured following market principles to increase their efficiency. The privatization and corporatization of water services have been consistently supported by the World Bank, the regional development banks, and the International Monetary Fund (IMF), and have often been a condition for developing countries to access international funds for water programmes or for loan reprogramming.[109] The question of privatization is particularly relevant in India. As discussed in other chapters in this book, the Indian government has,

105 *Ibid.*

106 Draft guiding principles, 'Extreme Poverty and Human Rights: The Rights of the Poor', Resolution 2006/9, Implementation of Existing Human Rights Norms and Standards in the Context of the Fight Against Extreme Poverty, in Report of the Sub-Commission on the Promotion and Protection of Human Rights on its Fifty-Eighth Session, UN Doc. A/HRC/2/2-A/HRC/Sub.1/58/36 (2006), para. 29 (emphasis added).

107 FBW Policy, note 62 above. The DWAF reported in its 2006–7 Annual Report that 74 per cent of the South African population have access to free basic water although the lack of basic infrastructure limits the provision of free water to the neediest communities. See South Africa, DWAF, Annual Report 2006–7, at 67–70. On the question of the amount of water provided for free, see *Lindiwe Mazibuko v. City of Johannesburg*, Case No. 06/13865, High Court of South Africa (Witswatersrand Local Division) (2008) and the text at note 63 above.

108 As noted elsewhere in this book, privatization can take different forms. See Chapter VII.

109 See Chapter V. See further, Jona Razzaque, 'Trading Water: The Human Factor', 13 *Rev. of European Community & Intl Envtl L.* 15, 22 (2004); Violeta Petrova, 'At the Frontiers of the Rush for Blue Gold: Water Privatization and the Human Right to Water' 31 *Brooklyn J. Intl L.* 577, 578 (2006).

during the last decade, increasingly been divesting itself in favour of the private sector including with respect to drinking water. In the arena of urban drinking water provision, for instance, domestic and foreign companies have emerged to undertake a variety of functions ranging from source development and augmentation to billing and metering.[110] Some initial attempts at privatization in urban water have, however, been aborted or at least postponed as a result of public protest.

The expansion of the privatization of water supply services the world over has undoubtedly been the most controversial aspect of reforms from the point of view of its implications for the right to water.[111] It has consequently been suggested that the right to water is indeed incompatible with privatization because privatization is often coupled with price increases and poor servicing of poor and/or rural neighbourhoods, and that consequently human rights obligations should prevent the private provision of these basic services.[112] While it is difficult to generalize on the impacts of the various kinds of privatization efforts on the realization of the right to water, and while violations of the right can also occur where water provision is in public hands, it is clear that delegating the management or ownership of water resources to the private sector can lead to abuses.[113] It thus cannot authorize states to disregard their international obligations.

There has generally been a lack of discussion and analysis in international arenas dealing with water regarding the controversial aspects of the involvement of the private sector in the water sector, in particular from the point of view of delivery and pricing questions. The approach of UN treaty and expert bodies has been limited to the statement that the human rights framework does not dictate a particular form of service delivery while leaving it to states to determine the best ways to implement their human rights obligations.[114] Despite this apparent neutrality as to the way in which states are to provide water services, and indeed the lack of prohibition of private sector involvement, human rights obligations do ask countries to regulate and monitor private water providers. General Comment No. 15 requires of states – both the hiring state and the state in which the company operates or is incorporated – that they ensure that private water providers do not compromise access to affordable, physically accessible, safe and acceptable water.[115]

110 See further Chapter VII.

111 See, e.g., the failures in Buenos Aires, Cochabamba, Jakarta or Manila.

112 See, e.g., Henri Smets, 'Economics of Water Services and the Right to Water', in Brown Weiss, Boisson de Chazournes and Bernasconi-Osterwalder eds, note 31 above, 173.

113 Draft Principles on Extreme Poverty, note 106 above para. 30 stating that 'the sale of water rights and privatization or management of water resources that results in a lack of access to drinkable water for groups of the population must be regarded as an encroachment upon this right'. See further Malgosia Fitzmaurice, 'The Right to Water' 18 *Fordham Envtl L. Rev.* 537, 557–68 (2007).

114 UNHCHR Report, note 7 above, para. 52.

115 General Comment No. 15, note 13 above, paras. 27 and 44; Sub-Commission Guidelines, note 41 above para. 2.3(e). See also UNHCHR, Consultation on Human Rights and Access to Water and Sanitation 52 (2007). See further Sub-Commission on the Promotion and Protection of Human Rights, Liberalization of trade in services and human rights (2002), UN Doc. E/CN.4/Sub.2/2002/9, para. 18.

One dimension of this duty to regulate is that full or partial privatization of water services should take place only when there is a clear and efficient regulatory framework.[116] This framework should include monitoring, participation, and imposition of penalties in cases of non-compliance. The High Commissioner also stresses that the right of individuals to participate in decision making is an especially important element when making decisions on private sector delivery.[117] In the Indian context, judicial judgments hold that the state cannot claim insufficient funds as a reason to not carry out its duties.[118] One critical question in the country, however, remains whether the state can have the assorted concessions, guarantees, rates of return and escrow mechanisms used to ensure private sector involvement when a lack of public resources is the major justification used for involvement of the private sector.

The consultation process conducted by the UNHCHR on the topic of the right of access to water identified the increasing societal expectation that companies ensure that their operations do not compromise access to safe drinking water.[119] Submissions received by the private sector during the consultation process highlighted the fact that various private water providers support recognition of access to safe drinking water as a human right. The responsibility of private corporations themselves under international human rights law is currently being explored by the HRC through the mandate of a Special Representative of the Secretary-General including in relation to the right to water.[120] Further than that, there has not been much attention given to the responsibility of states for acts of private water contractors, despite the fact that developing this issue would give states an incentive to enact more stringent legislation and other measures requiring more responsible conduct of water companies. It is clear that if a water contractor interferes with the right to water, a state may incur international responsibility for failure to prevent or punish the conduct of the private actor. The state may also incur international responsibility if the acts of the contractor are attributable to it.[121]

3. Due process and water service disconnections

In the context of a policy of cost recovery, limitations and disconnections of water services have appeared as logical options for water providers (whether public or

116 UNDP Human Development Report, note 1 above, 77.

117 UNHCHR Report, note 7 above, para. 53.

118 See Chapter VII.

119 UNHCHR Report, note 7 above, para. 31.

120 See, e.g., HRC, Report of the Special Representative of the Secretary-General on the issue of human rights and transnational corporations and other business enterprises, business and human rights: mapping international standards of responsibility and accountability for corporate acts, UN Doc. A/HRC/4/35 (2007).

121 With respect to companies involved in health care, see, e.g., Andrew Clapham, *Human Rights Obligations of Non-State Actors* 243 (New York: Oxford University Press, 2006).

private) in case of non-payment by users. Although in some countries water service disconnections are on the rise, there is a trend in state practice towards the prohibition of such measures due to the user's inability rather than unwillingness to pay.[122] An important feature of a number of regulatory frameworks for drinking water in India's urban areas is the possibility for the authority in charge to disconnect private water supply when the user fails to pay.[123] The question of whether such a vital resource as water for domestic uses can be legally interrupted and what are the due process requirements in such cases thus remains very pertinent.

Human rights law apparently falls short of prohibiting disconnections altogether despite the fact that the obligation to respect the right to water clearly prohibits states from unfairly interfering with people's access to water. Rather, current norms provide that disconnections must comprise the following procedural safeguards if they are applied:[124]

(a) timely and full disclosure of information on the proposed measures;
(b) reasonable notice of the proposed action;
(c) legal recourse and remedies for those affected; and
(d) legal assistance for obtaining legal remedies.

There is, however, a strong presumption against disconnecting or interrupting water services, in particular since in the words of the Committee on Economic, Social and Cultural Rights, '*[u]nder no circumstances* shall an individual be deprived of the minimum essential level of water'.[125] There is also a strong presumption that disconnections of institutions serving vulnerable groups, such as schools, hospitals and refugee camps, are prohibited. The High Commissioner further specifies that in addition to respecting commonly agreed principles of due process, disconnections for non-payment should not result in a person being denied access to a minimum amount of safe drinking water where that person provides that he or she is unable to pay for these basic services.[126] The Committee of Ministers of the Council of Europe, for instance, has recommended that social measures should be put in place to prevent the supply of water to destitute persons from being cut off.[127]

122 Prohibitions have been adopted in Argentina, Australia, Austria, Belgium, Brazil, Ireland, Luxembourg, Mexico, New Zealand, Norway, Spain, Sweden Switzerland, the United Kingdom and Ukraine.
123 See, e.g., Chapter VIII.
124 UNHCHR Report, note 7 above, para. 57.
125 General Comment No. 15, note 13 above, para. 56 (emphasis added).
126 *Ibid.*, para. 59.
127 Recommendation 14 Committee of Ministers to member states on the European Charter on Water Resources (2001), para. 5.

4. The role of local authorities and other communities of users

While the provision of water is primarily the responsibility of states, water law reforms are often characterized by the transfer of power and responsibilities from the state to regional or local authorities, and/or to other communities of users. It is of course very important that these bodies are made to comply with international requirements on the right to water. The obligations of states on the international level with the actions of other bodies that have responsibilities for water provision, as well as the obligations incumbent upon the bodies themselves, however, remain unclear. Emblematic of this trend, local authorities recognized the right to water at the Fourth World Water Forum in 2006.[128] With respect to local authorities that have been delegated the task of supplying drinking water, General Comment No. 15 provides that, as part of its obligation to protect the right to water, 'the State party still retains the responsibility to comply with its Covenant obligations, and therefore should ensure that these authorities have at their disposal sufficient resources to maintain and extend the necessary water services and facilities'.[129] States should thus ensure that local authorities guarantee equal access to affordable and physically accessible water of a good quality. This is part of their obligation to protect the right to water. Some indication of the responsibilities placed upon local authorities to whom responsibilities in water provision have been placed can be found in WHO guidelines to secure water quality.[130] There is, however, little guidance regarding the role and obligations of local authorities in water provision, and regarding the obligations of the state if the local authority does not comply with requirements on the right to water, beyond the general statement that it retains responsibility under international human rights law.

A second way in which water delivery is being vested away from the state is through structures that aim to decentralize decision making, by transferring water services operation to local communities of users.[131] This has been a particularly central aspect of water reforms in India, including with respect to urban and rural drinking water, and very often in the context of World Bank-funded projects.[132] There are no guidelines in international human rights instruments regarding decentralization. Further clarification is thus needed in terms of the standards to be applied by states when delegating water management to local communities of

128 Local Government Declaration on Water on the occasion of the Fourth World Water Forum, Mexico, 21 March 2006, available at www.worldwatercouncil.org/fileadmin/wwc/World_Water_Forum/WWF4/declarations/Local_Authorities_Declaration_english.pdf.

129 General Comment No. 15, note 13 above, para. 51.

130 WHO 2006, note 66 above, 11–12.

131 In the context of water sector reforms, 'decentralization' refers to the divesture of the state of many of its functions in water management and water services provision and, in many cases, their transfer to local authorities and users' associations, with the participation of the private sector. See further Philippe Cullet, 'Water Law – Evolving Regulatory Framework', in Cullet, Gowlland Gualtieri, Madhav and Ramanathan (eds.) note 62 above, 26 and Chapter V.

132 See Chapter VII.

users in order to ensure that the right to water is respected. These should include in particular adequate procedural safeguards such as public participation and transparency, non-discrimination, and accountability, as well as a clear division of responsibilities between the various actors.

5. *Prioritization of uses*

Issues of prioritization of uses have implications for water law reforms in that these reforms generally need to manage competing demands. In practice, essential domestic uses (such as drinking, washing clothes, personal hygiene, cooking and personal sanitation) are not consistently prioritized in relation to other uses such as agriculture or industry.[133] Indeed, domestic water use accounts for less than 5 per cent of total global water use, the balance being used by industry and agriculture.[134] Of this already limited amount, only a small proportion is for essential domestic uses. Additionally, there are often difficulties in making available a sufficient quantity of safe water because of unsustainability of existing water sources stemming from degradation of water catchment areas, pollution or over-exploitation.

For these reasons, the right to water requires of states that they include in laws and other acts clear prioritization of water uses, with primacy given to water for personal and domestic uses.[135] In India, the need for prioritization of drinking water is found in court cases as well as in legislation on groundwater.[136] In international documents, this basic principle is highlighted for instance in the Johannesburg Plan of Implementation, which calls for water to be allocated among competing uses in a way that gives priority to the satisfaction of basic human needs.[137] General Comment No. 15 notes that priority in the allocation of water should also be given to water resources to prevent starvation and disease, and that attention should be given to ensuring that disadvantaged and marginalized farmers have equitable access to water and water management systems.[138] It does not, however, adequately explore the existing and potential conflicts between basic uses and economic uses of water. The UNHCHR Report accordingly highlights that once a sufficient amount of safe drinking water to prevent disease has been secured for all, allocation of water among various uses – water for personal and domestic uses beyond this sufficient amount, water to produce food, water to sustain livelihoods, or water to ensure environmental hygiene – remains unclear, and calls upon the HRC to clarify obligations in this regard.[139]

133 *Ibid.*
134 UNDP Human Development Report, note 1 above, 2.
135 As reported by the UNHCHR, many countries have adopted legislation that addresses prioritization between uses. See UNHCHR Report, note 7 above, para. 40.
136 See references in Chapter IX.
137 Johannesburg Plan of Implementation, note 102 above, para. 26(c).
138 General Comment No. 15, note 13 above, para. 6.
139 UNHCHR Report, note 7 above, para. 62.

CONCLUSION

The perspectives examined in this chapter should be set within the broader framework of international law and policy applicable to water. Indeed, as seen in a previous chapter,[140] only certain aspects of water law have been developed at the international level and there is to date no international water law treaty. On the other hand, international economic and financial institutions have assumed a leading position in influencing national policies related to water. The human rights dimension of water accordingly needs to be integrated into a holistic approach to water on the international level that includes all its dimensions, including environmental ones. Human rights law has witnessed the evolution from sectoral approach – the right to water related to certain categories of vulnerable persons, such as children, women or war prisoners – to a universal and generally applicable right. The adoption of General Comment No. 15 in 2002 was thus noteworthy in that it responded to the need for water law to be addressed from the perspective of human rights law.

From a national perspective, an international rights-based approach for water has much to bring in contributing to the development of water law in the framework of water law reforms being conceived and/or implemented in various countries such as India. It is now clear that while these reforms address issues of cost recovery, efficiency, and regulation, they must not lose sight of the need to account for obligations related to the right to water in order to achieve the primary goal of basic access to safe water for all segments of the population. The recognition of the right to water should serve as a way to counter the effects of the economic approach to water by requesting of states that water is provided to all, including those who cannot afford it under the full cost recovery principle. Nevertheless, while providing firm grounding for the main components of a right to water, it appears that international standards often do not provide a comprehensive enough framework to address the most pressing and indeed controversial aspects of water law reforms. For instance, General Comment No. 15 does not engage in depth in the formulation of water policies or issues of water management, particularly with regard to the question of water privatization or the commodification of water, some of the most pressing issues in current water reforms.

International human rights law has a fundamental role to play, both in clarifying the nature of state obligations regarding access to water and providing individuals and non-state actors with redress mechanisms when their right of access to water is denied, both on the international and national levels. Not only are legally binding instruments needed in international human rights law that specifically address the right to water, but also the question of water remains

140 See Chapter III. Providing an overview of the international framework for water law, see also Philippe Cullet, 'Water Law – Evolving Regulatory Framework', in Cullet, Gowlland Gualtieri, Madhav and Ramanathan (eds.) note 62 above, 27, 28–31.

dominated by international economic law and policy rather than a human rights legal perspective. Accordingly, it has been noted that:

> specific, dedicated and sustained attention to safe drinking water [. . .] is currently lacking at the international level, given the broad range of issues that special procedures and treaty bodies have to address within their mandates and the specific questions that arise in relation to access to safe drinking water.[141]

The following issues are some amongst others that should be tackled in the context of a human right to water:

(1) The tensions existing between cost recovery policies and a human rights perspective, and the need for a more holistic approach to water law reforms. It is noteworthy that there is a similar lacuna in the Indian context, where many aspects of the current reforms are incompatible with the right to water with no legal guidance on how to address these disparities despite judicial and constitutional support of the right.
(2) The existence of severe inequalities and discrimination in the face of access to water, particularly as concerns the poor and rural populations, and particularly in view of the fact that financing from international institutions tends to favour big projects or almost exclusively focus on the involvement of private actors. In terms of pricing, all segments of the population are entitled to a water services system that takes into account capacity to pay when pricing and subsidies are put in place.
(3) The uncertainties surrounding the accountability of the state on the international and national levels in the face of the emergence of a range of new actors stemming both from decentralization and privatization efforts. In particular, more in-depth study is needed on international human rights requirements concerning the private provision of water services and the type of regulatory system that states must put in place in that respect.
(4) The lack of regulation regarding the extent to which responsibilities for water provision can be delegated by the state to other actors, whether private companies, local governments or different types of non-state bodies. For instance, can all critical tasks involved in providing water to people (such as setting of tariffs) be delegated, even when subject to regulation and monitoring?
(5) The difficulties that remain in the division of labour and responsibilities in water provision when different bodies and interests are involved at the local level.
(6) The gap in democratization linked to water law reforms and in particular the adoption of policies on privatization.

141 UNHCHR Report, note 7 above, para. 69.

XII. Water sector reforms and principles of international environmental law

David Takacs

INTRODUCTION

Over one billion people lack access to a basic supply of clean drinking water, defined by the World Health Organization (WHO) as (low) 20 litres/day within 1 km of the home. WHO estimates that 2.4 billion people lack adequate sanitation facilities. A child dies every fifteen seconds from diarrhoea, largely caused by contaminated water or poor sanitation; this death toll exceeds the ravages of HIV/AIDS on children.[1]

In India, according to UNICEF, 90 per cent of rural dwellers have access to a basic supply of safe drinking water, although only 48 per cent have access to adequate sanitation. WaterAid puts that figure at just 15 per cent.[2] The World Bank acknowledges the gap between official figures (94 per cent covered with water supply in rural areas and 90 per cent in urban areas) that 'are probably a better indication of the infrastructure that has been built than the services that are actually provided'.[3] WaterAid asserts that 8.7 per cent of Indian children die before reaching the age of five, mostly from preventable, often waterborne diseases.[4] Pollution is endemic, with the World Bank reporting that '[m]any rivers – even very large ones – have turned into fetid sewers'. According to the World Bank, about 15 per cent of India's aquifers 'are in critical condition', and this threatens to increase to 60 per cent by 2030. Climate change, with attendant droughts, severe storms, and melting glaciers, further portend threats to the

1 World Health Organization, The Right to Water 7, 12 (Geneva: WHO, 2003); World Health Organization, *The Global Water Supply and Sanitation Assessment 2000* at 1 (Geneva: WHO, 2000).

2 UNICEF, 'International Year of Sanitation 2008', available at www.unicef.org/india/wes_3950.htm; WaterAid, 'India', available at www.wateraid.org/international/what_we_do/where_we_work/india.

3 John Briscoe & R.P.S. Malik, *India's Water Economy: Bracing for a Turbulent Future* 35 (Oxford: Oxford University Press, 2006). Of course, the World Bank has policy reasons for underestimating or doubting the government's figures; were the government's figures to be correct, it would undermine the World Bank's claims that the government is a poor or inefficient provider of water services.

4 WaterAid, note 2, above.

stability of India's water supply.[5] Fifty per cent of India's seasonal precipitation falls in fifteen days and 90 per cent of river flows occur in four months, exacerbating seasonal water scarcity.[6]

As India's population soars over one billion, the needs of daily survival use, irrigated agriculture (often for export crops), and industry compete with each other, and with the needs of non-human species and ecosystem processes, and with the needs of unborn generations of humans and non-humans.

During the past two decades, powerful international actors have expressed deep concern about the growing global scarcity of water. To mitigate this scarcity, they have not turned to environmental politics, policy, and law. Instead, water sector reform (WSR) has focused on political restructuring based on economic principles. Under pressure, Southern governments increasingly abandon their roles as water providers, and turn over water services to private actors, who commodify and commercialize water. A changing regulatory environment may seriously impact the physical environment that supports human and non-human life.

How one conceives of 'environmental' problems and the law will shape how one seeks to solve these problems. Environmental law covers wide ground; in fact, some friction exists between environmental advocates who include or exclude humans as the locus of concern when discussing environmental problems. When framers of policy or law consider one environmental aspect disarticulated from the others, they may derive non-sustainable solutions, or those that may violate other precepts of domestic or international environmental law or international human rights law. Focusing exclusively on human needs may lead to pollution or degradation of ecosystem resources needed to sustain non-humans (which, in turn, may be needed to serve future generations of humans); focusing exclusively on non-human needs may ignore the real environmental needs of people today who are sick or dying because of insufficient or unclean water.

In this chapter, I offer a wide view: environmental problems happen when human action impacts assets from the non-human biological or physical world, creating difficulties for humans or non-humans. Thus 'water scarcity' is an 'environmental' problem because it concerns a resource from the physical world (water) that is adversely impacted by human action (overuse, inequitable distribution, pollution) with resulting problems for humans (shortage or contamination) and/or non-humans. Environmental laws may thus address a spectrum of concerns, from who or what gets to use shared watercourses, to how to protect non-human species, to the human right to a basic share of clean water, to who cannot arrogate and who must steward resources, to who gets to participate in decision making about how resources are distributed and managed.[7] Thus this

5 Briscoe and Malik, note 3, xv–xvi, 17–19, 67. They suggest at 19 that India could have a 'window of opportunity' to make use of the 'windfall' from melting glaciers, if it developed infrastructure (read: dams) to store this torrent of melting water.

6 *Ibid.*, at xv.

7 Sequentially: Convention on the Law of the Non-navigational Uses of International Watercourses, New York, 21 May 1997 (hereafter UN Watercourses Convention); Convention on Biological

chapter covers vast territory, and only skims the surface in addressing the disparate implications of current trends in WSR for environmental law.

In this chapter, I focus on WSR's cornerstone principles and resulting policy edicts. I focus in particular on the World Bank, which has driven economic and political solutions to problems of environmental scarcity, and explain how these principles and policy solutions have played out in India. To assess how these solutions do or do not adhere to principles of environmental law, I examine sources of that law including constitutions, international water law, and customary international environmental law. Throughout I offer ideas on how WSR might change to align solutions to water scarcity and inequitable distribution with principles of environmental law.

A. Water sector reform principles, strategies, and actors: Addressing environmental scarcity through economic and political restructuring

1. Agenda 21: A holistic approach to water sector reforms

Until 1990, most freshwater law was at the national level. Two seminal events in 1992 brought international water law to the fore, and helped set the agenda for an overarching philosophy of WSR that adopts non-environmental solutions to environmental scarcity.

The UN Conference on Environment and Development described multiple environmental threats creating dwindling or polluted water for humans and non-humans. Chapter 18 calls for 'holistic management of freshwater as a finite and vulnerable resource, and the integration of sectoral water plans and programmes within the framework of national economic and social policy'.[8] Indeed, its extensive documentation takes a broad approach to analyzing the problems of water scarcity and to solving those problems, and emphasizes that sustainable development should be at the core of watercourse management.[9] The strategy avers that:

Diversity, Rio de Janeiro, 5 June 1992 (hereafter Biodiversity Convention); Human Rights Council, Report of the United Nations High Commissioner for Human Rights on the Scope and Content of the Relevant Human Rights Obligations Related to Equitable Access to Safe Drinking Water and Sanitation Under International Human Rights Instruments, UN Doc. A/HRC/6/3 (2007), para. 66; Joseph L. Sax, 'The Public Trust Doctrine in Natural Resource Law: Effective Judicial Intervention', 68 *Michigan L. Rev.* 471 (1970); Convention on Access to Information, Public Participation in Decision Making and Access to Justice in Environmental Matters, Aarhus, 25 June 1998, UN Doc. ECE/CEP/43.

8 Agenda 21, Report of the UNCED, Rio de Janeiro, 3–14 June 1992, UN Doc. A/CONF.151/26/Rev.1 (Vol. 1, Annex II), Chapter 18.1, 18.2, and 18.6.

9 Patricia Wouters, The Legal Response to International Water Scarcity and Water Conflicts: The UN Watercourses Convention and Beyond 37 (Dundee, 1999), available at www.dundee.ac.uk/iwlri/Documents/Research/IWLRI%20Team/Wouters/GYIL.pdf.

> Water resources have to be protected, taking into account the functioning of aquatic ecosystems and the perenniality of the resource, in order to satisfy and reconcile needs for water in human activities. In developing and using water resources, priority has to be given to the satisfaction of basic needs and the safeguarding of ecosystems. Beyond these requirements, however, water users should be charged appropriately.[10]

While the strategy contemplates payment for water, it recognizes that ecosystems have no ability to pay, and that those satisfying 'basic needs' should not be charged. While the strategy encourages delegation of water management to 'the lowest appropriate level', it still retains multiple responsibilities for national governments. To protect water resources, water quality and aquatic ecosystems, the document both validates – and helps to harden as customary international environmental law – the polluter-pays principle and '[m]andatory impact assessment of all major water resource development projects potentially impairing water quality and aquatic ecosystems'.[11] The policy plan is comprehensive, and recognizes that an environmental problem requires an 'integrated' social, political, and economic response, and roots much of that response in reformulated environmental laws, policies, and responsibilities.

2. The Dublin Principles: Water as an economic good

The same year, attendees of the International Conference on Water and the Environment in Dublin produced a set of four Dublin Principles, which, while non-binding, have 'defined the basic guidelines for water resources management in the years to come'.[12] Principle 1 lays out the environmental imperative that ostensibly drives WSR, linking human and ecosystem needs to holistic management so all environmental needs are fulfilled:

> Freshwater is a finite and vulnerable resource, essential to sustain life, development, and the environment. Since water sustains life, effective management of water resources demands a holistic approach, linking social and economic development with protection of natural ecosystems.[13]

Principle 2 recommends that managers take a 'participatory approach [. . .] with

10 Agenda 21, note 8 above, chapter 18.8.

11 *Ibid.*, chapter 18.12, 18.40.

12 Dublin Statement on Water and Sustainable Development, International Conference on Water and the Environment, Dublin, 31 January 1992; Matthias Finger and Jeremy Allouche, *Water Privatization: Trans-National Corporations and the Re-Regulation of the Water Industry* 24 (London: Spon Press, 2002). See also, Chapters III and V for further background and interpretation of the Dublin Principles.

13 Dublin Statement on Water and Sustainable Development, International Conference on Water and the Environment, Dublin, 31 January 1992, Principle 1.

full public consultation and involvement of users in planning and implementation of water projects'. This will turn out to be code for reducing the role government plays in providing water.[14] Principle 3 notes the central role women play in water management, a principle that will be given little more than lip service in subsequent WSR.

Most importantly, principle 4 asserts that '[w]ater has an economic value in all its competing uses and should be recognized as an economic good'. While governments had generally provided water as a public service, here the Dublin Principles assert the 'basic right of all human beings to have access to clean water and sanitation at an affordable price' – not necessarily free.[15] The framers allege that 'failure to recognize the economic value of water' has led people to waste water and helped create the current environmental crisis. To mitigate the crisis, '[m]anaging water as an economic good is an important way of achieving efficient and equitable use, and of encouraging conservation and protection of water resources'.[16]

Thus in Dublin, the environmental crisis described in Agenda 21 (with its prescribed integrated, environmental solutions) yields a restricted economic and political solution: make people pay for water, take water provision out of government's hands, and people will subsequently manage and conserve the environmental resource they have carelessly despoiled and squandered. The authors do not suggest that an environmental crisis could be addressed through environmental law, even though 1992 also saw other environmental problems addressed through international environmental law solutions: the Convention on Biological Diversity addressed the biodiversity crisis, and the UN Framework Convention on Climate Change addressed the climate crisis.[17] Yet the 'environmental' crisis of scarce, unclean water will be a pretext for sweeping economic and political prescriptions with the potential to benefit the already wealthy and exacerbate the environmental crisis these policy prescriptions are supposed to mitigate. Agenda 21's emphasis on sustainable development has disappeared. Furthermore, these economic and management solutions to 'environmental' problems may violate environmental law.

3. International financial institutions' strategy to confront the 'environmental' crisis of water scarcity: Disinvest the state, decentralize, and privatize

Access to clean water comprises an environmental crisis: competing users lack sufficient uncontaminated water. The poor cannot fulfil their basic daily needs, and non-human entities and ecosystems cannot sustain themselves.

14 *Ibid.* See also Finger and Allouche, note 12 above, 31.

15 Dublin Statement on Water and Sustainable Development, International Conference on Water and the Environment, Dublin, 31 January 1992, Principle 4.

16 *Ibid.*

17 Biodiversity Convention, note 7 above and United Nations Framework Convention on Climate Change, 9 May 1992, 1771 UNTS 107.

To address this crisis, the World Bank and other international financial institutions (IFIs) have played a leading role in formulating, implementing, and imposing WSR throughout the world.[18] The World Bank has eagerly embraced the Dublin Principles, from which it explicitly derives its policies. For example, the World Bank points favourably to WSR in Mexico's new water management regime based on transferable property rights, management by users, and the elimination of distorting subsidies.[19] The World Bank has lent about $US 14 billion for water projects in India; it has contributed 72 per cent of external lending and grants for water there. To summarize what the World Bank feels is needed for India, John Briscoe writes that the end of an era of groundwater expansion is:

> going to require a dramatic transformation in the way in which public water services are provided to farmers, households and industries, in which the watchwords are water entitlements, financial sustainability, accountability, competition, regulation and entry of alternatives to government provision, including cooperatives and the private sector.[20]

The Bank's rules of engagement (a euphemism for conditionality) for water sector loans for India include (for irrigation) strengthening cost recovery, beneficiary participation and water entitlements; for urban water and sanitation, improving services to the poor and private sector participation; and a return to hydro-power with limited environmental and social impacts (although the World Bank does not explain how to limit those impacts).[21]

From the Dublin Principles – particularly from Principle 4 – the World Bank and other forces that stood to profit from water as an economic good promoted a seven-point set of policy prescriptions that comprise the hegemonic programme of WSR to remedy this environmental crisis, which I now review.

- *Water users waste water* because government has treated it as a public good to be given away. Because water has no economic value, waste leads to a problem of environmental scarcity.

Modern WSR operates from a paradigm that we will only respect, use carefully, and conserve things with economic value. Environmental scarcity of water comes from a common problem of a free, shared resource exploitable – and thus exploited – by all. As one author euphemistically puts it, '[w]asteful water use was

18 See Chapter V.

19 World Bank Operations Evaluation Department, Bridging Troubled Waters: Assessing the World Bank Water Resources Strategy 2, 21 (Washington, D.C.: World Bank, 2002).

20 Briscoe & Malik note 3 above, xx–xxi.

21 *Ibid.*, at 77–78.

also exacerbated by the absence of a payment culture'.[22] And, allegedly, if we do not pay for water, we will not value it, and we will waste and pollute it. Dharmadhikary argues that in India, this argument buttresses reforms in all sectors, including water.[23]

- In treating *water as a public good* to be given away (which leads to user waste), and in mismanaging the whole process by which it gives water away, *government creates the problem* of environmental scarcity of clean water.

According to the World Bank, governments under-price water (or simply give it away), thus promoting waste; monopolize service provision and thus eliminate beneficial competition; and fail to invest in maintaining water infrastructure. By seeing water as a public good, governments provide distorting subsidies that creates the environmental scarcity.[24] As part of the indictment, Briscoe writes that '[u]ser charges are negligible, resulting in lack of accountability and insufficient generation of revenue even for operation and maintenance'. The Indian government has failed at 'maintaining the stock of infrastructure or providing services [. . .] The implicit philosophy has been aptly described as Build-Neglect-Rebuild'.[25]

Shortly after the Dublin Principles were adopted, the World Bank highlighted how governments failed to provide various kinds of infrastructure, mentioning irrigation, water supply, and sanitation in particular. According to the World Bank, governments face no competition, and thus have no incentives to efficiency: they provide wasteful subsidies, they are not held accountable, and they lead citizens to overuse scarce resources.[26] India, in particular, is a state in disrepute because of how it has mismanaged its water resources.[27]

- To combat these intertwined factors of lack of economic value and government mismanagement, *water must be treated as an economic good*, to be allotted to the most economically efficient uses and paid for by consumers.

To view water as an economic good battles the philosophy which has informed Indian water policy and recent jurisprudence, that access to water is a human

22 Lilian Saade Hazin, 'Private Sector Participation in Water Supply and Sanitation: Realising Social and Environmental Objectives in Mexico', in Nick Johnstone and Libby Wood (eds.) *Private Firms and Public Water: Realising Social and Environmental Objectives in Developing Countries* 177 (Cheltenham: Edward Elgar, 2001).

23 Shripad Dharmadhikary, 'A Flawed Model for Water Regulation', *India Together*, 23 May 2007, available at www.indiatogether.org/2007/may/env-mwrra.htm.

24 Finger and Allouche, note 14 above, 102; World Bank, note 19 above, 21.

25 Briscoe and Malik, note 3 above, xvii.

26 World Bank, World Bank World Development Report 6–7 (New York: Oxford University Press, 1994).

27 Briscoe and Malik, note 3 above, 39.

right, and the government must protect this public trust resource.[28] Conceiving of water as an economic good fundamentally changes the laws and policies enacted to govern its distribution. If access to clean water is a human right, governments must design laws to ensure that all members of society receive a basic quantity. If water is governed under the public trust doctrine, then governments must steward the resource for present and future generations, and prevent private owners from arrogating the resource.[29]

But when water is a commodity, as McDonald and Ruiters express it, we experience another step of 'the transformation of all social relations to economic relations, subsumed by the logic of the market and reduced to the crude calculus of profit'. WSR realizes this commodification, as it promotes full cost recovery, where users must pay for all aspects of water infrastructure and management.[30]

The World Bank's approach for more than a decade has been, according to Coelho, 'an unwavering focus on commercial viability' as necessary for good service.[31] The Bank urges '[a]dopting pricing and incentive policies that achieve cost recovery, water conservation, and better allocation of water resources'.[32] While a Bank review of its own water policies discusses pollution of groundwater and salinization of aquifers, the World Bank believes this is *caused* by 'lack of regulation and effective water pricing',[33] and not, say, by Bank-endorsed (over) reliance on irrigated crops for export agriculture. A Bank analyst writes that '[m]ost of the revenues for these services are going to have to come from users, because [. . .] the urban authorities are going to have to invest massive amounts of public money in the sewerage systems needed to clean up the polluted rivers'.[34] How Briscoe comes up with this non sequitur leading from environmental despoliation to the mandatory requirement that users pay is not clear.

While an environmental crisis seems to precipitate these reforms, environmental solutions grounded in environmental law do not comprise the actual policies and actions that emerge. According to Briscoe, through water entitlements and cost recovery, 'the majority of water-related environmental problems in India would be ameliorated to a significant degree'.[35] Users on farms and in cities would stop wasting water, and would stop depleting aquifers. Government

28 Philippe Cullet, 'Water Law Reforms: Analysis of Recent Developments', 48/2 *J. Indian L. Institute* 206 (2006).

29 I will return to these points below.

30 David A. McDonald and Greg Ruiters, 'Introduction', in David A. McDonald and Greg Ruiters (eds.), *The Age of Commodity: Water Privatization in Southern Africa* 3 (London: Earthscan, 2005) and Cullet, note 28 above, 208.

31 Karen Coelho, 'The Slow Road to the Private – A Case Study of Neo-Liberal Water Reforms in Chennai', in Philippe Cullet, Alix Gowlland Gualtieri, Roopa Madhav and Usha Ramanathan (eds.) *Water Law at the Crossroads – National and International Perspectives with Special Emphasis on India* 81, 87 (New Delhi: Cambridge University Press, 2009).

32 Section 2(b), World Bank, Water Resources Management (OP 4.07, February 2000).

33 World Bank, note 19 above, 21.

34 Briscoe and Malik, note 3 above, 50.

35 *Ibid.*, at 60.

would transfer its attention from building and managing water infrastructure and would instead be engaged in creating fiscal space for investing in environmental quality and other public goods.[36] Cost recovery is thus a panacea that will cure environmental problems.

Of course, full costs are never really full, because users do not bear pollution externalities.[37] In the absence of government control, how those externalities will be paid for – and thus how environmental mandates would be respected – remains unexplained.

Northern advisors and donors encouraging full cost recovery is nothing new in India. In 1969, USAID believed that in India, '[w]ater supply systems should be considered as a public utility in the same way as electricity, gas, or telephone systems. The users or beneficiaries should pay the full cost of supplying water'.[38] Lee made a common-sense observation that in the West, to cover capital debt incurred by investment in water infrastructure, water users are encouraged to use more (and thus dispense more money for) water. Thus making water an economic good might lead to the drive to exploit *more* of it. Using water as a resource to be harnessed for industry, energy generation and corporate agriculture, which will fuel a nation's economy, requires more, not less water, thus exacerbating, not solving water scarcity. The National Water Policy, 2002 follows this script, calling for '[n]on-conventional methods' such as massive infrastructure projects like inter-basin water transfers and desalination. Of course, these projects and more conventional water development must be 'taking into account [. . .] environmental considerations', although what precisely that means is not explained or how that is to be accomplished when water goes to the highest industrial bidder is not established.[39]

While the National Water Policy, 2002 does not go all the way to full cost recovery, it does argue, 'water charges for various uses should be fixed in such a way that they cover at least the operation and maintenance charges of providing the service initially and a part of the capital costs subsequently'.[40] In India, the World Bank and Asian Development Bank promote transferable water rights, with free trading between entitlement holders.[41] As Dharmadhikary puts it, '[i]f the water sector is converted into a market with free trading of entitlements, it is only a matter of time before the control of water resources goes into the hands of those with economic power'.[42] This is the logical and probably desired outcome of water going to the highest bidder. How the poor will satisfy their basic right to water, or how non-human actors will participate in the bidding

36 *Ibid.*

37 Finger and Allouche, note 14 above, 77.

38 Terence R. Lee, *Residential Water Demand and Economic Development* 17 (Toronto: University of Toronto Press, 1969).

39 National Water Policy, 2002, available at www.ielrc.org/content/e0210.pdf, Sections 3.2 and 3.3.

40 *Ibid.*, Section 11.

41 Briscoe and Malik, note 3 above, 21–26.

42 Dharmadhikary, note 23 above.

process – in other words, how provisions of domestic and international environmental law would be fulfilled under water trading – remains unanswered.

- Water service delivery and management must be *decentralized* away from the government and towards private management entities.

Governments had their chance, and, according to critics, they failed. Through mismanagement and free giveaways of water, governments have created the problem of environmental scarcity of water. Conserving the role of the state in water provision does not conserve water, and so water provision must be decentralized.

But if the crisis of water scarcity results from the crisis of government mismanagement, how are those same governments supposed to be entrusted with regulating private entities that will now oversee this scarce environmental resource? According to Finger and Allouche, 'decentralization simultaneously requires a national organ responsible for inter-sectoral water resources management, as well as a regulatory framework. This dual process is not very clearly thought through in the various World Bank documents, given the fact that the World Bank adheres to a quite simplistic view of privatization'.[43] WSR requires these governments to represent the interests of the poor – many of whom will, presumably, be unable to afford to pay for water services. The government must also protect the non-human, but members of endangered species and representatives of ecosystems notably do not vote and do not pay taxes and cannot outbid others for the water they require.

- Decentralization demands *community participation* to manage the local resource.

In India, the National Water Policy, 2002 and the World Bank promote community participation.[44] It is also promoted in the World Bank-sponsored Swajal projects, which seek to put water service provision in local people's hands, who will get the water service delivery they are willing to pay for. Communities must pay 10 per cent of the upfront capital costs, and will be fully responsible to operate and maintain the water system.[45]

At first blush, these decentralized community participation mandates seem to contribute to a vibrant environmental democracy. But Cullet argues that this system bypasses the traditional, more democratic, constitutionally based panchayati raj system of governance and instead imposes a new layer of bureaucracy. Newly formed water users associations (WUAs) must manage the irrigation water they control. Furthermore, these WUAs are limited to landowners, and those who can pay for the initial infrastructure investments. They make no provisions for

43 Finger and Allouche, note 14 above, 104.

44 National Water Policy, 2002, Section 6.8; Cullet, note 28 above, 216; Briscoe and Malik, note 3 above, xx–xxi, 54–55.

45 See Chapter VIII.

persistent inequalities in caste system or for how to deter local elites from controlling local water decision making in their own interests. The central role of women (Dublin Principle 3) has faded away in the composition of these WUAs.[46] Underlying all of this is that WUAs had no choice about whether or not they wanted to take on these responsibilities, or back the World Bank-sanctioned WSR principles in the first place.

- Once removed from government's purview, private service providers can provide water most effectively.

Even if government has been inefficient and is guilty as critics charge, other alternatives remain: donors could invest in efforts that reform government to make it more effective and accountable.[47] Reforms could include careful guidelines for how to adhere to domestic and international environmental law principles that respect the human right to water and the needs of non-human entities. For example, in Southern Africa, 70 per cent of water goes to irrigate cropland, most of which goes to (water-hungry) export crops and not staple crops to feed local people.[48] Policy changes to address agricultural water use could address environmental needs.

Instead, recommended reforms, according to Dharmadhikary, 'started and ended with *financial* reforms'.[49] As Finger and Allouche put it, the World Bank 'ends up with a serious simplification: privatization becomes the answer to environmental concerns'.[50] Indeed, between 1990 and 2002, the World Bank made 86 of 276 water supply loans conditional on privatization.[51] In India, the National Water Policy, 2002 urges that '[p]rivate sector participation should be encouraged in planning, development and management of water resources projects for diverse uses, whenever possible'.[52] Privatization may take a variety of forms, including public–private partnerships that nonetheless change the 'underlying managerial ethos' of water from a public good to a profitable commodity.[53]

46 Cullet, note 28 above, 216 and Philippe Cullet, 'Water Law – Evolving Regulatory Framework', in Philippe Cullet, Alix Gowlland Gualtieri, Roopa Madhav and Usha Ramanathan (eds.) *Water Law at the Crossroads – National and International Perspectives with Special Emphasis on India* 26 (New Delhi: Cambridge University Press, 2009).

47 McDonald and Ruiters, note 30 above, 7.

48 Larry A. Swatuk, 'The New Water Architecture of SADC', in McDonald and Ruiters (eds.) note 30 above, 52 and Stephen C. McCaffrey, *The Law of International Watercourses* 7 (Oxford: Oxford University Press, 2007).

49 Dharmadhikary, note 23 above.

50 Finger and Allouche, note 14 above, 64.

51 Peter T. Muchlinski, *Multinational Enterprises and the Law* 552 (Oxford: Oxford University Press, 2007).

52 National Water Policy, 2002, Section 13.

53 Finger and Allouche, note 14 above, 81–82 for different schemes of privatization. See also McDonald and Ruiters, note 30 above, 2.

Privatization promoters argue that 'governments are corrupt, unaccountable, unimaginative and financially strapped, and unable to expand and upgrade water services on their own in a reliable and cost-effective manner'.[54] Privatization may result in 'unbundling', or handing over to the private sector whatever is profitable and leaving the state what is left over (namely, what is least profitable).[55] It may mean red-lining, or cutting out service to some poor communities completely.[56] Private providers may engage in 'cream-skimming', serving more wealthy water users, and leaving the state with fewer resources to fulfil its now more difficult mandates of providing water to the poor or engaging in environmental conservation.[57] This not only makes it harder for the state to fulfil its environmental law mandates, but also provides further ammunition to those who paint governments as inefficient or incapable of providing water services.

A study of Mexican water supply privatization makes the case that poor people buying from individuals pay five times more per unit of water than those who receive water from a private provider. Private providers can subsidize the poor by cross-tariffs. But, unfortunately, 'cross-subsidies only benefit households who are connected', and in Mexico, the very poor could not afford the connection fees in the first place.[58] While some case studies show that poor households benefit from privatization, this is not always the case, particularly where private service providers have no incentive to provide water to those who can't pay premium prices. Distributive justice is poorly served by unregulated markets.

Privatization means turning over water delivery to profiteers. Economists value the global market in water at US$1 trillion, and only 5–10 per cent of the market has been privatized: this is an enticing target for profit seekers.[59] As UNICEF notes:

> while privatization may bring many benefits in terms of incentives for efficient service delivery, it is not necessarily a comprehensive solution and risks further marginalizing the poor. First, there are few incentives to reach out to informal settlements in peri-urban and rural areas. Low incomes and limited 'voice' mean that the poor can bring little pressure to bear.[60]

So, for example, WSR in southern Africa has opened up water service to the private sector, and twenty-eight of the largest multinational enterprise water service providers have subsequently set up in the region.[61]

54 *Ibid.*, at 1.

55 Rudolf Nsorwine Amenga Etego and Sara Grusky, 'The New Face of Conditionalities: The World Bank and Water Privatization in Ghana' in McDonald and Ruiters (eds.) note 30 above, 279.

56 McDonald and Ruiters, note 30 above, 1.

57 Finger and Allouche, note 14 above, 64.

58 Hazin, note 22 above, 179.

59 McDonald and Ruiters, note 30 above, 31 and Vivienne Bennett, *The Politics of Water: Urban Protest, Gender, and Power in Monterrey, Mexico* 552 (Pittsburgh: University of Pittsburgh Press, 1995).

60 Quoted in Finger and Allouche, note 14 above, 55.

61 McDonald and Ruiters, note 30 above, 28.

Hilderling points out that privatization 'actually requires a strong government in order to regulate and control it and to enforce social and ecological conditions'.[62] Furthermore, governments are legally obliged to exert a strong regulatory hand to enforce environmental and human rights laws. Absent strict government regulation – that same government allegedly incompetent to manage water resources – private service providers who profit through providing water would have no necessary incentive to conserve, and every incentive to provide more water to those who can pay the most for it. So while WSR pushes privatization and full cost recovery as antidotes to the environmental problem of water scarcity, these policies may have precisely the opposite effect: private service providers will find new, possibly environmentally damaging sources of water (e.g., dams) that supply more water so that more water can be paid for, and encourage household waste so that households pay more.

- *Non-human* needs do not figure in the reforms.

Given that it allegedly stems from environmental imperatives, it is odd that WSR does not address human use of water's impact on non-human species or processes. One can judge the environmental impacts of these reforms not only on how they address environmental human needs, but also on how they address the needs of non-human individuals, communities, species, and ecosystem processes in present and future generations.

If water is a resource for non-human species and processes, states must enact and enforce laws that ensure that water remains available to sustain these species and processes. But when water is commodified, it is hard to see how any bit of water not used by the highest bidding humans would not be a waste.[63] The World Bank pays lip service to environmental priorities, such as 'restoring and preserving aquatic ecosystems and guarding against over-exploitation of groundwater resources, giving priority to the provision of adequate water and sanitation services for the poor'.[64] The Bank cites the former Chair of the Central Water Commission's critique of the National Water Policy, 2002: 'The need to balance the use of water with its deliberate non-use in order to maintain environmental balance of the riverine, estuarine, and the coastal ecosystems is negated.'[65]

Yet the same Bank report points favourably to huge new dam and other infrastructure projects.[66] While the World Bank recognizes that non-human nature suffers from current patterns of water allocation, its solutions – privatization, full cost recovery, local participation, large infrastructure – would seem to prioritize or

62 Antoinette Hilderling, *International Law, Sustainable Development and Water Management* 109 (Delft: Eburon, 2004).

63 Jayanta Bandyopadyay and Shama Perveen, 'Interlinking of Rivers in India: Assessing the Justifications', 39 *Economic & Political Weekly* 5307 (2004).

64 World Bank, note 32 above, Section 2(d).

65 Briscoe and Malik, note 3 above, 22–23.

66 *Ibid.*, at 23, 59.

encourage exclusive use by humans. How efficient pricing would help non-humans is also not addressed. Fish (and the river ecosystems in which they swim) can't pay for their own upkeep or participate in WUAs.

B. The legal framework: Examining whether water sector reform complies with environmental law

IFIs, private businesses, and political leaders have joined to forge the seven principles that guide modern WSR in India and elsewhere. Whether this approach to WSR will alleviate water scarcity and bring a reliable source of clean water to all citizens is questionable. Whether or not this approach to WSR is *legal* is also questionable.

Below I cite international water treaties, environmental human rights law, and emerging principles of customary international environmental law to examine whether or not the current, dominant approach to WSR adheres to environmental law. To the extent that WSRs actually provide more clean water to more citizens now and in the future, and create or preserve enough water to maintain functioning aquatic ecosystems, they will generally cleave international treaty-based, domestic, and customary international environmental law. To the extent WSRs undercut these goals, they violate the law.

1. Treaty law

UN Convention on the Law of the Non-Navigational Uses of International Watercourses

Much of WSR in India and elsewhere has implications for how neighbouring riparians may use shared waterways. Even when waterways or aquifers are not shared transboundary resources, principles that govern shared resources may influence domestic policy making.

The UN Convention on the Law of the Non-Navigational Uses of International Watercourses (UN Watercourses Convention) aims to ensure fair and equitable use of freshwater resources for current and future generations.[67] Overwhelmingly adopted by the UN General Assembly in 1997, only fourteen nations have signed it and it has not yet entered into force.[68] Nonetheless, it codifies three equity principles of customary international environmental law when applied to freshwater resources, and has served as a template for a growing number of regional water agreements.[69] While these environmental law principles

67 UN Watercourses Convention, note 7 above, and see Chapter III.

68 David Freestone and Salman M. A. Salman, 'Ocean and Freshwater Resources', in Daniel Bodansky, Jutta Brunnée, and Ellen Hey (eds.) *The Oxford Handbook of International Environmental Law* 350 (Oxford: Oxford University Press, 2007) and McCaffrey, note 48 above, 455.

69 Wouters, note 9 above, 42; McCaffrey, note 48 above, 376 and Freestone and Salman, note 68 above, 355.

apply to transboundary freshwater management, they can also be translated into ethical and legal guide rails for how neighbouring states within a nation, or simply neighbouring property owners, should jointly manage water resources to ensure that all may enjoy the right to water.

Article 5 stipulates that states must use shared watercourses in an 'equitable and reasonable manner'. To determine what is 'equitable and reasonable', some states must circumscribe their sovereign right to exploit their own resources, and consider the social and economic needs of the watercourse states concerned, the population dependent on the watercourse, and the availability of alternatives.[70] In *Hungary v. Slovakia*, citing the UN Watercourses Convention, the ICJ affirmed the right to equitable and reasonable sharing of transboundary waterways.[71]

Article 7 stipulates an '[o]bligation not to cause significant harm' to the quality or quantity of water for other states sharing the watercourse, and a duty to mitigate and compensate states once harm occurs. The Convention does not mention rights explicitly, but stipulates that when watercourse uses conflict, 'special regard' must be 'given to vital human needs'. Article 21 makes special note of the obligation to prevent pollution that may cause significant harm, 'including harm to human health and safety'. Finally, Articles 12 and 13 specify the need to notify and share information about activities that may adversely affect other riparian states.[72]

Thus the UN Watercourses Convention codifies emerging principles of equitable and reasonable utilization, no significant harm, and prior notification/cooperation at the intersection of environmental and human rights law. These principles work in tandem: states are obliged to prevent significant harm, and to give and receive information freely about water resource allocation and infrastructure development in order to ensure equitable and reasonable use.[73]

The Convention includes ecological considerations. Articles 20–23 require parties to protect and preserve ecosystems, prevent and control pollution, prevent the introduction of alien species, and protect the marine environment from polluted freshwater flows. Even further, 'equitable and reasonable utilization' and 'no significant harm' may be applied to non-human entities and processes that depend on shared waterways: Article 21 specifies that states must 'prevent, reduce and control pollution of an international watercourse that may cause significant harm to other watercourse States or to their environment'.[74]

One could argue that the South African reserve system makes domestic law

70 UN Watercourses Convention, note 7 above, Arts. 5 and 6.

71 *Case Concerning the Gabcikovo-Nagymaros Project (Hungary v. Slovakia)*, Judgment of 25 September 1997, ICJ. Rep 92, para. 85.

72 UN Watercourses Convention, note 7 above, Arts. 7, 10.2, 21.2, 13 and 14.

73 McCaffrey, note 48 above, 376, 402–3.

74 UN Watercourses Convention, note 7 above, Arts. 20–23 and McCaffrey, note 48 above, 376, citing these provisions as cementing as customary international environmental law the principle that states must protect ecosystems.

from these international legal principles with respect to non-human needs. South Africa's National Water Act mandates a reserve, whose requirements must be secured before any other water is allotted. The reserve must maintain enough water to fulfil basic human *and* non-human needs, which is the water required to protect the aquatic ecosystems of the water resource.[75]

Nothing similar exists in India. When users over-consume or pollute shared river basins, or build massive infrastructures (e.g., dams or river linking), they violate principles of customary international environmental law codified in the UN Watercourses Convention if they impact water flow into downstream riparians, which they would certainly do. Furthermore, while these widely accepted – and rather non-demanding – principles of customary international environmental law are available to officials or IFIs to adopt for managing water within the nation, they are absent from World Bank policy documents or India's National Water Policy, 2002. Backing down from its original position of absolute sovereignty over its shared watercourses, India has, since 1976, endorsed the principle of equitable and reasonable use. Its 1996 Ganges treaty with Bangladesh endorses this and the 'no significant harm' principle.[76]

Yet for the World Bank and the Indian government, dams are on the table. Despite the damage to non-human species and ecosystems, despite the potential violations of the UN Watercourses Convention for shared watercourses, and despite the internal upheavals and human rights violations that have occurred in India from past dam-building (which the World Bank acknowledges), the World Bank poses more dams as a certainty. Briscoe and Malik advocate 'willingness to consider financing high-return infrastructure that can be built to reasonable social and environmental standards'.[77] It is not clear what those standards are. Furthermore, without negotiating with Bangladesh over fair and equitable use, India is contemplating a river-linking project that would be the largest ever attempted, altering the courses of the Brahmaputra and Ganges before they reach Bangladesh. The project would construct 1,000 km of canals and 300 reservoirs, and flood 8,000 sq km of land currently housing three million people.[78] Obviously these new massive projects fulfilling WSR's goals would violate the transboundary rights of riparian nations or other Indian states to equitable and reasonable use, no significant harm, and prior notification/cooperation.

75 Alix Gowlland Gualtieri, 'South Africa's Water Law and Policy Framework: Implications for the Right to Water', in Philippe Cullet, Alix Gowlland Gualtieri, Roopa Madhav and Usha Ramanathan (eds.) *Water Law at the Crossroads – National and International Perspectives with Special Emphasis on India* 389 (New Delhi: Cambridge University Press, 2009); South Africa, Water Services Act, 1997 and South Africa, National Water Act, 1998, Chapter 3, Part 3.

76 UN GAOR, 31st Sess., Special Political Committee, 21st Meeting, 1976 2–3; cited in McCaffrey, note 48 above, 117, 292; Treaty on Sharing of the Ganges Waters at Farakka, New Delhi, 12 December 1996, 36 *ILM* 519 (1997).

77 Briscoe and Malik, note 3 above, xxii.

78 McCaffrey, note 48 above, 296.

Convention on Biological Diversity and Ramsar Convention on Wetlands

India has ratified both the Biodiversity Convention and the Ramsar Convention. The overarching goal of the Biodiversity Convention is the 'conservation of biological diversity, the sustainable use of its components, and the fair and equitable sharing of the benefits arising out of the utilization of genetic resources'. As in the UN Watercourses Convention, states have the sovereign right to exploit their own natural resources, but that right is circumscribed by the obligation not to cause damage to the environment of other states. Signatories must '[d]evelop national strategies, plans or programmes for the sustainable use of biological diversity' and 'integrate [. . .] the conservation and sustainable use into relevant sectoral, cross-sectoral plans, programs, and policies'. States are to promote '[i]n-situ conservation', where they are to '[r]egulate or manage biological resources important for the conservation of biological diversity whether within or outside protected areas, with a view to ensuring their conservation and sustainable use' and to '[p]romote the protection of ecosystems, natural habitats, and maintenance of viable populations of species in natural surroundings'.[79] The goal of the Ramsar Convention is 'the conservation and wise use of all wetlands through local, regional and national actions and international cooperation, as a contribution towards achieving sustainable development throughout the world'.[80]

India's National Water Policy, 2002 pays lip service to some of these principles. Water 'has to be treated as an essential environment for sustaining all life forms'. While planning and implementation of water involves 'socio-economic aspects and issues such as sustainability', nowhere in the National Water Policy, 2002 do the authors explain how to prioritize the resources named in the Biodiversity Convention or Ramsar. Ecology is the fourth of six (adjustable, depending on circumstances) priorities for allocating water. When planning projects, we know only that 'the preservation of the environment and the ecological balance should be a primary consideration'.[81]

How this will all happen when WSR promotes large-scale dams or irrigation projects that are clearly contemplated in the National Water Policy, 2002, and river-linking projects planned elsewhere, is not clear. It is axiomatic that dams and massive river-linking projects do great harm to functioning ecosystems, habitats, and non-human creatures that depend on riparian zones and other wetlands. Irrigated crops for agriculture, supported by IFI lending policies, demand great amounts of water and often lead to pollution from biocides and chemical fertilizers.[82] As discussed above, it is not clear how plans that promote the privatization of water resources, with concomitant full cost recovery, will fail to lead to *over*-development (not conservation) of water resources – thus more

79 Biodiversity Convention, note 7 above, Arts. 1, 3, 6, 8(c).

80 What is the Ramsar Convention on Wetlands? Available at www.ramsar.org/about/info2007-02-e.pdf.

81 National Water Policy, 2002, Sections 1.3, 5, and 6.2.

82 McCaffrey, note 48 above 7, 448.

pumping of rivers or aquifers, more irrigated export agriculture, and more pressure to build dams or link rivers. Principles of WSR give insufficient thought to environmental treaty provisions, and India and other nations who follow standard WSR prescriptions will likely violate their treaty obligations to steward their biological resources.

2. *Human right to water and right to a healthy environment*

Because water is a fundamental requirement of life for all humans and non-human creatures, it should be the source from which all WSR springs. The human right to water is both a starting point for policy that complies with international and domestic environmental and human rights law, and a performance measure against which one can measure the legality of government or IFI policy prescriptions. Each proposal for WSR should be measured against this question: How does this proposal further the right to a basic, domestic supply of clean water for present and future generations of humans? At the very least, the percentage of households who receive basic water supply before/after privatization, and the quantity, quality, and affordability of that water give a yardstick by which we can measure the human rights-based legality of WSR.

Chapter XI documents the growing acceptance that access to a sufficient amount of clean water is a basic human right. As party to the major human rights conventions, India has treaty-based responsibilities to respect, protect, and fulfil a host of rights from which the right to water may be derived.[83] The Indian Supreme Court has interpreted the '[p]rotection of life and personal liberty', enshrined as a fundamental right by Article 21 of the Constitution, to implicitly require a right to water.[84]

National governments that disinvest themselves of responsibility for providing clean drinking water do not simultaneously disinvest themselves of legal obligations at the intersection of environmental and human rights law. To the extent that India implements WSR that provides more clean, affordable water to more people, those reforms help India respect, protect, and fulfil the right to water – whether it's a treaty-based obligation, or one demanded by the Constitution's right to life.

India and the World Bank argue that its full cost recovery, privatization, and other WSR policies, including dam and river-linking projects, do advance the right to water for more Indian citizens. Briscoe claims that WSR makes water more affordable, and widens convenient access, which frees up more time for the poor to pursue other money-making opportunities. WSR promoters even pose dams as human rights-*enhancing*:

83 International Covenant on Civil and Political Rights, International Covenant on Economic, Social and Cultural Rights, and Convention on the Rights of the Child. See also Chapter XI.

84 Constitution of India, Art. 21; *F.K. Hussain v. Union of India*, AIR 1990 Ker 321 and *Venkatagiriyapps v. Karnataka Electricity Board & Others*, 1999 (4) Kar LJ 482.

> Much of the major water infrastructure which will be built in India in coming decades includes hydropower. Hydropower projects generate large revenues, and in most cases the number of people to be resettled by hydropower projects in India will be relatively small [. . .] It is therefore a doable task, with few difficult tradeoffs, to ensure that local people are major beneficiaries of such projects. This is not only ethically the right thing to do, but it means that costly delays in project implementation can be avoided.[85]

So framed, dams are ethically acceptable and promote human rights. Of course, according to Briscoe, 'India remains a country in which there are serious issues about how affected people – many of whom are from Scheduled Tribes – are dealt with in major infrastructure projects'. But human rights-enhancing opportunities abound! India has seen 'considerable progress, especially by modernizing hydropower companies, but there is still a long way to go before practice in India can compare favorably to practice in, say, China, where resettlement is considered to be «a development opportunity» rather than a cost'.[86] In the World Bank philosophy, dams are pro-people.

Despite proponents' claims, it is not clear whether WSR promotes the human right to water. Chapter XI reviews how WSR may impair state responsibility to respect, protect, and fulfil the right to water for its citizens. To 'respect' the right to water means to not actively prevent it from being realized. Damming rivers and displacing thousands of citizens directly violates the right to water, and damming or polluting shared watercourses may violate the right to water of downstream riparians.[87] Pollution run-off from large-scale, privatized, irrigated agriculture may destroy surface or groundwater supplies. Privatizing water may make the price of water beyond reach for the poor, with no incentive to provide the service for free (or cheap); it may result in a failure to hook up to water supplies for those who cannot afford the initial cost; it may mean direct disconnection from the water supply for those who fail to pay, as it has in various locales in Africa. For example, in one South African project, to enforce payment, a utility 'included water cut-offs, removing meters and portions of pipes to prevent illegal reconnections, reducing 24-hour supply to intermittent hours throughout the day and night, and installing restrictors in household taps to reduce the flow of water'.[88]

To protect water, according to General Comment No. 15, includes requiring

85 Briscoe and Malik, note 3 above, 59.

86 *Ibid.*

87 See, e.g., Philippe Cullet, 'The Sardar Sarovar Dam Project: An Overview', in Philippe Cullet, (ed.) *Sardar Sarovar Dam Project: Selected Documents* 1 (Aldershot: Ashgate 2007). McCaffrey, note 48 above, 371.

88 McDonald and Ruiters, note 30 above, 28; Sean Flynn and Danwood Mzikenge Chirwa, 'The Constitutional Implications of Commercializing Water in South Africa', in McDonald and Ruiters (eds.) note 30 above, 68; Laila Smith *et al.*, 'Public Money, Private Failure: Testing the Limits of Market Based Solutions for Water Delivery in Nelspruit', in McDonald and Ruiters (eds.) note 30 above, 139.

'State parties to prevent third parties from interfering in any way with the enjoyment of the right to water' and requiring adoption of 'necessary and effective legislative and other measures to restrain, for example, third parties from denying equal access to adequate water; and polluting and inequitably extracting from water sources, including natural sources, wells and other water distribution systems'. States must prevent private service providers from compromising equal, affordable, and physical access to sufficient, safe and acceptable water through an effective regulatory system, which includes independent monitoring, genuine public participation and imposition of penalties for non-compliance.[89] And to fulfil the right to water would mean that the government proactively promotes realization of the right

India and the World Bank (one of the third parties mentioned above) argue that WSR helps India to protect and fulfil basic human rights obligations; but absent close central government regulation (requiring, *inter alia*, no prepaid service requirements, no disconnections, cross-tariffs with strict standards to ensure water provision for all households) to make sure that the needs of all Indians are met, India may fail in its obligations. Treating water as a public service that the government provides, and educating people on this right, would ensure that India meets its requirement to protect and fulfil the right to water.

Finally, the right to water is a subset of a more general right to a clean and healthy environment. An Earthjustice report to the UN Human Rights Commission documents the 'repeated and increasing recognition of a human rights-based approach to environmental protection'.[90] The Indian Constitution guarantees: 'The State shall endeavour to protect and improve the environment and to safeguard the forests and wild life of the country.'[91]

The Indian government seems to be doing little to fulfil its constitutionally mandated responsibility to protect the environment as it implements WSR. As noted above, WSR in India (or elsewhere) demands nothing like the South African reserve system that maintains water for future human and current and future non-human use. Dams proposed to alleviate water scarcity (and benefit the forces behind privatization), and the pollution generated from irrigated agriculture (facilitated by dams) undermines the government's constitutional and international mandated duties to protect the environment and thus violate citizens' right to a healthy environment.

89 See Chapter XI and the Committee on Economic, Social and Cultural Rights, General Comment No. 15: The Right to Water (Articles 11 and 12 of the International Covenant on Economic, Social and Cultural Rights), UN Doc. E/C.12/2002/11 (2002), paras 23 and 24.

90 Earthjustice, Environmental Rights Report: Human Rights and the Environment 7 (Materials for the 61st Session of the United Nations Commission on Human Rights, Geneva, 14 March–22 April 2005), available at www.earthjustice.org/library/references/2005_ENVIRONMENTAL_RIGHTS_REPORTrev.pdf.

91 Constitution of India, Art. 48A.

3. Principles of customary international environmental law

Unlike hard law inscribed in national legislation, constitutions, or international treaties, some legal norms in international environmental law exist in a twilight realm: they may or may not legally require actors to follow them, and they may not be enforceable should actors violate them. Still, they carry moral and political force and should guide the decision making of policy makers crafting WSR.[92] According to Bodansky *et al.*, 'these norms operate as broad principles that frame legal discourse and diplomacy – so their formal legal status is of only limited practical significance. They play a significant role in the broader process of persuasion and justification that characterizes international environmental law, rather than as rules that govern behaviour'.[93]

These norms thus serve multiple functions in evaluating WSR. In some cases, enterprising public interest lawyers can use them as bases for litigation, as they have done with the public trust doctrine in India.[94] That is to say, violations of these principles may be actionable offences employed in a court of law to prevent plans or policies from coming into effect or continuing to operate. In other cases, to the extent that governments view these principles as *opinio juris*, they may be compelled to follow them, or shamed into not violating them. These norms may also inspire policy makers or NGO activists who can look to them – and to treaties or national policies that codify them – for overarching principles, concrete plans, or policy points they can attempt to implement when managing water.

Many of these principles are well developed for environmental problems as they pertain to human health and wellbeing. But part of the concordance with environmental law – be it treaty based or customary – means caring for and perpetuating non-human species and preserving functioning ecosystems in their own right. Thus as I explore the implications of WSR for environmental law, I discuss the implications for both humans and non-humans.

Sustainable development

Sustainable development consists of four core elements:

(1) Economic, social, and environmental policies must be integrated.
(2) The needs of present and future generations of humans must be considered (inter-generational equity).
(3) The needs of the poorest humans must be given first priority (intra-generational equity).
(4) 'Nature is the true infrastructure of society' and thus ecosystem services

92 Ulrich Beyerlin, 'Policies, Principles, and Rules' in Bodansky *et al.* (eds.), note 68 above, 447.
93 Daniel Bodansky, Jutta Brunnée and Ellen Hey, 'International Environmental Law: Mapping the Field', in Bodansky *et al.* (eds.), note 68 above, 23.
94 *MC Mehta v. Kamal Nath* 1997 (1) SCC 388.

must be preserved for the needs of present and future generations (usually of humans).[95]

Because these elements of sustainable development are vaguely defined, it is difficult to hold an actor legally responsible for fulfilling responsibilities, or for deriving policy standards against which one could measure an action in its name. Nonetheless, we can examine particular WSR policies to predict if they are likely to narrow the gap between rich and poor in the present and in future generations, and to see whether they are likely to degrade the natural environment upon which human communities depend.

WSR often grants lip service to the integration of economic, social, and environmental policies. But the reforms that emerged from the Dublin Principles – promoted by the World Bank, implemented in India – have seized the economic at the expense of the social and environmental. Efforts to restructure management to promote water as an economic good seldom make it clear how to ensure environmental safeguards for human or non-human uses of water, or how the needs of the poor will derive from the economic policy prescriptions: thus intragenerational equity is far from certain under devolved state responsibility, full cost recovery, unregulated user participation, and privatization. In the absence of safeguards, intra-generational inequity is more likely to occur as private actors get wealthier by selling water, or by buying water resources that they then use to generate even greater profits through generating energy, manufactured goods, or irrigated crops. Nothing like a South African reserve exists in WSR for India to protect future generations of humans and the current generations of non-humans. Water commodification will probably lead to greater exploitation and more use of scarce resources to foment profit in the present generation at the expense of future generations, thus impeding intergenerational equity. Finally, WSR key principles do not give priority to non-human species or processes; commodification that leads to more water use and huge infrastructure projects like dams, desalination plants, and river linking will do more harm than good to the non-human world.

Common but differentiated responsibility

Codified as Principle 7 of the Rio Declaration, common but differentiated responsibility (CBDR) means that nations who have created environmental problems, and as a result have acquired wealth that gives them the financial and technical means to mitigate those problems, are legally obliged to offer aid to help other states to cope with these problems.[96] It is the legal anchor of the Kyoto

95 Daniel Barstow Magraw and Lisa D. Hawke, 'Sustainable Development', in Bodansky *et al.* (eds.) note 68 above, 619–620.

96 Rio Declaration on Environment and Development, Rio de Janeiro, 14 June 1992, UN Doc. A/CONF.151/26/Rev.l (Vol. l), Principle 7; Philippe Cullet and Annie Patricia Kameri-Mbote,

Protocol, which stipulates the disproportionate responsibility of Northern nations to take steps to mitigate global climate change.[97] The UN Watercourses Convention requires common responsibility for riparian management, but emphasizes differing social and economic needs, and suggests compensation from polluter to polluted.[98] The Plan of Implementation of the World Summit on Sustainable Development stresses that states with greater wealth must help those with fewer means raise the proportion of people who have access to safe water.[99]

While CBDR promotes intra- and inter-generational equity, it is usually not applied in a domestic context (of course, it is always available as an ethical principle; the rich in any nation could follow its moral rationale and legal prescriptions and treat the poor accordingly). Instead, it usually pertains to the responsibility of Northern nations to Southern nations, because of the disproportionate share of environmental goods the North has appropriated, and the environmental pollution the North has emitted.[100] IFI lending, while not explicitly driven by CBDR, is touted as being driven by the desire of the North to help the South.

How does this pertain to WSR? Water reform in Southern Africa has opened up water provision to private sources, and twenty-eight large multinational water companies have set up shop in the region.[101] The World Bank adopted a French model of privatized water sources, which allowed French multinationals to exploit untapped water markets worldwide.[102] Not only is CBDR not operating as an equity principle within India or other developing nations; rather, Northern-based multinational water service providers are actually taking profits out of these nations, flowing capital from South to North, rather than vice versa as CBDR would mandate. To fulfil CBDR requirements, any privatization requires that capital flow from North to South, which is an unlikely result where profit is at stake and Northern actors have the technology, political clout, and business acumen to earn that profit.

'Activities Implemented Jointly in the Forestry Sector: Conceptual and Operational Fallacies', 10 *Georgetown Int'l Envtl. L. Rev.* 98, 102 (1997); Anita M. Halvorrsen, 'Common But Differentiated Commitments in the Future Climate Change Regime: Amending the Kyoto Protocol to Include Annex C and the Annex C Mitigation Fund', 18 *Colorado J. Int'l Envtl. L. & Pol'y* 247, 254 (2007).

97 Lavanya Rajamani, 'The Nature, Promise, and Limits of Differential Treatment in the Climate Regime', 16 *Yb. Int'l Envtl. L.* 81, 86 (2007) and Kyoto Protocol to the United Nations Framework Convention on Climate Change, 10 December 1997, Arts 10 and 12.

98 UN Watercourses Convention, note 7 above, Arts 6 and 7.

99 Plan of Implementation of the World Summit on Sustainable Development, Report of the World Summit on Sustainable Development, Johannesburg, 26 August–4 September 2002, UN Doc. A/CONF.199/20, para. 14.

100 Dinah Shelton, 'Equity', in Bodansky *et al.* (eds.) note 68 above, 657.

101 McDonald and Ruiters, note 30 above, 24.

102 Finger and Allouche, note 14 above, 91.

Polluter-pays principle

Principle 16 of the Rio Declaration states that '[n]ational authorities should endeavour to promote the internalization of environmental costs and use of economic instruments, taking into account the approach that the polluter should, in principle, bear the costs of pollution'.[103] This principle is also found in the Convention on the Protection and Use of Transboundary Watercourses and International Lakes, which sets principles for participating nations to govern transboundary water.[104] In conjunction with CBDR and the preventative principle, we find a moral and legal presumption that entities should first strive to prevent pollution; those who do pollute the global commons should internalize the cost of that pollution; and, in some versions of the polluter-pays principle, polluters also clean up and compensate those who will be most grievously affected by their pollution.[105] While the polluter-pays principle tends to apply to states, domestic laws – for instance, the United States' CERCLA/Superfund law – apply the principle to private actors.[106] The Indian Supreme Court has declared it to be a part of customary international environmental law, stating that:

> If a polluter refuses to pay compensation, his industry will be closed, and the compensation recovered as arrears of land revenue. If an industry sets up necessary pollution control devices now, it is still liable to pay for the past pollution it has generated.[107]

In India and elsewhere, principles of full cost recovery, privatization, and tradable water rights mean that, based on the most lucrative use of water, the highest bidder will obtain water, and highly polluting industries – industrial development and corporate, irrigated agriculture – will claim a disproportionate share of water rights. Despite the Supreme Court's ruling that the polluter-pays principle obtains in India, in the absence of strict court scrutiny or government regulation, it is not clear how the polluter-pays doctrine would be respected in India. The polluter will pay for the water, but nothing in WSR requires them to internalize the costs of pollution or pay for the externalities that the pollution causes![108] Thus WSR – derived from the environmental problem that people allegedly waste and

103 Rio Declaration on Environment and Development, Rio de Janeiro, 14 June 1992, UN Doc. A/CONF.151/26/Rev.l (Vol. l).

104 Convention on the Protection and Use of Transboundary Watercourses and International Lakes, Helsinki, 17 March 1992, UN Doc. ENWA/R.53, Art. 2(5)(b).

105 Hilderling, note 62 above, 210, 159; United Nations Development Programme, *Human Development Report 2007/2008: Fighting Climate Change: Human Solidarity in a Divided World* 41 (Basingstoke: Palgrave Macmillan, 2007) and Sumudu A. Atapattu, *Emerging Principles of International Environmental Law* 438–9, 463 (Ardsley, NY: Transnational Publishers, 2006).

106 US Comprehensive Environmental Response, Compensation, and Liability Act, Title 42, Chapter 103.

107 *Vellore Citizens' Welfare Forum v. Union of India* (1996) 5 SCC 647.

108 Hilderling, note 62 above, 104.

contaminate water because it is *not* an economic good – imposes a managerial solution that may well produce *more* pollution, where the polluter will *not* pay, but the poor will pay in terms of contaminated water sources. This is an empirical prediction that can be used as a standard to test the prudence – and the environmental legality – of WSR. For example, water quality can be tested before and after WSR is implemented in a region to see the effects of commodification and privatization of water. And if pollution increases, international law dictates that the polluter should not only be required to prevent it, but should be required to clean up after itself.

Public trust doctrine

Fifteen hundred years ago, the Roman Empire's leading jurists' wisdom became codified in the *Corpus Juris Civilis*. The Emperor Justinian added these words to one section: 'By the law of nature these things are common to all mankind, the air, running water, the sea and consequently the shores of the sea.'[109] This notion – the public trust doctrine – suggests that certain resources (usually water) are shared property of all citizens that the state must protect in perpetuity.[110] The notion that certain resources are fundamental to human life, and that governments must steward these resources, has spread throughout civil and common law legal systems.

In 1997, the Indian Supreme Court installed the public trust doctrine as law. In that case, the Minister of the Environment impermissibly allowed a motel to be built at the mouth of a river, and allowed the motel to change the river's course, creating floods in nearby villages, in violation of the public trust doctrine. Before invoking the public trust doctrine, the Court alludes to:

> the classic struggle between those members of the public who would preserve our rivers, forests, parks and open lands in their pristine purity and those charged with administrative responsibilities who, under the pressures of the changing needs of an increasingly complex society, find it necessary to encroach to some extent upon open lands heretofore considered inviolate to change.[111]

The public trust doctrine is a sound legal footing on which to base any WSR. In India, of course, it is the law of the land. But it also encompasses many other environmental law principles: the human right to water (the government must steward water resources to meet the basic needs of all citizens); fair and equitable use; impermissibility of significant harm (to be policed by the state steward);

109 Thomas Glyn Watkin, *An Historical Introduction to Modern Civil Law* 63 (Aldershot: Ashgate, 1999) and Mark Dowie, 'In Law we Trust', 22 *Orion* 18, 21 (2003). *See also* Atapattu, note 105 above, 438.

110 Dowie, note 109 above, 20.

111 *MC Mehta v. Kamal Nath* 1997 (1) SCC 388.

polluter-pays (the government must be vigilant about private users spoiling the common resource); and environmental democracy (a fundamental procedural right for affected members of the public to appeal directly against government decisions that compromised resources held in the public trust).[112] Policies implemented under WSR that allow the government to abandon its stewardship responsibilities and allow private interests to arrogate water resources violate the public trust doctrine and should be illegal in India.

Environmental democracy

Environmental democracy, namely the right to meaningful participation in environmental matters, is an emerging customary international environmental law norm. Principle 10 of the Rio Declaration asserts that concerned citizens should have the right to access all pertinent information about environmental decision making, to participate in all decision-making processes, and to access all judicial and administrative proceedings, including redress and remedy.[113] We might also include the requirement of environmental impact assessments with full public input and participation, and a general right of access to just governance.[114] The UNECE Aarhus Convention (open to all nations to join) is at the forefront of attempts to codify and implement these principles of environmental democracy, but they are finding their way into various multilateral environmental agreements and into national constitutions or statutes. The European Court of Human Rights has asserted that the right to access information about environmental hazards is fundamentally linked to respect for privacy and family life. The African Convention on the Conservation of Nature and Natural Resources also guarantees access to justice in these issues.[115]

The National Water Policy, 2002 promotes public participation 'from the project planning stage itself'.[116] At first blush, the WSR principle that calls for user participation seems to promote environmental democracy in WUAs that govern irrigation, or as embodied in the World Bank-backed Swajaldhara Guidelines that require local water users to take responsibility for operating water systems.[117] But Cullet claims that this participation is actually imposed on users; both WUAs and the Swajaldhara Guidelines (which are not part of any democratically passed law) require beneficiaries to manage and fund the maintenance of the water

112 Joseph L. Sax, *Defending the Environment: A Strategy for Citizen Action* 235 (New York: Borzoi Books, 1971).

113 Rio Declaration on Environment and Development, Rio de Janeiro, 14 June 1992, UN Doc. A/CONF.151/26/Rev.l (Vol. l) and Atapattu, note 105 above, 13.

114 Agenda 21, Report of the UNCED, Rio de Janeiro, 3–14 June 1992, UN Doc. A/CONF.151/26/Rev.1 (Vol. 1, Annex II), Chapter 18. Called for in Convention on the Protection and Use of Transboundary Watercourses and International Lakes, Helsinki, 17 March 1992, UN Doc. ENWA/R.53. See also Atapattu, note 105 above, 305–6.

115 Jonas Ebbesson, 'Public Participation' in Bodansky *et al.* (eds.) note 68 above, 699, 701.

116 National Water Policy, 2002, Section 6(8).

117 See Chapter VI.

infrastructure – whether they want to or not.[118] User participation is not a model of participatory democracy; participation in both schemes is often limited to water users, and those who can't afford to pay or those who are not landowners may be excluded from participating. Furthermore, these groups are not drawn from constitutionally mandated, democratic panchayati raj associations and may simply mirror local power arrangements where women or lower caste members lack proportional representation.[119] The right to access all judicial and administrative proceedings, including redress and remedy, is lacking in WUAs or the Swajaldhara Guidelines. No environmental impact assessments are monitoring the prospective environmental impacts of policy changes that disinvest the government from providing water as a basic right. Despite rhetoric about local community involvement and participation, both WUAs and communities operating under Swajaldhara face onerous, non-democratic, possibly discriminatory systems of water management.

CONCLUSION

WSR ought to follow the precepts of domestic and international environmental law; it does not always do so. In particular, the drive to disinvest the state in favour of private service providers with some form of (circumscribed) community involvement may run counter to the dictates of environmental law to preserve species and ecosystems, respect and protect the human right to clean water, promote sustainable development, foment environmental democracy, and preserve water resources in the public trust for vital uses of present and future generations.

Water sector reforms are moving away from, not towards, compliance with international environmental law and the domestic environmental law of many nations, including India. The World Bank and other powerful actors promoting WSRs do not feel the need to heed codified or customary environmental law. Rather than adhering to the dictates of sustainable development, or conserving and stewarding resources, or prioritizing the rights of the poor and the non-human (be they in current or future generations), WSRs further marginalize them. Allegedly implemented to solve a problem of environmental scarcity and contamination, WSRs seem more likely to exacerbate those problems. Rather than seeing the public trust doctrine or the human right to water as legal rules rooted in moral philosophies protecting the dignity of human life, and basing policies on those principles, WSRs see water resources as exploitable and available to the highest bidder. The poor and the non-human are unlikely to emerge better off as government devolves its responsibility for water management to private sources that profit by charging the full cost of water.

118 See Chapter VI in this volume; Cullet, note 28 above, 216; Ministry of Rural Development, Guidelines on Swajaldhara, 2003.

119 Cullet, note 28 above, 216 and Chapter VI.

XIII. Water, health and water quality regulation

Sujith Koonan and Adil Hasan Khan

INTRODUCTION

The obvious link between the protection and promotion of public health and access to safe and adequate water is indisputable.[1] Access to safe water is essential to health which is a basic human right.[2] Safe drinking water plays a major role in the overall health of human beings. It is estimated that about 70 to 80 per cent of illnesses are water and sanitation related. It must be noted that women and children suffer most from lack of access to safe water. The total number of deaths caused by malnutrition due to unsafe water, inadequate sanitation and insufficient hygiene is estimated at 860,000 per year in children under five years of age.[3] Unsafe drinking water has also been identified as one of the major health problems in India. It is estimated that water and sanitation-related infections and diseases constitute 60 to 80 per cent of all illnesses in India.[4]

This link and implications could be considered as one of the major reasons why governments across the world have played and continue to play a key role in providing water supply to all. Added to that, access to safe and adequate water is also perceived as an essential input to the overall development of the nation, both

1 Some studies expose the staggering health implications of unsafe and inadequate access to water, which are estimated to account for 1.73 million deaths each year due to diarrhoeal diseases. For details, see Guy Howard Jamie and Bartram, Domestic Water Quantity, Service Level and Health (Geneva: World Health Organization, 2003); Annette Pruss-Ustun *et al.*, Safer Water, Better Health: Costs, Benefits and Sustainability of Interventions to Protect and Promote Health (Geneva: World Health Organization, 2008) and Anil Agarwal *et al.*, *Water, Sanitation, Health – for All?: Prospects for the International Drinking Water Supply and Sanitation Decade, 1981–1990* (London: Earthscan, 1981).

2 World Health Organization, Guidelines on Drinking Water Quality (Geneva: WHO, 3rd edn 2006).

3 Pruss-Ustun *et al.*, note 1 above, 7.

4 Government of India, Implementation Manual on National Rural Water Quality Monitoring and Surveillance Programme 18 (New Delhi: Government of India, Ministry of Rural Development, 2004).

in social and economic aspects.[5] The development of policy and institutional frameworks *vis-à-vis* drinking water supply demonstrates this health concern. For instance, concern for public health was the major reason for bringing water supply under the direct and complete control of the government in Europe and North America during the nineteenth and early twentieth centuries.[6]

Public health concerns are also one of the most important reasons, if not the primary reason, behind the setting up of water supply systems in particular and water resource regulation and management in general. The evolution of a number of rules, regulations and standards in the international and Indian contexts seem to testify to this observation. In the international context, the World Health Organization (WHO) has formulated several guidelines and standards regarding water quality and quantity primarily to address health issues.[7] In the Indian context, one important example in this regard is the Manual on Water Supply and Treatment formulated by the Government of India.[8]

Despite the presence of a number of binding and non-binding instruments, the problem of unsafe and inadequate water supply and its implications for public health is often highlighted as a persisting critical issue.[9] This raises, importantly, the question of the legal nature of the existing framework and the effectiveness of implementation. In this context, this chapter tracks the existing law and policy framework in India and at the international level, addressing health concerns in water resource regulation and management. This chapter further seeks to address the reasons for any ineffectiveness in the implementation.

The scope of this chapter is limited to the examination of the regulatory framework, mainly in India, addressing water quality. This essentially includes analysis of water quality standards and guidelines. Unavoidably, this chapter also aims to examine the framework provided under the Water (Prevention and Control of Pollution) Act, 1974. Since an examination of water quality regulation in isolation would be highly artificial and incomplete, a brief account of its relation and link with relevant human rights (right to health, right to sanitation and human right to water) and the nature of the consequent government obligations are also examined.

5 N. Praveen *et al.*, 'Drinking Water Quality in Urban India: Why and How it is Getting Worse – A Case Study of South Indian Cities', in K.V. Raju (ed.) *Elixir of Life: The Socio-Ecological Governance of Drinking Water* 353 (Bangalore: Books for Change, 2006).

6 Werner Troesken, 'Typhoid Rates and the Public Acquisition of Private Waterworks – 1880–1920', 59/4 *Journal of Economic History* 927, 929 (1999).

7 See, e.g., World Health Organization, note 2 above.

8 Government of India, Manual on Water Supply and Treatment (New Delhi: Government of India, Ministry of Urban Development, 3rd edn 1999). See also Ministry of Rural Development, Agenda Notes for State Secretaries' Conference on Rural Drinking Water and Sanitation (New Delhi: Government of India, Ministry of Rural Development, 2008).

9 Central Pollution Control Board, Status of Water Quality in India – 2007 at 19 (New Delhi: Central Pollution Control Board, 2008). See also Smita Misra and Bishwanath Goldar, 'Likely Impact of Reforming Water Supply and Sewerage Services in Delhi', 43(41) *Economic & Political Weekly* 57 (2008).

A. Health and water: Historical context and development

1. General historical context

The importance of the quantity of water available for human consumption has been a major concern for a long time. However, the emphasis on quality came to light only after the sources and effects of drinking water contaminants came within human understanding in the latter half of the nineteenth century. During this period, it was proven by scientists that diseases such as cholera could be eradicated by providing safe water for all.[10] This understanding seems to have triggered government efforts to set water quality standards and to promote water treatment mechanisms. In fact, the idea and practice of water treatment is regarded as one of the most significant public health advances of the twentieth century.[11]

Water quality concerns seem to have been a major reason why governments in the United States and the United Kingdom took responsibility for water supply, particularly in urban areas. For instance, the outbreak of cholera in London in 1840 is considered as a major impetus for the enactment of the Metropolis Water Act, 1852. The government had begun to exercise significant control over suppliers by virtue of this legislation and required quality requirements such as filtering of water.[12] The concern of the government even resulted in the taking over of the operational assets of some private water supply companies.[13]

In the United States, the major highlighted rationale for institutional changes in water supply in the late nineteenth and the first quarter of the twentieth century was the protection of public health through preventing waterborne epidemics.[14] The emphasis on health concerns resulted in the wide promotion of water treatment methods such as filtration and chlorination.[15] The regulatory framework was also developed to ensure safety. Legal regulation of drinking water quality in the United States began in the early twentieth century. The federal regulation of drinking water was initiated in 1914 and set standards for the bacterial quality of drinking water. This initiative ultimately culminated in the enactment of Safe Drinking Water Act, 1974.[16]

Ever since the government started taking major initiatives in providing water supply, the main impetus was the perceived connection between unsafe water and

10 United States Environmental Protection Agency (USEPA), The History of Drinking Water Treatment, EPA-816-F-00-006, February 2006.

11 *Ibid.*

12 James Salzman, 'Thirst: A Short History of Drinking Water', Duke Law School Legal Studies Research Paper No. 92, December 2005, available at http://papers.ssrn.com/sol3/papers.cfm?abstract_id=869970.

13 *Ibid.*, at 21.

14 Troesken, note 6 above, 929.

15 *Ibid.*, at 930.

16 USEPA, note 10 above, 18.

epidemic diseases. Hence, the major focus was on 'safety'. Indeed, the regulatory effort in this regard in the West has successfully eradicated epidemics such as cholera.[17] This shows the crucial role of water quality regulation in public health advancement, and thereby overall social and economic development.

2. Developments at the international level

There is no binding legal instrument at the international level regulating water quality standards. One of the reasons for this could be the unfeasibility of having a universal regulatory approach and standards. It has been argued that no single approach regarding water quality regulation is universally applicable. The regulation largely depends upon such factors as needs, regulatory potential and the capability of individual countries.[18] Another reason that has been highlighted is the advantage provided by the use of a risk–benefit approach in the establishment of national standards and regulations.[19]

However, international instruments also play a crucial role in the regulation of water quality, particularly those prepared under the auspices of the WHO. Special mention in this regard should be given to the drafting and periodical updating of the Guidelines for Drinking Water Quality (hereafter the WHO Guidelines). The third edition in this series was published in 2006.[20] The role of the WHO as a knowledge producer is also significant because the WHO produces a number of educative and informative publications that could be a help or guidance for individual countries.[21]

The WHO initiatives can be considered as major developments at the international level directly addressing the issue of water quality regulation. However, there are other international instruments, though broad in scope, which recognize the close link between water quality regulation and public health. For instance, some documents expressly state the protection of human health and wellbeing through prevention, controlling and reducing water-related diseases.[22] The need for emphasis on water quality regulation can also be seen in a number of policy statements at the international level.[23] Some instruments at the regional level such

17 Maggie Black and Rupert Talbot, *Water – A Matter of Life and Health* 18 (New Delhi: Oxford, 2005).

18 World Health Organization, note 2 above, 2.

19 *Ibid.*, at 2.

20 *Ibid.*

21 For example, the WHO in collaboration with IWA Publishing brings about the *Journal of Water Health*. This journal covers a range of issues related to water and health.

22 See, e.g., Protocol on Water and Health to the 1992 Convention on the Protection and Use of Transboundary Watercourses and International Lakes, London, 17 June 1999, UN Doc. MP.WAT/AC.1/1999/1.

23 See, e.g., Agenda 21, Report of the UNCED, Rio de Janeiro, 3–14 June 1992, UN Doc. A/CONF.151/26/Rev.1 (Vol. 1, Annex II); Local Actions for a Global Challenge, the Ministerial Declaration of Fourth World Water Forum, 21–22 March 2006; UN General Assembly Resolution 58/217, International Decade of Action 'Water For Life', UN Doc. A/RES/58/217 (2004); United Nations Water Conference – Resolutions, in Report of the United Nations Water

as the Helsinki Declaration on Health and Environment, 1994 address this issue directly.[24]

These documents at the international and regional levels generally serve the purpose of highlighting the health impacts of unsafe and inadequate water and urging all member countries to take appropriate action. While the normative value of these instruments can be acknowledged, specific norms on water quality regulation are clearly missing at the international level. It could further be said that existing instruments are not more than mere policy statements by the international community broadly acknowledging the need for water quality regulation.

3. Developments in India

Water quality problem is a critical issue in India that needs to be addressed legally. For instance, fluoride and arsenic contamination of groundwater is widely cited as one of the major challenges in drinking water quality regulation. The reported population directly affected and at risk in case of fluoride and arsenic contamination is estimated as 25.1 and 71.3 million respectively.[25] The alarming situation of water quality in India is also confirmed by the Planning Commission of India in one of its documents. It is estimated that there are about 217,000 habitations suffering from water quality problems, with 118,088 habitations suffering from excess iron, 31,306 from excess fluoride, 23,495 from excess salinity, 13,958 from excess nitrate and 5,029 from excess arsenic.[26]

The issue of water quality gained the attention of the government long back. Water pollution was one of the earliest environmental issues on the agenda of the Indian State. Government initiatives to formulate a legal framework to ensure the quality of water resources started in the early 1960s. As early as 1962, the Ministry of Health had appointed an expert committee on water pollution.[27] This initiative eventually culminated in the enactment of the Water (Prevention and Control of Pollution) Act, 1974 (hereafter the Water Act).

Conference, Mardel Plata, 14–25 March 1977, United Nations publication, Sales No. E.77.II.A.12, available at www.ielrc.org/content/e7701.pdf; UN General Assembly Resolution 55/2, United Nations Millennium Declaration, UN Doc. A/RES/55/2 (2000).

24 Declaration on Action for Environment and Health in Europe adopted by the Second European Conference on Environment and Health, Helsinki, 20–22 June 1994.

25 Ramesh Chandra Panda, 'Water Quality Scenario in India – An Overview', in International Life Science Institute-India, *Water Quality Management: South Asian Perspective Vision 2025* at 37 (New Delhi: International Life Science Institute-India, 2002). In the case of Gujarat, it has been observed that water in as many as 38 per cent of habitations was found not to comply with WHO guidelines with respect to fluoride content and 23.6 per cent were above the maximum permissible limit. See Indira Hirway, 'Ensuring Drinking Water to All: A Study in Gujarat', in K.V. Raju ed., note 5 above, 78.

26 Planning Commission – Government of India, *Eleventh Five-Year Plan 2007–12 – Volume II – Social Sector* (New Delhi: Oxford University Press, 2008).

27 Bharat Desai, *Water Pollution in India: Law and Enforcement* 45 (New Delhi: Lancer Books, 1990).

The enactment of the Water Act can be seen as a landmark in the water quality regulatory framework in India. Apart from the Water Act, there are other legal instruments addressing water quality issues such as the Food Safety and Standards Act, 2006 and the Environmental (Protection) Act, 1986. For instance, the Food Safety and Standards Act, 2006 includes the quality regulation of packaged 'drinking water' and 'water used in the food during its manufacture' within its scope.[28]

In addition to these enactments, water quality issues are directly and expressly addressed through certain instruments at the national level. Key instruments in this regard are the water quality specifications prepared by the Bureau of Indian Standards (BIS)[29] and the Manual on Water Supply and Treatment issued by the Central Public Health and Environmental Engineering Organization (CPHEEO).[30] While BIS specifications seek to address water quality issues in the rural water supply, the CPHEEO Manual addresses the quality aspects of the urban supply. However, these instruments are not mandatory in nature and therefore not legally enforceable.

Government initiative towards water quality management and regulation is not confined to the above mentioned key documents. The concern of the government to promote public heath by ensuring safe water can be found in various other efforts and initiatives. There has been a constant effort on the part of the government towards ensuring the quality of drinking water, particularly through the promotion of several drinking water schemes, missions, programmes and the establishment of various institutional mechanisms to address the issue of safe and sufficient drinking water for all.

The Accelerated Rural Water Supply Programme (ARWSP) can be considered as a pioneer effort by the government in this regard. The ARWSP was introduced in 1972–1973 to provide financial assistance to state governments to ensure the supply of drinking water in rural India. Significantly, the ARWSP guidelines mention quality concerns. One of the primary objectives of the ARWSP is to preserve the quality of water. This concern is further reflected in the classification of rural habitations with water resources of inadequate quality as 'not covered/no safe source' habitations.[31] The increasing concern for water quality can also be seen in the recent enhancement of funds under the programme to address water quality problems. Financial assistance to tackle the quality problems has been enhanced from 15 per cent to 20 per cent of the total ARWSP fund.[32]

28 Food Safety and Standards Act, 2006, s. 2(j).

29 BIS is a statutory institution established under the Bureau of Indian Standards Act, 1986 to promote harmonious development of the activities of standardization, marking and quality certification of goods and attending to connected matters in India.

30 The CPHEEO is a Technical Wing of the Ministry of Urban Development, Government of India, and deals with the matters related to Urban Water Supply and Sanitation Including Solid Waste Management in the country.

31 Government of India, Accelerated Rural Water Supply Programme Guidelines, available at http://ddws.gov.in/popups/arwsp_pop.htm.

32 Ministry of Rural Development, Annual Report 2007–08 at 185 (New Delhi, Government of India, Ministry of Rural Development, 2008).

The ARWSP was given a mission approach with the launch of the National Drinking Water Mission in 1986, which was later renamed the Rajiv Gandhi Drinking Water Mission in 1991.[33] Water quality concerns continue to get significant attention in the Mission. The primary objectives of the Mission include: monitoring water quality; resolving identified problems through science and technology to ensure that available water is of acceptable quality; and ensuring the sustainability of water quantity and quality on a long-term basis through proper water management techniques and implementation of management information systems. Water quality was accorded high priority in the Mission. Parameters and safety standards on water quality have clearly been laid down in the Mission document. A separate document was prepared for this purpose, namely the Executive Guidelines for the Implementation of Water Quality Testing Laboratories.[34]

Initiatives taken by the Government of India since the 1980s show that emphasis is placed on wider health implications in relation to water quality regulation. Major initiatives include 55 Mini-Missions and five Sub-Missions formed under the National Drinking Water Mission, 1986. The major object of these Mini-Missions was to develop an integrated approach and cost-effective technological packages to tackle water quality issues.[35] The initiative by the central government has continued in subsequent years (e.g., the constitution of a Sub-Mission to deal with the problem of arsenic in West Bengal in 1994, and the Expert Committee on Rural Water Supply Program, 1994).[36]

The Constitution of National Rural Drinking Water Quality Advisory Committee under the auspices of the Ministry of Rural Development, Government of India in 2006 is a recent initiative which shows the constantly increasing concern about water quality.[37] The Mandate of the Committee includes the enhancement of drinking water quality monitoring and surveillance. Promotion of the standardization of water quality is also a major task.[38]

Altogether, governmental efforts towards establishing a water quality regulatory framework have existed since the early 1960s. From a legal point of view, there are a number of instruments – enactments, administrative regulations and guidelines – addressing water quality issues. However, the scope and application of these instruments are different. This reveals a scenario of a water quality regulatory framework in India which is highly fragmented and complex.

33 Government of India, Water Quality – An Overview (New Delhi: Government of India, Department of Drinking Water Supply), available at www.ddws.gov.in/water_quality_pop.htm.

34 *Ibid.*

35 Som Pal, 'Key Issues in Water Quality Management', in International Life Science Institute-India, note 25 above, 18.

36 *Ibid.*, at 19.

37 The notification on the Constitution of National Rural Drinking Water Quality Advisory Committee by the Ministry of Rural Development, Government of India, No. W-11027/4/96-TM II (Pt.), 8 March 2006, available at www.ddws.gov.in/WQAdvisoryCommittee.doc.

38 *Ibid.*, at 2.

B. Water quality regulation in the human rights context

The relationship between water quality regulation and human rights jurisprudence is very significant. This is particularly because human rights jurisprudence is an important legal basis from which a rights–duty discourse on water quality regulation emanates. There are at least three human rights which can be considered as highly relevant in this context: the right to health, the right to sanitation, and the right to water.[39] This close link can be explained in the following way. Firstly, water quality concerns form an important part of the normative content of these human rights. Secondly, safe and potable drinking water is generally regarded and recognized as an indispensable element for their realization. Thirdly, the human rights context is critical and relevant given the obligation it casts on the state to ensure proper and effective water quality regulation.

1. The right to health

International human rights law expressly recognizes the right to health, which is found in a number of legal instruments. For instance, the Universal Declaration of Human Rights recognizes a right to a 'standard of living adequate for the health of himself and of his family'.[40] The International Covenant on Economic, Social and Cultural Rights (ICESCR) provides the most comprehensive provision on the right to health. The ICESCR recognizes the 'right of everyone to enjoy the highest attainable standard of physical and mental health'.[41] This right has been given an elaborate form and content through General Comment No. 14.[42]

General Comment No. 14 recognizes health as a fundamental human right indispensable for the exercise of other human rights. It further provides that 'every human being is entitled to enjoy the highest attainable standard of health conducive to living a life in dignity'.[43] It is expressly recognized that 'safe and

39 The list of human rights having legal implications for water quality regulation is not exhaustive. For instance, right to a clean and healthy environment as constructed by the Supreme Court of India could also be a legal basis to argue for the responsibility of the government to undertake efforts to prevent and control water pollution. Indeed, this right has been invoked at least in some cases to this effect. For details, *M.C. Mehta v. Union of India*, AIR 1988 SC 1115.

40 Universal Declaration of Human Rights, UN General Assembly Resolution 217 A (III), 10 December 1948, Art. 25(1). For a brief survey of other instruments at the international level, *see* Mihir Desai and Kamayani Bali Mahabal, Health Care Case Law in India: A Reader 5–11 (Mumbai: Centre for Enquiry into Health and Allied Themes and India Centre for Human Rights & Law, 2007).

41 Art. 12.1 of the International Covenant on Economic, Social and Cultural Rights, New York, 16 December 1966, UN Doc. A/6316 (1966).

42 Committee on Economic, Social and Cultural Rights, General Comment No. 14, The Right to the Highest Attainable Standard of Health (Art. 12 of the International Covenant on Economic, Social and Cultural Rights), UN Doc. No. E/C.12/2000/4 (2000).

43 *Ibid.*, at para.1.

potable water and adequate sanitation' is an important 'underlying determinant' of the right to health.[44]

Another important feature of the right to health under international human rights law is that it has been recognized as 'a right to enjoy a variety of facilities, goods, services and conditions necessary for the realization of the highest attainable standard of health'.[45] This expression implicitly gives reference to drinking water supply and water supply for other domestic purposes. The quality concern has also been expressly recognized by providing that 'as well as being culturally acceptable, health facilities, goods and services must also be scientifically and medically appropriate and of good quality'.[46]

Two aspects of the right to health are particularly relevant in the context of water quality regulation. First, the right to health expressly comprises 'safe and potable' water as an important underlying determinant. Second, the right casts specific obligations upon state parties. This includes the obligation of state parties to take appropriate legislative measures for the progressive realization of human rights.[47] In the context of water quality regulation, it could be said that state parties are in principle obligated to formulate and implement a water quality regulatory framework.

A similar rationale can be applied in the Indian context. The right to life under Article 21 of the Constitution has been interpreted very liberally and progressively by the higher judiciary in India during the past decades to include apparently all related aspects of life such as livelihood, health and education. The right to health has also been read under Article 21 by the Supreme Court of India in a few cases.[48] Hence, it could be argued that the obligation of the state *vis-à-vis* water quality regulation could be constructed from the right to health in the international and Indian legal contexts.

2. *The right to sanitation*

Sanitation plays a significant role in achieving public health goals. Inadequate sanitation facilities, unsafe sanitation practices and lack of environmental hygiene are considered major routes to water-related epidemics. This obvious link between sanitation and public health has elevated sanitation as one of the major

44 *Ibid.*, at para. 11.

45 *Ibid.*, at para. 9.

46 *Ibid.*, at para. 12(d).

47 For details on obligations of state parties, Committee on Economic, Social and Cultural Rights, note 42 above and Committee on Economic, Social and Cultural Rights, General Comment No. 3, The Nature of State Parties Obligation (Art. 2, para. 1 of the International Covenant on Economic, Social and Cultural Rights), UN Doc. E/1991/23 (1990).

48 *Consumer Education and Research Centre v. Union of India* (1995) 3 SCC 42. *See also Virender Gaur v. State of Haryana* (1995) 2 SCC 577; *Kirloskar Brothers v. ESI Corporation* (1996) 2 SCC 682 and *State of Punjab v. Mahinder Singh Chawla* (1997) 2 SCC 83. For an account of the relationship between Art. 21 and the right to health, see Manoj Kumar Sinha, *Enforcement of Economic, Social and Cultural Rights: International and National Perspectives* 268–277 (New Delhi: Manak Publications, 2006).

concerns at the international and national levels. The growing concern of the international community can be seen in the declaration of 2008 as the International Year of Sanitation.[49] Moreover, virtually all member states of the United Nations have recognized the right to sanitation in at least one political declaration.[50] This concern further emphasizes the need for treating sanitation along with water in human rights terms.

Even though sanitation as a single point intervention has the potential to support the realization of other human rights, human rights instruments offer little guidance as to the scope and content of the term 'sanitation'. A UN report acknowledges that 'the contour of the human rights framework for sanitation remains imprecise'.[51] However, express reference to the 'right to sanitation' can be found in several soft law instruments and implicit reference could be deduced from several, if not all, human rights treaties.[52] For instance, a document at international level refers to the right to sanitation and defines it as 'the right of everyone to have access to adequate and safe sanitation that is conducive to the protection of public health and the environment'.[53] The Delhi Declaration of the Third South Asian Conference on Sanitation expressly recognizes access to basic sanitation as a human right.[54] Accordingly, the broad normative contents of the right to sanitation have been constructed, at least in principle, by emphasizing physical accessibility, affordability, adequacy, cultural acceptability and physical security.[55]

The legal conceptualization of right to sanitation in human rights terms is highly significant in the context of water quality regulation. This is mainly because inadequate sanitation is a major root cause of unsafe drinking water. Moreover, water quality regulation in isolation would be unable to achieve public health goals given the close link between water quality management and public health goals on the one hand and access to sanitation on the other.[56] Therefore, any approach which treats access to safe water and sanitation differently and in a compartmentalized manner would be highly artificial and counterproductive.

49 UN General Assembly Resolution 61/192, International Year of Sanitation, 2008, UN Doc. A/RES/61/192 (2007).

50 Centre on Housing Rights and Evictions (COHRE), Sanitation: A Human Rights Imperative (Geneva: Centre on Housing Rights and Evictions, 2008).

51 Report of the United Nations High Commissioner for Human Rights on the Scope and Content of the Relevant Human Rights Obligations Related to Equitable Access to Safe Drinking Water and Sanitation Under International Human Rights Instruments, UN Doc. A/HRC/6/3 (2007).

52 *Ibid.*, Annexes I and II.

53 Sub-Commission on the Promotion and Protection of Human Rights, Report of the Special Rapporteur on the Realization of the Right to Drinking Water and Sanitation, UN Doc. No. E/CN.4/Sub.2/2005/25 (2005).

54 Delhi Declaration on 'Sanitation for Dignity and Health' by the Third South Asian Conference on Sanitation held in New Delhi on 16–21 November 2008, available at www.ielrc.org/content/e0802.pdf.

55 United Nations High Commissioner for Human Rights, note 51 above, 12.

56 Black and Talbot, note 17 above, 96.

Recognition and realization of human right to sanitation, in this regard, could be considered as a complementary legal step towards an effective water quality regulation. This is particularly relevant in the Indian context given the abysmal scenario of access to sanitation. For instance, some studies show that only 200 of India's 400 major cities partially have sewers and only three per cent of effluent is treated.[57]

3. The right to water[58]

At international level, there are several binding and non-binding instruments establishing the right to water. Added to that, some of the recent developments have emphasized or re-emphasized the nature and scope of the right and the kind of obligation it casts upon the state. For instance, General Comment No. 15 adopted by the Committee on Economic, Social and Cultural Rights expressly provides that everyone is entitled to 'sufficient, safe, acceptable, physically accessible and affordable water for personal and domestic use'.[59] The legal concept of the human right to water also entails obligations for governments to take all possible measures towards its realization. As it has been conceptualized in international law, the right to water can be seen to provide for express references to a quality criterion.

The Indian legal system has also, at least in principle, recognized the right to water. There are a number of cases wherein the higher judiciary has interpreted the right to life under Article 21 of the Constitution to include the right to water. A reference, howsoever abstract it may be, to water quality and the obligation of the government to ensure it, can be read in these judgments. In one case the Supreme Court of India has recognized that the right to water under Article 21 imposes a duty upon the State to provide its citizens with 'clean drinking water'.[60] In another case, the Andhra Pradesh High Court expressed a well-articulated enunciation of the right with respect to the quality element by stating that under Article 21 the state has a duty to provide all its citizens with 'clean and adequate drinking water'.[61] Therefore, the obligation of the government to formulate and implement water quality regulation could be considered as directly emanating from the concept of the human right to water also.

57 Manoj Nadkarni, 'Drowning in Human Excreta', 10(19) *Down to Earth* (2001), available at www.indiaenvironmentportal.org.in/node/9699.

58 This section briefly explores the link between the right to water and water quality regulation. For details on the legal concept of the right to water, see Chapter XI.

59 Committee on Economic, Social and Cultural Rights, General Comment No. 15: The Right to Water (Articles 11 and 12 of the International Covenant on Economic, Social and Cultural Rights), UN Doc. E/C.12/2002/11 (2002), para. 2.

60 *AP Pollution Control Board (II) v. Prof. M.V. Nayudu*, 2000 (3) SCALE 354, available at www.ielrc.org/content/e0010.pdf.

61 *P.R. Subas Chandran* v. *Government of Andhra Pradesh* 2001 (5) ALD 771.

C. Regulatory framework

1. International regulatory framework

It has already been observed that there is no binding legal instrument at the international level which lays down water quality standards. Existing instruments, including at the regional level, remain largely limited to recognizing the need for dealing with water quality regulation to achieve public health goals.

In this legal vacuum, the WHO Guidelines represent an important instrument which expressly provides for norms and values regarding water quality regulation. The significance of the WHO Guidelines is illustrated by the broad acceptance of the standards enshrined in the Guidelines at the national level, although there is no binding legal obligation for countries to implement them. For instance, the BIS standards on drinking water in India more or less follow the WHO Guidelines. On this basis, the present section briefly examines the content of the Guidelines.

The WHO Guidelines are a comprehensive document which covers almost all aspects of water quality regulation. Most importantly, they describe the acceptable limit of various substances in drinking water, which is referred to as the 'guideline value'.[62] The guideline value is specified on the basis of separate consideration of various aspects of water quality, such as the microbial, chemical and radiological aspects.

In addition to the incorporation of scientifically approved water quality standards, the WHO Guidelines also cover desired institutional mechanisms and management approaches. They set out the role of various governmental agencies, at both the central and local levels, in water quality regulation and management. It is important to note that the Guidelines recognize and emphasize the crucial role of local authorities in water quality management and regulation, and in educating the public on water quality management.[63] They also underline the important role of other actors such as water vendors, water suppliers and the public in water quality management.[64]

Another important feature of the WHO Guidelines is their emphasis on pragmatism. They state that national water quality standards may differ 'appreciably' from the standards contained in the Guidelines after taking into account a variety of environmental, social, cultural, economic, dietary and other conditions affecting potential exposure.[65] It is further stated that:

> A programme based on modest but realistic goals – including fewer water quality parameters of priority health concern at attainable levels consistent

62 A 'guideline value' is defined as representing 'the concentration of a constituent that does not exceed tolerable risk to health of the consumer over a lifetime of consumption'. See World Health Organization, note 2 above, 31.

63 *Ibid.*, at 11–12.

64 *Ibid.*, at 15.

65 *Ibid.*, at 33.

> with providing a reasonable degree of public health protection in terms of reduction of disease or reduced risk of disease within the population – may achieve more than an overambitious one, especially if targets are upgraded periodically.[66]

It could be stated that the WHO Guidelines occupy a significant position in that they are a valid reference source for individual countries intending to frame or modify water quality standards. They also serve an educative purpose by constantly promoting emphasis on water quality management and regulation, mainly through knowledge production and dissemination processes.[67]

2. *Regulatory framework in India*

The water quality regulatory framework in India consists of different legal and institutional arrangements addressing water quality issues at different angles. This multi-faceted regime includes statutory frameworks, rules, administrative regulations and guidelines. Adding to this complexity, the nature and scope of these regimes are significantly different. While some are binding in nature, others are not. Some frameworks address water quality issues at a national level, while others address them at the local level. Some regimes focus specifically on drinking water quality issues. At the same time, some regimes influence water quality issues by addressing generally the environmental quality or the protection of public health. Having noted the complexity and multiplicity of the Indian regulatory framework, this section focuses primarily on instruments providing for water quality standards.

National-level regulation

There are a number of instruments at the national level which address water quality issues directly or indirectly. It is not the purpose of this chapter to mention and analyze each and every instrument separately. Generally, this section examines the regulatory approach enshrined in the regulatory framework at the national level as a whole, with special mention and reference given to certain key instruments.

The regulation of water quality has two main aspects. The first is the setting up of water quality parameters (i.e., the provision of acceptable limits of various constituents that could cause health hazards). The second aspect is the establishment of institutional mechanisms to effectively implement regulations. Institutional mechanisms essentially include the establishment of agencies at various levels for monitoring and surveillance.

66 *Ibid.*

67 WHO guidelines have been used as a valuable reference in various instruments at the national level. In the Indian context, see Ministry of Rural Development, Draft Guidelines for the Preparation of Legislation for Framing Drinking Water Regulation (New Delhi: Ministry of Rural Development, 2007).

Key instruments related to water quality standards in India are the Bureau of Indian Standards (BIS) Specification IS:10500 and the CPHEEO Manual on Drinking Water Quality. Both these instruments follow a similar regulatory approach, at least with regard to water quality parameters.

Under BIS IS:10500, there are two quality characteristics of drinking water, namely essential and desirable characteristics. All essential characteristics are to be examined routinely (crucially, there is no exact indication as to what 'routinely' entails), while all desirable characteristics should be examined either when a doubt arises as to water quality or when the potability of water from a new source is to be established. The acceptable limit of different constituents also follows the same classification (i.e., a desirable limit and a permissible limit). The idea of a permissible limit is subject to the condition of the 'absence of an alternate source'.

The objective of BIS IS:10500 was to assess the quality of water resources, and to monitor the effectiveness of water treatment and supply by the concerned authorities. Other instruments adopted under the auspices of the BIS address water quality issues which are not covered under BIS IS: 10500. For instance, the quality of packaged drinking water is covered under BIS IS: 14543 of 2004 and the quality of packaged natural mineral water is covered under BIS IS: 13428 of 2005.

The CPHEEO Manual also follows a similar approach in the case of the prescription of parameters. The only difference resides in the usage of terminology. The CPHEEO Manual uses the term 'acceptable' in the place of 'desirable' and 'cause for rejection' instead of 'permissible'. The parameters included in the CPHEEO Manual are also similar to those contained in BIS IS:10500.

Another important aspect of water quality regulation is the protection of water at the source point. This can be achieved mainly through preventive measures taken to control the pollution of water sources. It is a fact that the major source of water pollution in India is the discharge of untreated or partially treated effluent into water resources.[68] To address this issue, the central government enacted the Water (Prevention and Control of Pollution) Act of 1974 (hereafter the Water Act). The Water Act lays down various means and procedures to ensure the quality of water resources in the interest of public health and the environment.[69]

The Water Act focuses mainly on the control of the discharge of various effluents into rivers and streams. Most importantly, it provides for two kinds of regulatory tools. The first is the permit system or the consent procedure. This means that a person must obtain consent from the State Pollution Control Board before taking steps to establish any industry, operation or process, any treatment

68 *Ibid.*, at 39.

69 The health concern is expressly recognised in the Water Act. The definition of 'pollution' explicitly states 'public health and safety' as a concern. See Water (Prevention and Control of Pollution) Act, 1974, s. 2(e) available at www.ielrc.org/content/e7402.pdf.

and disposal system or any extension or addition to such a system which might result in the discharge of sewage or trade effluents into a stream, well or sewer or onto land.[70] The permit or consent may contain conditions related to the location, construction and use of the outlet as well as the nature and composition of the new discharges. The second regulatory tool is the prescription of standards for the discharge of effluents. The Water Act enables concerned State Pollution Control Boards to prescribe such standards. For instance, the Kerala Pollution Control Board has prescribed quality parameters with respect to groundwater[71] and the Utter Pradesh Pollution Control Board has prescribed quality parameters for river water.[72]

The Water Act also provides for various measures to enforce regulations. The State Pollution Control Boards are authorized under section 33A to issue orders to any person, officer or authority, including orders to close, prohibit or regulate any industry, operation or process, and to stop or regulate the supply or water, electricity or any other service.[73] Until 1988, the enforcement tool of the State Pollution Control Board was criminal prosecution. This was revised by the 1988 amendment to the Water Act of 1974. The Board now has the power to close non-compliant companies or to cut their water and power supply.

The second key aspect of water quality regulation consists in the institutional mechanism which is complementary to the substantive framework. A few institutions at national level are addressing water quality regulatory issues directly. It is a fact that the implementation of water quality regulation needs to be done at the local level, and that modes of implementation largely depend upon the nature of water resources and their prevailing uses. Therefore, powers and functions of the national level institutions are mostly directive in nature, which need to be followed by agencies such as public health departments, State Pollution Control Boards, or water supply agencies at the state and municipal levels.

The Central Pollution Control Board (CPCB) plays a key role in water quality management and regulation at the national level. One significant example of its role in this regard is the Guidelines for Water Quality Management, 2008. This comprehensive document covers different aspects of water quality regulation such as prescription of quality parameters, frequency of sampling and guidance as to the containers to be used for the collection of samples.[74] The CPCB also plays a significant role in the Union Territories since the mandate of

70 *Ibid.*, s. 24.

71 The Kerala Pollution Control Board at least provides for parameters on chemical contaminants consistently with the BIS IS: 10500 of 1991. See Kerala Pollution Control Board, Water Quality Monitoring (Thiruvananthapuram: Kerala Pollution Control Board), available at www.keralapcb.org/standards/standards_municipal.htm#mun6.

72 Uttar Pradesh Pollution Control Board, Primary Water Quality Criteria for Designated Best Use Classes, available at www.uppcb.com/river_quality.htm.

73 Water (Prevention and Control of Pollution) Act, 1974, s. 33A.

74 Central Pollution Control Board (CPCB), Guidelines for Water Quality Management (Delhi: Central Pollution Control Board, 2008).

water quality regulation in the Union Territories comes directly within its purview.[75]

Other key agencies at the national level include the Department of Drinking Water Supply under the Ministry of Rural Development, the Central Ground Water Board and the Water Quality Assessment Authority (WQAA) constituted under the Environmental (Protection) Act, 1986.[76] The Ministry of the Environment and Forests (MoEF) also plays an important role in water quality regulation on the basis of its power derived from the Water Act and the Environmental (Protection) Act. The Uniform Protocol on Water Quality Monitoring Order, 2005 can be considered as a good example of the role of MoEF in this regard.[77]

State-level regulation

Under the Constitution of India, water supply and public health fall within the legislative competency of the concerned state governments.[78] Consequently, water quality regulation falls primarily within their mandate. The regulatory framework at the central level makes little sense if the standards that it sets out are not respected and recognized through legal (or at least policy) instruments at the state level. There are exceptions to this general observation, one of which is the Food and Safety Act, 2006, which includes packaged drinking water.[79] Moreover, the effectiveness of water quality regulation is ultimately tested by examining the quality at the consumption point. Therefore, a legal framework for water supply systems at the state level is crucial insofar as quality regulation is concerned.

The water supply system in India is highly complex and consists of different legal and institutional frameworks. The responsibility for rural water supply in most of the States lies with public health engineering departments, water authorities, locally constituted statutory agencies or panchayati raj institutions. Urban water supply is mostly entrusted to municipal corporations, municipal authorities, water authorities, water boards or developmental authorities. For instance, in Andhra Pradesh, water supply and sanitation is the exclusive responsibility of the

75 See, e.g., the notification which lays down standards for compliance by small-scale industries located in the Union Territories, CPCB Notification No. 1/2 (71)87 Plg, 7 April 1988, available at http://envfor.nic.in/legis/water/no1-2-71-87.html.

76 Ministry of Environment and Forests, Water Quality Assessment Authority Order, Order No. S.Q. 583 (E), 29 May 2001. The WQAA has taken some significant steps in water quality regulation in India. Examples include the role played by the WQAA, together with the Water Quality Monitoring Committee constituted by the MoEF, in bringing about the Uniform Protocol on Water Quality Monitoring Order and the constitution of thirty-three state-level water quality review committees. *See* Action Taken/Status Note on the minutes of the meeting held on 5th September 2006 to discuss the 'National River Conservation Programme', available at http://nac.nic.in/atrs/river_conservation_5sept.pdf.

77 Ministry of Environment and Forests, Uniform Protocol on Water Quality Monitoring Order, 2005, Notification No. S.Q. 2151, 17 June 2005.

78 Constitution of India, Art. 246(3) together with Schedule 7, List II, Entry 6 and 17.

79 Food Safety and Standards Act, 2006, s. 2(j).

Panchayat Raj Engineering Department. Whereas in Delhi, the agency responsible for water supply is the Delhi Water Board constituted under the Delhi Water Board Act, 1998.

These different agencies derive powers and responsibilities from different sources. For instance, the Panchayat Raj Engineering Department in Andhra Pradesh functions as a separate department directly under the control of the state government. Water supply agencies in urban areas derive their powers mainly from municipal acts or separate statutes constituting water boards. For instance, the Delhi Water Board Act, 1998 sets forth all norms and procedures regarding water supply in Delhi. Therefore, water quality regulation at the local level depends primarily upon the quality criteria provided in Municipal Acts.

Municipal Acts use a variety of criteria applicable to the levels of water quality and quantity that need to be ensured; these include the terms 'proper and sufficient', 'insufficiency and unwholesomeness', 'fit for human consumption', 'sufficient supply of pure and wholesome', 'pure and fit for human consumption' and 'defective and insufficient'.[80] Certain Acts actually empower the State governments to adopt rules specifying exact quantity or quality standards – and the same would be binding upon the municipalities.[81]

These criteria suffer from at least two important shortcomings. Firstly, they are highly discretionary and vague, and their meaning and scope can easily be contested. It is unlikely that application of these criteria could create any concrete and legally enforceable duty upon water supply agencies, and it can at least be said that they are of a minimally binding nature. Moreover, such criteria cannot be considered as an adequate alternative to a precise code containing scientific and legally binding parameters and procedures.

Secondly, the criteria and the duties they create are subject to exceptions, with regard to which the cost factor is critical. Thus, while several of the Acts exhort municipalities to provide water of adequate amount and quality, this has to be done keeping in mind 'reasonable costs'. Costs and the non-performance of the statutory duties of municipalities have been at the core of numerous cases argued before the courts in the last few decades. In principle, the Supreme Court of India has rejected the defence based on financial constraints at least in a couple of cases.[82] However, this 'reasonable cost' exception continues to exist and there does not seem to be any clarity regarding its nature and scope.

In the case of rural water supply, the BIS specifications are generally referred to as the 'recommended standards' for drinking water supply. For instance, the Madhya Pradesh Public Health Engineering Department – the agency responsible for the supply of safe, potable and pleasant water – refers to water quality criteria provided in the BIS specification. Some documents at the national level

80 See, e.g., New Delhi Municipal Council Act, 1994, ss. 11(d)(ii) and 147(4) and Cantonments Act, 2006, s. 186(2).

81 Bengal Municipalities Act, 1932, s. 311 and Bihar and Orissa Municipal Act, 1922, s. 314.

82 *Municipal Council, Ratlam v. Vardichand Ors* (1980) 4 SCC 162 and *Dr. BL Wadhera v. Union of India and Ors.* (1996) 2 SCC 594.

also refer to the BIS IS 10:500 insofar as the monitoring of the concentration of contaminants is concerned.[83]

Another instrument relevant in the context of rural drinking water supply is the CPCB Guideline.[84] The Guideline essentially covers water at the source point and it inevitably includes groundwater resources. This is particularly relevant given the fact that groundwater is a major source of drinking water, and that groundwater resources are in a critical situation mainly due to contamination and depletion.[85]

Hence, it could be stated that the water quality regulatory framework in India consists mainly of legally non-binding instruments at the national level and some rudimentary provisions in Municipal Acts. The major limitation of the existing framework is that it does not confer any significant legally enforceable rights and duties.

D. Monitoring and surveillance: Practice in India

Substantive aspects such as standardization and rules related to sampling form only one part of the regulatory regime for water quality. Equally important are the monitoring and surveillance elements, which represent perhaps the most critical part of the regime since they require substantive investments in financial and human resources as well as a considerable infrastructure. In a simple sense, monitoring and surveillance refer to the means and processes for ensuring compliance with substantive norms. The effectiveness of the water quality regulatory framework depends upon the effectiveness of the monitoring and surveillance mechanisms.

In this context, this section broadly examines certain key aspects of water quality monitoring and surveillance, but does not intend to elaborate upon the details and effectiveness of all monitoring and surveillance programmes in India. In fact, this can be done only through specific case studies. The scope is limited to examining the broad spectrum of the monitoring and surveillance framework in a legal perspective by referring to certain key programmes.

1. Monitoring and surveillance

By definition, the concept of monitoring and surveillance are different. In relation to drinking water quality control, 'surveillance' refers to constant supervision,

83 Government of India, Guidelines for National Drinking Water Quality Monitoring and Surveillance Programme 3 (New Delhi: Ministry of Rural Development, 2006).

84 Central Pollution Control Board, note 74 above.

85 Planning Commission of India, Report of the Expert Group on 'Ground Water Management and Ownership' (Delhi: Government of India, Planning Commission, 2007); M. Dinesh Kumar and Tushar Shah, Groundwater Pollution and Contamination in India: The Emerging Challenge (IWMI-TATA Water Policy Briefing Paper, 2006); National Academy of Agricultural Sciences, Emerging Issues in Water Management – The Question of Ownership (New Delhi: National Academy of Agricultural Sciences, Policy Paper No. 32, 2005); Prabir Naik, Drinking Water Problem in Rural India, 94(8) *Current Science* 964 (2008).

from a public health point of view, of the safety and acceptability of drinking water supplies. It requires a continuous and systematic programme of sanitary inspection and water quality testing, carried out at different points of the water distribution system.[86] 'Monitoring' involves laboratory and/or spot testing of water samples collected from different locations in the water supply system, including sources, water treatment plants, distribution systems and house reservoirs.[87] However, the two concepts are complementary to each other and could be considered as indispensable parts of the water quality regulatory framework.

Monitoring and surveillance mechanisms generally seek to address such issues as the frequency of sampling, the location from where samples need to be taken and particular parameters which need to be tested in particular locations. They also comprise remedial measures to be taken in cases of the existence of risks posed to public health due to water contamination or threat of contamination. Hence, an effective monitoring and surveillance mechanism leads to pollution prevention, and prompt and timely action in case of identified health risks.

Having noted the importance and the role of water quality monitoring and surveillance in the protection of public health, some critical questions need to be addressed from a legal point of view: who is responsible for carrying out monitoring and surveillance? If there is a responsible agency, what is the legal framework which it must abide by? And what is the legal nature of such responsibility?

The key legal framework addressing the issue of water pollution in India is the Water Act, 1974. The CPCB and State Pollution Control Boards (SPCBs) are the agencies responsible for implementing the Water Act.[88] The CPCB and SPCBs were constituted to maintain or restore the wholesomeness of water resources in the country.[89] The responsibility of water quality monitoring and surveillance can also be read into the broad mandate of the CPCB and SPCBs as provided in the Water Act.

Indeed, the CPCB has taken some initiatives in water quality monitoring and surveillance. The CPCB together with the SPCBs have formed a network of monitoring stations across the country. The present network comprises 1,245 stations in twenty-seven States and six Union Territories spread over the country. The monitoring is done on a monthly or quarterly basis for surface waters and on a half-yearly basis in the case of groundwater.[90] The monitoring network covers 50 rivers, 78 lakes, six tanks, twenty-six ponds, eight creeks, nineteen canals,

86 Government of India, Draft Guidelines for Preparation of Legislation for Framing Drinking Water Regulations (New Delhi: Government of India, Ministry of Rural Development, 2007).

87 *Ibid.*, at 39–40.

88 The major instruments remain inadequate insofar as monitoring and surveillance of water quality are concerned. For instance, the BIS Standards use only guiding phrases such as 'routine' (in case of essential characteristics) and 'to be tested when pollution is suspected' (in case of toxic substances). These kinds of provisions are unlikely to have major practical impact.

89 Statement of Objects and Reasons of the Water (Prevention and Control of Pollution) Act, 1974.

90 Central Pollution Control Board, note 9 above, 10.

nineteen drains and 382 wells. Among the 1,245 stations, 695 are on rivers, 86 on lakes, nineteen on drains, nineteen on canals, six on tanks, twelve on creeks/seawater, twenty-six on ponds and 382 on groundwater stations.[91] Presently the inland water-quality monitoring network is operated under a three-tier programme – the Global Environmental Monitoring System (GEMS), Monitoring of Indian National Aquatic Resources System and Yamuna Action Plan.[92] Water samples are being analyzed for twenty-eight parameters consisting of physico-chemical and bacteriological parameters for ambient water samples apart from field observations.[93]

The CPCB follows a use-based approach in water quality monitoring and surveillance. This means the water quality will be regulated according to the use that a particular water source is put to. This has been termed as the 'designated best use' approach. As per this approach, out of several uses a particular water body is put to, those that demand the highest quality of water are 'designated best uses' and quality will be regulated accordingly. The CPCB guidelines lay down quality criteria for different uses. For instance, 'A' class water body means a water source which is used for drinking without any treatment. Quality parameters prescribed are high for this class of water bodies.[94]

The monitoring process promoted by the CPCB follows a further classification according to the element of human influence on water resources. It follows a classification of 'baseline station' and 'trend station'. The former refers to a station where there is no human activity influence. All sources other than 'baseline stations' are classified as 'trend stations'.[95] The frequency of monitoring needs to be based on this classification. All trend stations will be monitored with increased frequency. For instance, the CPCB Guidelines provide that all baseline stations, in the case of groundwater sources, will be monitored once a year, whereas all trend stations will be monitored four times a year.[96]

It appears that the key focus of the mandate of the CPCB and SPCBs is on monitoring water resources at the source point, for instance rivers, streams, ponds and groundwater resources. This forms only one part of the total monitoring and surveillance spectrum. This role of PCBs can be considered as little less than sufficient to ensure safe drinking water for those who collect and directly use surface water or groundwater resources.[97] However, the focus of the CPCB does not seem to cover the monitoring and surveillance of water treatment, distribution networks and individual storage systems. These aspects are particularly critical in urban water supply and piped rural water supply where unsafe

91 *Ibid.*, at 10.

92 *Ibid.*

93 *Ibid.*, at 11.

94 Central Pollution Control Board, note 74 above, Annex I.

95 *Ibid.*, Annex II.

96 *Ibid.*

97 The quality monitoring of groundwater resources also comes within the purview of groundwater authorities envisaged under groundwater laws. For details, see Chapter IX.

distribution network and individual storage systems can pose a threat to public health.[98]

It has already been observed that agencies responsible for urban water supply are mainly municipal bodies or separate water boards. The legal framework that determines the powers and functions of these agencies contains only minimal provisions regarding water quality regulation. Most of the Municipal Acts do not contain specific legal provisions as to water quality monitoring and surveillance. In practice, water quality monitoring and surveillance in urban water supply are undertaken either inadequately or not undertaken at all.

A survey conducted by the National Institute of Urban Affairs shows that around 57 per cent of urban centres do not monitor raw water at all.[99] The study also shows that in seven metropolitan cities, the situation is similar.[100] The periodicity of water quality monitoring varies from alternate days to once every six months. The study also shows that a majority of urban centres do not monitor distribution networks. The main reasons highlighted for this situation are lack of infrastructure and of human resources.[101]

Hence, it can be said that quality monitoring and surveillance in urban water supply in India is highly localized, and that as a consequence there is no uniformity in practice among water supply agencies across the country. Legal provisions in this regard are minimal and do not seem to have any practical impacts. The situation is similar with regard to rural water supply in that monitoring and surveillance of rural water supply quality have been identified as 'extremely inadequate'.[102] It has also been observed that there is no institutionalized quality monitoring and surveillance system in India.[103] To overcome this, a new management approach is being promoted by the central government, particularly with regard to rural water supply, which gives a more important role to local authorities as well as local communities.[104]

2. *Legal aspects: A critique*

At the outset, it should be noted that there is no specific and mandatory legal framework in India laying down rules and procedures for water quality regulation. The monitoring and surveillance mechanisms, which have been described above, appear highly fragmented and discretionary. The problem of fragmentation appears as a result of the existence of a number of institutional mechanisms

98 For an express recognition of the importance of water quality monitoring and surveillance of water distribution systems, Ministry of Rural Development, note 67 above, 40.

99 National Institute of Urban Affairs, Status of Water Supply, Sanitation and Solid Waste Management in Urban Areas 45 (New Delhi: National Institute of Urban Affairs, 2005).

100 *Ibid.*, at 46.

101 *Ibid.*, at 47 and 51.

102 Government of India, note 4 above, 3.

103 *Ibid.*, at 4.

104 Government of India, Guidelines for National Drinking Water Quality Monitoring and Surveillance Programme (New Delhi: Ministry of Rural Development, 2006).

which can be identified as responsible for monitoring and surveillance. Despite the number of institutions, there is hardly any legal provision which makes it mandatory for these agencies to conduct proper, effective and periodical monitoring and surveillance. The existing documents laying down the processes and procedures for water quality monitoring and surveillance are optional and therefore not legally binding at all.

Resource constraints are another main problem often highlighted as contributing to the inadequacy of monitoring and surveillance mechanisms. These constraints include limited infrastructure, as well as financial and human resources. The financial constraint argument makes very little sense while comparing the public health cost of inadequate monitoring and surveillance. Moreover, the financial constraint argument is limited by the fact that the protection of public health is supposed to be a matter of high priority for the government. Not only that, a pollution-free environment in general, and access to adequate water in particular, are fundamental rights of every individual as per the interpretation of Article 21 of the Constitution by the higher judiciary.[105] Such an interpretation confers a duty upon the government to ensure that all citizens are provided with 'clean and adequate drinking water'.[106] Though limited in scope, the phrase 'clean and adequate' emphasizes the quality criterion.

In order to achieve effective water quality regulation, a well-equipped institutional mechanism responsible for monitoring and surveillance is essential. The institutional mechanism should be buttressed by a legal framework, preferably at the central level, prescribing the necessary elements of monitoring and surveillance.[107] A legal framework at the central level should establish legally binding rules and aim to significantly harmonize existing regulations. As many water quality problems are local in nature, a decentralized organizational structure such as village-level and district-level units would be most effective and thus preferable. Over and above, an adequate legal framework for water quality regulation would be a significant step towards the realization of human rights in general and the rights to health and water in particular.

CONCLUSION

In India, public health concerns have been a major impetus behind the number of initiatives undertaken to improve drinking water supply. Attaining a safe and

105 See, e.g., *Virender Gaur v. State of Haryana* (1995-1) 109 PLR 591 (S.C.); *Subhash Kumar v. State of Bihar* AIR 1991 SC 420; *V. Lakshmipaty v. State* AIR 1992 Kant. 57 and *Damodar Rao v. SO Municipal Corporation, Hyderabad* AIR 1987 AP 170.

106 *PR Subash Chandran v. Government of Andhra Pradesh* 2001 (5) ALD 771.

107 Under the Constitution of India, the power to legislate on drinking water lies with the concerned state government. However, the Constitution allows the central government to legislate upon this matter provided that such powers have been conferred by state governments. Constitution of India, Art. 252.

adequate water supply has been an important item on the government's agenda at least since the introduction of five-year plans. It cannot be stated that these initiatives have failed completely; in fact, they have produced some significant results, one example being the increased coverage of drinking water supply. Another major achievement of government initiatives is 100 per cent eradication of guinea worm disease.[108]

However, water quality regulation, in a strict legal sense, is still at an early and rudimentary stage, primarily because water quality regulatory frameworks in India are largely dominated by non-binding guidelines, codes and manuals. Even though Municipal Acts contain reference to water quality, they are minimal and as such not likely to bring about any significant results. This reveals the fact that water quality regulation in India is highly obscure and complex. Adding to this complexity the responsibility for water supply is fragmented among a number of agencies. This often leads to a situation wherein water quality monitoring and pollution prevention happens primarily during the outbreak of epidemics.[109] In all probability, this situation has severe public health implications.

One of the important legal consequences of the present obscurity and complexity of water quality regulatory frameworks is the absence of clearly defined right–duty norms. In other words, opportunities for legal action against a water supplying agency because of inadequate water quality are very rare or unclear. In addition, because of the lack of a legal framework prescribing quality parameters and other procedures, different agencies follow different norms. For instance, some agencies monitor water quality every month, while others do not monitor it at all.

This complex and unclear scenario points to the need for the adoption of a comprehensive drinking water legislation, preferably at the central level, prescribing mandatory quality standards and rules related to monitoring and surveillance. In fact, some initiatives have already been taken in this direction. For instance, in 2007 the Ministry of Rural Development formulated 'Draft Guidelines for the Preparation of Legislation for Framing Drinking Water Regulation'.[110]

The legal responsibility of the supplier should be clearly and expressly defined in such a legal framework so that any violation results in legal action. The standard of duty of care that the supplier should comply with also needs to be stated expressly. A proper articulation of 'right–duty' aspects can bring about effective results in water quality regulation. The burden of proof should also be preferably on the supplier. Water quality parameters, as well as requirements for monitoring

108 Panda, note 25 above, 37.

109 *Ibid.*, at 37.

110 The draft guidelines are currently being discussed by various central government departments and ministries (including the Department of Drinking Water Supply, the Central Groundwater Board, the Department of Legal Affairs and the Ministry of Urban Development) and relevant state government departments (including the Public Health Engineering Departments, Public Health Departments, State Groundwater Boards and Rural Water Supply and Sanitation Departments) as well as certain international agencies (UNICEF, WHO and the WSP).

and surveillance, can be modelled on existing documents such as the BIS Code IS: 10500, CPHEEO Manual and Uniform Protocol on Water Quality Monitoring Order, 2005. Alternatively, these relevant documents could be annexed to any future drinking water legislation.

An effective legal framework to ensure water quality requires an effective institutional mechanism to periodically monitor water quality. This should be buttressed by effective cooperation between the water supplier and the monitoring agency. The aim of surveillance should be to undertake routine, independent monitoring of the water supply from a public health point of view. Implementation should be at the local level. Therefore, the institutional mechanism should be decentralized as this would enhance the effectiveness and reduce the cost.

All states should have a clear-cut water surveillance policy, which includes monitoring of water quality at the source point and user's end. The efficiency of these policies will depend upon adequate institutional mechanisms sustained by sufficient resources, such as mobile testing laboratories in all district headquarters. The responsibility of the government *vis-à-vis* water quality regulation needs to be addressed primarily from the point of view of human rights to establish a comprehensive and binding legal framework regulating water quality.

However, this regulatory framework can only ensure quality up to the point of consumption. To achieve comprehensive results, it should be complemented by government initiatives to educate the public to promote in-house practices such as hygiene, storage, and use. This is critical because individual behaviour such as boiling drinking water and the proper washing of hands are a crucial determining factor in eradicating water-related diseases.

XIV. Final remarks

Philippe Cullet, Alix Gowlland Gualtieri and Roopa Madhav

Recent trends in water sector reforms across the world highlight the importance of changes to water policy and legal frameworks. On the one hand, water resource use and water conflicts – both national and international – have been the subject of extensive research. On the other hand, reforms being carried out to legal frameworks, as a crucial component of broader water sector reforms, have not received adequate attention. This book contributes to a deeper understanding of the multi-faceted water-related legal, policy, and regulatory frameworks that govern the use of water resources at local, national and international levels.

The scope and importance of water law reforms have dramatically expanded in the last decade. The focus of this book has been to provide a theoretical understanding of water law reforms, both international and national, more particularly in India. Overall, in mapping the reform process, the primary task of the researchers has been to uncover the motivations and principles informing reforms, to better understand tensions between water as an economic good and water as a human right, and to explore the interface and impact of international discourses at the national level.

In this context, this book has attempted to analyze aspects associated with the changing legal framework for water use from a range of perspectives, ranging from human rights to agriculture, from development to trade, and from the local to the national and international levels. Existing water law is, on the whole, often patchy and incomplete. This is the case at the international level where water law is narrowly conceived, as well as in India. In the Indian context, an in-depth analysis of changes introduced to legal and institutional frameworks has been attempted in a sectoral manner, in keeping with the sectoral nature of the reforms.

Overall, international water law is well developed with regard to cooperation among states concerning issues and activities that are clearly transboundary in scope, such as navigation on international watercourses. In recent decades, the importance of collaboration on non-navigational aspects of international watercourses has rapidly grown and is now recognized as a core objective of international water law. However, international water law is yet to be effectively developed with regard to cooperation on issues related to water found within national boundaries. Further, while international water law has at least started integrating an environmental perspective, the social and human rights aspects of water remain largely absent. The absence of a human rights perspective in water law has now been addressed from the perspective of human rights law through the adoption of General Comment 15 and through the inclusion of the right to water on the agenda of the Human Rights Council. This constitutes an important conceptual framework which needs to be integrated in water law instruments – both international and national – if it is to be effectively implemented in practice.

Ongoing law reforms focus on water management as the main answer to existing water-related issues. This framework is insufficient as it does not take into account new important legal principles that should be integrated into water law. For instance, the precautionary principle which is the object of much discussion in the context of new technologies like biotechnology is also of relevance in the context of water, since science is not able to accurately predict all the consequences for the water system triggered either by climate change or by more direct human intervention such as the proposed interlinking of rivers in India. The principle of integration of environment and development which calls on the state to ensure that development is compatible with the need to protect and improve the environment is also of immediate relevance since water policies have, for instance, often been tilted more towards increasing agricultural production through irrigation than focused on achieving higher yields while minimizing environmental damage. These and other principles need to be more substantively integrated into water law.

The main thrust of the reforms is an attempt to turn water into an economic good, which is to be managed like any other commodity. This, however, is the more articulated and vigorously debated aspect of the reforms. The other crucial aspect that requires greater enquiry is the recasting of the state from a primary provider of services to a facilitator or regulator of services in the water sector, with far-reaching consequences. Other aspects, such as the legal framework over groundwater, decentralization and greater participation of water users and notably, the land–water linkage having a direct impact on access and control over water resources, have also largely been sidelined in popular debates around water sector reforms. The chapters in this book bring to the fore these concerns in a more holistic manner.

Research for this book largely focused on evolving legal, institutional and policy frameworks. It is thus largely pre-determined by the agenda of reforms being introduced on the ground. Understanding these largely episodic developments as a cohesive whole has proved to be a challenge. Also, research within India on water law has been relatively scarce and hence the literature on various emerging topics is limited. Besides, research has been limited to selected areas that have seen extensive legal and institutional changes. Several areas that are also seeing extensive changes such as watershed management, rainwater harvesting or even tank rejuvenation have therefore not been explored in any great depth.

The limits identified in existing reforms provide the basis for a series of recommendations for the future, both in terms of further water law reforms and research. While reforms to the water sector are certainly necessary to improve control over and access to water resources in India, they have to date remained inadequate in fulfilling the overarching goal of providing access to water for all in accordance with national and international human rights requirements. Thus, the need to map out future directions for legal reforms is vital for the water sector in the country, especially given its important complexities. While the reforms being carried out certainly appear challenging for policy makers, they also provide an opportunity to feed into the need to diverge from current perspectives and to adopt a more holistic approach. Four main issues can be drawn from the analysis carried out in the chapters constituting this volume, since they touch upon a broad range of issues related to water law reforms from both a national and an international perspective.

The first observation is that, overall, the present water framework in India is highly fragmented and lacks cohesiveness, both from a substantive and a structural point of view. The set of laws and policies inherited from the colonial period were characterized by a sectoral approach which has to date not been remedied by the present set of reforms. Water law reforms undertaken thus need to be buttressed by a comprehensive approach to water; one that represents the multi-faceted dimensions of the resource and one that reflects these dimensions in a single national legal framework for water.

The second important point to emerge from the research undertaken is that the initial phase of the reforms has been influenced by a clearly identifiable policy thrust backed largely by external financial institutions such as the World Bank and the ADB. The increased IFI presence in the water sector has undoubtedly had impacts in terms of the priorities which dominate in the reforms, namely those of financial cost recovery and efficiency, establishing of water rights, reducing the role of the state and increasing the cost for water provision. The need to counter the domination of an economic-based approach to water thus has to be addressed at the root of the problem. On the one hand, at the national level, a solid legal framework must be introduced. This should set out substantive aspects like the human right to water, as well as procedural ones like transparency and public participation, which can act as a leverage against IFI conditionalities. This framework should also serve to limit, or constrain, private sector involvement in water provision. On the other hand, at the international level, there is a need to strengthen the scope and content of the human right to water, and to develop state obligations relating to the environmental protection of water resources and in particular drinking water sources. In parallel, the duties of international institutions in this regard need to be clearly defined, and IFI policies need to incorporate strong safeguards regarding the right to water and environmental requirements in conditionalities for loans made to countries of the South.

Thirdly, there has been a major shift in the role of the state with regard to water provision, which is increasingly reflected in water law reforms. The new policies introduced in India over the past decade have imposed upon communities an increasingly important responsibility for their own water supply, while in parallel significantly reducing the role of the state in providing drinking water. The effect is that the state is now viewed as a regulator rather than as a provider of water. This has the potential to counter established measures of equity, and to directly or indirectly lead to violations of the human right to water. It appears that if access to water becomes dependent mainly on the financial capacity of specific individuals or specific panchayats, the poor and rural dwellers in particular will be the biggest losers in terms of inequalities in the face of access to water. For instance, the imposition of operation and maintenance costs to individual villages will lead to reduced access to water in those areas less well-endowed in the resource. Such policies need to be reversed because water is far too fundamental for human life. In view of the increased corporatization of public sector water utilities and private sector involvement, there is a need to retain state responsibility in water provision in order for it to comply with human rights obligations, at least in terms of monitoring and regulating these other actors.

Fourthly, there is a fundamental, and more global, need for a radical reorientation in the perspectives applicable to access to water. As mentioned above, the role of international donors has meant that water sector reforms in India have focused primarily on governance, financial efficiency, cost recovery and institutional issues, and have only marginally touched upon concerns of equity and sustainability. This has meant that many aspects of the current reforms are simply incompatible with the realization of the right to water, and that there is as yet little legal guidance on how to address these disparities despite judicial and constitutional support of the right. Moreover, widespread pollution of freshwater sources such as lakes, tanks and rivers are on the rise as aquifers are degraded through human use and saline intrusion. Deteriorating water quality not only impacts the human population (for instance in terms of water quality) but also ecological and biological diversity. These tensions between an economic perspective on water on the one hand, and a human rights and/or environmental one on the other reflect a more general trend in international policy towards the commodification of water, which is increasingly found in water sector reforms initiated in other countries. Since the dominant perception is that water needs to be regulated and managed on a commercial basis like any other market good, all water-related institutions are called upon to 'manage' water on commercial lines. This implies that any pre-existing social perspective on the provision of water is largely sidelined. The next phase of reforms and policy changes thus need to address in a more concrete way the tensions existing between the different dimensions of water, and particularly to restore the balance in favour of a more sustainable and equitable approach to the water sector.

On this basis, the following selected, specific proposals can be made for the water sector and water law reforms taking place in India:

(1) The right to water has been only implicitly derived from the Constitution under the right to life clause in Article 21. A constitutional amendment incorporating an explicit right to water would help to ensure that water law reforms are undertaken with the ultimate goal of achieving the realization of the right.

(2) While decentralization needs to be a continued focus, the challenges of inter-sectoral allocation of water and issues around inter-state water sharing need

a clear mandate, which can only be moved with concerted intervention at the Union level.

(3) The role of panchayati raj institutions (PRIs) needs to be effectively woven into the reform efforts to introduce decentralization and participation through user associations in irrigation and rural drinking water schemes. Presently, the fact that water users associations are established independently from PRIs implies that their basic structure is unrelated to the democratic structure of the latter and cannot lead to reducing existing inequalities in access to water.

(4) Proposed legislation and efforts to regulate groundwater use do not go far enough in seeking to provide equitable and socially just sharing of existing groundwater resources. Further, new laws do not include sufficient provisions to ensure that poorer users of groundwater will not be harmed by the new controls and charges. While there is a progressive withdrawal of the state from water-related functions, the future agenda of reforms for the water sector should take into account the need for a governmental role in regulating groundwater use in the context of the increasing importance of groundwater for all water uses.

(5) There should be a global review or study of current policy initiatives so as to re-align them with constitutional right to water guarantees, and with basic minimum (core) rights of access to water including safe drinking water, water for sanitation and health. This should incorporate an assessment of the implementation of the principles and measures proposed in the new laws, especially in terms of assessing whether these laws are successful according to the framework they set for themselves as well as in the broader context of existing laws and principles in which they arise.

(6) Finally, an in-depth determination of the scope of India's commitments under international human rights and environmental law is required. In particular, as a party to the ICESCR, the government is bound to progressively implement protected human rights including the right to water. The progressive withdrawal of the state from the provision of water supply services in rural areas and the imposition of a growing share of the capital and maintenance costs of drinking water to villagers constitute, in all likelihood, backward steps. Reform policies also need to be reconsidered in the light of international law obligations.

Future work on water law reforms needs to explore issues related to the implementation of recently adopted laws such as legislation setting up independent water regulatory authorities and issues that have not yet been given sufficient attention.

Work related to the implementation of already adopted laws will be very important to map out and understand the true extent of the reforms introduced since a number of consequences are only likely to arise in the course of their application on the ground. This is due in part to the fact that regulatory authorities have broad-ranging powers whose actual extent will only become apparent with time and will be partly linked to the willingness of existing government departments to relinquish their own existing powers. Further, updated work will be necessary to take into account the conceptual developments that can be identified between an earlier version of an act (e.g., the Maharashtra Water Resources Regulatory Authority Act, 2005) and a later version, such as the Uttar Pradesh Water Management and Regulatory Commission Act, 2008.

Beyond the reforms analyzed in this book, a number of other issues need to be given further attention. Firstly, the framework of analysis needs to be broadened to include, for instance, sanitation. While water supply and sanitation can be examined separately up to a certain point, the intrinsic links between the two call for recognizing the synergies that exist between the human right to water and the human right to sanitation. Further, there are a number of synergies between the regulation of water supply and sanitation. Indeed, both fall in principle under the remit of the states and both are covered by the decentralization provisions of the 73rd and 74th constitutional amendments.

Secondly, the analysis needs to be brought down to much more specific contexts, in particular with regard to urban water supply. Indeed, the fragmented nature of the legal frameworks governing water supply in cities makes any general analysis a difficult endeavour. The necessity to have a broader picture of the different regimes governing access in cities is nevertheless increasingly important in view of the rapid urbanization and of the significant urban reform projects that also include the water sector, such as the Jawaharlal Nehru National Urban Renewal Mission and the Urban Infrastructure Development Scheme for Small and Medium Towns. The fact that individual cities can be governed by different legal regimes also calls for specific attention in view of the constitutional imperative to fulfil the fundamental rights for all citizens equally. This is, for instance, the case in Delhi that is divided in three different zones for reasons that are neither hydrological nor social. The division, which sees New Delhi and the cantonment area governed by different legal regimes, and the reality which sees the New Delhi area get an allocation of water per capita that is vastly superior to any other area of Delhi, cannot be justified on the basis of the fundamental right to water. The separate treatment meted out to different residents of a similar city, to residents of different cities linked to the size of the city they inhabit – with a higher per capita allocation for bigger cities and the broad distinction between urban and rural citizens with regard to water provision – all necessitate further work. This is necessary to better understand the relationship of these regimes with the basic principles of water law, such as the fundamental right to water, and to examine these regimes in the context of the proposed and ongoing reforms for urban areas that will in all likelihood fundamentally change the basic principles on which urban water supply will be based in the future.

Bibliography

Agarwal, A. and S. Narain, *Dying Wisdom: The Rise, Fall and Potential of India's Traditional Water Harvesting Systems* (New Delhi: Centre for Science and Environment, 2001).

—— and Indira Khurana, *Making Water Everybody's Business* (New Delhi: Centre for Science and Environment, 2001).

Agarwal, Anil *et al.*, *Water, Sanitation, Health – for All?: Prospects for the International Drinking Water Supply and Sanitation Decade, 1981–1990* (London: Earthscan, 1981).

Alagh, Yoginder K., Ganesh Pangare and Biksham Gujja, *Interlinking of Rivers in India* (New Delhi: Academic Foundation, 2006).

Allan, J.A., 'Overall Perspectives on Countries and Regions', in Peter Rogers and Peter Lydon (eds.) *Water in the Arab World: Perspectives and Prognoses* (Cambridge, MA: Harvard University Press, 1994).

—— 'Virtual Water – The Water, Food, and Trade Nexus: Useful Concept or Misleading Metaphor?', 28 *Water Int'l* 4 (2003).

Allan, Tony, 'Watersheds and Problemsheds: Explaining the Absence of Armed Conflict over Water in the Middle East', 2(1) *Middle East Rev Int'l Affairs J.* 49 (1998).

Anonymous, 'What Price for the Priceless?: Implementing the Justiciability of the Right to Water', 120 *Harvard L. Rev.* 1067 (2007).

Arul, Carolin, *Water Laws and Rights in India* (New Delhi: Indian Society for Promoting Christian Knowledge, 2005).

Atapattu, Sumudu A., *Emerging Principles of International Environmental Law* (Ardsley, NY: Transnational Publishers, 2006).

Bajpai, P. and L. Bhandari, 'Ensuring Access to Water in Urban Households', 36/39 *Economic & Political Weekly* 3774 (2001).

Bandyopadyay, Jayanta and Shama Perveen, 'Interlinking of Rivers in India: Assessing the Justifications, 39/50 *Economic & Political Weekly* 5307 (2004).

Barandat, Joerg, 'Wasserrecht – Keine Trockene Juristerei', 17(1) *Vierteljahresschrift fuer Sicherheit und Frieden* 37 (1999).

—— and Aytul Kaplan, 'International Water Law: Regulations for Cooperation and the Discussion of the International Water Convention', in Waltina Scheumann and Manuel Schiffler (eds.) *Water in the Middle East* 13 (Berlin and New York: Springer, 1998).

Baumann, Cynthia, 'Water Wars: Canada's Upstream Battle to Ban Bulk Water Export', 10 *Minnesota J. Global Trade* 114 (2001).

Bennett, Vivienne, *The Politics of Water: Urban Protest, Gender, and Power in Monterrey, Mexico* (Pittsburgh: University of Pittsburgh Press, 1995).

Bernasconi-Osterwalder, Nathalie, 'Water, Agriculture, and Subsidies in the International Trading System', in Edith Brown Weiss, Laurence Boisson de Chazournes and Nathalie

Bernasconi-Osterwalder (eds.) *Fresh Water and International Economic Law* 207 (Oxford: Oxford University Press, 2005).

Black, Maggie and Rupert Talbot, *Water – A Matter of Life and Health* (New Delhi: Oxford University Press, 2005).

Bluemel, Erik B., 'The Implications of Formulating a Human Right to Water', 31 *Ecology LQ* 957 (2004).

Bodansky, Daniel, Jutta Brunnée and Ellen Hey, 'International Environmental Law: Mapping the Field', in Daniel Bodansky, Jutta Brunnée and Ellen Hey (eds.) *The Oxford Handbook of International Environmental Law* 1 (Oxford: Oxford University Press, 2007).

Boisson de Chazournes, Laurence and Salman M.A. Salman (eds.), *Water Resources and International Law* (Leiden: Martinus Nijhoff, 2005).

Brewer, J. *et al.* (eds.), *Irrigation Management Transfer in India: Policies, Processes and Performance* (New Delhi: Oxford and IBH Publishing, 1999).

Briscoe, John and R.P.S. Malik, *India's Water Economy: Bracing for a Turbulent Future* (Oxford: Oxford University Press, 2006).

—— (eds.), *Handbook of Water Resources in India* (New Delhi: Oxford University Press, 2007).

Caponera, Dante A., *Principles of Water Law and Administration* (London: Taylor and Francis Group, 2nd edn, 2007).

Chilton, J., 'Groundwater', in Deborah Chapman, *Water Quality Assessments – A Guide to Use of Biota, Sediments and Water in Environmental Monitoring* 412 (London: Taylor & Francis, 2nd edn, 1996).

Clapham, Andrew, *Human Rights Obligations of Non-State Actors* (New York: Oxford University Press, 2006).

Clark, Robert Emmet (ed.), *Water and Water Rights*, Vol. I (Indiana: The Allen Smith Company Publishers, 1967).

Cossy, Mireille, 'Le statut de l'eau en droit international économique – Principaux aspects au regard des règles de l'Organisation mondiale du commerce', in Laurence Boisson de Chazournes and Salman M.A. Salman (eds.). *Water Resources and International Law* 169 (Leiden: Martinus Nijhoff, 2005).

Cossy, Mireille, 'Water Services at the WTO', in Edith Brown Weiss, Laurence Boisson de Chazournes and Nathalie Bernasconi-Osterwalder (eds.) *Fresh Water and International Economic Law* 117 (Oxford: Oxford University Press, 2005).

Coward, Walter Jr., 'Irrigation Development: Institutional and Organisational Issues', in E. Walter Coward Jr. (ed.), *Irrigation and Agricultural Development in Asia: Perspectives from the Social Sciences* 27 (Ithaca and London: Cornell University Press, 1980).

Crook, John R. and Stephen McCaffrey, 'The United Nations Starts Work on a Watercourses Convention', 91 *American J. Intl L.* 374 (1997).

Cullet, Philippe, 'Human Rights and Displacement: The Indian Supreme Court Decision of Sardar Sarovar in International Perspective', 50 *Int'l & Comp. LQ* 973 (2001).

—— 'Water Law Reforms: Analysis of Recent Developments', 48/2 *Journal of the Indian Law Institute* 206 (2006).

—— 'Water Law – Evolving Regulatory Framework', in Philippe Cullet, Alix Gowlland Gualtieri, Roopa Madhav and Usha Ramanathan (eds.) *Water Law at the Crossroads – National and International Perspectives with Special Emphasis on India* 26 (New Delhi: Cambridge University Press, 2009).

—— (ed.), *Sardar Sarovar Dam Project: Selected Documents* (Aldershot: Ashgate, 2007).

—— and Alix Gowlland Gualtieri, 'Local Communities and Water Investments', in Edith Brown Weiss, Laurence Boisson de Chazournes and Nathalie Bernasconi-Osterwalder (eds.) *Fresh Water and International Economic Law* 303 (Oxford: Oxford University Press, 2005).

—— and Annie Patricia Kameri-Mbote, 'Activities Implemented Jointly in the Forestry Sector: Conceptual and Operational Fallacies', 10 *Georgetown Int'l. Envtl. L. Rev. 98* (1997).

—— Alix Gowlland Gualtieri, Roopa Madhav and Usha Ramanathan (eds.) *Water Law at the Crossroads – National and International Perspectives with Special Emphasis on India* (New Delhi: Cambridge University Press, 2009).

Daseking, Christina *et al.*, Lessons from the Crisis in Argentina (Washington D.C.: IMF Occasional Paper No. 236, 2005).

Dellapenna, Joseph W., 'The Berlin Rules on Water Resources: The New Paradigm for International Water Law', in Randall Graham (ed.) *The World Environmental and Water Resource Congress 2006: Examining the Confluence of Environmental and Water Concerns* (Omaha: American Society of Civil Engineers, 2006), available at www.ualg.pt/5cigpa/comunicacoes/Berlin%20Rules%20Summary.doc.

—— 'Transboundary Watersharing and the Need for Public Management', 133(5) *Journal for Water Resources Planning and Management* 397 (2007).

Desai, Bharat, *Water Pollution in India: Law and Enforcement* (New Delhi: Lancer Books, 1990).

Desai, Mihir and Kamayani Bali Mahabal, Health Care Case Law in India: A Reader (Mumbai: Centre for Enquiry into Health and Allied Themes and India Centre for Human Rights & Law, 2007).

Dowie, Mark, 'In Law we Trust', 22 *Orion* 18 (2003).

Dubash, Navroz K, *Tubewell Capitalism: Groundwater Development and Agrarian Change in Gujarat* (New Delhi: Oxford University Press, 2002).

—— The Electricity-Groundwater Conundrum: Case for a Political Solution to a Political Problem', 42(52) *Economic & Political Weekly* 45 (2007).

—— 'Independent Regulatory Agencies: A Theoretical Review with Reference to Electricity and Water in India', 43(40) *Economic & Political Weekly* 43 (2008).

Dupuy, Pierre-Marie, 'Le droit à l'eau, un droit international?' in Guillaume Grisel (ed.), *La mise en oeuvre du droit à l'eau – Actes du XXIXème Congrès ordinaire de l'IDEF* (Genève: Schulthess, 2006).

Dwivedi, Gaurav, Rehmat, and Shripad Dharmadhikary, *Water: Private, Limited* (Badwani: Manthan Adhyayan Kendra, 2006).

Etego, Rudolf Nsorwine Amenga and Sara Grusky, 'The New Face of Conditionalities: The World Bank and Water Privatization in Ghana', in David A. McDonald and Greg Ruiters (eds.) *The Age of Commodity: Water Privatization in Southern Africa* 275 (London: Earthscan, 2005).

Finger, Matthias and Jeremy Allouche, *Water Privatisation: Trans-National Corporations and the Re-Regulation of the Water Industry* (London and New York: Spon Press, 2002).

Fitzmaurice, Malgosia, 'General Principles Governing the Cooperation between States in Relation to Non-Navigational Uses of International Watercourses', 14 *Yb. Int'l Envtl. L.* 3 (2005).

—— 'The Right to Water', 18 *Fordham Envtl L. Rev.* 537 (2007).

Flynn, Sean and Danwood Mzikenge Chirwa, 'The Constitutional Implications of Commercializing Water in South Africa', in David A. McDonald and Greg Ruiters (eds.) *The Age of Commodity: Water Privatization in Southern Africa* 59 (London: Earthscan, 2005).

Food and Agricultural Organization, Law and Sustainable Development since Rio: Legal Trends in Agriculture and Natural Resource Management (Rome: FAO, FAO Legislative Studies 2002).

Freestone, David and Salman M. A. Salman, 'Ocean and Freshwater Resources', in Daniel Bodansky, Jutta Brunnée, and Ellen Hey (eds.) *The Oxford Handbook of International Environmental Law* 337 (Oxford: Oxford University Press, 2007).

Gangadhar, V., 'Riches of Rayalaseema', in Anil Agarwal, Sunita Narain and Indira Khurana, *Making Water Everybody's Business* 33 (New Delhi: Centre for Science and Environment, 2001).

Girouard, Robert, 'Water Export Restrictions: A Case Study of WTO Dispute Settlement Strategies and Outcomes', 15 *Georgetown Int'l. Envtl. L. Rev.* 247 (2003).

Gleick, Peter, 'The Human Right to Water', 1 *Water Policy* 487 (1999).

Gleick, Peter H. *et al.*, *The World's Water 2002–2003: The Biennial Report on Freshwater Resources* (Island Press: Washington, 2002).

Gold, Joseph, 'Use of Fund's Resources', in Margaret Garrisen de Vries and J. Keith Horsefield (eds.) *The International Monetary Fund 1945–1965, Volume II: Analysis* 522 (Washington D.C.: International Monetary Fund, 1969).

—— Conditionality (Washington D.C.: IMF Pamphlet Series No. 31, 1979).

Gowlland Gualtieri, Alix, 'South Africa's Law and Policy Water Framework: Implications for the Right to Water', in Philippe Cullet, Alix Gowlland Gualtieri, Roopa Madhav and Usha Ramanathan (eds.) *Water Law at the Crossroads – National and International Perspectives with Special Emphasis on India* 389 (New Delhi: Cambridge University Press, 2009).

Groendfelt, David and Mark Svendson (eds.), Case Studies in Participatory Irrigation Management (Washington: World Bank Institute, 2000).

Haan, Esther de, 'Balancing Free Trade in Water and the Protection of Water Resources in GATT', in Edward H.P. Brans *et al.* (eds.) *The Scarcity of Water* 245 (London: Kluwer Law International, 1997).

Halvorrsen, Anita M, 'Common But Differentiated Commitments in the Future Climate Change Regime: Amending the Kyoto Protocol to Include Annex C and the Annex C Mitigation Fund', 18 *Colorado J. Int'l Envtl. L. and Pol'y* 247 (2007).

Hazin, Lilian Saade, 'Private Sector Participation in Water Supply and Sanitation: Realising Social and Environmental Objectives in Mexico', in Nick Johnstone and Libby Wood (eds.) *Private Firms and Public Water: Realising Social and Environmental Objectives in Developing Countries* 177 (Cheltenham: Edward Elgar, 2001).

Hey, Ellen, 'The Watercourses Convention: To What Extent Does it Provide a Basis for Regulating Uses of International Watercourses?', 7 *Rev. European Community & Int'l Envtl L.* 297 (1998).

Hildering, Antoinette, *International Law, Sustainable Development and Water Management* (Delft: Eburon Academic Publishers, 2004).

Hirway, Indira, 'Ensuring Drinking Water to All: A Study in Gujarat', in K.V. Raju (ed.) *Elixir of Life – The Socio-Ecological Governance of Drinking Water* 74 (Bangalore: Books for Change, 2007).

Hodgson, Stephen, Legislation on Water Users' Associations (Rome: Food and Agricultural Organization, Legislative Study 79, 2003).

Hoekstra, Arjen Y. and Ashok K. Chapagain, 'Water Footprints of Nations: Water used by People as a Function of their Consumption Pattern', 21 *Water Resources Management* 35 (2007).

Hoering Uwe and Ann-Kathrin Schneider, King Customer? The World Bank's 'New' Water Policy and its Implementation in Sri Lanka and India (Brot für die Welt, 2004).

Holland, Ann-Christin Sjolander, *The Water Business: Corporations versus People* (Bangalore: Books for Change, 2005).

Hooja, Rakesh, 'Below the Third Tier: Water Users Associations and Participatory

Irrigation Management in India', 1 *Indian Journal of Federal Studies* (2004), available at www.jamiahamdard.edu/cfs/jour4-1_4.htm.

—— *Management of Water for Agriculture* (New Delhi: Rawat Publications, 2006).

Howard, Guy and James Bartram, Domestic Water Quantity, Service Level and Health, (Geneva: World Health Organization, 2003).

Jackson, John H., *The Jurisprudence of GATT and WTO* (Cambridge: Cambridge University Press, 2000).

Jacob, Alice and S.N. Singh, *Law Relating to Irrigation* (Bombay: N.M. Tripathi, 1972).

Jairath, Jasveen. *Water User Associations in Andhra Pradesh* (New Delhi: Concept Publishing Company, 2001).

James, A.J., 'From Sector Reform to Swajaldhara – Scaling up in India', 23/2 *Waterlines* 11 (2004).

Janakarajan, S. and Marcus Moench, 'Are Wells a Potential Threat to Farmer's Well Being? Case of Deteriorating Groundwater Irrigation in Tamil Nadu', 41(37) *Economic & Political Weekly* 3977 (2006).

Kathpalia, G.N. and R. Kapoor, Water Policy and Action Plan for India 2020: An Alternative (Delhi: Alternative Futures, 2002).

Kidd, Michael, 'Not a Drop to Drink: Disconnection of Water Services for Non-Payment and the Right of Access to Water', 20 *South African J. Human Rts* 119 (2004).

Kiss, Alexandre, Environmental Law Principles (Geneva: UNITAR, 2005).

Koonan, Sujith, 'Groundwater: Legal Aspects of the Plachimada Dispute', in Philippe Cullet, Alix Gowlland Gualtieri, Roopa Madhav and Usha Ramanathan (eds.) *Water Law at the Crossroads – National and International Perspectives with Special Emphasis on India* 159 (New Delhi: Cambridge University Press, 2009).

Koppen, Barbara van, A Gender Performance Indicator for Irrigation: Concepts, Tools and Applications (Colombo: International Water Management Institute, IWMI Research Report 59, 2002).

Kumar, M. Dinesh, 'Impact of Electricity Prices and Volumetric Water Allocation on Energy and Groundwater Demand: Analysis from Western India', in V. Ratna Reddy and S. Mahendra Dev (eds.) *Managing Water Resources: Policies, Institutions, and Technology* 118 (New Delhi: Oxford University Press, 2006).

—— and O.P. Singh, 'Virtual Water in Global Food and Water Policy Making: Is There a Need for Rethinking?', 19 *Water Resources Management* 759 (2005).

Langford, Malcolm, 'Ambition that Overleaps Itself? A Response to Stephen Tully's "Critique" of the General Comment on the Right to Water', 26 *Netherlands Quarterly of Human Rts* 433 (2006).

Langston, Craig and Grace Ding, *Sustainable Practices in the Built Environment* (Oxford: Butterworth and Heinemann, 2nd edn, 2001).

Lee, Terence R, *Residential Water Demand and Economic Development* (Toronto: University of Toronto Press, 1969).

Liebenberg, Sandra, 'The Value of Human Dignity in Interpreting Socio-Economic Rights', 21 *South African J. Human Rts* 1 (2005).

Little, Scott P, 'Canada's Capacity to Control the Flow: Water Export and the North American Free Trade Agreement', 8 *Pace Int'l L. Rev.* 127 (1996).

Lorenz, Frederick and Edward Erickson, *The Euphrates Triangle: Security Implications of the Southeastern Anatolia Project* (Washington D.C.: National Defense University Press, 1999).

MacDonnell, Lawrence J, 'Rules Guiding Groundwater Use in the United States', 1 *Indian Juridical Review* 43 (2005).

Madhav, Roopa, 'Tirupur Water Supply and Sanitation Project: A Revolution in Water Resource Management?', in Philippe Cullet, Alix Gowlland Gualtieri, Roopa Madhav and Usha Ramanathan *Water Law at the Crossroads – National and International Perspectives with Special Emphasis on India* 201 (New Delhi: Cambridge University Press, 2009).

Magraw, Daniel Barstow and Lisa D. Hawke, 'Sustainable Development' in Daniel Bodansky, Jutta Brunnée, and Ellen Hey (eds.) *The Oxford Handbook of International Environmental Law* 613 (Oxford: Oxford University Press, 2007).

Mahapatra, Sushanta Kumar, 'Functioning of Water Users Association or Pani Panchayat in Orissa: Principle, Procedure, Performance, and Prospects', 3/2 *Law, Environment and Development Journal* 126 (2007).

Mansuri, Ghazala and Vijayendra Rao, Community-Based (and Driven) Development: A Critical Review (Washington D.C.: World Bank, Policy Research Working Paper No. 3209, 2004).

Mariño, Manuel and Karin E. Kemper (eds.), Institutional Frameworks in Successful Water Markets: Brazil, Spain and Colorado, USA (Washington D.C.: World Bank, Technical Paper No. 427, 1999).

Mason, Simon *et al.*, 'Linkages Between Sub-national and International Water Conflicts: The Eastern Nile Basin', in Hans-Guenther Brauch *et al.* (eds.) *Facing Global Environmental Change: Environmental, Human, Energy, Food, Health and Water Security Concepts* 325 (Berlin: Springer Verlag, 2007).

McCaffrey, Stephen, 'A Human Right to Water: Domestic and International Implications', 5 *Georgetown Int'l. Envtl. L. Rev.* 1 (1992).

—— 'The Human Right to Water', in Edith Brown Weiss, Laurence Boisson de Chazournes and Nathalie Bernasconi-Osterwalder (eds.) *Fresh Water and International Economic Law* 93 (Oxford: Oxford University Press, 2005).

—— *The Law of International Watercourses* (Oxford: Oxford University Press, 2007).

McDonald, David A. and Greg Ruiters, 'Introduction', in David A. McDonald and Greg Ruiters *The Age of Commodity: Water Privatization in Southern Africa* 3 (London: Earthscan, 2005).

McMahon, J.A., *The WTO Agreement on Agriculture: A Commentary* (Oxford: Oxford University Press, 2006).

Mehta, Lyla, 'Problems of Publicness and Access Rights: Perspectives from the Water Domain', in Inge Kaul *et al.* (eds.) *Providing Global Public Goods – Managing Globalization* 556 (Oxford: Oxford University Press, 2003).

Misra, Smita and Bishwanath Goldar, 'Likely Impact of Reforming Water Supply and Sewerage Services in Delhi', 43(41) *Economic & Political Weekly* 57 (2008).

Mollinga, Peter P. and Alex Bolding (eds.), *The Politics of Irrigation Reform: Contested Policy Formulation and Implementation in Asia, Africa and Latin America* (Aldershot: Ashgate, 2004).

Mori, Katsuhiko, 'Virtual Water Trade in Global Governance', in Arjen Y. Hoekstra (ed.) Virtual Water Trade: Proceedings of the International Expert Meeting on Virtual Water Trade (Delft: IHE, Value of Water Research Report Series No. 12, 2003).

Mosse, David, *The Rule of Water: Statecraft, Ecology and Collective Action in South India* (New Delhi: Oxford University Press, 2003).

Muchlinski, Peter T., *Multinational Enterprises and the Law* (Oxford: Oxford University Press, 2007).

Muralidhar, S., 'The Right to Water: An Overview of the Indian Legal Regime', in Eibe Riedel and Peter Rothen (eds.) *The Human Right to Water* 65 (Berlin: Berliner Wissenschafts-Verlag, 2006).

Nadkarni, Manoj, 'Drowning in Human Excreta', 10 (19) *Down to Earth* (2001), available at www.indiaenvironmentportal.org.in/node/9699.

Narasimhan, T.N., 'Groundwater Management and Ownership', 43(7) *Economic & Political Weekly* 21 (2008).

Narayanamoorthy A. and R.S. Deshpande, *Where Water Seeps! Towards a New Phase in India's Irrigation Reforms* (New Delhi: Academic Foundation, 2005).

National Institute of Rural Development, *India Panchayati Raj Report 2001: Four Decades of Centralized Governance in Rural India, Volume 1* (Hyderabad: National Institute of Rural Development, 2002).

Olleta, Andrés, 'The World Bank's Influence on Water Privatisation in Argentina – The Experience of the City of Buenos Aires' in Philippe Cullet, Alix Gowlland Gualtieri, Roopa Madhav and Usha Ramanathan (eds.) *Water Law at the Crossroads – National and International Perspectives With Special Emphasis on India* 230 (New Delhi: Cambridge University Press, 2009).

Pal, Som, 'Key Issues in Water Quality Management', in International Life Science Institute-India, *Water Quality Management: South Asian Perspective Vision 2025* at 18 (New Delhi: International Life Science Institute-India, 2002).

Panda, Ramesh Chandra, 'Water Quality Scenario in India – An Overview', in International Life Science Institute-India, *Water Quality Management: South Asian Perspective Vision 2025* at 37 (New Delhi: International Life Science Institute-India, 2002).

Pant, Niranjan, 'Some Issues in Participatory Irrigation Management', 43(1) *Economic & Political Weekly* 30 (2008).

Parveen, Saila and I.M. Faisal, 'Trading Virtual Water between Bangladesh and India: A Politico-economic Dilemma', 6 *Water Policy* 1 (2004).

Praveen, N. *et al.*, 'Drinking Water Quality in Urban India: Why and How it is Getting Worse – A Case Study of South Indian Cities', in K.V. Raju (ed.) *Elixir of Life: The Socio-Ecological Governance of Drinking Water* 353 (Bangalore: Books for Change, 2006).

Perry, Christopher John, Michael Rock and David Seckler, Water as an Economic Good: A Solution, or a Problem? (Colombo: International Irrigation Management Institute, Research Report No. 14, 1994).

Petrova, Violeta, 'At the Frontiers of the Rush for Blue Gold: Water Privatization and the Human Right to Water', 31 *Brooklyn J. Intl L.* 577 (2006).

Phansalkar, Sanjiv and Vivek Kher, 'A Decade of the Maharashtra Groundwater Legislation: Analysis of the Implementation Process', 2(1) *Law, Environment and Development Journal* 67 (2006).

Postel, Sandra L. and Aaron T. Wolf, 'Dehydrating Conflict', 126 *Foreign Policy* 60 (2001).

Prasad, Kamta, *Farmers' Participation in Irrigation Management in India* (New Delhi: Uppal Publishing House, 2001).

—— (ed.), *Water Resources and Sustainable Development* (New Delhi: Shipra, 2003).

Pruss-Ustun, Annette *et al.*, Safer Water, Better Health: Costs, Benefits and Sustainability of Interventions to Protect and Promote Health (Geneva: World Health Organization, 2008).

Public Citizen, *Water Privatization Fiascos: Broken Promises and Social Turmoil* (Washington D.C.: Public Citizen, 2003).

Purohit R.C. and Virendra Kumar, 'Ownership of Ground Water and its Pricing: Rajasthan Perspective', in Saleem Romani *et al.* (eds.), *Groundwater Governance – Ownership of Groundwater and its Pricing* 354 (New Delhi: Capital Publishing Company, 2007).

Rajamani, Lavanya, 'The Nature, Promise, and Limits of Differential Treatment in the Climate Regime', *16 Yb. Int'l Envtl. L.* 81 (2007).

Raju, K.V., Keshab Das and S. Manasi, 'Emerging Trends in Rural Water Supply: A Comparative Analysis of Karnataka and Gujarat', in K.V. Raju (ed.) *Elixir of Life – The Socio-Ecological Governance of Drinking Water* 1 (Bangalore: Books for Change, 2007).

Rao, V. Sitarama, *Law Relating to Water Rights* (Hyderabad: Asia Law House, 1996).

Rathore, M.S., 'Water Rights and Other Alternatives to Groundwater Management in India', in Saleem Romani *et al.* (eds.) *Groundwater Governance – Ownership of Groundwater and its Pricing* 333 (New Delhi: Capital Publishing Company, 2007).

Ratner, Steven R., 'Corporations and Human Rights: A Theory of Legal Responsibility', 111 *Yale L.J.* 443 (2001).

Razzaque, Jona, 'Trading Water: The Human Factor', 13 *Rev. of European Community & Int'l Envtl. L.* 15 (2004).

Riedel, Eibe, 'The Human Right to Water and General Comment No. 15', in Eibe Riedel and Peter Rothen, *The Human Right to Water* 19 (Berlin: Berliner Wissenschafts-Verlag, 2006).

Rosegrant, Mark and Hans Binswanger, 'Markets in Tradable Water Rights: Potential for Efficiency Gains in Developing Country Irrigation', 11 *World Development* 22 (1994).

Sadeleer, Nicolas de, *Environmental Principles: From Political Slogans to Legal Rules* (Oxford: Oxford University Press, 2002).

Saleth, Maria R., 'Groundwater Markets in India: A Legal and Institutional Perspective', 29(2) *Indian Economic Review* 157 (1994).

—— 'Irrigation Privatization in India: Options, Issues and Experience', 34(26) *Economic & Political Weekly* A-86 (1999).

—— 'Water Rights and Entitlements', in John Briscoe and R.P.S. Malik (eds.) *Handbook of Water Resources in India* 282 (New Delhi: Oxford University Press, 2007).

—— and Ariel Dinar, *The Institutional Economics of Water: A Cross-Country Analysis of Institutions and Performance* (Cheltenham: Edward Elgar, 2004).

Salman, Salman M.A. and Siobhán Alice McInerey-Lankford, The Human Right to Water: Legal and Policy Dimensions (Washington, D.C.: The World Bank, 2004).

Sampat, Preeti, "Swa'-jal-dhara or 'Pay'-jal-dhara – Sector Reform and the Right to Drinking Water in Rajasthan and Maharashtra", 3/2 *Law, Environment and Development Journal* 101 (2007).

Sangameswaran, Priya, Roopa Madhav and Clifton D'Rosario, '24/7 Privatization and Water Reform: Insights from Hubli-Dharward', 43(14) *Economic & Political Weekly* 60 (2008).

Sax, Joseph L., 'The Public Trust Doctrine in Natural Resource Law: Effective Judicial Intervention', 68 *Michigan L. Rev.* 471 (1970)

—— *Defending the Environment: A Strategy for Citizen Action* (New York: Borzoi Books, 1971).

Shaban, Abdul and R.N. Sharma, 'Water Consumption Patterns in Domestic Households in Major Cities', 42/23 *Economic & Political Weekly* 2190 (2007).

Shah, Tushaar, 'India's Master Plan for Groundwater Recharge: An Assessment and Some Suggestions for Revision', 43/51 *Economic & Political Weekly* 41 (2008).

Shah, Tushaar *et al.*, 'Energy-Irrigation Nexus in South Asia: Improving Groundwater Conservation and Power Sector Viability', in Mark Giordano and Karen G. Villholth (eds.) *The Agricultural Groundwater Revolution: Opportunities and Threats to Development* 211 (Oxfordshire: CABI International, 2007).

Siddiqui, Iqbal Ahmed, 'History of Water Laws in India', in Chhatrapati Singh (ed.) *Water Law in India* 289 (Delhi: Indian Law Institute, 1992).

Singh, Chhatrapati, *Water Rights and Principles of Water Resource Management* (Bombay: N.M. Tripathi, 1991).

—— (ed.), *Water Law in India* (New Delhi: Indian Law Institute, 1992).
Singh, K.K. (ed.), *Amin and Shastri's The Law of Easements* (Lucknow: Eastern Book Company, 4th edn, 1970).
—— *Farmers in the Management of Irrigation Systems* (New Delhi: Sterling Publishers, 1991).
Singh, O.P. *et al.*, 'Virtual Water Trade in Dairy Economy: Irrigation Water Productivity in Gujarat', 39/31 *Economic & Political Weekly* 3492 (2004).
Sinha, Manoj Kumar, *Enforcement of Economic, Social and Cultural Rights: International and National Perspectives* (New Delhi: Manak Publications, 2006).
Smets, Henri, 'Economics of Water Services and the Right to Water', in Edith Brown Weiss, Laurence Boisson de Chazournes and Nathalie Bernasconi-Osterwalder (eds.) *Fresh Water and International Economic Law* 173 (Oxford: Oxford University Press, 2005).
Smith, Laila *et al.*, 'Public Money, Private Failure: Testing the Limits of Market Based Solutions for Water Delivery in Nelspruit', in David A. McDonald and Greg Ruiters (eds.) *The Age of Commodity: Water Privatization in Southern Africa* 130 (London: Earthscan, 2005).
Sorel, Jean-Marc, 'Sur quelques aspects juridiques de la conditionalité du FMI et leur conséquences', 7 *European J. Int'l L.* 1 (1996).
Stiglitz, Joseph E, *Globalization and its Discontents* (New York: W.W. Norton, 2003).
Swart, Mia, 'Left out in the Cold? Crafting Constitutional Remedies for the Poorest of the Poor', 21 *South African J. Human Rts* 215 (2005).
Swatuk, Larry A., 'The New Water Architecture of SADC', in David A. McDonald and Greg Ruiters (eds.) *The Age of Commodity: Water Privatization in Southern Africa* 43 (London: Earthscan, 2005).
Tanzi, Attila and Maurizio Arcari, *The United Nations Convention on the Law of International Watercourses* (The Hague: Kluwer Law International, 2001).
Thobani, Mateen, *Tradable Property Rights to Water* (Washington D.C.: World Bank, 1995).
Tuerk, Elisabeth, Aaron Ostrovsky and Robert Speed, 'GATS and its Impact on Private Sector Participation in Water Services', in Edith Brown Weiss, Laurence Boisson de Chazournes and Nathalie Bernasconi-Osterwalder *Fresh Water and International Economic Law* 143 (Oxford: Oxford University Press, 2005).
Troesken, Werner, 'Typhoid Rates and the Public Acquisition of Private Waterworks – 1880–1920', 59(4) *Journal of Economic History* 927 (1999).
Tully, Stephen, 'A Human Right to Access Water? A Critique of General Comment No. 15', 23 *Netherlands Quarterly of Human Rts* 35 (2005).
United Nations, *Water for People – Water for Life* (Paris: UNESCO, 2003).
—— *Water – A Shared Responsibility* (Paris: UNESCO, 2006).
United Nations Development Programme, *Human Development Report 2006 – Beyond Scarcity: Power, Poverty and the Global Water Crisis* (New York: UNDP, 2006).
Upadhyay, Sanjay and Videh Upadhyay, *Handbook on Environmental Law Volume III: Environmental Protection, Land and Energy Laws* (New Delhi: LexisNexis, 2002).
Upadhyay, Videh, 'Water Management and Village Groups: Role of Law,' 37(49) *Economic & Political Weekly* 4907 (2002).
—— 'Canal Irrigation, Water User Associations and Law in India – Emerging Trends in Rights-based Perspective', in Philippe Cullet, Alix Gowlland Gualtieri, Roopa Madhav and Usha Ramanathan (eds.) *Water Law at the Crossroads – National and International Perspectives with Special Emphasis on India* 111 (New Delhi: Cambridge University Press, 2009).
Vaidyanathan, A. *Water Resource Management: Institutions and Irrigation Development in India* (Delhi: Oxford University Press, 1999).
—— 'Role of Law in Water Resources Management', 1 *Indian Juridical Review* 12 (2004).

—— *India's Water Resources, Contemporary Issues on Irrigation* (New Delhi: Oxford University Press, 2006).

Vani, M.S., *Role of Panchayat Institutions in Irrigation Management* (New Delhi: Indian Law Institute, 1992).

Venkateswarlu, Davuluri, 'Politics of Irrigation Management Reforms in Andhra Pradesh', *in* Rakesh Hooja, Ganesh Pangare and K.V. Raju (eds.), *Users in Water Management: The Andhra Model and its Replicability in India* 159 (New Delhi: Rawat Publications, 2002).

Watkin, Thomas Glyn, *An Historical Introduction to Modern Civil Law* (Aldershot: Ashgate, 1999).

Werksman, Jacob, 'Greening Bretton Woods', in Philippe Sands (ed.) *Greening International Law* 73 (London: Earthscan, 1993).

World Bank, Natural Disasters: Counting the Cost (Washington, D.C.: World Bank, 2004).

—— Water Resources Sector Strategy: Strategic Directions for World Bank Engagement (Washington D.C.: World Bank, 2004).

World Bank Operations Evaluation Department, Bridging Troubled Waters: Assessing the World Bank Water Resources Strategy (Washington, D.C.: World Bank, 2002).

—— Efficient, Sustainable Services for All? – An OED Review of the World Bank's Assistance to Water Supply and Sanitation (Washington D.C.: World Bank, OED Report No. 26443, 2003).

World Health Organization, The Right to Water (Geneva: WHO, 2003).

Index